Handbook of Research on Waste Diversion and Minimization Technologies for the Industrial Sector

Ashok K. Rathoure
M/s Akone Services, Lucknow, India

IGI Global
PUBLISHER of TIMELY KNOWLEDGE

A volume in the Advances in Environmental Engineering and Green Technologies (AEEGT) Book Series

Published in the United States of America by
IGI Global
Engineering Science Reference (an imprint of IGI Global)
701 E. Chocolate Avenue
Hershey PA, USA 17033
Tel: 717-533-8845
Fax: 717-533-8661
E-mail: cust@igi-global.com
Web site: http://www.igi-global.com

Library of Congress Cataloging-in-Publication Data

Names: Rathoure, Ashok K., 1983- editor.
Title: Handbook of Research on waste diversion and minimization technologies for the industrial sector / Ashok K. Rathoure, editor.
Description: Hershey, PA : Engineering Science Reference, an imprint of IGI Global, [2021] | Includes bibliographical references and index. | Summary: "This book explores efficient and cost-effective waste diversion and waste minimization strategies for the industrial sector"-- Provided by publisher.
Identifiers: LCCN 2020009702 (print) | LCCN 2020009703 (ebook) | ISBN 9781799849216 (hardback) | ISBN 9781799855972 (paperback) | ISBN 9781799849223 (ebook)
Subjects: LCSH: Factory and trade waste--Management. | Waste minimization.
Classification: LCC TD897 .W33 2021 (print) | LCC TD897 (ebook) | DDC 628.4--dc23
LC record available at https://lccn.loc.gov/2020009702
LC ebook record available at https://lccn.loc.gov/2020009703

This book is published in the IGI Global book series Advances in Environmental Engineering and Green Technologies (AEEGT) (ISSN: 2326-9162; eISSN: 2326-9170)

British Cataloguing in Publication Data
A Cataloguing in Publication record for this book is available from the British Library.

Advances in Environmental Engineering and Green Technologies (AEEGT) Book Series

Sang-Bing Tsai
Zhongshan Institute, University of Electronic Science and Technology of China, China & Wuyi University, China
Ming-Lang Tseng
Lunghwa University of Science and Technology, Taiwan
Yuchi Wang
University of Electronic Science and Technology of China Zhongshan Institute, China

ISSN:2326-9162
EISSN:2326-9170

Mission

Growing awareness and an increased focus on environmental issues such as climate change, energy use, and loss of non-renewable resources have brought about a greater need for research that provides potential solutions to these problems. Research in environmental science and engineering continues to play a vital role in uncovering new opportunities for a "green" future.

The **Advances in Environmental Engineering and Green Technologies (AEEGT)** book series is a mouthpiece for research in all aspects of environmental science, earth science, and green initiatives. This series supports the ongoing research in this field through publishing books that discuss topics within environmental engineering or that deal with the interdisciplinary field of green technologies.

Coverage

- Green Technology
- Radioactive Waste Treatment
- Biofilters and Biofiltration
- Waste Management
- Air Quality
- Contaminated Site Remediation
- Pollution Management
- Water Supply and Treatment
- Sustainable Communities
- Cleantech

IGI Global is currently accepting manuscripts for publication within this series. To submit a proposal for a volume in this series, please contact our Acquisition Editors at Acquisitions@igi-global.com or visit: http://www.igi-global.com/publish/.

The Advances in Environmental Engineering and Green Technologies (AEEGT) Book Series (ISSN 2326-9162) is published by IGI Global, 701 E. Chocolate Avenue, Hershey, PA 17033-1240, USA, www.igi-global.com. This series is composed of titles available for purchase individually; each title is edited to be contextually exclusive from any other title within the series. For pricing and ordering information please visit http://www.igi-global.com/book-series/advances-environmental-engineering-green-technologies/73679. Postmaster: Send all address changes to above address.

Titles in this Series

For a list of additional titles in this series, please visit:
http://www.igi-global.com/book-series/advances-environmental-engineering-green-technologies/73679

Solar Concentrating Modules With Louvered Heliostats Emerging Research and Opportunities
Dmitry Strebkov (Federal State Budget Scientific Institution, Russia & Federal Scientific Agroengineering Center VIM, Russia) Natalya Filippchenkova (Federal State Budget Scientific Institution, Russia & Federal Scientific Agroengineering Center VIM, Russia) and Anatoly Irodionov (Federal State Budget Scientific Institution, Russia & Federal Scientific Agroengineering Center VIM, Russia)
Engineering Science Reference • © 2021 • 267pp • H/C (ISBN: 9781799842767) • US $190.00

Smart Agricultural Services Using Deep Learning, Big Data, and IoT
Amit Kumar Gupta (Amity University Jaipur, India) Dinesh Goyal (Poornima Institute of Engineering and Technology, India) Vijander Singh (Manipal University Jaipur, India) and Harish Sharma (Rajasthan Technical University, India)
Engineering Science Reference • © 2021 • 280pp • H/C (ISBN: 9781799850038) • US $195.00

Role of IoT in Green Energy Systems
Vasaki Ponnusamy (Universiti Tunku Abdul Rahman, Malaysia) Noor Zaman (Taylor's University, Malaysia) Low Tang Jung (Universiti Teknologi PETRONAS, Malaysia) and Anang Hudaya Muhamad Amin (Higher Colleges of Technology, UAE)
Engineering Science Reference • © 2021 • 305pp • H/C (ISBN: 9781799867098) • US $195.00

Eco-Friendly Energy Processes and Technologies for Achieving Sustainable Development
Mir Sayed Shah Danish (University of the Ryukyus, Japan) and Tomonobu Shah Senjyu (University of the Ryukyus, Japan)
Engineering Science Reference • © 2021 • 340pp • H/C (ISBN: 9781799849155) • US $195.00

Precision Agriculture Technologies for Food Security and Sustainability
Sherine M. Abd El-Kader (Electronics Research Institute, Cairo, Egypt) and Basma M. Mohammad El-Basioni (Electronics Research Institute, Cairo, Egypt)
Engineering Science Reference • © 2021 • 437pp • H/C (ISBN: 9781799850007) • US $195.00

Energy-Efficient Underwater Wireless Communications and Networking
Nitin Goyal (Chitkara University Institute of Engineering and Technology, Chitkara University, Punjab, India) Luxmi Sapra (Chitkara University Institute of Engineering and Technology, Chitkara University, Punjab, India) and Jasminder Kaur Sandhu (Chitkara University Institute of Engineering and Technology, Chitkara University, Punjab, India)
Engineering Science Reference • © 2021 • 339pp • H/C (ISBN: 9781799836407) • US $195.00

IGI Global
PUBLISHER of TIMELY KNOWLEDGE

701 East Chocolate Avenue, Hershey, PA 17033, USA
Tel: 717-533-8845 x100 • Fax: 717-533-8661
E-Mail: cust@igi-global.com • www.igi-global.com

Editorial Advisory Board

List of Contributors

Table of Contents

Detailed Table of Contents

Any technology that is developed and used will have certain components or constituting mechanisms such as materials used, procedure, or technological process involved. The quality of life on planet with respect to environment such as fresh air to breathe, clean water for drinking, and other purposes are essential for all forms of life on earth. Therefore, whatever developments are undertaken should not degrade our environment. It is essential to bear in mind that whatever development, be it construction, new power plants, transportation modes, etc., should be carefully weighed in terms of its contribution to the environment, especially if there are any negative aspects that emerge from use of machinery, equipment, technologies, etc. We should analyze any technology that we use in terms of its cost-benefit study. Here in this chapter, the authors have discussed environmentally friendly technologies for our sustainability to use the natural resources wisely with some case studies of natural as well as manmade disasters.

This chapter uses the DPSIR framework to examine logistical, infrastructural, and operational challenges in SA's SWM system using Buffalo City Municipality as case study. Findings suggest that SWM is an endemic problem in the municipality characterised by a preference to landfill disposal, minimal recycling tendencies, poor waste services, non-enforcement of existent waste regulation, inadequate funding, and non-involvement of urbanites and informal sector in the process. To counter these challenges, BCM has enhanced its financial resources to facilitate scientific landfilling and transit to perceptions of solid waste as a useful good. The municipality has collaborated with local industries and international organizations to synergize efforts to implement state and local waste management plans. For an integrated model of SWM, this review suggests SA and BCM's need to plan innovatively, adopt landfill-mining, plan

and biodegradable biopolymers that are environmentally friendly in wastewater treatment. The critical parameter to determine the effectiveness of pollutants removal is chemical oxygen demand, colour, and total suspended solids. In this aspect, this chapter also discussed some of the significant findings done in previous studies to provide proper understandings and implications on this topic.

Unplanned and unmonitored growth of civilization and industries recorded in the previous decades have resulted in the introduction of several pollutants of emerging concern like pharmaceuticals and antibiotic drugs, endocrine disrupters, personal care products, etc. in adjacent aquatic environments. Moreover, the primary considerations of recent wastewater treatment strategies include energy neutrality and efficient recovery of valuable products as end/byproducts of the concerned process. The present study reviews such recent advances in designs, development, and implementation of algae based photobioreactors for simultaneous CO2 capture and treatment of effluents containing various types of emerging contaminants with negligible energy input. It also discusses the subsequent usage of algal biomass for production of value-added products like bioethanol, alginate, etc. This study has primarily addressed the potential and limitations of such photobioreactors for simultaneous achievement of wastewater remediation, CO2 sequestration, and bioenergy production with a zero-discharge concept.

The world over, efforts are being made for the safe disposal through development of cost-effective technologies to convert fast growing solid waste into useful products. In India, one such solid waste causing concern is the automobile windshield waste. The main sources of waste are automobile windshield manufacturers, dismantlers, recyclers, processors, etc. Automobile windshield waste emanating from various sources is a mixture of glass, PVB, additives, etc. The glass layer is fused with the PVB plastic film. Automobile windshield plastic film cannot be separated and is also not being recycled. At present, the reclaiming of PVB materials from automobile windshield waste is not being carried out in India. It is presently being thrown in dumpsites and land fill sites. It has negative impact both on human health and environment.

In contemporary society, the production and use of plastics is growing due to their assorted applications. However, these tendencies result to production of plastic waste whose effects to the environment are destructive. This chapter explores the production tendencies of plastics and plastic-waste chronologically,

one for the last few decades. So, these heavy metals must be removed from the environment by proper techniques to save the society. The chapter deals with the different toxic material removal technologies and mathematical modeling.

Soumya Saswat Priyadarshi, College of Engineering and Technology, Bhubaneswar, India

Six different concentrations of Chlorella vulgaris microalgae culture are incorporated with raw wastewater to improve treatment efficiency in three phases. In phase one of study, 30% and 40% microalgae were incorporated with raw wastewater with 24 hrs retention period for 18 days in a batch mode and these conditions are made same for all phases. During phase two and phase three of study, 20%, 25% and 35%, 45% micro-algae concentrations used to treat raw wastewater respectively. Amongst all the six concentrations, 30% microalgae behaves most efficiently and is able to reduce 34 to 81% COD, 43 to 75% nitrate, and 36 to 99% of phosphorous. Overall ammonia nitrogen reduction varies from 80% to 99%. No attempt was made to control temperature or evaporation losses. These attempts can definitely increase the growth process of microalgae and increase the removal efficiency of nutrients in wastewater.

Rachana Dubey, ICAR Research Complex for Eastern Region, India
Arbind Kumar Choudhary, ICAR Research Complex for Eastern Region, India
Janki Sharan Mishra, ICAR Research Complex for Eastern Region, India
Ashutosh Upadhyaya, ICAR Research Complex for Eastern Region, India
Sharad Kumar Dwivedi, ICAR Research Complex for Eastern Region, India
Surajit Mondal, ICAR Research Complex for Eastern Region, India
Karnena Koteswara Rao, ICAR Research Complex for Eastern Region, India
Abhishek Kumar Dubey, ICAR Research Complex for Eastern Region, India

Population growth, industrialisation, urbanisation, and climate change have created huge pressure on freshwater resources to fulfil the demand. Approx. 70-80% of the freshwater supply returns as wastewater, which is difficult to tackle and manage. We need to tackle the freshwater demand from different sectors like domestic, industrial, and agriculture. Most important is how to use the wastewater safely in agriculture. Therefore, it is an apt time to refocus on ways to recycle water especially in sectors like agriculture and for ecosystem services. The major concern in using wastewater in agriculture is its quality as the wastewater may carry pathogens, heavy metals, and many other pollutants, which might reach to human beings and animals via food chain. A solution to wastewater reuse is through bioremediation techniques. Bioremediation should be considered as a feasible and futuristic technology for safe use of wastewater in agriculture as it will reduce the burden on centralised water treatment system as well as it being economic and eco-friendly.

Foreword

Waste is the unwanted or useless solid materials generated from combined residential, industrial and commercial activities in a given area. It may be categorized according to its origin (domestic, industrial, commercial, construction or institutional); according to its contents (organic material, glass, metal, plastic paper etc); according to hazard potential (toxic, non-toxin, flammable, radioactive, infectious etc). Waste minimization is an important element of sustainable development because it allows improvements to business performance to be created concomitantly with improvements to environmental performance. Elimination of waste followed by reduction at source and then recycling form the preferred hierarchy of options. Reduction at source techniques includes good practice as well as changes to raw materials, processes, and products. Recycling on-site is preferred to recycling off-site, and of the four levels in the hierarchy of recycling, direct re-use is preferred since the material to be recycled is homogeneous, and from a known and reliable source. Systematic approaches to design and change can be implemented by considering first the service that needs to be provided, then the products needed to provide the service, and then the processes in which the products are made. Such a logical approach allows wastes arising from a variety of activities to be incorporated into integrated waste minimization studies. Such risings include product waste, process waste, and utility waste, the latter being created in the utility systems that drive manufacturing processes. There are 7Rs viz. Refuse, Reduce, Repurpose, Reuse, Recycle, Rot, Rethink. Every year, millions of tons of waste are generated from both households and building construction, most of which end up in landfills with a small percentage being recycled. Thus, there is a great need for waste minimization as this will not only have a huge environmental impact but also present substantial economic and social benefits.

Waste minimization entails limiting the amount of waste that is generated, thereby helping to eliminate the production of persistent and harmful wastes, effectively supporting efforts that promote a society that is sustainable. Thus, waste minimization involves a change of societal patterns that relate to production and consumption as well as redesigning products to eliminate the generation of waste. Waste Minimization is reduction in the quantity of hazardous wastes achieved through a conscientious application of innovative or alternative procedures. Simple adjustments to a process producing wastes may be the only requirement to achieve some results. However, looking at the broader picture in the University environment, it is often difficult to recognize waste reductions due to the complex and changing growth patterns within the campus community. Reductions are often offset by increased staff and student growth and/or building construction. Reuse of waste means any operation by which products or components that are not waste are used again for the same purpose for which they were conceived.

Here, in the current book, Dr. Ashok K. Rathoure has done a great job for the same in the area of waste minimization techniques especially for industrial sector. This will be a milestone in the current era where the waste is a major global problem.

Vishnu D. Rajput
Southern Federal University, Rostov-on-Don, Russia

Preface

Uncontrolled and rapid industrialization and urbanization is the major cause of rapid environmental degradation. Use of nature's freely available resources in an uncontrolled manner and throwing of waste into the environment has caused more stress to the assimilative capacity to dilute and disperse the wastes and recover the used natural resources. The depletion of resources has threatened the life of future generations. Increasing numbers of industries and production of various categories of wastes cause major environmental pollution. These problems have forced governments to control the industrial activities of waste discharge directly into natural environments. Formation of effluent and emission standards and regulation for wastes to be discharged into natural bodies of water, air and soil has forced industries to establish waste treatment facilities in their own industrial premises. This was the concept of end-of-pipe (EOP) treatment, in which the basic production process was the same with addition of waste treatment facilities. The basic concept of the installation of treatment units was to bring down the concentration and load of the effluent within the allowed limits before discharging into natural resources.

Rapid depletion of resources was not the focal point in EOP treatment process. But, with time, government bodies were asked to look towards solutions to control the rapid depletion of resources. The EOP treatment was unable to improve the environmental condition as expected, although the degradation speed was slowed down. Increasing pressure from the public about environmental condition has forced governments to implement more stringent standards. This has resulted in the concept of waste minimization in terms of quality and quantity.

Living standards and quality of life in most of the developed countries have increased since the Second World War, due to rapid industrialization and urbanization. This growth and development, however, have involved a nearly thirty-fold increase in use of fossil fuels over the past century and an over fifty-fold increase in industrial production for the same period. Most of the consumption of fossil fuel and production has taken place since 1950. It is estimated that between 1955 to 1988 world industrial output increased approximately six times, the World's population only doubled. The result of rapid industrialization is the rapid depletion of resources, and environmental impacts due to production of enormous waste exceeding assimilation capacity of the environment. At the beginning of this industrialization process, not enough attention was given about the state of the environment; assumptions were made that the nature has the capacity to restore the resources we extract and accept what we throw into it. These wastes have now become pollutants due to their enormous volume and multiple characteristics.

Based on the effects of pollutants in the environment, pollutants can be classified either as threshold (dangerous even at low concentration) or non-threshold (dangerous when present at a certain level) pollutants. Waste generated from industries is nonhomogeneous, with differing characteristics and composition. Similarly, these wastes are multimedia in nature with wastes disposed in all three media namely air, water, and soil. Effect of mixed pollutants of different characteristics can be synergistic and antagonistic depending upon the resultant effects. For example, the combined effects of sulfur dioxide in the air could produce acid rain. On the other hand, it is advantageous to have wastes with two different pH ranges (acidic and alkaline) which could eliminate the neutralization unit completely.

Continuous degradation of earth's environment mainly due to industrial pollution is one of the major concerns of many environmental regulators. In addition to this, public attitude towards environmental protection has led governments to control industries disposing high load waste in to the environment. Pollution control standards defined first for ambient air and water and soil quality standards were mostly derived from public health considerations and ecological justifications. Public concern over industrial wastes led the application of more stringent regulations, authorization, and procedures.

ORGANIZATION OF THE BOOK

The book is organized into 18 chapters. A brief description of each of the chapters follows:

Chapter 1 representing the Environment Friendly Technologies: Concept and Need for Sustainability. Any technology that is developed and used, will have certain components or constituting mechanisms such as Materials used, procedure or technological process involved, the effect on life around. The quality of life on planet with respect to environment such as fresh air to breathe, clean water for drinking and other purposes etc. are essential for all forms of life on earth. Therefore, whatever developments are undertaken should not degrade our environment. It is essential to bear in mind that whatever development, be it constructions, new power plants, transportation modes etc. should be carefully weighed in terms of its contribution to the environment, especially if there are any negative aspects that emerge from use of machinery, equipment, technologies etc. We should analyse any technology that we use in terms of its Cost Benefit study. Here in this chapter, author has discussed about environment friendly technologies for our sustainability to use the natural resources wisely with some case studies of natural as well as man-made disasters.

Chapter 2 reviews then Solid Waste Management Using the DPSIR Framework in a Southern Africa Case Study. This book chapter uses the DPSIR framework to examine logistical, infrastructural and operational challenges in SA's SWM system using Buffalo City Municipality as case study. Findings suggest that SWM is an endemic problem in the municipality characterised by a preference to landfill disposal, minimal recycling tendencies, poor waste services, non-enforcement of existent waste regulation, inadequate funding and non-involvement of urbanites and informal sector in the process. To counter these challenges, BCM has enhanced its financial resources to facilitate scientific landfilling and transit to perceptions of solid waste as a useful good. The municipality has collaborated with local industries and international organizations to synergize efforts to implement state and local waste management plans. For an integrated model of SWM, this review suggests SA and BCM's need to plan innovatively, adopt landfill-mining, plan strategically on all waste cycle aspects, and modify existing regulations to accommodate urbanites and informal waste pickers in the process.

Chapter 3 shows the Solid Waste, Treatment Technologies and Environmental Sustainability: Solid Wastes and their Sustainable Management Practices. Wastes has a long history of association with humans and other organisms. It is inevitable and has different sources right from crop residues (Agriculture), food production and its movement through food chain, industries and their processes. Waste has different environmental implications like leachate can contaminate groundwater, emissions from waste burning cause air pollution and its dumping in soil cause soil pollution. There are different methods and technologies being used in the world to minimize waste. All the technologies being used are reviewed and presented in it with their pros and cons of all the technologies. This chapter will include all these technologies and associated environmental concerns.

Chapter 4 is speaking about Assessment of phytoremediation efficiency of Coriandrum sativum in metal polluted soil and sludge samples: A green approach. A greenhouse pot experiment was conducted to evaluate the phytotoxic effect of heavy metals (Zn, Cu, Ni and Cr) on the growth (plant height, plant and seed weight) of Coriandrum sativum and uptake of metals. For this purpose, the polluted soil and sludge samples were collected from three sites of Moradabad: Karula nala (KS), Dhauri nala (DS) and Karula nala sludge (KSL). Metal content in the plant and soil was determined by AAS technique. The plants were also hydrodistilled using Clevenger apparatus and the extracted oil was analysed by GC and GC/MS. The content of copper in KS and KSL, Zn in KS, Ni in DS and KSL and Cr in KS was above the permissible limits according to Indian standards. Plant weight was significantly affected in the different soil samples. Among all the collected polluted soil and sludge samples, DS sample showed the highest metal accumulation, while in the KSL sample, the plant could not survive. The study revealed that coriander can be used to remediate the contaminated soil with economic return and metal free final product, essential oil.

Chapter 5 demonstrates the ability of Previous, Contemporary and Prospects of e-waste and its management. The contemporary world is driven by electronic gadgets without which the survival of mankind is perceived to be incomplete. The uncontrolled dependence of mankind on electronic gadgets has resulted in enhanced production of these gadgets leading to the accumulation of e-waste. Both technological innovation and market expansion have played an important role in electronic waste (e-waste). Owing to hazardous material composition, electronic waste causes environmental problems during the waste management phase if not properly pre-treated. Growing attention is being given to the impacts of these hazardous components from e-waste on the environment. Many countries have drafted legislation to improve the reuse, recycling, and other forms of recovery of such wastes so as to reduce disposal problems. The purpose of this chapter is to present an overview of electronic waste, the current status of management of electronic waste, and recycling technologies for the recovery of metals from end-of-life electronic equipment.

Chapter 6 illustrates the technological interventions in management of Design and Development of Small-Scale Oil Palm Waste Shredder for Achieving Environmental Sustainability. This chapter presents the design and development of an economically and technically feasible small-scale oil palm waste shredder to contribute to oil palm waste management. Most available oil palm waste shredders are unaffordable due to high initial investment cost, large, heavy and un-portable, which have been the drawbacks for small and medium enterprises to participate in the industry that relates with oil palm waste management. To address this issue, a low cost and small-scale oil palm waste shredder were designed and produced. The central design concept deduced is to compromise acceptable efficiency with an affordable price by integrating all mandatory components of oil palm waste shredder into a comparatively small and portable support structure. The innovations made in the design are size reduction and adoption of double-shafts cutters with seven jaws technology. Preliminary calculations supported the design. The cutting force produced is up to 19.45kN with safety factor of 4.5. This design can handle other oil palm wastes as oil palm frond is harder than other oil palm wastes.

Chapter 7 is about microalgae Performance of Chitosan and Polyglutamic Acid in Palm Oil Mill Effluent Treatment. This chapter presents the study on pollutant removal efficiency in palm oil mill effluent using chitosan and polyglutamic acid (PGA). Up until today, palm oil mill effluent (POME) considered one of the significant sources of environmental pollution. The characteristics of POME include contaminating the source of drinking water, which also harmful to the aquatic ecosystem by creating a highly acidic environment or causing eutrophication. With increasing public awareness of environmental pollution, it creates the need to address this issue. Both chitosan and PGA are non-polluting food-based anionic and biodegradable biopolymers that are environmentally friendly in wastewater treatment. The critical parameter to determine the effectiveness of pollutants removal is chemical oxygen demand, colour, and total suspended solids.

Chapter 8 presents the phytoremediation Microalgal photobioreactors as an integrated approach for simultaneous wastewater treatment, carbon sequestration and recovery of valuable resources. Unplanned and unmonitored growth of civilization and industries recorded in the previous decades have resulted in the introduction of several pollutants of emerging concern like pharmaceuticals and antibiotic drugs, endocrine disrupters, personal care products etc. in adjacent aquatic environments. Moreover, the primary considerations of recent wastewater treatment strategies include energy neutrality and efficient recovery of valuable products as end/by-products of the concerned process. The present study reviews such recent advances in designs, development and implementation of algae based photobioreactors for simultaneous CO2 capture and treatment of effluents containing various types of emerging contaminants with negligible energy input. It also discusses the subsequent usage of algal biomass for production of value-added products like bioethanol, alginate, etc. This study has primarily addressed the potential and limitations of such photobioreactors for simultaneous achievement of wastewater remediation, CO_2 sequestration and bioenergy production with a zero-discharge concept.

In Chapter 9, the author has discussed about the Research and Development in Automobile Windshield Waste (AWW). World over, efforts are being made for the safe disposal through development of cost-effective technologies to convert fast growing solid waste into useful products. In India, one such solid waste causing concern is the automobile windshield waste. The main sources of waste are: automobile windshield manufacturers, dismantlers, recyclers, processors etc. Automobile windshield waste emanating from various sources is a mixture of glass, PVB, additives etc. The glass layer is fused with the PVB plastic film. Automobile windshield plastic film cannot be separated and is also not being recycled. At present, the reclaiming of PVB materials from automobile windshield waste is not being carried out in India. It is presently being thrown in dumpsites and land fill sites. It has negative impact both on human health and environment.

Chapter 10 indicates the interaction of Plastic Waste Menace and Approaches to its Management through Biodegradation. In contemporary society, the production and use of plastics is growing due to their assorted applications. However, these tendencies result to production of plastic waste whose effects to the environment are destructive. This book chapter explores the production tendencies of plastics and plastic-waste chronologically, discusses its negative environmental effects and proposes the adoption of biodegradation as a management approach. A number of effects including death, retarded growth and development, emergence of cancers and altered physiology are associated with ingestion of plastics in animals and humans. Water and air pollution are associated with unscientific disposal of plastic waste resulting to leaching and release of greenhouse gases. The use of microbes such as bacteria and fungi that have inherent enzymes such as lipases, esterases, hydrolases and amylases is discussed as a sustainable solution to the plastic menace. The successful use of biodegradation must be coupled with scientific research to optimise the factors influencing the process.

Chapter 11 is about the research innovation in Plastic Waste Disposal and Role of Microorganisms in Bioremediation of Plastics. Plastics are the most commonly used polymers for conventional applications. Plastic wastes accumulating the environment are presenting an ever-increasing ecological threat. Low Density Polyethylene is major cause of long-term environmental pollution. An eco-friendly approach to resolve this ever-growing persistent menace is Bioremediation. The diverse metabolic capability of microbes can be effectively exploited for breakdown of plastic wastes. Another approach is to synthesise biodegradable or organic plastics which is soluble in the natural environment. There are several advantages associated with these biodegradable plastics or plastic degrading microbes. They can be composted with organic wastes and returned to enrich the soil. Their use will not only reduce stress and loss of habitat of wild animals caused by dumping of conventional plastics but will also lessen the labour expenses for the removal of plastic wastes in the environment because they degrade naturally. This review describes bioremediation/biodegradation process, production, types of and advantages bio-plastics.

Chapter 12 presents the sustainable solid waste management via biological treatment- A critical study: Biological treatment of solid waste. The solid wastes generated as a result of human activity have become very complex. Unscientific management of such wastes has resulted in environmental pollution; thereby putting a serious risk to the ecosystem and human health. A sustainable waste management is required for reduced pollution levels as well as for safeguarding the environment for future generation. Biological methods have gained in importance mainly because of its environmentally friendly option for recovering materials and energy. This is true as per the guidelines of "Hierarchy of Sustainable Waste Management". Composting and anaerobic digestion are two treatment options that is commonly followed in both developing as well as developed countries. The aim of this chapter is thus to address the technical background, advantages as well shortfalls related to the waste treatment via biological methods. An up-to-date literature thus presented may be beneficial to the scientists and researchers for working towards further improvement of the present system of waste management.

Chapter 13 is a review on isolation of Acetobacter and production of vinegar from pineapple and papaya waste (peels) using *Acetobacter aceti*. The diverse agro-chemical zones create India as the second largest producer of contemporary fruits and vegetables. These are used for production of processed fruit and vegetable products such as juices, preserved jams that leads to generation of large amount of wastes, both solid and liquid. This waste is incautiously or carelessly being thrown without any treatment promoting environmental deterioration. It deals with how these fruit waste especially pineapple and papaya peels which are usually discarded during the processing or consumption of the fruit are useful in preparation of vinegar which is termed as a useful product from waste. it is also discussed about the general characteristics of Acetic acid bacteria, importance of the bacteria and how it can be isolated by natural sources which is cost effective. So, the waste peels generated after consuming the fruits and vegetables need not be discarded as waste instead it can be used in so many ways to produce useful products.

Chapter 14 is about bioremediation and biodegradation technologies for removal of pollutants and wastes from environment. Bioremediation is defined as the process whereby organic wastes are biologically degraded under controlled conditions to an innocuous state, or to levels below concentration limits established by regulatory authorities for bioremediation to be effective, microorganisms must enzymatically attack the pollutants and convert them to harmless products. As bioremediation can be effective only where environmental conditions permit microbial growth and activity, its application often involves the manipulation of environmental parameters to allow microbial growth and degradation to proceed at a faster rate. Bioremediation techniques are typically more economical than traditional methods such as incineration, and some pollutants can be treated on site, thus reducing exposure risks for clean-up personnel, or potentially wider exposure as a result of transportation accidents. Bioremediation is based on natural attenuation the public considers it more acceptable than other technologies.

Chapter 15 is about use of wastage biomaterial/biomass for wastewater treatment and modeling: Mathematical modeling for adsorption process. Recently, the existence of toxic materials in our environment causes severe concern to the living organism. The main causes for the contamination of such toxic materials in the environment are rapid industrialization and agricultural activities. The different toxic entities present in the environment mainly depend on the nature of the industry. However, some of the common generated toxic pollutants present in the water bodies are heavy metal ions, phenol, dyes, pesticides, insecticides, detergents and a wide spectrum of aromatic compounds. Among those substance heavy metals have been reported the major one for the last few decades. So, these types of heavy metals must be removing from the environment by proper technique to save the society. The present chapter deals with the different toxic material removal technologies and mathematical modelling.

Chapter 16 is focusing on wastewater treatment using microalgae: Nutrient Removal. Six different concentrations of Chlorella vulgaris microalgae culture is incorporated with raw wastewater to improve treatment efficiency in three phases. In phase one of study, 30% and 40% microalgae were incorporated with raw wastewater with 24 hrs retention period for 18 days in a batch mode and these conditions are made same for all phases. During phase two and phase three of study, 20%, 25% and 35%, 45% microalgae concentrations used to treat raw wastewater respectively. Amongst all the six concentration 30% microalgae behaves most efficiently and able to reduce 34 to 81% COD, 43 to 75% Nitrate and 36 to 99% of phosphorous, Overall ammonia nitrogen reduction varies from 80%to 99%. No attempt made to control Temperature or evaporation losses. These attempts can be definitely increasing the growth process of Microalgae and increase the removal efficiency of nutrients in wastewater.

Chapter 17 represent the safe use of wastewater in agriculture through bioremediation process. population growth, industrialisation, urbanisation and climate change has created huge pressure on freshwater resources to fulfil the demand. Approx. 70-80% of the freshwater supply returns as wastewater which is difficult to tackle and manage. We need to tackle the freshwater demand from different sectors like domestic, industrial and agriculture. Most important is how to use the wastewater safely in agriculture therefore it is an apt time to refocus on ways to recycle water especially in sectors like agriculture and for ecosystem services. The major concern in using wastewater in agriculture is its quality as the wastewater may carry pathogens, heavy metal and many other pollutants which might reach to human beings and animals via food chain. A solution to wastewater reuse is through bioremediation techniques. Bioremediation should be considered as a feasible and futuristic technology for safe use of wastewater in agriculture as it will reduce the burden on centralised water treatment system as well as it is economical and eco-friendly.

Chapter 18 has explored the possibilities of characterization and reuse of wood waste. Wood is the most versatile natural material, it can be used to obtain sawn wood, wood for civil construction, naval purpose, packaging, furniture, wood flooring, composites, bioenergy, pulp and paper. More recent the nanotechnology and the biorefinery processes creates news possibilities for wood use, however the wood processing still generates a large amount of waste. The wood can be processed mechanically or chemically, the mechanical processing of the wood has residues like sawdust, cuttings and wood chips, from the chemical processing of wood the main residue of this processing is black liquor.

CONCLUSION

The current scenario where we are reaching to next level of increased population (8 billion) and due to increased demand of supplies the waste generation is so high. Hence, we need to adopt zero waste policy everywhere. In this connection, the current waste minimization practices required to modify and adopt some new techniques for waste minimization at source for better environment and future sustainability.

Ashok K. Rathoure
M/s Akone Services, Lucknow, India

Acknowledgment

This book is the result of the dedicated effort of numerous individuals. We would like to acknowledge the help of all the people involved in this project and, more specifically, to the authors and reviewers that took part in the review process. Without their support, this book would not have become a reality.

First, we would like to thank each one of the authors for their contributions. Our sincere gratitude goes to the chapter's authors who contributed their time and expertise to this book.

Second, we wish to acknowledge the valuable contributions of the reviewers regarding the improvement of quality, coherence and content presentation of chapters. Most of the authors also served as referees; we highly appreciate their double task.

Additionally, we would like to thank our Editorial Advisory Board.

Our Production Coordinator skillfully managed the production of the textbook. We wish to thank the publisher for the sustained interest shown by him during the entire work. A sincere expression of the thanks goes to the entire staff of IGI Global.

Generous participation of faculties and students to eliminate errors in the text and to eliminate errors in the text and to refine the presentation will be greatly acknowledged.

Last and not least, we beg forgiveness of all those who have been with us over the course of the years and whose names we have failed to mention.

Ashok K. Rathoure
M/s Akone Services, Lucknow, India

Chapter 1
Environmentally Friendly Technologies:
Concept and Need for Sustainability

Ashok K. Rathoure
https://orcid.org/0000-0001-9131-1346
M/s Akone Services, Lucknow, India

ABSTRACT

Any technology that is developed and used will have certain components or constituting mechanisms such as materials used, procedure, or technological process involved. The quality of life on planet with respect to environment such as fresh air to breathe, clean water for drinking, and other purposes are essential for all forms of life on earth. Therefore, whatever developments are undertaken should not degrade our environment. It is essential to bear in mind that whatever development, be it construction, new power plants, transportation modes, etc., should be carefully weighed in terms of its contribution to the environment, especially if there are any negative aspects that emerge from use of machinery, equipment, technologies, etc. We should analyze any technology that we use in terms of its cost-benefit study. Here in this chapter, the authors have discussed environmentally friendly technologies for our sustainability to use the natural resources wisely with some case studies of natural as well as manmade disasters.

BACKGROUND

Environment is basically, the support system that upholds life on earth. It provides the basic requirements to support life such as fresh air that living organisms respire or breathe, food to eat, fresh water for plants & animals and also land where living beings live. There is a rhythm or pattern or natural order in the environment, which keeps the environment in a self-sustaining mode. However, man is in constant interaction with the environment. As a result of developments of civilization over the years; the industrial development; and new breakthroughs in various fields, he has had interventions with Nature. Many of the self-sustaining in-built order in nature are being threatened by way of various human and industrial activities.

DOI: 10.4018/978-1-7998-4921-6.ch001

ENERGY REQUIREMENTS AND DEVELOPMENT

Energy is a pre-requisite or in fact essential for meeting our developmental needs or accomplishing the development goals for any economy. While moving from a solely agrarian economy and further on the coming of the industrial revolution and later on, with the further developments in Information and Communication Technologies, mankind has come to increasingly depend on various sources for power to run the machinery and equipments, gadgets etc. in our homes and in industrial units.

Renewable Energy and Non-renewable Energy

We can broadly categories our energy sources into two - namely, Renewable Energy and Non-renewable Energy. Renewable Energy refers to the energy that can be generated from renewable sources. Renewable sources are basically those which are in nature and which gets renewed or replenished by the natural processes in the environment. For instance, Water. Water is a natural resource that gets replenished in nature. There is lot of fresh water resources on earth such as the rivers, streams, pounds, springs etc. Apart from this, there are oceans and sea as well. By the process of evaporation, water gets evaporated. This water vapor gets condensed to form clouds. Further, by way of rain or snow etc., it comes down to the earth's surface. Thus, water changes to water vapor form and again to water during this cycle. Water in fact, is getting continuously circulated by this natural process. That is why, we say, water is a renewable resource provided by nature. Another renewable resource we have is from the sun, the solar energy. Sun's light and heat fall on the earth's surface daily and can be considered by us for meeting the energy needs. Yet another one, is the Wind energy, which can be harnessed in areas where there is potential for wind energy.

Now talking about Non-renewable Energy, Non-renewable Energy, on the other hand, it refers to the energy from sources, which cannot be easily replenished, as is the case of renewable energy. Non-renewable energy comes from depleting sources. This means that we need to be concerned about such sources as it cannot be made available for all ages. Therefore, non-renewable sources are not sustainable energy sources and when we use it in the present generation, it is by limiting the future generations to such energy sources. As a big measure towards environment sustainability and environment management, it is imperative that we turn to Renewable sources of energy to meet our energy needs. Every region should explore the scope for use of renewable energy there. This means the scope of using renewable sources of energy needs to be explored. Areas which are abundant with water sources, solar energy, wind etc. are to be assessed for harnessing renewable energy.

Further, it is matter of concern that even now, electricity is not available in many parts of a country, as in case of India. It is required that it is made available to all parts of a nation so that development takes place everywhere. In case of more populous countries, the challenge is higher. It has to be well taken care of, so that infrastructural bottlenecks do not hamper the developmental goals. Depending only a single source of power will have limitations, if the supply-demand gap in power sector is high. Also, the institutional arrangements that are put in place for power needs may not suffice to meet requirements. Environment is the hub of rich natural resources. The rich flora and fauna, streams, rivers, oceans, mountain ranges are not only sources of natural beauty but also are sources of abundant energy. These natural resources found on our planet Earth is the source of rich energy sources which are of great use for mankind. The use of any technology is for a stated purpose. The technological product or the equipment would have been designed to perform certain stated predefined or predetermined tasks that would serve to enhance the efficiency of operations or scale up operations or even be an innovation that can help radically re-engineer the current processes and systems of practice followed in undertaking any activity.

Technological products, by way of its day-to-day use, entail an interaction. By interaction, the technological product operates in an environment - that is a sum total of the surroundings; inclusive of the user of the technology, the inputs or such resources that go into the use or running of the technological product, the output generated out of the technological product. The use of a technological product will have an effect on the environment – this effect can be in three lines - (1) positive effect (2) neutral effect or (3) negative effect. When the product used, adds to the richness of the environment, by any means, then it has a positive effect. When the product used, does not harm or add any richness, then it will of neutral nature. And of more concern, is when the product used has an ill-effect, then it can be said to be of negative effect. This negative effect can be - of a nature of direct effect or of an indirect effect. The negative effect on the environment, can be - of being of a main effect or of the nature of a side effect. Technologies that are designed and developed, taking into cognizance, that the environment in which it operates, is also as important as the use of any such technology and in fact much more - are those designed as Environment friendly Technologies.

THE CONCEPT OF EFT

There are some features or aspects which help us to discern whether the technologies that are in use in our societies and even in our day-to-day lives, are environment friendly or on the other hand, environment harmful. Some of the aspects that are found to be characteristic in environment friendly units are:

- Use environment friendly mechanisms
- Use natural resources
- Preserve or protect our environment
- Maintaining the Ecosystem
- Sustainable development

Any technology that is developed and used, will have certain components or constituting mechanisms such as Materials used, procedure or technological process involved, the effect on life around. That is, basically, there will be some inputs or a set of inputs. The set of specified inputs in the required combination or configuration which will be subject to a specified process – that is, some technological process and further, outputs are generated. So, if we look at the components for managing our environment in an environment friendly approach, we need to focus on Environment friendly technologies. So, we need to look even beyond inputs, process and outputs. Retake So the components will encompass for Environment friendly technologies the following:

- Inputs (materials, equipment)
- Process
- Outputs
- Uses
- Beneficiaries - people, industry, flora and fauna, environment
- Environmentally Sustainable.

Inputs - When we talk about inputs for an Environment friendly technology, it will include the materials required, the equipment which will be used, the spare parts/ other items etc. The inputs that are used in a EFT should be environment friendly, that is it should by itself, be an input that is Environment friendly. That is why when we are using renewable energy sources, those are environmentally sustainable and hence, Environment friendly. The Materials that we use, would be positive and harmless for our environment, if that used are products which are EF or such products used are such which can be recycled and reused, or the used products are biodegradable waste creation or by-products, as a result of the use of a technological product and is not harmful to society, that is, waste management/ by-product should be an environment friendly mechanism

Process - The process entailed in the use of EFT should not be harming the environment. It should be environment friendly. By way of use of any process, there can be changes or implications in the immediate environment in the short run or in the long run. This needs to be taken into cognizance when are looking at technologies or designing solutions, that we aim to make environment friendly.

Noise pollution, air pollution, harmful radiation / other health hazards etc are some concerns that need to considered in this context. The effect of such processes on living beings - humans, flora, fauna, aquatic life etc. should be examined.

Outputs - The main output in use of EFT, of course, is the desired or intended output for which the EFT has been developed for. For instance, in case of wind turbines, the electrical energy that we are able to harness is the desired output from use of the wind power. While designing EFTs, it is also imperative to consider the possible uses of such EFTs, the beneficiaries of such EFTs, which can be people, industry, flora and fauna, and even the environment itself.

We should analyze any technology that we use in terms of its Cost Benefit study. That is what all costs go into the making and use of the technology. The cost will include not only cost of raw materials, consumables etc. but also cost to the environment where it will be made operational. Another term that is referred in the context of setting up of a new project or plant, is the Environment Impact Analysis. That is, in what way and to what extent the new project, if set up, will impact our environment.

NEED FOR ENVIRONMENT FRIENDLY TECHNOLOGIES (EFT)

We have made great progress in use of industrial equipment, machinery, electronic gadgets, ICT equipment, latest technologies etc. We have also built a greater number of factories, plants, industrial units etc. We have explored and made considerable interventions into the hitherto pristine environment at many places. This has resulted in man-environment interactions that has in some aspects, already affected our environment. For instance, air pollution levels have increased. Non-biodegradable waste has gone into such enhanced levels of which we need to be concerned about, water bodies affected with pollution is noticed at many places. The biodiversity that we have on earth is another important aspect. No doubt, we need to preserve the rich biodiversity we have on planet earth. The different varieties and variants, we have of the different flora especially in the forests, valleys, mountains are to be protected. Activities and developments undertaken by us should be such it does not affect or harm the plant life. Similarly, deforestation, soil erosion, landslides etc. as a result of man's interventions into nature can result in loss of natural vegetation on earth. Plants and trees are very important to maintain the ecosystem and also is life giving for the animal life on earth. Similarly, the fauna on earth is also to be preserved. Deforestation and the resulting reduced forest cover, will affect the wildlife. Many species found only in specific forests and in mountainous regions are being reported to have falling numbers. The quality of life on planet with respect to environment such as fresh air to breathe, clean water for drinking and other purposes etc. are essential for all forms of life on earth. Therefore, whatever developments are undertaken should not degrade our environment. It is essential to bear in mind that whatever development, be it constructions, new power plants, transportation modes etc. should be carefully weighed in terms of its contribution to the environment, especially if there are any negative aspects that emerge from use of machinery, equipment, technologies etc.

Footprints that we leave on the earth will have bearing on the future generations as well. The extent to which we should use the resources on earth, should be taking into consideration the need of the future generations also who will inhabit the same planet. Sustainable development precisely refers to this concept wherein we do not compromise on the needs of the future generations by focusing on the needs of only the current generation. As part of increasing consciousness on the use of various resources, we have started to measure how much of a resource we utilize.

EFT AND SENSITIVITY

Solar Energy is being used in many institutions such as hospitals, educational institutions, business firms etc. to generate electricity. The energy from the sun is being sourced to meet their power requirements. It is interesting to note that they have got clear advantages in implementing energy obtained from renewable sources.

Solar for Transportation

When we talk about solar powered devices, we also need to know that today solar powered transportation is also possible. Solar energy has been considered for use in the transportation sector such as solar buses, solar boats, airports that use solar energy to power etc. These are initiatives which will help us to move towards sustainable transport. This is significant in the context that transportation systems entail huge energy consumption and emissions. So, in terms of air pollution/ water pollution etc. these need to be carefully weighed for environment sustainability. Greenhouse gas emissions have been a matter of concern. As per the UNEP (United Nations Environment Programme), the transportation sector is major source of greenhouse gas emissions and also other forms of pollution; and that 95% of the world's transport energy is from fossil fuels. So, it is a major concern and efforts required in place to mitigate greenhouse gases emission. Therefore, moving towards green transport and clean energy is the need. Engaging in environment friendly practices and technologies are essential. We will discuss about some of the environment friendly technologies installed for the transportation sector and also what can be our role in mitigating environment pollution in subsequent modules.

Case Study of Cochin International Airport Limited

Cochin International Airport Limited (CIAL) is the first airport in the world to be solar powered. The foray into solar power began with the setting up of a pilot plant of a capacity of 100kWp in the year 2013. Further, it went on to installation of higher capacity of 1 MW. Considering the huge power requirements on a daily basis and the electricity tariffs to the State Electricity Board, CIAL decided to install solar power plants of capacity adequate to meet its power needs. In this effort, CIAL after taking into consideration the requirements arrived at the designed capacity of 12 MWp, apart from this, a provision to enhance by 2.4 MWp. The project was done in the year 2015 within a short time period of six months. This plant is the major solar plant and generates around 45000 units on an average daily basis, and by this CIAL adorned the rarest of rare distinctions, of being - the first fully solar powered airport in the world. It is a magnificent sight to see the solar panels laid on around 45 acres of land, suitably configured to generate power of desired capacity. You will recall from our previous modules where we discussed about the principle of working of photovoltaic technology that solar panels convert the solar energy into electricity which is DC or direct current. Inverters are used here too, to convert the DC to AC. Transformers are used to step up and further, connected to the grid. CIAL has gone in for availing the banking facility provided by the State Electricity Board. As per this, instead of having a mechanism of storage using battery, the excess power generated can be given to the grid and such excess power provided can be sourced from the grid. Further, the excess units provided to grid will be paid the electricity board. So here it has the scope for not only being self-sufficient and thereby grid neutral, but also being a contributor towards meeting power requirements outside. The extent of power generation through a renewable source such as solar, is considerable when it is to meet the needs of an airport. So, it only once again highlights to us about the great scope that lies in harnessing solar energy for our needs. Further, in tune with its efforts to remain grid neutral with further expansion plans and new developments such as international terminal, CIAL had taken up setting of additional solar plants in its premises such as the canal top solar project. The carport solar project which is an interesting case where car parking was suitably designed to house solar panels and generate power. This is especially a commendable contribution in a State such as Kerala where the State purchases a good proportion of its power requirements from the Central Grid.

Case Study of Metro Rail System in India

The Metro rail system is being increasingly considered as a good mode of urban transport. The metro rail system has certain advantages such as produces less noise and unlike road transport vehicles, emissions are not a matter of concern here. In India also, in many of the big cities, metro rail system has been implemented. Even the metro rail system can be best utilized to generate power using renewable source such as the solar power. The space available along the metro and the roof tops can be considered for optimum utilization.

Case Study of Solar Ferry Service

ADITYA, India's first solar ferry which was inaugurated on 12th Jan 2017. As per its designer, Sandith Thandashery, whom we just listened to, it scores very well on the five important parameters for a passenger ferry which are (Thandasherry, 2018):

1. Safety,
2. Comfort,
3. Pleasure,
4. Cost-effectiveness
5. Eco-friendliness.

It is stated that by annually using ADITYA, the solar ferry, the saving is, in terms of 94 tons of CO2 and 35000 liters of diesel. So, it is clearly indicating that the solar ferry is a greener & better option for water transport. Further, in terms of cost, it is saving of 22 lakh rupees. This boat has been operational for a year now on the Vaikom -thavanakadavu route and passengers commute via this ferry service. When we talk about this solar ferry boat, the eco-friendly aspect is a big advantage with respect to water transport, primarily due to following aspects. The conventional ferry boats which are diesel powered cause pollution namely- air pollution and water pollution. Emissions from such diesel engine powered boats include CO^2 & other such matter. The gases emitted will affect health of those who operate the boat and that of the passengers. A solar boat helps to not only have avoid diesel but also associated emission or production of CO_2 & such gases. With regard to diesel powered boats, water pollution also is a problem. The oil which is used, may leak from engine, and machinery system during the course of operation of such boats, and this can eventually get easily discharge into the water. Similarly, fuel that is unburned can also be getting released into the water body. This will pollute the water body and also will be dangerous to the aquatic life (Thandasherry, 2018). This institution also has begun using solar energy for the power needs of its office. Few solar panels have been provided on the roof top of the firm is being used to meet some of their power needs. So, as an organization, the Navalt is employing solar power and is a good example for many more institutions to follow environment friendly measures.

SENSITIVITY TOWARDS ENVIRONMENTAL ISSUES

Degradation of the natural environment, extinction of many varieties of flora and fauna and their impact on human lives etc. are now visible all over the world. Though it is enriched with natural resources the situation in India is even more precarious as it is a densely populated country. The sustainable development of the country is facing big challenges due to the environmental pollution, those especially linked to land, water and air. Hence it is a necessity to sensitize these issues in the realm of Environmental Education. Let us have a look into the contemporary environmental issues experienced by us locally and which will later help us to think and act globally.

- Environmental pollution- Land pollution, Air Pollution, Water pollution and Sound pollution.
- Deforestation
- Spoilage of natural resources
- Climate change

Environmental Pollution

Air pollution is one of the worst conditions India faces. The International Energy Agency (IEA) reports that by 2040, India has to face about 9 lakh premature deaths in the country due to the shooting up of air pollution only. As India is at the 11th position out of 20 in pollution rate the life expectancy of newborn children may go down about 15 months. Also, it is a threatening statistic according to the Environmental Performance Index of 14th rank of India among the 180 most polluted countries in the world. Local observations and experiences in the villages of Kerala thrown light into a very big phenomena happening in the flowing water bodies. The word river brings forth an image of flowing water among all the people without the difference of urban or rural, time and space. The traditional concept of running river has been changed now. Almost all the small rivers in the urban places and a few in the rural places (areas) are stagnant with a lot of solid dumps, pollutants and other wastages. But the astonishing seen we see all around is the stagnant water bodies with a layer of precipitated pollutants and solid wastes in dark colors and seen as the habitat of varieties of mosquitoes and microbial organisms. The health mission of Kerala like Arogya Keralam is placing warning boards every nook and corner to avoid stagnation of water to prevent mosquitoes and other microorganisms which cause various kinds of communicable and chronic diseases. Basically, these warning boards are not meant for rivers and such other water resources. They meant the house hold surroundings, drainages and water holding things like utensils, bottles etc. But in the present situation more than this house hold surroundings it is applicable to the water bodies like rivers and ponds. This is a serious issue which seeks immediate solutions from the part of the public and also from the government.

Environmental Degradation

The rapid growing population, urbanization, economic development, industrialization etc. cause very serious environmental degradation. The natural demographic phenomena like birth, death and human migration taken place and continues its way as usual. The varying and increasing population density demands a lot of infrastructural developments and which results in the huge amount of utilization of natural resources. The technologies and scientific interventions for life facilitations cannot be avoided by the common people. Because convenience factor is the prime concern of our time and for that we compromise a lot with all environmental issues without looking forward and without leaving anything behind except these challenges. India is growing very fast in population to overcome China, the most populous country of the world. In the same manner the imbalance in ecological factors are also growing. In its true sense, we must agree that the general public is not at all bothered about the environmental concerns. It is high time to think and act properly to stick on a developing and sustainable path. Heavy loss of natural resources, heavy pressure on land (land degradation), deforestation, habitat destruction and loss of bio diversity are some of the crucial issues happen due to urbanization. The changing consumption pattern cause rising demand of energy which results in spoilage of natural resources.The infrastructural developments, increase in number of branded expensive motor vehicles, increase in trade and businesses etc. are the economic development indicators of the present days. Even in educational field the status of institutions is indicated by the luxurious facilities provided by the management. Also, it is a trend to rate the quality of the schools on the basis of the amount paid by the parents as donation and other yearly expenses to meet the education of their children. This trend is seen everywhere and in all sectors of life.

Several studies have been conducted on environmental costs and developmental costs and it is found that the majority of the public are of the opinion of increasing developmental costs than environmental costs. It is a clearly evident social truth and common tendency among the public that they always follow superficial and immediate developmental strategies than the philosophical and futuristic dimensions of sustainable development. Environmental quality and sustainability are considered only after the fulfillment of basic needs and sometimes after the fulfillment of all the livelihood requirements even by the government. At the part of government there might have been a number of social development commitments like poverty eradication, employment opportunities, and utilization of development funds from agencies. Also, it is matter of political and democratic governance and policy of the ruling party for getting approval and appreciation from the general public and their voters. We have to agree with the policies taken for eradicating of poverty and to provide a shelter with basic requirements to all. In a civilized society nobody can object these fundamental rights of the citizens in the cost of environmental protection. But sustainability cannot be compromised. Some developed countries advocate this policy as a better thing to attain a developed state of economy in the countries first and then implement developmental activities for environmental sustainability. Earth with pure air, fertile land, clear and pure water, diversity of flora and fauna, cool atmosphere is needed to everyone. Our development should not harm our young one's existence. That is the concern of being educated and educate others environmentally. Education is the best route and destination of the journey of sustainable development and teacher education is the junction for enrooting the technical support and guidelines through interdisciplinary approaches. Thus, the integrity of work education in the process of sustainable development can be assured through environmental education.

Unfortunately, the environmental concerns are being awakened only at the time of crisis management. When a crisis arises, the circumstances will immediately be altered. Behind every movement against environmental threats there will be a learned group committed to educate the public at any cost. We call them activists. But the policy makers and bureaucrats ignore environmental costs to meet their objectives of economic development. Sustainability is not a prime concern for them. Then who can assure environmental sustainability? Solutions have to be found and implemented at personal level and local level. Several studies have been conducted all over the world to determine the reasons of environmental degradation and climate change those having high correlation in different magnitudes. That is just like the relation of deforestation with global warming, carbon emission, water scarcity etc.

Among them global warming becomes very crucial and prime reason for almost all the other challenges. There are many studies which reveal the results of threatening truth of the climate variation due to global warming in India. The highest temperature noticed in India during the summer of 2016 was in Rajasthan, 51 degree Celsius- a real indicator of the increasing rate of global warming over the past five years. The interior areas near the forest boarders are also facing such a challenge. We can again observe the example of the fragile ecological area of Western Ghats of Idukki, one of the most beautiful tourist places of South India. In my own experience when my life started there in the midst of a cold summer- yes it was a cold summer 21 years back in the high range areas- it was not possible to take bath without warm water. There was no fan in any of the houses and it felt as a symbol of luxury only. Now the story has been changed. In summer it is not even to think to sleep at nights without a fan. Then no need to tell about days! This is the change story of a period of almost 21-22 year's duration. The changes of the recent years are not slow as such. The drastic change occurs in the atmospheric temperature is being unpredictable. Another issue India face is the melting of Himalayan glaciers in an alarming rate and its increase in frequency which results in floods and the deadly natural calamities happens in the fertile river banks where the rich yields help us to sustain in food production. The frequent forest fire in summers doubles the risk of global warming. The imbalanced state of layers of earth again threatens us by earthquakes. Totally life becomes a big challenge and a chance. Everything happens only because of our negligence not because of ignorance. This is the context where our role is essential by imparting a fruitful practical education to make us and others to practice on environmental sustainability. In olden days people were planning their livelihood according to the seasons and climate. Now we are not able to predict even the monsoon. We all experience the unpredictable variations and extreme end of weather and thus weather forecasting is being a satire in social media. It is not the fault of the scientific inventions and technology but the repulsion of the extreme level of environmental degradation. Our height of intellect is being defeated by the intensity of natural calamities even though we have several technologies to predict and measure it but no technology to resist it. Hence the only way to be aware is the process of education and the only way to resist is the product of education.

Kerala Floods of 2018- A Paradigm Shift

The Kerala floods of 2018 were a great leveler. All and sundry were put under the same roof- the roof of the rehabilitation camps. Man, machines and government policies all contributed to the disaster. There was allegation of policy paralysis at all levels. One of the fundamental issues that the floods brought into focus was the fact that hard environmental decisions could no longer be postponed. The Gadgil committee had brought into sharp attention the devastation that environmental profligacy would bring on the Western Ghats. Therefore, the report had categorically stated that mining, quarrying and other debilitating activities should be foreclosed in the ghat areas. But profit motive combined with an utter lack of concern for our fellow beings had prevented any concrete action as mentioned in the report. The policy makers also found it convenient to follow the diluted recommendations of the Kasturirangan report. The results are there for all to see. Another constant source of concern is the big dams built on the Ghat areas. The ecology of the Ghat area is least suited for big dams such as the Idukki dam. The result of any mishap associated with the dam is unimaginable. The recipe for the disaster was prepared by an alleged miscalculation about the intensity of the forthcoming rains which prevented authorities from opening the dam's flood gates till the last moment. When they were forced to open the dams at last, a massive deluge resulted that engulfed areas from Idukki to Trivandrum. The fact that massive deforestation had taken place in the ghats made sure that there was little holding capacity for the slopes, hastening the downward flow of water. Illegal mining of sand had left little holding capacity of lands of the plains as well.

As usual, hot political discussions had been aroused. Reasons were highlighted in various dimensions. Our concern in this particular platform for environmental education is focusing on the power of nature and its eternal truth. For the knowledge driven society, it is in the form of scientific evidences. Even though the belief on the power of Nature cannot be denied as it is based on Natural Sciences. It is a fact that no scientific and technologic developments are capable enough to over throw nature's power. India had a fair share of Calamities, natural and manmade over these several decades. Disasters resulting in loss of lives and other assets are still increasing. This is sharply pointing towards the emerging need of sustainable thoughts and practices inculcated through the commitments and love on nature.

Floods, hurricanes, tornadoes, volcanic eruptions, earthquakes, tsunamis, and other geologic processes are natural disasters. Some examples for manmade calamities happened in India:

1. Bhopal Gas Tragedy,
2. Flyover Collapse happened in Kolkotta Girish Park,
3. Building collapse happened in Lalitha park,
4. Maha Kumbh Mela Tragey happened in Allahabad.

Reasons for the Manmade Disasters

- Innocence of the people
- Lack of awareness
- Carelessness on environmental matters

- Negligence of the common people
- Lack of scientific knowledge at the end of the people who have not sufficiently educated
- some others who are of orthodox mentality may not have scientific attitude
- Lack of foresight
- Avoidance due to laziness
- Lack of concern with social and environmental matters
- Irresponsibility
- Some may not have any kind of concern to anything and least or little bothered of their accountability

In its broad sense, behind almost all the natural disasters happened in this century, there is a hidden hand of human involvement. That is the main reason behind many of the political and environmental discussions going on in media along with or immediately after the disasters. Whoever is blamed, government or the public is not at all a concern on the environmental point of view. But the facts regarding the intensity of the destruction and its root cause and the limitations of the already existing technological and logistic measures are to be considered with prime concern. It is a general tendency of human being to overcome the difficulties and worries of any kind of disasters immediately after its relief, if the difficulties are of common nature and are affected by many. The theoretical studies done on Anthropological perspectives with regard to environmental sustainability become relevant in this scenario. The theories are Environmental determinism and environmental possibilism. Determinism works on the already determined societal norms. That means it works according to the nature and norms of the society. The nature and norms according to environmental determinism is formed by the determination of type of settlements, their trade, culture and dressing, agriculture etc. Hence, we can say that the life style of such societies is decided according to the physical environment. But in possibilism it is the perception of the individuals and has to be changed according to the availability of resources and other facilities. Environmental possibilism is the theoretical approach advocates the possibilities adaptive technologies and alternative measures according to the nature of environment. It satisfies the environmental demands in connection with the changing situations. It is against the earlier mentioned environmental determinism. In the context of disasters natural or manmade whatever it is environmental possibilism find its way as a measure to overcome the consequences.

ENVIRONMENTAL POSSIBILISM

- Integrate environmental education as an interdisciplinary approach in all levels of education
- Attitude towards environment has to be developed in a positive way
- Try to inculcate the self-responsibility of the individuals in each and every deed in our practical life.
- Develop scientific attitude of the people by adopting apt environmental education techniques
- The marginal public have to be educated and initiated for the imparting of awareness on sustainable development
- The effect of sustainable development technologies has to be spread among the public at its maximum
- The funding allotments done for the grass root level agencies have to be utilized in a transparent manner. Absence of transparency in the processes associated with planning, sanctioning and

implementation very often results in debilitation of the overall scheme and ultimately defeating the very purpose for which it was initiated.

- The enthusiasm shown at the first phase of implementation of benign environmental schemes is hardly reflected at subsequent stages. The grandiose elaborations made at the drawing boards is found missing when the implementation takes place. The erosion of substance in such schemes ultimately becomes self-defeating and often works at cross purposes.
- Though laws of the country are often strongly worded, in practice and implementation they are found extremely wanting. This was actively highlighted in landmark cases such as the Ratlam Municipality case where the apex court has attested to this fact in no uncertain terms. When implementation is carried out in letter and spirit, there is no doubt that we shall be beneficiaries of a vibrant environment.
- Development has to go hand in hand with environmental protection. An area of concern is the building norms that exist in our country. Even the existing norms are observed in their violation as is seen in the Mudappa Case. There is an urgent need to combine building norms with protection norms for the environment. Waste management, energy conservation and water resources management have to be integrated into building norms to effectively protect the environment.
- Water recharging through rain water harvesting, though a part of present-day town planning has to be bolstered. The scale and magnitude of its implementation needs extension. The ground loss of water currently experiences and the periodic droughts faced by the country can become a thing of the past if effectively implemented.
- Solar energy is an undying source of clean fuel. The policy makers have taken baby steps towards making solar energy the cornerstone of energy policy, but much needs to be done. The adoption of solar power as the source of energy by Kochi Metro and Cochin International Airport needs commendation. The same needs to be implemented in every house of the country.
- The use of eco-friendly products for everyday use needs promotion. For instance, during the recent floods one of the major pollutants found in Kerala is sodium acrylate used majorly in the production of sanitary napkins and diapers. Needless to state contaminants of this nature change the hue and texture of our ecosystem irretrievably.

CONCLUSION

Calamities unify mankind, perhaps due to the realization that once the material gains have been eroded by an unforgiving nature, everyone is under and even roof. We have witnessed various major calamities. If even an iota of the efforts put out at the time of calamity had been made from time to time, a catastrophe of this scale could have been avoided altogether. Therein lies the message for the future.

REFERENCES

Thandasherry, S. (2018, Jan.). High Fives for Solar Powered Ferry Ops. *Ship & Boat International*, 18-24.

Chapter 2
A Review on Solid Waste Management Using the DPSIR Framework in a Southern Africa Case Study

Joan Nyika
https://orcid.org/0000-0001-8300-6990
University of South Africa, South Africa

Ednah Onyari
https://orcid.org/0000-0001-6606-2675
University of South Africa, South Africa

ABSTRACT

This chapter uses the DPSIR framework to examine logistical, infrastructural, and operational challenges in SA's SWM system using Buffalo City Municipality as case study. Findings suggest that SWM is an endemic problem in the municipality characterised by a preference to landfill disposal, minimal recycling tendencies, poor waste services, non-enforcement of existent waste regulation, inadequate funding, and non-involvement of urbanites and informal sector in the process. To counter these challenges, BCM has enhanced its financial resources to facilitate scientific landfilling and transit to perceptions of solid waste as a useful good. The municipality has collaborated with local industries and international organizations to synergize efforts to implement state and local waste management plans. For an integrated model of SWM, this review suggests SA and BCM's need to plan innovatively, adopt landfill-mining, plan strategically on all waste cycle aspects, and modify existing regulations to accommodate urbanites and informal waste pickers in the process.

DOI: 10.4018/978-1-7998-4921-6.ch002

BACKGROUND

In the last two decades, SA has had unprecedented population growth, which has affected urbanisation trends, waste generation and management. A World Bank survey documented by the South African Institute of Race Relations (2013) details that 62% of the population resides in urban areas which is a rise from 52% in 1990s. This population outburst has overwhelmed municipality authorities reducing their capacity to provide waste services and leading to unscientific handling (Samson, 2010; Muzenda et al., 2012; Katisiimeh et al., 2013; Nyika et al., 2020). In SA, more than 42 million cubic metres of solid waste is generated annually, which translates to 0.7 kg per individual daily (Nkosi *et al.*, 2013). In Eastern Cape Province, approximately 1.6 Million tonnes of solid waste are generated annually from industrial, commercial and residential sources (Nyika et al., 2020). The Buffalo City Municipality (BCM) in Eastern Cape is the largest generator at 619, 099 tonnes annually compared to neighbouring municipalities though Solid Waste Management (SWM) is poorly funded and uncoordinated (DEDEA, 2009). Fiehn and Ball (2005) observed a similar trend citing inadequate waste services for majority of the population, unlicensed and illegal management procedures, unclear waste-related regulations and inadequate recycling action plans as major issues in SWM. The Integrated Pollution and Waste Management Policy noted SWM as SA's biggest twenty-first century problem (Simelane & Mohee, 2012).

At municipality level as is the case in Nelson Mandela Bay and BCM, Integrated Waste Management Plans (IWMP) to oversee the implementation of NWMS have been formulated though not successfully. Ali (2010) and Dlamini et al. (2019) argued that the willingness to better waste services is there but the capacity and infrastructure could be an impediment. On a different viewpoint, Fobil et al. (2010) as well as Rasmeni and Madyira (2019) claimed that it is the lack of a well thought SWM plan that precedence stakeholder participation especially the urbanites, which ails the capacity of municipalities in waste service provision. Senkoro (2003) and Cobbinah et al. (2017) support these suppositions claiming that poor stakeholder involvement causes South African municipalities to spend 20-50% of their revenue on SWM but only about 35% have access to effective waste services. Additionally, existent literature lacks holism in that it focuses on government's failures and recommends improvements without assessing the role of urbanites and private sector in the waste cycle (Oteng-Ababio, 2010; Oteng-Ababio et al., 2013; Cobbinah et al. 2015, 2017; Amoah & Kasoe, 2014; Addaney & Oppong, 2015; Dlamini et al., 2019; Nyika et al., 2020). In view of this situation, it is necessary to assess the current situation of SWM, implement and formulate effective measures that synergise development of municipalities countrywide (Dlamini & Simatele, 2016). This books focuses on this issue using the BCM case study.

SWM IN SA AND BCM USING THE DPSIR FRAMEWORK

A Brief Introduction of SA and BCM

SA is one of the largest multiracial countries of Africa and consists of North West, Northern Cape, Mpumalanga, Limpopo, KwaZulu-Natal, Gauteng, Free State, Eastern, and Western Cape provinces (Statistics SA, 2011). It has a population of approximately 57 million and Gauteng is the most populous province followed by KwaZulu-Natal and Eastern Cape. BCM is located in Amathole District of Eastern Cape Province. It neighbours Ngqushwa, Nkonkobe and Amahlathi local municipalities as shown in Figure 1. The area has a total of 223,568 households and 755,200 people that is 11.5% of Eastern

Cape's total population and a population density of 300 people every square kilometre (Statistics SA, 2011). It is the second largest economy in Eastern Cape following Nelson Mandela Bay Metropolitan Municipality (NMBMM) with a gross value of R29.7B (DEDEAT, 2013). Economic sectors include finance, community services, manufacturing, trade, transport, mining and agriculture rated at 29%, 28%, 17%, 13%, 8%, 0.4% and 1%, respectively (BCMM, 2011).

Figure 1. Map of Buffalo city metropolitan (BCMM, 2011)

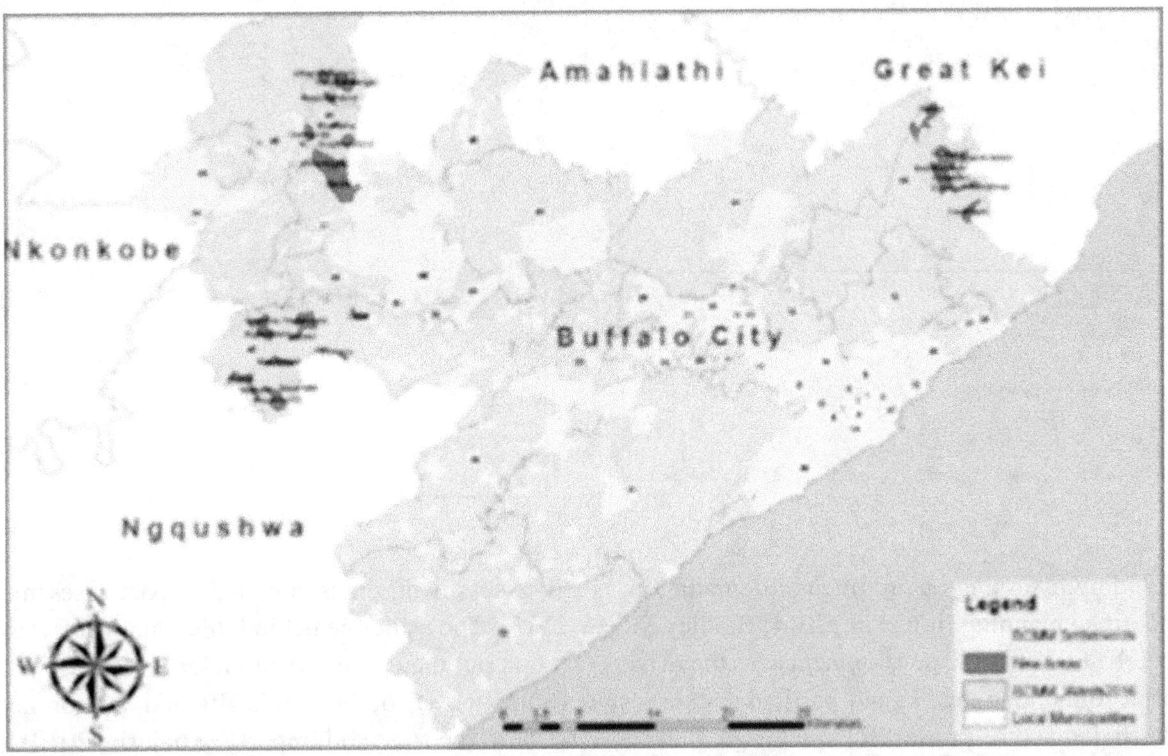

DPSIR Framework

The DPSIR framework (Figure 2) is a methodology for analysing environmental problems, which was modified from the Pressure-State-Response (PSR) framework of the Organisation for Economic Cooperation and Development (OECD) and United Nations Environmental Planning (UNEP) (Kazuva et al., 2018). Its goal is to evaluate efficiency and effectiveness of policy responses to environmental problems.

The initial model PSR was criticised for being narrow, unidirectional and linear since it did not show the forces that exert pressure to induce a state of environmental change and did not show the direct-indirect and short-long term impacts of resultants changes (Cao, 2005; Svarstad et al., 2008; Linster & Fletcher, 2012). However, in 1994 the OECD added the driving force and impact indices in the framework resulting to DPSIR, which was adopted by European Environmental Agency (EEA) in 1999 (Kazuva et al., 2018). Following this move, the model has been successfully adopted by the United Nations (UN) and United States Environmental Protection Agency (USEPA) in various environmental management initiatives

Figure 2. SWM factors in BCM according to DPSIR framework

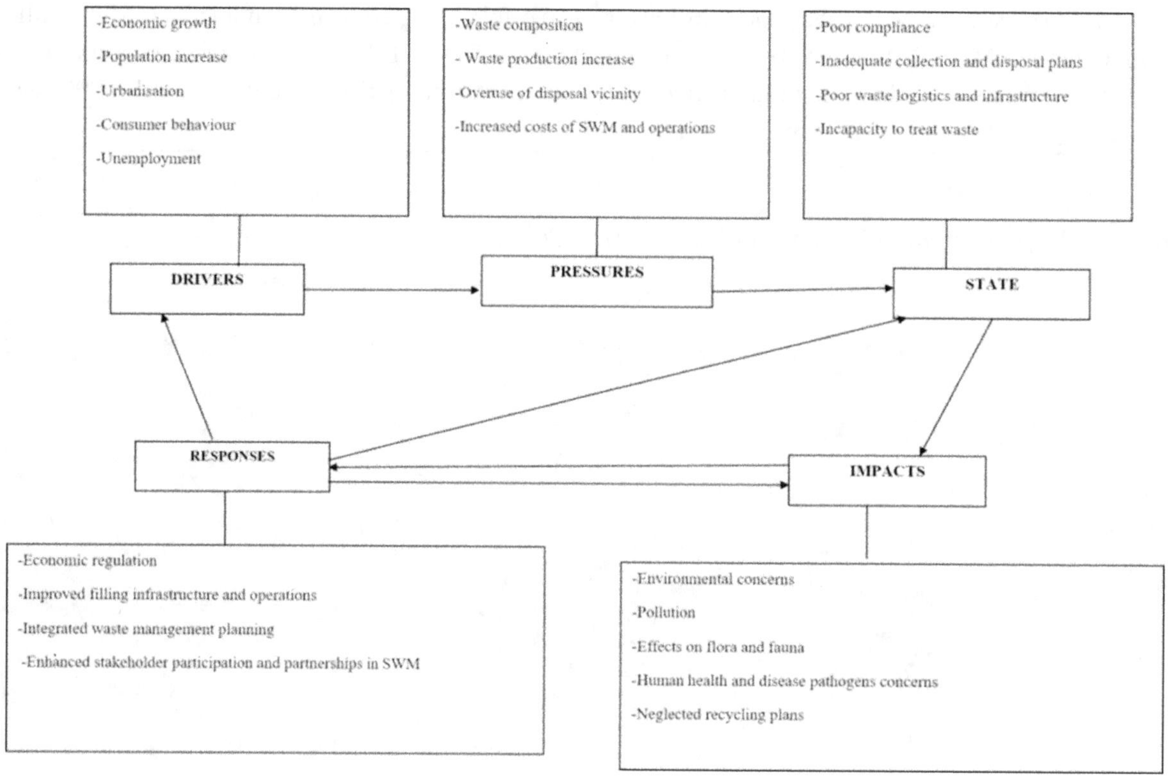

including assessing environmental degradation, health assessment, environmental impact assessment and urban planning (Liu et al., 2014; Bradley & Yee, 2015). The principle behind this framework is that social and economic developments are the drivers (D) that put the environment under pressure (P) and result to its poor state (S). A particular state results to impacts (I) on human health and environment, which demand responses (R) to manage, prevent or adapt to the state and impacts (Khajuria, 2010).

In SWM, the DPSIR framework has been applied to solve a number of issues. In Dar es Salaam city of Tanzania, DPSIR was used to identify indicators of environmental risks associated with solid waste (Kazuva et al., 2018). The model was used to predict waste generation by formulating and adjusting indicators of such tendencies in Czech Republic (Hrebicek et al., 2017). In developing countries of South Asia, the model was used to assess SWM and correlate economic growth in the countries with resultant environmental risk (Khajuria, 2010). In China's Ningbo city, DPSIR was used to demystify the reasons behind waste over-generation by analysing the intricate interactions of stakeholders in the SWM sector (Ke'er, 2010). These studies demonstrate the wide application of DPSIR framework in SWM initiatives globally.

CONCEPTUALISING DPSIR FRAMEWORK IN SWM OF SA AND BCM

Drivers

SWM in SA and BCM using the DPSIR framework is represented in Figure 2. Population is a major driving force to SWM. According to the United State Secretariat (2010), SA has about 50 million people up from 14 million in 1950. In the last 20 years, it has had unprecedented growth in population characterised by relocation of 62% of the entire population to urban areas (Samson, 2010; Muzenda et al., 2012; Nyika et al., 2020). Statistics SA (2003) rated the number higher claiming that 55.1% of South Africans lived in urban areas by 1990s but the number grew to 64.8% and 68.5% in 1996 and 2001, respectively. Urbanisation is prompted by positive economic growth and the search for employment. Urbanites mostly agglomerate in informal settlements, which limit waste disposal and treatment options due to space reductions and increase the demand for goods' replacement thereby affecting household consumption rates (Hrebicek et al., 2017; Gutberlet, 2018). In BCM, the scenario is evident where a 0.6% population growth and a 1.67% economic growth rate compounded by urbanisation promoted waste generation (DEDEAT, 2013). United States Secretariat (2010) rated the municipality as the seventh largest in SA with a population of more than 0.8 million people. This trend directly affects SWM in that unemployed residents remain in their residential areas all-day conducting activities that lead to waste generation (Hrebicek et al., 2017).

Consumer behaviour affects waste generation in aspects of packaging, willingness to pay and attitudes towards waste. In SA and BCM, SWM perceptions must change from landfill disposal tendencies to waste reduction and reuse trends. Landfilling is the most preferred method of SWM in SA since 90% of the 108 million tonnes of waste generated is disposed off (DEA, 2012; Oelofse, 2018). A similar trend is evident in BCM where most of the waste is diposed off in its Roundhill and King Williams landfills. Urbanites of developing countries assume that local and national governments are responsible for waste cycling, do not adhere to the polluter's pay principle and lack caution during generation (Amoah & Kasoe, 2014; Cobbinah et al., 2017). In characterising consumer behaviour, Dlamini et al. (2017) and Cobbinah et al. (2017) noted that most South Africans and Ghanaians, respectively lack technical knowledge on waste segregation and reuse despite their predominant perceptions of dissatisfaction to local government due to inadequate waste services. Additionally, they consume packaged and pre-prepared foods, fresh fruits and vegetables, which enhance waste production (Dlamini et al., 2017). The trend shows disconnect of residents in the waste management processes despite the fact that they are the generators. Nyika et al. (2020) made a similar observation in a review of SWM tendencies in most cities of SA.

The high rate of unemployment in SA and BCM is a driver in SWM. According to Statistics SA (2011), unemployment rate for the country was at 29.8% while that of BCM was 35%. Unemployment enhances waste generation and informal picking (Oteng-Abbabio et al., 2013; Linzner & Lange, 2016; Niekerk & Weghmann, 2019). The DEA (2016) reported the existence of 62,147 waste pickers operating as trolley pushers or at landfills across SA though the sector is not formalised in the mainstream SWM system. This form of recycling in urban areas is on ad hoc basis propagated by the private sector mainly informal waste pickers at kerbsides (Nkosi et al., 2013; Oelofse, 2018). Their motive is waste valorisation and subsequently, getting a livelihood and the practice has a significant role in driving SWM (Cobbinah et al., 2013, Oteng-Abbabio et al., 2013; Linzner & Lange, 2016; DEA, 2016; Hettiarachchi et al., 2018). In Roundhill landfill of BCM, about 50 informal waste pickers were identified (Infrastructure News, 2016).

Pressures

Solid waste composition varies in municipalities and between nations based on industrial structures, management regulations in place, economic situation, socio-cultural behaviour and lifestyle (Abdel-Shafy & Mansour, 2018). Composition is a critical aspect in determining appropriate techniques to handle wastes and in waste-to-energy processes particularly gasification and incineration (ASME, 2014). SA and BCM's waste composition is as shown in Table 1 whereby majority of the waste is from food scrap, ash and vegetable waste while metal and glass form the minority (FWSC, 2014). The distribution is similar to existent literature that documented the composition to be typical in suburban areas of developing countries (Ramachandra & Bachanda, 2007; Nabegu, 2010; Mudau, 2012; Gumbi, 2016). The USEPA (2013) had a similar view claiming that 55-80% of wastes in developing countries are from households while commercial and market areas contribute 10-30%. The latter consists of institutional and industrial waste. This composition depicts a lot of heterogeneity as well as divergent physicochemical characteristics in the waste owing to its different sources. This nature is a setback during sorting to facilitate effective reuse and recycling as Abdel-Shafy and Mansour (2018) noted and advised on the need to fractionate waste at generation stage before meaningful pre-treatment is done.

Table 1. Waste composition in municipalities of developing countries (FWSC, 2014)

Waste type	Percentage
Food scrap	40
Ash, dirt	21
Vegetable	20
Paper cardboard	5
Plastics	6
Textile rubber	4
Metal	3
Glass	1

Landfill disposal is the most preferred SWM approach in many South African cities such as Cape Town, City of Johannesburg, City of Tshwane, Ekurhuleni, Mangaung and Msunduzi municipalities (SACN, 2014; Gumbi, 2016; Godfrey, 2019). However, these municipalities suffer from limited airspace. In BCM, the two operational landfills, Round Hill and King Williams can hold limited waste, a situation that has forced the municipality to embark on waste reduction strategies including recycling and separation as alternatives. The municipality closed some general and garden-waste disposal areas such as James Pearce Park, Highgate, Mdantsane, NU 12, Ducats, West Bank, Beacon Bay, Ilitha, Phakamisa and Zwelitsha in the 1980s due to diminishing filling space (FWSC, 2014). All municipalities in SA report a similar trend, which is compounded by increased waste generation trends as Taljaard (2011) observed. A similar scenario is reported in Ghana where increased generation and disposal of even untreated hazardous waste without sorting or reuse has limited available filling space (Cobbinah and Erdiaw-Kwasie, 2016). Challenges associated with diminishing disposal space have therefore triggered contemporary SWM trends to focus on diverting landfill waste to other products as reported in developed countries such as Netherlands and United States to manage rising generation levels (Oteng-Ababio et al., 2013; Cobbinah & Erdiaw-Kwasie, 2016; Cobbinah et al., 2017). It is from the prerequisite to have adequate open space and environmental risks associated to landfill disposal that the practice is discouraged in Europe (Aracil et al., 2018).

SWM services in SA are functions of municipalities through the user-pay principle. According to Statistics SA (2009), the total income from municipalities in 2007 and 2008 was 110,123, 127,630 million Rand, respectively, which is very low, compared to the costs incurred. This revenue is low compared to the capital investment of commissioning a landfill. Limited SWM funding and income could be attributed to non-adherence to the polluter pay principle according to Nahman and Godfrey (2010). Goldblatt (2009) and DEA (2016, 2019) shared similar views claiming that most South African municipalities had low budgetary allocations for SWM that varied among municipalities making average costing impossible to realize. BCM for instance invested more than R800 million to rehabilitate its operational landfills and improve the SWM cycle in the last five years yet SWM is not up-to-standard (BCMM, 2018). SWM drivers increase operational and infrastructural costs in BCM just as many South African municipalities according Nahman and Godfrey (2010). In Vienna of Austria, Dhaka of Senegal and Damascus of Syria, a similar trend where SWM was crippled by low budgetary allocations amidst a high demand for operational and maintenance cost was reported (Brunner & Fellner, 2007).

State

BCM has a SWM Department (SWMD) that oversees environmental sustainability and in particular, solid waste cycling (BCMM 2018). The municipality just like the country relies on several pertinent pre-determined legal frameworks in SWM planning. These are mainly oriented towards Command and Control (CAC) according to Nahman and Godfrey (2010) and are aimed at promoting adherence and enforcement and whose failure results to penalties. These legislations include the National Environmental Management Act (NEMA) of 1998 that puts measures to reduce and eliminate pollution causatives including solid waste and the National Environmental Management Waste Act (NEMWA) of 2008 that outlines the National Waste Management Strategy (NWMS) focusing on waste reduction and minimisation rather than the current disposal trends.

BCM operates under these regulations particularly the NWMS that advocates on the theme of 'cradle to grave' SWM that perceives landfill disposal as the last resort in the management hierarchy (DEDEA, 2009). At the municipality level, BCM relies on the Municipal By-Laws, Municipal Structures, Local Government Municipal Systems and Municipal Financial Management Acts in SWM. By-laws facilitate waste management at local level; financial management act oversees proper use of resources to deliver waste services while the municipal structures and local government systems acts facilitate development of integrated development plans that inculcate waste management. The municipality's Integrated Waste Management Plan (IWMP) guides the operations of SWMD by providing predictions of waste generated and best management trends (BCMM, 2018). At the local level, most of the by-laws are outdated and do not supplement modern waste management practices and legislations at national level. Similarly, literature documented weak regulations, poor municipality governance and ignored checks and balances as drawbacks to effective SWM at municipality level (Simatele & Etambakonga, 2015; Simelane & Mohee, 2015; Simatele et al., 2017).

Non-compliance in BCM is evident in the municipality landfills. In the case of Roundhill landfill, both untreated general and hazardous waste is received daily to the tune of 600, 000 tonnes. The landfill has had challenges to do with non-compliance in recent times owing to increased waste traffic and unscientific disposal characterised by uncontrolled tipping in unlined cells (Dookhi, 2016). A similar scenario is reported in King Williams whereby the landfill is overwhelmed with treated and untreated waste often disposed unscientifically (BCMM, 2018). This is typical of landfills in developing cities such as Makassar city of Indonesia which is overwhelmed by increased waste disposal most of which is not sorted or pre-treated (Permana et al., 2015). Hazardous and general waste sites that were closed including Old Everite, James Pearce Park, Berlin and Dimbaza landfills of BCM have not adhered to the Department of Water and Sanitation (DWS) minimum requirements. Furthermore, municipality residents dump waste in the sites even after closure because monitoring is never done (Dookhi et al., 2015; Jewaskiewitz & Dookhi, 2017). Unpermitted dumpsites such as Kayser's Beach and Chiselhurst also receive waste from BCM alluding to high levels of non-compliance and poor disposal practices as FWSC (2014) reported.

BCM suffers of limited resources, poor operational techniques and standards, which make SWM for the municipality strenuous (FWSC, 2014). Most residents do not have access to designated refuse transfer stations hence littering and illegal dumping are widespread in the municipality (Riddin, 2018). Simatele et al. (2017) and Cobbinnah et al. (2017) reported similar trends in Johannesburg and Berekum towns of SA whereby waste services were below par and authorities' management capacity was overwhelmed. According to Statistics SA (2011) waste collection services by BCM was rated as a main approach to SWM while construction of communal dumps, self-made pits and illegal dumping contributed the minority trend as shown in Figure 3. Of concern in the trend are the environmental risks that unscientific SWM practices pose even at minimal levels even with overwhelmed capacity of SWMD to better the current situation. Oteng-Ababio et al. (2013) demystified the situation claiming that the lack of or limited access to waste services among the urban poor results to waste heaping and some residents opt to reduce it by burning while others practice illegal dumping.

SWM infrastructure of municipalities in SA faces a challenge of unparalleled waste levels and BCM is one such example. Efficient collection depends on the availability of vehicles equivalent to waste generated. Riddin (2013) reported that only 10 refuse compactor trucks of the 26 available in BCM were operational, a situation attributed to their old age and their constant breakdowns. This has greatly affected refuse collection and its disposal to Round Hill landfill of Berlin. Residential garbage from hotels that should be collected twice a week like in the case of Kennaway Court accumulates for more than a fortnight, forcing the management to seek the services of private companies to dispose the waste at added costs (Riddin, 2013). This is typical among poor urban communities of developing countries that have to put up with uncollected waste and even sometimes seek alternatives to manage it such as private waste services according to Oteng-Ababio et al. (2013).

The poor infrastructure state in BCM is compounded by the lack of suitable plan for waste fractionation before transportation for easier management, which according to Abdel-Shafy and Mansour (2018) is a crucial aspect in effective SWM. Evenly, the transporting trucks cannot accommodate this modification even if separation could be possible at generation sites. According to FWSC (2014), only 61% of BCM households had access to weekly kerbside refuse collection too. This percentage has been fluctuating over time as shown in Table 2, which could be due to logistical, infrastructural and separation challenges in the SWMD of the municipality (LGHSA, 2017). The situation is characteristic in many urban areas of developing countries where financial and institutional constraints result to incapacity to handle waste

Figure 3. Waste disposal options in BCM (LGHSA, 2017)

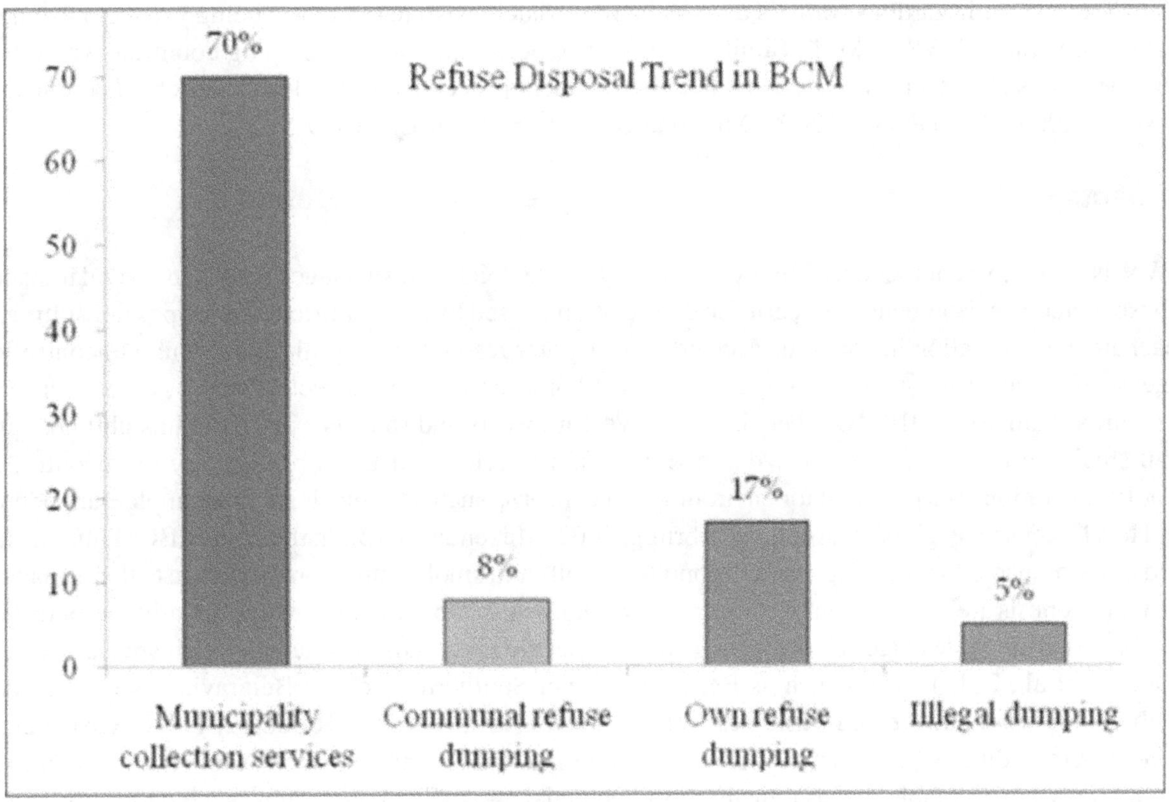

and provide such services to residents (UN-Habitat, 2010; Okot-Okumu & Nyenje, 2011; Abderrahmane et al., 2014; Okot-Okumu, 2015; Dlamini et al., 2019, Nyika et al., 2020).

The current capacity to handle and treat waste in the municipality is inadequate. The two operational landfills are overwhelmed as they are receiving more waste that their capacity to handle. A similar scenario was reported in Pudong, China whereby waste composting superseded the landfill daily handling and treatment capacity (Minghua et al., 2009). To be environmentally acceptable, the capacity to handle waste should be supplemented by data collection throughout the waste cycle (Hazra & Goel, 2009; Gumbi, 2016; Godfrey, 2018). Inadequate data collection and records on composition of collected waste, aspects of sorting and separation, illegal dumping and enforcement actions or incidents further compounds the logistical management approaches to use in streamlining landfill disposal tendencies in SA and BCM to acceptable levels (BCMM, 2018). In Kolkata, India, a similar scenario was reported whereby waste logistics data was unavailable and deterred effective planning on waste handling and treatment (Hazra &

Table 2. Trend of weekly refuse collection services (LGHSA, 2017)

Year	2001	2007	2011	2017
% Weekly refuse collection services	71.2	60	70.4	98

Goel, 2009; Blaauw, 2017). Unscientific disposal is a common scenario in the municipality since only 13, 000 tonnes of hazardous waste is disposed appropriately despite the figure being way below annual productions rates (FWSC, 2014). Similar trends are reported in many developing countries where filling operations are never monitored and bypass treatment processes, which lead to leachate percolation (Asnani, 2005; Hazra & Goel, 2009; Minghua et al., 2009; Godfrey, 2018).

Impacts

SWM is a sensitive component of environmental hygiene, which if mismanaged results to several impacts. These include environmental concerns on degradation caused by unscientific waste disposal resulting to leachate contamination in soils, surface and groundwater resources, air pollution through uncontrolled release of methane and waste burning and the spread of diseases through vectors such as rodents, insects and birds (Kubanza, 2010; Nkosi et al., 2013). Vector insects and rodents exacerbate unsightliness and foul smells in heaped and illegally dumped wastes. The release of gases particularly in uncontrolled landfill operations is a potential for environmental concerns such as explosions, fires, unpleasant odours and to a larger scope, global warming (Zurbrugg, 2002; Muzenda, 2012). In the case of BCM, uncovered and non-compacted waste heaps even beyond lined cells and smoke guns from burnt waste that enhances methanogenesis are environmental concerns (Jewaskiewitz & Dookhi, 2017). Additionally, uncontrolled leachate runoff and leakages on the storage system that are potentials of water pollution are also evident (Dookhi et al., 2015). Areas such as Berlin, Gonubie, Southernwood and Belgravia have to contend with odours from uncollected waste heaps that last even up to a month (Riddin, 2013). Environmental concerns regarding SWM practices if unregulated often lead to pollution of water resources and air. Continually unacceptable disposal methods in unofficial and official dumpsites worsen the situation in SA as Kubanza (2010) and Dlamini et al. (2019) observed. Pollution increases environmental costs by inflicting negative effects on community health and through diminution of already scarce land space and renewable resources degradation. In BCM, water and soil pollution has been a major concern in areas near dumpsites as according to Dookhi (2016). Jewaskiewitz and Dookhi (2017) have also reported air pollution in the form of smog from burnt waste in Roundhill landfill.

Poor SWM in SA has negative effects on flora and fauna as reported by Khumalo (2016). The burning of waste to ash facilitates the leaching of harmful heavy metals and are taken up by plants even in low concentrations (Khumalo 2016). The elements pollute soils and suppress the growth of some plants such as mushrooms (Adefemi & Awokunmi, 2009). In Uthukela district of KwaZulu-Natal, plants exposed to waste ash that had heavy metals depicted retarded growth, as the elements were still active even after burning (Khumalo, 2016). Poor SWM tendencies also affect soil organisms that are important in biogeochemical cycling. The cycling of heavy metals from waste to soil results to death of soil organisms and has a direct effect on primary and secondary consumers (Adefemi & Awokunmi, 2009). In BCM, plant populations were adversely affected by physical disturbances during Roundhill landfill construction while the fauna are under pressure from landfill expansion and unscientific SWM according to Kalule (n.d). Additionally, poor SWM is associated to high prevalence of respiratory, infectious and tropical diseases (Wolfville & Scotia, 2002; Kubanza, 2010).

Considering SWM challenges including airspace limitations in municipalities of SA, landfill disposal is not sustainable but can be reduced through waste reuse and recycling. Effectiveness of recycling depends on stakeholder participation, efficiency and effectiveness of infrastructure and equipment. Manaf et al. (2009) had a similar suggestion claiming that municipalities with inconsistent collection services, inadequate stakeholder equipment and limited infrastructure for SWM usual have a recycling challenge. The NWMS of SA hails the uptake of recycling as a shift from traditional end of pipe solution to reductive approaches of waste management but in practice, it is neglected. DEAT (2007) and DEA (2019) reported that 80% of SA residents were pursuing some form of recycling though it did not specify the recycled waste, which alludes to the practice being localised rather than formal. Recycling trends are driven by unemployment and the need for alternative livelihoods in SA just as witnessed in Indian (Sharholy et al., 2008) and Kenyan (Rotich et al., 2006) municipalities whereby many informal waste pickers are poor unemployed urban dwellers.

In BCM, the uptake of recycling is limited as there are inadequate data records on the practice, and if available, they are not updated. According to FWSC (2014), recycling of all kinds of solid waste remains elusive compared to the produced quantities. As shown in Table 3, only 77%, 18%, 38% and 13% of metal, plastic, paper and glass, respectively is recycled while the rest is disposed off. In areas where recycling is practiced, it is on ad hoc basis not included in the BCM management system. These shortfalls are associated with limited, recycling markets and professionals, non-recognition of informal waste pickers and limited financial support for recycling initiatives and their associated infrastructure by the government (Nissim & Inbar, 2005; Rotich et al., 2006; Minghua et al., 2009; Gutberlet, 2018).

Responses

Table 3. Status of recycling in BCM (FWSC, 2014)

Type of waste	Recycled Tonnes	Produced Tonnes	% of Recycled waste
Metal	22,000	28,600	77
Plastic	3,300	18,300	18
Paper	16,000	42,800	38
Glass	1,500	11,760	13

Sustainable SWM involves tactical and strategic decision-making at operational level particularly in selection of filling and waste treatment sites (Wang et al., 2009), improving capacity to reuse and recycle (He et al., 2009; Nyika et al., 2020), in service zoning (Mourao et al., 2009) and in planning for collection and separation (Chu et al. 2009; Dlamini et al., 2019). To achieve these, financial, technical and institutional constraints must be overcome at local levels with collaboration of the private sector (UN-Habitat, 2010; Oteng-Ababio et al., 2013). Following the recognition of the aforementioned challenges in SWM, SA and BCM are taking several correctional responses to facilitate sustainability in the sector as summarised in Table 5.

Traditionally, SWM just as other environmental challenges were tackled using the CAC policies, which control behaviour directly through stipulated rules that must be adhered to or enforce conformity by imposing fines and penalties (Perman et al., 2003). Economic Instruments (EI) that focus on indirect behaviour change through environmental subsidies and taxes are preferred to ineffective CAC policies (UNEP, 2005; Nahman & Godfrey, 2010; Godfrey & Oelofse, 2017; Godfrey, 2018). Some of the EI used in SWM in SA and BCM are shown in Table 4. The aim of these incentives is to abate waste generation by service providers, consumers and producers while encouraging recovery, reuse and recycling (Inter-American Development Bank, 2003). Several documentations report of the increased demand for EI use compared to CAC owing to their acceptability among waste generators (Pearce & Turner, 1994; Bell & Russell, 2002; Inter-American Development Bank, 2003). EI are advantageous in that they are incentives for novelty and potential avenues for revenue generation that promote economic efficiency (National Treasury, 2006). However, their successful implementation requires political will from authorities, institutional capacity for monitoring, compliance evaluation and sound-functioning markets (Inter-American Development Bank, 2003; UNDP, 2005). These preconditions must be strengthened in developing countries including SA where SWM policies are skewed to CAC mechanisms (Bell & Russell, 2002; Russell & Vaughan, 2003).

In tackling challenges associated with unscientific landfill disposal, the municipality is enhancing the capacity of operational Round Hill and King Williams Town landfills. In the latter, the municipality hired Envitech Solutions Limited for specialist consulting services and a turn-around strategy implementation to enhance compliance and improve the operations and infrastructure of the landfill in a project worth R136 million (Dookhi, 2016). Turn-around operations in the landfill included repairs and rehabilitating the lining and leachate collection system, re-capping and compacting waste to accommodate more and deter uncontrolled tipping (Jewaskiewitz & Dookhi, 2017). Infrastructure News (2016) reported improved compliance levels from 30% to 80% following these rehabilitative initiatives. However, the situation needs to be monitored and evaluated to realize full compliance.

Additionally, the municipality is focused on enhancing compliance and logistical operations in King Williams Landfill to ensure hazardous waste received is treated and disposed off within the cells as BCMM (2018) noted. BCM is in the process of reviving closed Second Creek, Mdantsane and Nu 2 landfills to avail more disposal space within existent landfill regulations. The municipal government has injected R2, R8 and R2 million to procure waste skips, mobile transfer stations and additional street litterbins, respectively to enhance waste logistics (BCMM, 2018). According to The LGHSA (2017), these efforts have borne fruits as weekly refuse collection services in the municipality were rated at 98% in 2017 but require follow up for sustainability. Additionally, the municipality had budgeted for 20 new refuse collection trucks to be purchased in the 2018/2019 financial year to improve waste services

Table 4. EI in SWM in SA and BCM

Instrument	Incentives Involved	Execution	References
Waste Collection charges based on quantity	Put a price on every unit of waste collected for landfilling to minimize generation and promote reuse/recycling	At municipality level	Choe & Fraser, 1998; United Nations Environmental Program, 2005
Deposit-refund scheme	A deposit is incurred after purchase and refunded on return of the used product for repackaging. Discourages disposal	Through public-private sector collaborations	Reschovsky & Stone, 1994; Inter-American Development Bank, 2003
Product and input tax	Hike prices for environmentally influential inputs/ products to discourage use and resultant waste generation	At national level	Pearce & Turner, 1994; Forum for Economic and the Environment, 2002
Polluter Pay Principle (PPP)	Waste generators incur direct costs of production including collection, disposal, treatment and resultant externalities such a health and environmental effects	At national level	DEAT, 1999

in uncovered areas (Riddin, 2013). These measures encourage landfill disposal and depict a persistence of traditional SWM approaches as noted in Makassar municipality of Indonesia whereby management responses were skewed to disposal tendencies (Permana et al., 2015). Similarly, in Vienna of Austria, Dhaka of Senegal and Damascus of Syria, filling tendencies were preferred to other alternatives due to limited municipal SWM budgets (Brunner & Fellner, 2007). The need to re-evaluate alternative SWM measures that are sustainable in SA and BCM is indispensable.

Following the above considerations, BCM formulated and endorsed an integrated waste management plan (IWMP) in 2002 to ensure legal compliance of transfer stations and landfills, improve waste services and reduce the quantities of disposal by 35% through buy -backs for recycling in a 5-year period (DEDEA, 2009). Other municipalities in SA such as Ekurhuleni, e-Thekwini, Cape Town and Johannesburg have undertaken this move but in most cases, the plans become outdated and do not serve as effective guidelines to SWM (SACN, 2014). The effectiveness BCM's 2002 plan remains questionable and undocumented, which could be attributed to weak planning departments of the municipality despite the area's extensive urban geography. A similar scenario is reported in Tshwane municipality of SA (Simatele & Etambakonga, 2015; Gumbi, 2016; Niekerk & Weghmann, 2019) and Kinshasa municipality of Congo (Samson, 2008) whereby IWMP are informal and outdated most often due to limited municipal finances for SWM. As such, the plans fail to appreciate and tackle intricate contemporary SWM challenges holistically. SA and BCM are currently re-evaluating their strategic plans and interventions to improve SWM as SACN (2014) highlighted.

SWM is intricate and demands integrated and comprehensive models that allow municipalities to perfect in localised waste management and concurrently maximise on its valorisation and environmental benefits with minimal economic costs (Anschutz et al., 2004; Nyika et al., 2020). With these factors in mind, BCM management recently held a waste 'Indaba' (conference) to enhance stakeholder participation and partnership in resolving SWM challenges (BCMM, 2018). The use of partnerships in SWM in developing countries is commended for increased investment in the sector with low public expenditure towards sustainable development (Bartone & Leitman, 1994; Post, 1999). Oteng-Ababio et al. (2013) however shared a different opinion claiming that partnerships for SWM promote tactical win-win technologies and are a form of neo-colonialism where global actors dominate in exploiting developing countries. Baud et al. (2001) claimed that partnerships involving local and global actors help in demystifying the multi-faceted nature of MSW irrespective of the actors' motive.

Table 5. Measures in place for SWM in BCM

Category	Specific Measures
Landfill disposal	-Repairs and rehabilitation of operational Roundhill and King Williams landfill for enhanced operation capacity -Enhanced compliance to landfill regulations -Increased funding to build transfer stations and street litterbins -Reviving closing landfills -Buy new transport and compacting trucks -Specialist consulting services to improve landfill infrastructure -Informal waste scavenging, reuse and recycling
Planning	-Developing IWMP
Partnerships	-BCM, ELIDZ and SEPA -BCM and BKCOB -BCMDA and DEA

BCM collaborated with the Swedish Environmental Protection Agency (SEPA) to analyse the waste generated and facilitate separation at the source (Reputation Matters, 2014). In a campaign dabbed, "All hands on waste", waste separation is done using colour coded bins to enable appropriate reuse and disposal. This initiative involved the East London Industrial Development Zone (ELIDZ) in identifying waste economy opportunities by formalising waste separation sector, encouraging composting plant and urban agriculture to create sustainable livelihoods for the poor in the municipality through reuse and recycling. Separation at source has already started in BCM through the partnership. Furthermore, BCM has used R6 million to make operational buyback centres at Unit 17, Sunny Ridge and Cafferian Heights in Mdantsane (BCMM, 2014). A similar initiative is documented in a partnership between Ghana's Accra Metropolitan Assembly and Canada's Chagon City and County Waste Company for purchase of waste compacting trucks, improvement of waste services and recycling of disposed waste (Oteng-Ababio et al., 2013).

CONCLUSION

Using the DPSIR framework, this book chapter establishes that population growth and economic development that promote urbanisation, unemployment, consumer behaviour skewed to negative tendencies and perceptions to waste are the drivers to SWM in SA and BCM. These drivers have a direct effect on the composition and volumes of waste since more space is used for disposal and waste management costs become inflated. These pressures are evident in the current state of SWM in SA and BCM where the entire waste cycle is disrupted and compliance to filling regulations, collection, transport, disposal and treatment operations are inadequate. Consequently, the greater effects of these failures are felt in form of environmental concerns such as pollution, disruption of animal and plant growth, neglect of waste minimisation plans and increased prevalence of communicable diseases. This book chapter recommends the need to rethink SWM innovatively through enhanced landfill mining and the formalising of informal waste pickers, the use of EI to streamline SWM behaviour and improved political will of all stakeholders of the sector.

REFERENCES

Abdel-Shafy, H., & Mansour, M. (2018). Solid waste issue: Sources, composition, disposal, recycling and valorisation. *Egyptian Journal of Petroleum*, *27*(4), 1275–1290. doi:10.1016/j.ejpe.2018.07.003

Abderrahmane, Y., Okkacha, Y., & Hassibaa, B. (2014). Sustainable solid waste management in the city of Mecheria (Western Algeria). *Energy Procedia*, *50*, 953–959. doi:10.1016/j.egypro.2014.06.114

Addaney, M., & Oppong, R. (2015). Critical issues of municipal solid waste management in Ghana. *Journal of Energy and Natural Resource Management*, *2*, 30–36.

Adefemi, E., & Awokunmi, E. (2009). The impact of municipal solid waste disposal in Ado-Ekiti metropolis, Ekiti-state, Nigeria. *African Journal of Environmental Science and Technology*, *3*(8), 186–189. doi:10.5897/AJEST09.075

Ali, A. (2010). Wasting time on solid waste in developing countries. *Waste Management (New York, N.Y.)*, *30*(8-9), 1437–1438. doi:10.1016/j.wasman.2010.04.025 PMID:20493678

American Society of Mechanical Engineers (ASME). (2014). *Waste to Energy, Summary Report*. Available Online: http://www.asme.org

Amoah, S. T., & Kosoe, E. A. (2014). Solid waste management in urban areas of Ghana: Issues and experiences. *Wa Journal of Environmental Pollution and Human Health*, *2*, 110–117.

Anschutz, J., Ijgosse, J., & Scheinberg, A. (2004). *Putting integrated solid waste management to practice*. Waste.

Aracil, C., Haro, P., Fuentes-Cano, D., & Gomez-Barea, A. (2018). Implementation of waste-to-energy options in landfill-dominated countries: Economic evaluation and GHG impact. *Waste Management (New York, N.Y.)*, *76*, 443–456. doi:10.1016/j.wasman.2018.03.039 PMID:29610061

Arendse, L., & Godfrey, L. (2010). *Waste management indicators for national state of environment reporting*. Available Online: http://www.unep.or.jp/ietc/kms/data/2010.pdf

Asnani, P. U. (2005). *United States Asia Environmental Partnership Report, United States Agency for International Development, Centre for Environmental Planning and Technology, Ahmedabad; 2005. Technical Committee Report, West Bengal SWM Mission 2005*. Government of West Bengal.

Bartone, C., & Leitman, J. (1994). *Towards environmental strategies for cities, urban infrastructure notes, urban No. UE-11*. World Bank.

Baud, I., Grafakos, S., Hordijk, M., & Post, J. (2001). Quality of life and alliances in solid waste management contributions to urban sustainable development. *Cities (London, England)*, *18*(1), 3–12. doi:10.1016/S0264-2751(00)00049-4

Bell, R., & Russell, C. (2002). Environmental policy for developing countries. *Issues in Science and Technology*, *18*, 63–70.

Bello, I., Ismail, M., & Kabbashi, N. (2016). Solid waste management in Africa: Integrated development plan review: A city growing with you, Eastern Cape, SA. A review. *International Journal of Waste Resources*, *6*, 1–4.

Blaauw, P. (2017). Informal employment in South Africa: Still missing pieces in the vulnerability puzzle. *Southern African Business Review*, *21*, 339–360.

Bradley, P., & Yee, S. (2015). *Using the DPSIR framework to develop a conceptual model: Technical support document*. USEPA.

Brunner, P., & Fellner, J. (2007). Setting priorities for waste management strategies in developing countries. *Waste Management & Research*, *25*(3), 234–240. doi:10.1177/0734242X07078296 PMID:17612323

Buffalo City Metropolitan Municipality. (2011). *Buffalo City Metropolitan Integrated Development Five-Year Plan: 2011 to 2016. Adopted by Council on 30th June 2011*. Buffalo City Metropolitan Municipality.

Buffalo City Metropolitan Municipality. (2018). *Solid waste management department*. Available Online: https://www.buffalocitymetro.gov.za/Municipality/MSWaste

Buffalo City Metropolitan Municipality (BCMM). (2014). *Metro waste management forum to be launched.* Available Online: http://www.bcmmnews.co.za/blog.php?id=8

Cao, H. (2005). An initial study on DPSIR model. *Environmental Science & Technology*, *28*, 110–111.

Choe, C., & Fraser, I. (1998). The economics of household waste management: A review. *The Australian Journal of Agricultural and Resource Economics*, *42*(3), 269–302. doi:10.1111/1467-8489.00052

Chu, F., Labadi, N., & Prins, C. (2006). A scatter search for the periodic capacitated arc routing problem. *European Journal of Operational Research*, *169*(2), 586–605. doi:10.1016/j.ejor.2004.08.017

Cobbinah, P. (2015). Local attitudes towards natural resources management in rural Ghana. *Management of Environmental Quality*, *26*(3), 423–436. doi:10.1108/MEQ-04-2014-0061

Cobbinah, P., Addaney, M., & Agyeman, K. (2017). Locating the role of urbanites in solid waste management in Ghana. *Environmental Development*, *24*, 9–12. doi:10.1016/j.envdev.2017.06.004

Cobbinah, P., Erdiaw-Kwasie, M., & Amoateng, P. (2015). Africa's urbanisation: Implications for sustainable development. *Cities (London, England)*, *47*, 62–72. doi:10.1016/j.cities.2015.03.013

DEA. (2019). *National waste management strategy*. Available online: https://www.environment.gov.za/documents/strategicdocuments/wastemanagement

DEAT. (1999). *National waste management strategy*. Available Online: http://www.deat.gov.za/ProjProg/WasteMgmt/waste.html

DEAT. (2007). Assessment of the status of waste service delivery and capacity at the local government level. DEAT: General waste management directorate and Palmer Development group, Pretoria, SA.

Department of Economic Development and Environmental Affairs (DEDEA). (2009). *Draft integrated general waste management plan*. GIBB Engineering, Province of the Eastern Cape.

Department of Economic Development, Environmental Affairs and Tourism (DEDEAT). (2013). *2013-14 Policy Statement: Honourable MEC Mcebisi Jonas, Eastern Cape Provincial Government.* Available Online: http://www.dedea.gov.za/Speeches/Policy%20Speech%202013%20-%20MEC%20Jonas.pdf

Department of Environmental Affairs and Technology (DEAT). (2008). The capacity audit and needs analysis survey for environmental impact administrators. Department of Environmental Affairs and Tourism. Undertaken by Regenesys, Pretoria, SA.

Department of Environmental Affairs (DEA). (2016). *Report on the Determination of the Extent and Role of Waste Picking in SA; DEA: Pretoria, SA, 2016*. Available online: http://sawic.environment.gov.za/documents/5413.pdf

Dlamini, B., Rampedi, I., & Ifegbesan, A. (2017). Community resident's opinions and perceptions on the effectiveness of waste management and recycling potential in the Umkhanyakude and Zululand district municipalities in the KwaZulu-Natal province of SA. *Sustainability*, *9*(10), 1–19. doi:10.3390u9101835

Dlamini, S., & Simatele, D. (2016). Unrecognized informal solid waste recycling in an emerging African megacity: A study of Johannesburg, SA. *WIT Transactions on Ecology and the Environment*, *202*, 1–3.

Dlamini, S., Simatele, M., & Kubanza, N. (2019). Municipal solid waste management in South Africa: From waste to energy recovery through waste-to-energy technologies in Johannesburg. *Local Environment, 24*(3), 249–257. doi:10.1080/13549839.2018.1561656

Dookhi, A. (2016). Engineering compliance- East London regional waste disposal site- A case study. *Proceedings of the 23rd WasteCon Conference*. Available Online: http://www.infrastructurene.ws/wp-content/uploads/sites/4/2017/10/17.-Dookhi-AS.pdf

Dookhi, N., Jewskiewitz, S., & Jewaskiewitz, B. (2015). East London Regional (Roundhill) waste disposal site- wasting no time on the road to compliance. *Civil Engineering (New York, N.Y.), 23*, 35–41.

Fiehn, H. & Ball, J. (2005). *Integrated waste management: background research paper produced for the SA Environment Outlook report on behalf of the Department of Environmental Affairs and Tourism*. Jarrod Ball and Associates.

Fobil, J., Armah, N., Hogarh, J., & Carboo, D. (2008). The influence of institutions and organizations on urban waste collection systems: An analysis of waste collection system in Accra, Ghana. *Journal of Environmental Management, 86*(1), 262–271. doi:10.1016/j.jenvman.2006.12.038 PMID:17291677

Future Works Sustainability Consulting (FWSC). (2014). *Review of the Buffalo City Metropolitan Municipality integrated environmental management plan and coastal zone management plan: Draft state of the environment report*. FWSC.

Godfrey, L. (2019). Waste plastic, the challenge facing developing countries-ban it, change it, collect it? *Recycling, 4*(3), 1–6. doi:10.3390/recycling4010003

Godfrey, L., & Oelofse, S. (2017). Historical Review of Waste Management and Recycling in South Africa. *Resources, 6*(4), 57. doi:10.3390/resources6040057

Goldblatt, M. (2009). *Research paper: Macro-economic trends and economic instruments*. Department of Environmental Affairs, Pretoria.

Gumbi, S. (2016). *Current waste management and minimisation patterns and practices: an exploratory study on the Ekurhuleni metropolitan municipality in South Africa* (MSc Thesis). University of South Africa, RSA.

Gumbo, T., & Simelane, T. (2015). Innovations in municipal solid waste management: Experiences from eThekwini municipality, SA. In T. Simelane & R. Mohee (Eds.), *Future directions of municipal solid waste management in Africa* (pp. 202–217). Africa Institute of SA. doi:10.2307/j.ctvh8r2sj.18

Gutberlet, J. (2018). *Waste in the city: Challenges and opportunities for urban agglomerations, urban agglomeration Mustafa Ergen*. IntechOpen. Retrieved 09 July 2020 from: https://www.intechopen.com/books/urban-agglomera

Hazra, T., & Goel, S. (2009). Solid waste management in Kolkata, India: Practices and challenges. *Waste Management (New York, N.Y.), 29*(1), 470–478. doi:10.1016/j.wasman.2008.01.023 PMID:18434129

He, L., Huang, G., Zeng, G., & Hu, H. (2009). Identifying optimal regional solid waste management strategies through an inexact integer programming model containing infinite objectives and constraints. *Journal of Waste Management, 29*(1), 21–31. doi:10.1016/j.wasman.2008.02.003 PMID:18406594

Hettiarachchi, H., Meegoda, J., & Ryu, S. (2018). Organic waste buyback as a viable method to enhance sustainable municipal solid waste management in developing countries. *International Journal of Environmental Research and Public Health, 15*(11), 1–15. doi:10.3390/ijerph15112483 PMID:30405058

Hrebicek, J., Kalina, J., Soukopova, J., Prasek, J., & Valta, J. (2017). *The forecasting waste generation model based on linked open data and the DPSIR framework: Case study concerning municipal waste in the Czech Republic.* Available Online: https://www.muni.cz/en/research/publications/1348999

Infrastructure News. (2016). *Roundhill rehabilitation reaches new heights*. Available Online: http://www.infrastructurene.ws/2016/08/02/roundhill-rehabilitation-reaches-new-heights/

Inter-American Development Bank. (2003). *Economic instruments for solid waste management: global review and applications for Latin America and the Caribbean*. D.C. Inter-American Development Bank.

Jewaskiewitz, S. & Dookhi, N. (2017). Roundhill restoration. *Sustainable Landfills,* 16-19.

Kalule, S. (n.d.). *A case study for developing a framework for ecological risk assessment for landfill sites in SA*. Available Online: https://www.academia.edu/4558844/EGM_826_assignment_4_Manuscript_Kitumba_Kalule

Katisiimeh, M. W., Burger, K., & Mol, A. P. (2013). Informal waste collection and its co-existence with the formal waste sector: The case of Kampala, Uganda. *Habitat International, 38,* 1–9. doi:10.1016/j.habitatint.2012.09.002

Kazuva, E., Zhang, J., Tong, Z., Si, A., & Na, L. (2018). The DPSIR model for environmental risk assessment of municipal solid waste in Dar es Salaam city, Tanzania. *International Journal of Environmental Research and Public Health, 15*(8), 1–30. doi:10.3390/ijerph15081692 PMID:30096829

Ke'er, C. (2010). *Controlling municipal solid waste (MSW) over-generation with a "waste disposal fee" based on an efficient stakeholder partnership* (Msc Thesis). Lund University.

Khajuria, A. (2010). *Application of the DPSIR framework for municipal solid waste management in south Asian developing countries* (PhD Dissertation). Osaka University.

Khumalo, S. (2016). *Environmental impact of household solid waste disposal practices on plant growth in rural areas of KwaZulu-Natal: A case study of UThukela district municipality* (Msc. Dissertation). University of SA.

Kubanza, S. (2010). *Perception and issues of solid waste management in SA, Johannesburg case study* (Msc Thesis). University of Witwatersrand.

Linster, M., & Fletcher, L. (2012). *Using the pressure-state-response model to develop indicators of sustainability, USA*. Available Online: http://documentacion.ideam.gov.co/openbiblio/bvirtual/017931/DocumentosIndicadores/Temasvarios/Docum26.pdf

Linzner, R., & Lange, U. (2013). Role and size of informal sector in waste management—A review. *Water Resources Management, 166,* 69–83.

Liu, Y., Hao, M., Guo, H., Gao, W., & Xiang, N. (2014). Establishment of assessment indicator system for environment-friendly society based on DPSIR model. *Environmental Protection Science, 16,* 14–21.

Local Government Handbook of SA (LGHSA). (2017). *Buffalo City Metropolitan Municipality. 117 Oxford Street, Cnr North and Oxford Streets*. Trust Centre.

Manaf, L., Samah, M., & Zukki, N. (2009). Municipal solid waste management in Malaysia: Practices and challenges. *Journal of Waste Management, 29*(11), 2902–2906. doi:10.1016/j.wasman.2008.07.015 PMID:19540745

Minghua, Z., Xiumin, F., Rovetta, A., Qichang, H., Vicentini, F., Bingkai, L., Giusti, A., & Yi, L. (2009). Municipal solid waste management in Pudong New area, China. *Waste Management (New York, N.Y.), 29*(3), 1227–1233. doi:10.1016/j.wasman.2008.07.016 PMID:18951780

Mourao, M. C., Nunes, A. C., & Prins, C. (2009). Heuristic methods for the sectoring arc routing problem. *European Journal of Operational Research, 196*(3), 856–868. doi:10.1016/j.ejor.2008.04.025

Mudau, A. (2012). *A laboratory investigation of the effects of water content and waste composition on leachate and gas generation from simulated MSW* (MSc Thesis). University of the Witwatersrand, SA.

Municipalities of SA. (2018). *Buffalo City Metropolitan Municipality*. Available Online: https://municipalities.co.za/map/7/buffalo-city-metropolitan-municipality

Muzenda, E., Ntuli, F., & Pilusa, T. J. (2012). Waste management strategies and situation in SA. *Engineering and Technology, 68*, 309–312.

Nabegu, A. (2010). An Analysis of Municipal Solid Waste in Kano Metropolis, Nigeria. *Journal of Human Ecology (Delhi, India), 31*(2), 111–119. doi:10.1080/09709274.2010.11906301

Nahman, A., & Godfrey, L. (2010). Economic instruments for solid waste management in SA: Opportunities and constraints. *Resources, Conservation and Recycling, 54*(8), 521–531. doi:10.1016/j.resconrec.2009.10.009

National Treasury. (2006). *A framework for considering market-based instruments to support environmental fiscal reform in SA*. National Treasury.

Nemerow, N. L. (2009). *Environmental Engineering: Environmental Health and Safety for Municipal Infrastructure, Land Use and Planning, and Industry* (6th ed.). Wiley.

Niekerk, S., & Weghmann, V. (2019). *Municipal solid waste management services in Africa*. Working Paper, Public Service International, Johannesburg, RSA.

Nissim, I., & Inbar, S. (2005). From dumping to sanitary landfills-solid waste management in Israel. *Waste Management (New York, N.Y.), 25*(3), 323–327. doi:10.1016/j.wasman.2004.06.004 PMID:15823747

Nkosi, N., Muzenda, E., Zvimba, J., & Pilusa, J. (2013). *The current waste generation and management trends in SA: A review. International Conference on Integrated Waste Management and Green Energy Engineering*, Johannesburg, SA.

Nyika, J., Onyari, E., Mishra, S., & Dinka, M. (2020). Waste management in South Africa. In A. Pariatamby, F. Shahul, & M. Bhatti (Eds.), *Sustainable waste management challenges in developing countries*. IGI Global. doi:10.4018/978-1-7998-0198-6.ch014

Oelofse, S. (2018). Opportunities and challenges in waste management. *CSIR*. Available Online: http://sustainabilityweek.co.za/assets/files/Day%203%20%20CSIR%20-%20Waste.pdf

Okot-Okumu, J. (2015). Solid waste management in Uganda; challenges and options. In T. Simelane & R. Mohee (Eds.), *Future directions of municipal solid waste management in Africa* (pp. 107–135). Africa Institute of South Africa. doi:10.2307/j.ctvh8r2sj.14

Okot-Okumu, J., & Nyenje, R. (2011). Municipal solid waste management under decentralisation in Uganda. *Habitat International*, *35*(4), 537–543. doi:10.1016/j.habitatint.2011.03.003

Oteng-Ababio, M. (2010). Private sector involvement in solid waste management in Ghana: The case of the Greater Accra Metropolitan Area (GAMA). *Waste Management & Research*, *28*(4), 322–329. doi:10.1177/0734242X09350247 PMID:19854814

Oteng-Ababio, M. (2011). Governance crisis or attitudinal challenges: collection, storage and transportation of solid waste in Ghana. *Integrated Waste Management, 1, 3-22*.

Oteng-Ababio, M., Arguello, J., & Gabbay, O. (2013). Solid waste management in African countries: Sorting the facts from the fads in Accra, Ghana. *Habitat International*, *39*, 96–104. doi:10.1016/j.habitatint.2012.10.010

Pearce, D., & Turner, R. K. (1994). *Economics and solid waste management in the developing world*. CSERGE working paper WM 1994-05. Centre for Social and Economic Research on the Global Environment.

Perman, R., Ma, Y., McGilvray, J., & Common, M. (2003). *Natural resource and environmental economics*. Pearson Education.

Permana, A., Towolioe, S., Aziz, N., & Ho, C. (2015). Sustainable solid waste management practices and perceived cleanliness in a low-income city. *Habitat International*, *49*, 197–205. doi:10.1016/j.habitatint.2015.05.028

Post, J. (1999). The problems and potentials of privatizing solid waste management in Kumasi, Ghana. *Habitat International*, *23*(2), 201–216. doi:10.1016/S0197-3975(98)00045-9

Ramachandra, T., & Bachamanda, S. (2007). Environmental audit of Municipal Solid Waste Management. *International Journal of Environmental Technology and Management*, *7*(3/4), 369–391. doi:10.1504/IJETM.2007.015152

Rasmeni, Z., & Madyira, D. (2019). A review of the current municipal solid waste management practices in Johannesburg city townships. *Procedia Manufacturing*, *35*, 1025–1031. doi:10.1016/j.promfg.2019.06.052

Reputation Matters. (2014). *Buffalo City Metropolitan Municipality calls for all hands on waste*. Available Online: http://reputationmatters.co.za/2014/02/17/buffalo-city-metropolitan-municipality-calls-for-all-hands-on-waste/

Reschovsky, J., & Stone, S. (1994). Market incentives to encourage household waste recycling: Paying for what you throw away. *Journal of Policy Analysis and Management*, *13*(1), 120–139. doi:10.2307/3325093

Rich, C., Gronow, J., & Voulvoulis, N. (2008). The potential for aeration of MSW landfills to accelerate completion. *Waste Management (New York, N.Y.)*, *28*(6), 1039–1048. doi:10.1016/j.wasman.2007.03.022 PMID:17531464

Riddin, T. (2018). *BCM solid waste operates on 10 of 26 refuse trucks*. Available Online: https://www.dispatchlive.co.za/news/2018-04-10-bcm-solid-waste-operates-on-10-of-26-refuse-trucks/

Rotich, H. K., Yongsheng, Z., & Jun, D. (2006). Municipal solid waste management challenges in developing countries e Kenyan case study. *Waste Management (New York, N.Y.)*, *26*(1), 92–100. doi:10.1016/j.wasman.2005.03.007 PMID:16006111

Russell, C. S., & Vaughan, W. J. (2003). *The choice of pollution control policy instruments in developing countries: arguments, evidence and suggestions*. Edward Elgar.

Samson, M. (2010). *Organising reclaimers in Tshwane, SA-Lessons from Topdown and bottom up experiences. Women in informal employment: Globalizing and Organizing*. WIEGO.

Schübeler, P. (1996). Conceptual framework for municipal solid waste management in low-income countries. In K. Wehrle & J. Christen (Eds.). UNDP/UNCHS/World Bank/SDC Collaborative Programme on Municipal Solid Waste Management in Low-Income Countries. Academic Press.

Senkoro, H. (2003). *Solid Waste Management in Africa: A WHO/AFRO Perspective*. Paper 1, Presented in Dares Salaam at the CWG Workshop.

Sentime, K. (2011). Profiling solid waste pickers: A case study of Braamfontein-Greater Johannesburg. *Africanus*, *41*, 96–111.

Senzige, J., Makinde, D., Njau, K., & Nkansah-Gyeke, Y. (2014). Factors influencing solid waste generation and composition in urban areas of Tanzania: The case of Dar-es- Salaam. *American Journal of Environmental Protection*, *3*(4), 172–178. doi:10.11648/j.ajep.20140304.11

Sharholy, M., Ahmad, K., Mahmood, G., & Trivedi, R. (2008). Municipal solid waste management in Indian cities-A review. *Waste Management (New York, N.Y.)*, *28*(2), 459–467. doi:10.1016/j.wasman.2007.02.008 PMID:17433664

Simatele, D., & Etambakonga, C. L. (2015). Scavenging for solid waste in Kinshasa: A livelihood strategy for the urban poor in the Democratic Republic of Congo. *Habitat International*, *49*, 266–274. doi:10.1016/j.habitatint.2015.05.029

Simelane, T., & Mohee, R. (2012). Future directions of municipal solid waste management in Africa. *Policy Brief. Briefing*, *81*, 1–5.

Simelane, T., & Mohee, R. (2015). *Future directions of municipal solid waste management in Africa*. Africa Institute of SA.

South African Cities Network (SACN). (2014). State of waste management in cities-Phase 2: Modelling the effects of landfilling as a disposal method. SACN.

South African Institute of Race Relations (SAIRR). (2013). *SA 'two-thirds urbanised*. Available at: www.southafrica.info/news/urbanisation-240113.htm

Statistics, S. A. (2003). *Investigations into appropriate definitions of urban and rural in SA*. Report No. 03-02-21.

Statistics, S. A. (2009). *Non-financial municipal census of municipalities for the year ending June 2008*. Statistics SA.

Statistics, S. A. (2011). *Buffalo City Municipality, Census 2011*. Available Online: http://beta2.statssa.gov.za/?page_id=1021andid=buffalo-city-municipality

Svarstad, H., Petersen, L. K., Rothman, D., Siepel, H., & Watzold, F. (2008). Discursive biases of the environmental research framework DPSIR. *Land Use Policy*, *25*(1), 116–125. doi:10.1016/j.landusepol.2007.03.005

Taljaard, A. (2011). *Critical perspectives on the definition of wastes in SA- experiences within the steel-making industry in environmental management*. North West University, SA.

Tukahirwa, J. T., Mol, A., & Oosterveer, P. (2011). Access of urban poor to NGO/CBO-supplied sanitation and solid waste services in Uganda: The role of social proximity. *Habitat International*, *35*(4), 582–591. doi:10.1016/j.habitatint.2011.03.006

UN-Habitat. (2010). Solid waste management in the world's cities e Water and sanitation in the world's cities 2010. London: United Nations Human Settlements Programme (UN-Habitat), Earthscan.

UNEP-United Nations Environment Programme. (2005). *Integrated Waste Management Scoreboard: a tool to measure performance in municipal solid waste management*. UNEP.

United Nations Environment Program (UNEP). (2005). *Solid waste management*. Available Online: http://www.unep.or.jp/ietc/Publications/spc/SolidWaste Management/index.asp

United Nations Secretariat. (2010). *Population division of the department of economic and social affairs: World population prospects*. Author.

United States Environmental Protection Agency. (2013). *Wastes-non-hazardous waste-municipal solid waste*. Author.

Wang, G., Qin, L., Li, G., & Chen, L. (2009). Landfill site selection using spatial in- formation technologies and AHP: A case study in Beijing, China. *Journal of Environmental Management*, *90*(8), 2414–2421. doi:10.1016/j.jenvman.2008.12.008 PMID:19375842

Wolfville, P., & Scotia, N. (2002). *Study on solid waste management options for Africa. Project Report Final Draft Version*. The African Development Bank.

Zurbrugg, C. (2002). Urban solid waste management in low income countries of Asia: How to cope with the garbage crisis: scientific committee on problems of the environment (SCOPE). Urban solid waste management review session, Durban, SA.

Zurbrügg, C., Caniato, M., & Vaccari, M. (2014). How assessment methods can support solid waste management in developing countries—A critical review. *Sustainability*, *6*(2), 545–570. doi:10.3390u6020545

Chapter 3
Solid Waste, Treatment Technologies, and Environmental Sustainability:
Solid Wastes and Their Sustainable Management Practices

Zia Ur Rahman Farooqi
https://orcid.org/0000-0003-2398-1895
University of Agriculture, Faisalabad, Pakistan

Abdul Kareem
University of Agriculture, Faisalabad, Pakistan

Faizan Rafi
University of Agriculture, Faisalabad, Pakistan

Shujahat Ali
https://orcid.org/0000-0002-3760-9042
Mirpur University of Science and Technology (MUST), Pakistan

ABSTRACT

Waste has a long history of association with humans and other organisms. It is inevitable and has different sources right from crop residues (agriculture), food production and its movement through food chain, industries and their processes. Waste has different environmental implications like leachate and can contaminate groundwater. Emissions from waste burning cause air pollution, and its dumping in soil cause soil pollution. There are different methods and technologies being used in the world to minimize waste. All the technologies being used are reviewed and presented in it with their pros and cons of all the technologies. This chapter will include all these technologies and associated environmental concerns.

DOI: 10.4018/978-1-7998-4921-6.ch003

INTRODUCTION

Solid waste is every type of solid substance which arises from human and other organisms and considered as unwanted. Solid wastes can be organic or inorganic fractions and generally do not carry any value to any organism. Wastes can be kitchen refuse, packaging, clippings, cloth, bottles, paper, cans, and dry batteries. There are different types of solid wastes on the basis of biodegradability, inertness, e-wastes, composites and domestic. All these wastes are generated from industrial processes and domestic activities (Kumar et al., 2018).

BACKGROUND

Expanding population, growing economies and fast urbanization has incredibly accelerated the metropolitan waste generation rate in developing nations (Guerrero et al., 2013). Cities typically are responsible for waste generation (Burnley, 2007; Sujauddin et al., 2008). Solid waste is one of the important byproducts of urban lifestyle, which is growing faster due to rapid urbanization. In the last 1 year, the world population increased from 2.9 to 3 billion, but waste generation increased from 0.68 to 1.3 billion tons, and per capita generation increased from 0.64 to 1.2 kg/day, which become almost double within 10 years. A prediction was made on the basis of increasing solid waste generation; about 4.3 of urban population will generate 1.42 kg/capita/day of municipal solid waste by 2025. However, per capita solid waste generation varied in different countries and different cities within a country (Mian et al., 2017).

There are different associated risks of solid wastes and their treatment technologies. Like leachates in landfill sites cause soil and groundwater pollution, gaseous emissions from combustion, incineration, pyrolysis and waste to energy processes cause air pollution and climate change, solid waste smell and its presence in any area creates nuisance (De and Debnath, 2016).

MAIN FOCUS OF THE CHAPTER

To avoid all the risks associated with solid wastes, numerous technologies are devised to manage them. Some are old and conventional (recycling, open burning, dumping into the seas) and pose further environmental risks, some are new (landfilling, biomethanation, pyrolysis, composting, plasma gasification, waste to energy and incineration). These methods also have some minor environmental implications like pyrolysis, combustion etc. have gaseous emissions problems. So, there was a need of some ways to manage solid wastes with sustainable way. This purpose was accomplished when 3Rs and agricultural wastes disposal with agricultural wastes residues incorporation into the fields, composting of municipal wastes, production of biofuels and biogas was invented. These practices are sustainable as well as profitable because they do not only reduce wastes but also generate biofuels, energy and electricity as well.

SOLID WASTES, TYPES, AND SOURCES:

There are different types of solid waste that are released from different sources and they are described briefly in the following;

Biodegradable Waste

This waste type include any waste organic material which can be degraded into carbon dioxide, water, methane or simple organic molecules by micro-organisms and other living things. These wastes are decomposed by composting, aerobic and in-aerobic digestion. These wastes are decomposed by the above processes domestically as well as commercially with the purposes of reduction of waste volumes with the aim to reduce harm to human health and impact on the environment and production of useful products and use them wherever possible. It is important to choose correctly the composition of various biodegradable wastes and ensure proper temperature regime (Kliopova et al., 2019).

Biodegradable waste is commonly damped, combusted or buried into the soil. As the biodegradation rate of organic wastes is low, and also leads to the emission of gas which cannot be captures and contribute to the greenhouse effect. The amount of gasses emitted through this process is estimated about 30% of the global methane emissions (Nowicki et al., 2016). To control the gaseous emissions from degradation of biodegradable wastes, production of activated carbon adsorbents from this type of waste has become a very interesting alternative to its combustion, recycling or composting (Biedenbach et al., 2015). Activated carbon act as efficient adsorbents used for purification of gases and liquids because they have large surface area and micro-porous structure, ensuring very good sorption capacity towards a wide range of gasses and other substances. It also advantage that activated carbon also produced from any waste substance containing carbon in it. Therefore, taking into regard the high cost of activated carbon production from fossil coals along with the economic and ecological aspects, recently much attention has been paid to the use of organic waste for this purpose. The post-agricultural production waste is easily available, renewable and inexpensive, that is why the activated carbons obtained from this type of waste have been subjected to adsorption tests from gas and liquid phases.

Recyclable

Any waste material which can be used again for beneficial purpose is often called as Recyclable waste. It is the method of renovating waste materials into new and useful materials and objects or reusing them as a whole. It is best alternative to conventional waste disposal method as it saves natural resource materials and help in lowering greenhouse gasses. It can prevent the wastage of potentially useful resources and materials, and reduce the consumption of fresh raw materials, thereby reducing energy usage, air pollution (from incineration), and water pollution (from landfilling). Recycling is a major constituent of modern waste reduction strategy of 3Rs (Reduce, Reuse, and Recycle). Thus, it aims at environmental sustainability by substituting raw material inputs. International Standardization Organization also set some standards related to recycling such as ISO 15270:2008 for plastics waste and ISO 14001:2015 for environmental management control of recycling practice (Hohenthal et al., 2019; Roychoudhuri and Debnath, 2020).

Materials which can be recycled include glass, paper, cardboard, metal, plastic, tires, batteries, and electronic equipments. These materials are either delivered to recycling center and sorted, cleaned, and reprocessed into new materials destined for manufacturing new products. Due to rapid economic growth and population increase, problems of environmental pollution, climate change, and resource depletion have become increasingly serious in the world. Recyclable waste recycling is becoming one of the key approaches to simultaneously respond to the above issues, and governments have begun to regulate and promote source separation and urban waste recycling in recent years. There are three main types of recycling: primary, secondary, and tertiary. In primary recycling, wastes are recycled into same things i.e. like paper into more paper. In secondary recycling process, wastes are modified to use for any-other way i.e. cutting plastic bottle to use it in garden, and tertiary recycling include complete transformation of the waste into a new product by chemical processes, burning or any-other process (Zhang et al., 2016; Zhao et al., 2018; Zink et al., 2018; Shi et al., 2019).

Inert

Any type of waste which cannot be decomposed biologically or chemically reactive and does not decomposed. It is the major part of the waste and covers up to 50% of the total generated wastes which poses a challenge of its management. It also has major contribution in affecting human health and the environment as it contains many types of secondary pollutants trapped in it. Inert wastes include demolition waste, plastic, rubber, ceramic and debris etc. Its management only include its recycling as filling and base material for floors, concrete aggregate, building material and tiles, etc. (Mendoza et al., 2017; Tatsuno et al., 2019).

Many Governments has set up storage facilities for inert waste management i.e. French Ministry of Environment workgroup defined three types of storage facilities according to the nature of the inert waste viz:

1. Type F facility stores plaster, asbestos-cement and certain inert industrial wastes.
2. Type G facility stores demolition waste mixed with certain industrial wastes.
3. Type H facility stores non-polluted soil from works and sorted demolition wastes and technically separable plaster certain industrial wastes (Emmanuel et al., 2002).

E-waste

It is electronic waste of discarded electrical devices. Used electronics are refurbished, reuse, resale, or metals used in their mother and circuit boards are recovered for reuse in new equipments. Casual management of e-waste in developing countries can lead to environmental pollution and adverse human health effects as electronic scrap components such as CPUs contain potentially harmful materials such as lead, cadmium, beryllium, or brominated flame retardants. Recycling and disposal of e-waste may involve significant risk to the health of workers and their communities. Electronic Equipment Directive has classified e-waste into ten categories: Large household appliances (including cooling and freezing equipments), small household appliances (computers), consumer electronics (television), lamps, toys, tools and medical devices, monitoring and control instruments, and Automatic dispensers. On the other hand, some categorize e-wastes into six categories, namely: (1) Temperature exchange equipment (air conditioners, freezers), (2) Screen and monitors, (3) Lamps, (4) Large equipment (washing machines, electric stoves), (5) Small equipment (microwave, electric shaver), and (6) Small IT and telecommunication equipment (mobile phones, printers). Products in each category vary in longevity profile, impact, and collection methods, among other differences (Ilankoon et al., 2018; Narbón-Perpiñá and Prior, 2020). In 2016, Asia imported significant amount of the most extensive volume of e-waste (18.2 Mt), accompanied by Europe (12.3 Mt), America (11.3 Mt), Africa (2.2 Mt), and Oceania (0.7 Mt). The smallest in terms of total e-waste made, Oceania was the largest generator of e-waste per citizen (17.3 kg), with hardly 6% of e-waste gathered and recycled. Europe is the second largest e-waste generator per citizen with an average of 16.6 kg. America generates 11.6 kg and solicits only 17% of the e-waste caused in the provinces, which is commensurate to the assortment count in Asia (15%) (Kumar et al., 2017; Dutta and Goel, 2017).

Composite

Composite is aggregate of two or more materials. This type of waste creates tons of problems as it is hard to deal with multiple type of wastes at a time. To date, there has been no successful attempt to recycle the composite materials and hence the waste is increasing day by day. Composite waste include liquid waste, organic waste, solid and recyclable rubbish, and hazardous waste.

Composite recycling has been an issue in research over the last decade and some methods were proposed: pyrolysis, oxidation in a fluidized bed, and chemical recycling. Among them, pyrolysis is currently the only process with commercial-scale implementations (Hu et al., 2017; Li et al., 2017).

Domestic

Domestic waste means any no putrescible waste, consisting of organic and in-organic materials, such as paper, cardboard, yard clippings, wood, or similar materials, generated in a dwelling, including the real property upon which it is situated, containing four living units or less. Household waste, also known as domestic waste or residential waste, is disposable materials generated by households. This waste can be comprised of hazardous and non-hazardous waste. Non-hazardous waste can include food scraps, paper, bottles, etc. which can be recycled or composted (Mian et la., 2017). Domestic waste is waste that is generated as a result of the ordinary day-to-day use of a domestic premise and is either: collected by or on behalf of a local government as part of a waste collection and disposal system. During the last decade, a growing interest in recycling of domestic waste has emerged, and action plans to increase the recycling of domestic waste have been agreed by many governments. The implementation of new systems and equipment for the collection of domestic waste has been separated at the source. However, only limited information exists on possible occupational health problems related to such new systems. Occupational accidents are very frequent among waste collectors. Based on current knowledge, it appears that the risk factors should be considered as an integrated entity, i.e. technical factors (poor accessibility to the waste, design of equipment) may act in concert with high working rate, visual fatigue due to poor illumination and perhaps muscle fatigue due to high workload (Ihedioha et al., 2017). Typical waste management process used around the world is explained in the fig. 1.

SOLID WASTE GENERATION AND ASSOCIATED ENVIRONMENTAL RISKS

Around the world, waste generation rates are rising. In 2016, 2.01 billion tons of solid waste was generated with the rate of 0.74 kg/person/day. It is estimated that annual waste generation will be increased up to 70% with rapid increase in population in 2050. In Pakistan, 5.5 million USD are invested in a composting project in Lahore. Due to this facility, 150,000 tons of carbon dioxide equivalents are reduced, and compost production increased from 300 to 1000 tons/day (World Bank, 2019).

Solid waste is generated from 0.283 to 0.612 kg/capita/day in Pakistan and its growth rate is 2.4%/year (Nintendo et al., 2019).

Figure 1. Typical Solid Waste Management Process Flow

The following are the main problems regarding solid waste management in Pakistan:

- No existence of waste collection system.
- Waste is dumped or thrown into the streets.
- Wastes are not categorized properly.
- No existence of enough sanitary landfill sites.
- No awareness among citizens about waste management and associated environmental risks.

These all factors contributes in spreading human health diseases and environmental pollution.

Health Hazards

There are numerous environmental and human health impacts associated with solid wastes as described;

- Gasses originated from wastes cause skin and eye infections.
- Dust generated during wastes handling cause breathing problems.
- Bees and flies breed on rotting garbage and spread diseases like diarrhea, dysentery, typhoid, hepatitis and cholera.
- Mosquitoes also breed on waste piles and transmit malaria and yellow fever.
- Dogs, cats and rats dwelling on waste spread diseases including plague and flea born fever.
- Waste collectors face different diseases like intestinal, parasitic and skin diseases (Lal et al., 2016; Ihedioha et al., 2017; Mustafa et al., 2017).

SOLID WASTES MANAGEMENT TECHNOLOGIES AND THEIR TYPES

Solid waste management is regarded as a centralized process and this is mostly based on the economic status of countries (Srivastava et al., 2014). The first step in solid waste management is making strategy and its monitoring. The modern waste management strategies are expensive and mostly used by developed countries (Hassan et al., 2016). The following solid wastes management technologies are used generally and divided into two types i.e., ancient and modern solid waste management technologies.

Ancient

The old solid waste management technologies include recycling, open burning, dumping into seas and rivers, and plowing crop residues directly into the fields.

Recycling

One of the most sustainable means of solid waste management is recycling, as stated in the literature. This process does not adversely affect human health, social conditions, and the environment. This sustained technique does not diminish the quality of life due to poor economic conditions (Troschinetz & Mihelcic, 2009). There is a clear difference between the research conducted on recycling in developed countries as compared to the developing countries. The recycling facilities in the developed countries are heavily industrialized that are produced as a result of the daily life activities of a citizen. The technical applications are focused on the developed countries, including tools and models (Daskalopoulos et al., 1998). In developing countries, the focus is on the research on recycling for direct factors, practical, and elements related to waste management. The competency is the best indicator actual behavior of society towards the different strategies of solid waste management. If the concept and the proper way to recycle the waste is understood by any community, it is easier to persuade them to adopt it (Corral-Verdugo, 1997).

The plastic has contributed a lot in water and land pollution. The problem with it is its non-degradability and not degraded well. The primary recycling is employed to recycle plastic. First cycling is the reintroduction of scrap, single-use plastic edges, and its parts to the extrusion cycle to make products like the plastic. A material from scrap plastic has a similar feature to the original ones. This type of recycling must be only done with semi-clean scrap, that's why it is not mostly selected by the recyclers. A significant number of our daily life products are produced by recycling like pipes, door profiles, windows, grocery shoppers, and blinds (Al-Salem et al., 2009).

Figure 2. Solid Wastes Management Technologies

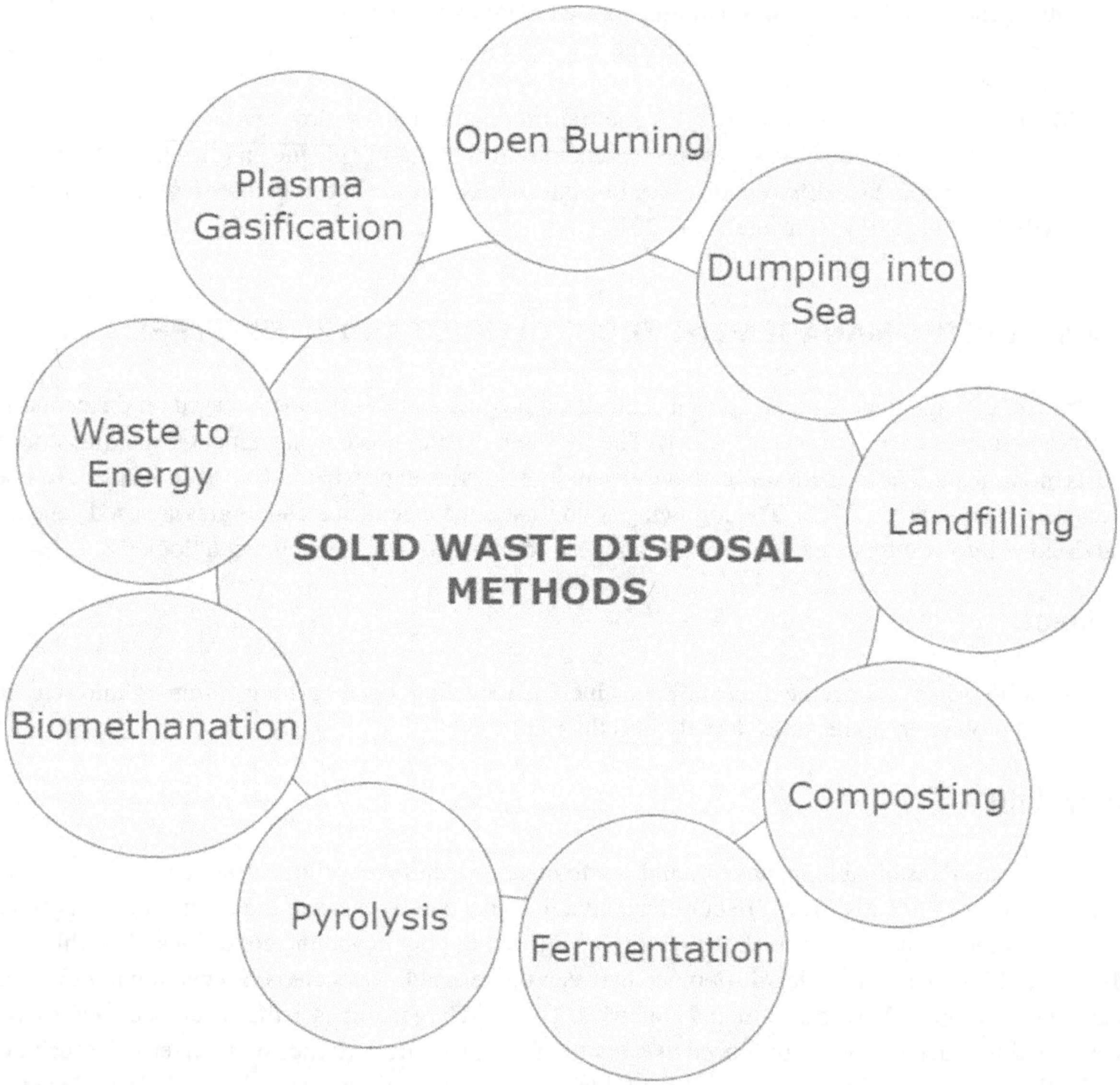

There is another type of recycling known as chemical recycling. This is feasible for those heterogeneous and contaminated waste materials for those which cannot be separated, or their separation is economically not suitable. This is one of the best ways to protect the environment from the plastics. It may be done in closed-loop recycling or open-loop recycling based on the products made from different types of plastics. In closed-loop recycling, the same products are manufactured from which they were made. It may be made entirely from recycled products or a mixture of them. The PET packaging products are its example. In open-loop, recycled plastics are utilized for different products than the one they were made from (Ragaert et al., 2017).

Open Burning

Open burning is one of the most common strategies to manage solid waste in the developing countries of the world. This is mostly done in open places. The masses are not aware of its ill effects on humans. The people find it the easiest way to get rid of solid waste. The problem of waste has been aggravated by an increase in solid waste in Thailand. This problem was mainly due to change in client behavior, growth in population, inadequate facilities to tackle waste, and economic expansion. These are the main reasons causing every house to adopt their own techniques in the elimination of this waste. Open burning is one of the main strategies adopted by many people (Pansuk et al., 2018).

One of the main concerns with open burning is its effects on the environment and human health. There are a lot of emissions from open burning, which are very toxic to human health. The open burning has a direct role in air pollution and climate change due to greenhouse gases from solid waste due to open burning. For example, in Thailand, 4.09 kilo tones/year solid waste is burnt by open burning and emit 0.61 kilo tones of nitrous oxide, 499 kilo tones carbon dioxide, and 26 kilo tones/year of methane in 2016. Sulfur dioxide, carbon monoxide, and many other air pollutants are added in the environment by open burning in Thailand (Pansuk et al., 2018).

Dumping Into Seas and Rivers

It is a common practice to dump solid waste into the rivers and seas. This has played havoc for the marine and freshwater ecosystems. There is a significant concern about the pollutants being added in these systems by the solid waste. This is mostly seen in developing countries where the authorities do not have any mechanism to check the composition of solid waste. The only source for this kind of dumping is the anthropogenic activities. The huge population growth along the seas and rivers have resulted in increased spill in them. In a short-term period, the degradation of these activities has reached the health of the population across the coastal areas. The pollutants included in them are heavy metals, hydrocarbons from oil, nitrate and phosphates, fertilizers, pesticides, and insecticides. Eutrophication is caused by the nutrients in the waters and estuaries near the shore. In the river, the ecosystem is maintained by the steady discharge of the water from the river with that of salty water of the sea in the estuary. The sediment load from the river is balanced by ocean erosion in the delta and its adjacent coastal areas. Due to the industrial zones, many persistent organic pollutants are added in the solid waste. This solid waste, when dumped in the seas or rivers, directly affects the life in the water systems. The other significant source of pollutants in the seas and rivers is sewage waste. The solid waste, along with the sewage sludge, is also very damaging to the sea and river life. The sandstorms are one of the unique ways to pollute water systems. The pollutants are carried by the strong winds into the river or the sea (Daoji & Daler, 2004).

The solid waste being dumped in the rivers also composed of papers, files, bottles, cans, bags, cigarette packaging, and many other products. Most of the waste from the construction is thrown in nearby water reservoirs. The primary mode of transmission of solid waste in rivers, canals, streams, and seas is the drainage system of different cities. It is also observed that they may get stuck in the plantings along the beaches. Most of it is dumped in rivers and seas. The drainage system is also blocked by the solid waste leading to the more production of sediments due to absence of protection of surfaces, and the natural discharge is maximized, the degradation of water quality and the entrance of pollutants in the food chain (Salles et al., 2012).

Ploughing in the Fields

Ploughing in the fields is related to waste management in rural areas. The ploughing is the physical manipulation of soil in the field. Firstly, it is essential to note the type of solid waste being added in the fields. It may be crop residues or any chemical wastes. Crop residues or any organic substance buried in the fields by ploughing these in soils. The wastes added in soils should be degradable necessarily. Ploughing and tillage is done in the fields after crop harvesting. It is also done by collecting wastes from other areas and ploughing in fields. Many farmers in the developing countries also burn the crop stubbles in fields that is the primary cause of smog in these areas. The major drawback of using ploughing is that they mostly take out heavy objects from the field (Abdoli et al., 2008).

Modern

Many modern strategies are being applied to deal with solid waste. These technologies are mostly used in the developed countries of the world. The advanced waste management strategies are presented in fig. 1 and are discussed below;

3Rs

The 3Rs (Reduce, Reuse, and Recycle) strategy of waste management is widely used in developing countries. It is the strategy that is sustainable and acceptable in society. The developed countries are focusing on the development of sustainable solid waste management. This is one of the best ways to conserve the environment. But this is not adopted by the developing countries due to their expensive instrumentation and installation. There are many concerns in the developing world on which they focus. 3Rs are way more sophisticated systems as compared to ancient strategies. First, the awareness of a reduction in solid waste is being created in developed countries. People are encouraged to use of disposable things. Many policies are made to address the problem of reduction in solid waste. There are many analyses being done by the researchers of this field. The reduction policy is made by keeping in view the waste generated from a single household to the factory. The economic advantage of reducing waste may be an essential point to raise the importance of this issue in the world. Reusing has been used for many years in the developed world. The reusing must be done such that there is minimal use of the physical and chemical modifications. There are a number of economic, environmental, and social advantages of reuse. The reusing of certain materials must be done under the proper conditions because they may require special requirements for it. Recycling of the materials is the last approach in 3Rs. This is the most utilized technique for solid waste treatment being applied in many developing countries (Mena et al., 2011).

Following are the benefits of using 3Rs

- The developing countries can attract foreign investment by emission reduction credits.
- The organic gas can be produced by adopting these methods, which can be used for energy production purposes.
- The power generation can be efficiently done.
- It is eco-friendly approach and replenishes the nutrients in the soil.
- Reduced greenhouse gas emissions.

- These approaches may increase the GDP of the country by creating new jobs (Chowdhury et al., 2014).

Composting

As the sizes of the cities and urban areas are increasing the solid waste production is also growing. This has led to the development of modern strategies. Composting is one of the most critical strategy to tackle waste. The organic waste produced by the landfills can be made into such a product that may be used to enhance agricultural productivity. The volume of the waste is considerably reduced by composting. It is a biological process that involves the conversion of much heterogeneous organic waste into organic matter rich substances under the controlled conditions of oxygen, moisture, and temperature along with the microbial population. The organic substances such as vegetables, leaves, fruits, manure, sludge, and other waste are converted by microorganisms into humus like materials. These materials can be used as an organic fertilizer in the fields. Fungi, bacteria, and actinomycetes are the decomposers employed in the process of composting. The basic concept of composting is the microbial decomposition of organic substances. The decomposers release enzymes which convert complex materials into simpler ones. The composting, done in the presence of oxygen is called aerobic and the composting done in the absence of oxygen is called anaerobic composting. Vermicomposting, in-vessel composting, biomineralization, open windrows, controlled microbial composting and aerated static pile composting are other composting systems which are employed to convert wastes into useful soil fertility enhancing products (Atalia et al., 2015).

Landfilling

Landfilling is one of the most economical strategies for managing solid waste. This technique is often confused with open dumping. To avoid this confusion, scientists classified some conditions. The following conditions must be fulfilled by a site to be called as landfill site;

- There should be anaerobic environment.
- There should be one daily cover at least.
- There should be connected pipe system for collection of leachates.
- There should be natural ventilation facility (Hassan et al., 2000).

The second system of classification is based on operational purposes. Level (I) include controlled tipping. Level (II) has a landfill having embankment and daily covering of soil. Level (III) has a leachate recirculation system. Level (IV) contains a treatment system for leachate (Idris et al., 2004).

Incineration

It is energy producing process which is carried-out in engineered facility and employs thermal decomposition at high temperatures to destroy and reduce waste volumes. Generally, combustible wastes or wastes with significant organic contents are considered suitable for incineration. However, any solid waste which pose hazards to environment can be incinerated. Different incineration technologies have been developed for handling various types and physical forms of hazardous wastes. The significant sub-systems which may be incorporated into a solid waste incineration system are waste preparation and feeding, combustion chamber, air pollution control, and residual/ash handling facility. The selection of an appropriate system combination of these components is primarily a function of the physical and chemical properties of the waste stream or streams to be incinerated (Oppelt, 1987).

Fermentation

The organic fractions of the municipal solid wastes and other biodegradable solid substrates like agricultural wastes have to be conveniently treated to reduce their environmental impact and to recover energy and material while massive disposal treatments e.g., landfill or incineration should be avoided. An obvious choice to achieve this goal is given by the anaerobic digestion and co-digestion processes for methane production. However, anaerobic fermentation processes can be used for the production of valuable products. In fact, aerobic fermentation is an effective bioprocess for the production of volatile fatty acids and other low weight organic compounds such as lactic acid and alcohols. These can used for the production of methyl and ethyl-esters to be added to gasoline because of their high octane number (between 103 and 118), as a low cost external carbon source for the production of biopolymers, like poly-hydroxy-alkanoates, for sustainment of the biological processes for nutrient removal in waste water treatment plants, still, for anaerobic fermentation with a high hydrolysis rate being the first step of the anaerobic digestion process for biogas production when treating organic wastes with a high biodegradability (Bolzonella et al., 2005).

Pyrolysis

It is one of the most efficient and environmental-friendly strategies. It also has many operational advantages over other approaches. If it is adopted at a large industrial scale, it has proven to be very economical in terms of fuel pricing. The pyrolysis is defined as a procedure in which the degradation is thermally achieved in an inert environment of long-chain organic substances either in the presence of a catalyst knows as catalytic pyrolysis or in the absence of catalysis called Thermal Pyrolysis. The long-chain complex organic compounds are converted into simpler molecules by control the flow of heat and pressure. The by-products of this procedure are known as tars. The long residence time is required to operate this process. The most common solid waste materials used in the pyrolysis are plastics. Due to the utilization of plastic, it has gained much fame in the market. About 80% of daily life plastic is composed of thermosets and thermoplastics. These are converted products like those of fossil fuels. The setup of the pyrolysis may range from a micro-industry level to a large scale. It is dependent on the type of reactors installed, and the kind of plastic added. The factors affecting the rate of pyrolysis are the pressure, temperature, type of catalyst, heat transfer limit, and residence time (Al-Salem et al., 2017).

Biomethanation

It is the process of extraction of methane gas from waste burning (Yan et al., 2019). Different types of feed-stock like rice straw and other crop residues, livestock dung and fruit and vegetable leaves remains that contain high amounts of organic material in them, are used for methane generation and biogas production (Negi et al., 2018). Anaerobic digestion is commonly used to extract methane gas from different types of wastes (Donoso-Bravo et al., 2016).

Waste to Energy/ Refused Derived Fuel

Due to energy shortage and increased demands and rapid depletion of fossil fuels resources, researchers are doing work on cheap, reliable and sustainable sources of energy. Different solid wastes are used for energy production through burning and digestion (Kothari et al., 2010). As food is necessity of every human, the wastes produced during food production, processing and consumption. About one third of food produced globally is left as waste which allows the sustainable production of energy from the food wastes through combustion, digestion and fermentation processes (Uçkun Kiran et al., 2014). In USA, energy production from municipal solid waste management consists of recycling, composting and combustion with energy production from them (Psomopoulos et al., 2009).

Figure 3. Process Flow of Biofuels and Energy Production from Solid Wastes here

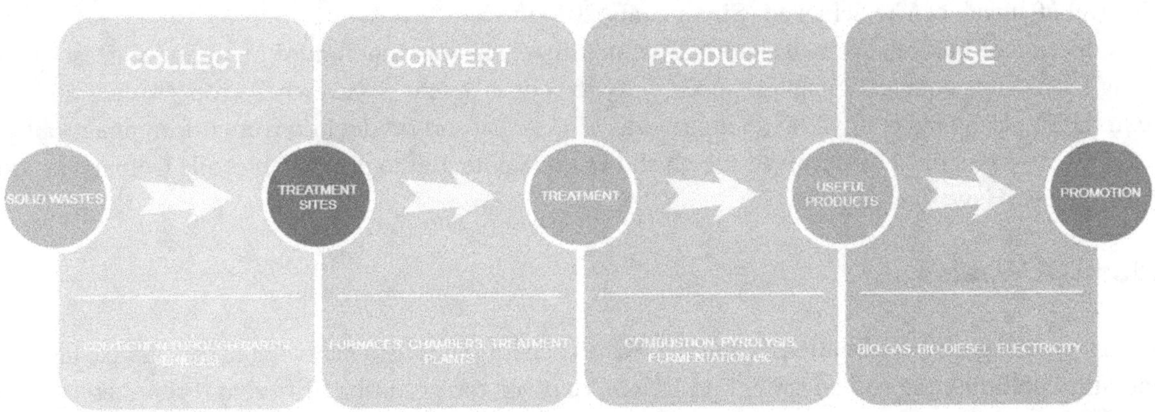

Plasma Gasification

Plasma gasification is comparatively a newer advancement in existing technologies in which solid wastes are changed to energy-filled products (Sanlisoy and Carpinlioglu, 2017). This technology is capable of providing sustainable disposal for various wastes as well as providing the energy requirements of the world (Favas et al., 2017). Different types of wastes like industrial and municipal wastes are used in this process for energy generation. This process also have the advantages of reducing the chances of soil and groundwater pollution because wastes are burned and used for energy generation. By using this technology, up to 71.8% energy recovery can be obtained from feed-stock (Messerle et al., 2016; Wang et al., 2019). Fig. 3 shows the waste collection and its conversion to useful products in a sustainable way.

SUSTAINABLE WASTE MANAGEMENT

There is an increased concern about the wastes management whose production is increasing day by day. Increasing environmental concerns like soil and groundwater pollution, it is necessary to remove the solid wastes or use them for any purposeful use. For this, different solid waste conversion to energy technologies are being used for sustainable waste management and energy production.

Agricultural Wastes

Due to increased agricultural crop residues production, their management is necessary because these are produced in huge amounts all over the world. Agricultural wastes are conventionally burned in the fields, used as fuels in homes, or incorporated in the fields after chopping by rotavators. Agricultural wastes contain different types of useful substances in them which can be used for different purposes like biochar production for use as biosorbent, carbon sequestration enhancement, biofuels production like biodiesel, and recovery of cellulose, fibers and biopolymers etc. which are used in medicine and food industry (Mostafa et al., 2018). These wastes also used as removal agents of dyes from industrial effluents. As depicted from the various experiments that up to 90% of removal efficiency of dyes from industrial effluents can be achieved (Singh et al., 2017).

Production of biofuels from agricultural wastes also helps in reduction of greenhouse gasses by reducing fossil fuels combustion and reduced emissions of different gasses when crop residues are set to fire in the fields (Wang et al., 2018; Yilmaz et al., 2018). Bio-oil production from different agricultural wastes are also practiced by pyrolysis which shows good potential to produce bio-oils from agricultural crop residues (Hawash et al., 2017).

Industrial Wastes

Due to increasing industrialization, industrial wastes generation is also increasing. This increasing waste generation polluting the soil and water pollution. To avoid the pollution, different technologies are being used for effective removal and management of these wastes. Bricks manufacturing from industrial wastes is now being used as alternate of conventional bricks (Andreola et al., 2016; Wiemes et al., 2017). By using combustion and digestion, energy is also produced from different types of industrial wastes. Electricity is produced from sugarcane bagasse, different sludges and used in the industrial as well as supplied for domestic use (Contreras-Lisperguer et al., 2018; Hassan et al., 2019). Green cements are also produced by converting industrial wastes into geopolymers and using the product as replacement of conventional cement. Geopolymers derived cement product is eco-friendly (reduction of 60% consumption and carbon emissions as compared to conventional cement production) and also has the advantage of industrial waste disposal (Zhang et al., 2017).

Domestic Wastes

Rapid urbanization has resulted in large amounts of municipal wastes generation. Increase in municipal solid wastes results in increased pollution levels in environment, which must be managed accordingly. So, there must be any effective way of municipal solid waste disposal (Polyakova et al., 2018).

Some strategies of their disposal are using them as fuel by mixing with other wastes, use them as soil conditioners and fertilizers (Tıkız and Pehlivan, 2018). These wastes are also used as biosorbents for heavy metals (Quintas et al., 2019), processing for production of composts (Jiya et al., 2019), use of different fruit peel extracts for bio-control agents (El-Gendy et al., 2016) and production of electricity (Ayodele et al., 2017).

E-Wastes

Electronic waste or E-waste is one of the growing waste stream all over the world. It posse different hazardous environmental impacts. Heavy metals in e-waste cause soil and water pollution, piles of e-wastes present a bad image of the area. In the world, 41 million tons of e-waste was produced in 2014. It is increasing with the rate of 3-5% each year. This poses a challenge to the researchers to devise a proper method to cope with this problem (Kumar et al., 2017; Tansel, 2017).

Recycling is the strategy which is being used since the ages. Recycling of e-waste is a necessity to address the pollution problems and shortage of mineral resources for electronics industry as well as to reduce the environmental pollution and human health risks (Zeng et al., 2017). Recycling and demolishing is also done to recover the precious rare earth elements like metal recovery from printed circuit boards, hard disc drives and displays (Garlapati, 2016; Işıldar et al., 2018).

COST TO BENEFIT RATIOS OF DIFFERENT WASTE MANAGEMENT TECHNOLOGIES

Fuel Production

Energy production from different solid wastes is now widely considered for a sustainable approach for energy production and waste disposal simultaneously. This approach is very cost-effective as feedstock is usually the wastes which are headache otherwise. Cost-benefit analysis for waste to energy was done and it was revealed that it has more benefits than fossil fuels in combination with reducing global warming potential of greenhouse potential (Wang et al., 2016). Waste cooking oil is used for energy generation for production of biodiesel. By spending 0.946 USD, we can obtain benefits of 1.298 USD (Mohammadshirazi et al., 2014; Singhabhandhu and Tezuka, 2010; Zhao et al., 2016). Landfilling and energy recovery from the wastes is a beneficial industry. It can be depicted in the study in which authors showed an estimated benefits of 1.92 million USD to 16.63 million USD earning in a single year by a single landfill site in China (Zhou et al., 2015).

Recycling

Recycling is a win-win strategy in which, used or scrap products are reused or sold to the areas where these products are needed. It has more benefits than the costs (Pickin, 2008). By recycling, different types of resources are used again and again. Developed countries send their old equipments to developing countries adopting new technology by giving old technology to developing countries is a mode of recycling e-waste. For example, old models of equipment's are sold at cheaper prices to developing countries. By adopting this method, they get rid of their e-wastes as well as earn some money by selling their e-waste (Tanskanen, 2013).

SOLUTION AND RECOMMENDATIONS

As indicated in the chapter, there are numerous ways by which solid wastes can be disposed-off in a sustainable way. Wastes to energy, pyrolysis, biomethanation, combustion, incineration and fermentation techniques are extensively used for energy and biofuels production. Generally, above mentioned technologies are thought to be useful and eco-friendly. But, there are some issues associated with these, like air pollution levels may increase near pyrolysis, incineration and combustion facilities. Gasses emissions can contribute in climate change and increase health problems in nearby areas. Some techniques require sophisticated and costly equipment and setups like Biomethanation, pyrolysis and fermentation, and some require labor to run the processes. These issues should be scientifically studied and addressed accordingly.

FUTURE RESEARCH DIRECTIONS

Sustainable solid wastes management is an innovative approach by which solid wastes are collected, treated, reused or processes for fuel and energy production. But due to the issues of air pollution in the case of combustion, pyrolysis etc., research should be done to invent or modify the ways of fuels and energy production without harming the air as it is already polluted due to excessive carbon and other greenhouse gas emissions. Landfilling cause soil and groundwater contamination by leaching of heavy metals and other contaminants. All these issues must be addressed to minimize these risks with continuous and wise use of available resources for environmental health and its improvements along with fuel and energy production.

CONCLUSION

Different types of solid wastes pose environmental threats. They cause environmental pollution by contaminating soil, water air and other biotic components. Solid wastes are of different types like agricultural wastes (crop residues, animal dung, fruit peels, thinning and pruning wastes), industrial wastes (sewage sludge, wastewater, ash, and miscellaneous solid wastes) and domestic solid wastes. Their management is of great concern in modern era. There are different approaches which are used for solid waste management like dumping into sea, landfilling, and open burning. But these management strategies are not eco-friendly and poses environmental threats. With the progress in science and technology, scientists invented various sustainable methods to cope with problems associated with solid waste management and energy shortage collectively. They devised the methods of fuel production from crop residues, sewage sludge and animal dung. They produced biogas, biodiesel, and electricity from solid wastes by processes like incineration, digestion, methanation, pyrolysis and plasma gasification etc. These methods are well studied and cost effective. Future studies are being done to assess the more advance method of energy recovery from wastes having less emission so that we can totally eradicate the climate change problems originated by use of fossil fuels.

REFERENCES

Abdoli, M., Samieifard, R., & Jalili, G. M. (2008). *Rural solid waste management*. Academic Press.

Al-Salem, S., Antelava, A., Constantinou, A., Manos, G., & Dutta, A. (2017). A review on thermal and catalytic pyrolysis of plastic solid waste (PSW). *Journal of Environmental Management*, *197*, 177–198. doi:10.1016/j.jenvman.2017.03.084 PMID:28384612

Al-Salem, S., Lettieri, P., & Baeyens, J. (2009). Recycling and recovery routes of plastic solid waste (PSW): A review. *Waste Management (New York, N.Y.)*, *29*(10), 2625–2643. doi:10.1016/j.wasman.2009.06.004 PMID:19577459

Andreola, F., Barbieri, L., Lancellotti, I., Leonelli, C., & Manfredini, T. (2016). Recycling of industrial wastes in ceramic manufacturing: State of art and glass case studies. *Ceramics International*, *42*(12), 13333–13338. doi:10.1016/j.ceramint.2016.05.205

Atalia, K., Buha, D., Bhavsar, K., & Shah, N. (2015). A review on composting of municipal solid waste. *Journal of Environmental Science. Toxicology and Food Technology*, *9*(5), 20–29.

Ayodele, T., Ogunjuyigbe, A., & Alao, M. (2017). Life cycle assessment of waste-to-energy (WtE) technologies for electricity generation using municipal solid waste in Nigeria. *Applied Energy*, *201*, 200–218. doi:10.1016/j.apenergy.2017.05.097

Biedenbach, D. J., Kazmierczak, K., Bouchillon, S. K., Sahm, D. F., & Bradford, P. A. (2015). In vitro activity of aztreonam-avibactam against a global collection of Gram-negative pathogens from 2012 and 2013. *Antimicrobial Agents and Chemotherapy*, *59*(7), 4239–4248. doi:10.1128/AAC.00206-15 PMID:25963984

Bolzonella, D., Fatone, F., Pavan, P., & Cecchi, F. (2005). Anaerobic fermentation of organic municipal solid wastes for the production of soluble organic compounds. *Industrial & Engineering Chemistry Research*, *44*(10), 3412–3418. doi:10.1021/ie048937m

Burnley, S. J. (2007). A review of municipal solid waste composition in the United Kingdom. *Waste Management (New York, N.Y.)*, *27*(10), 1274–1285. doi:10.1016/j.wasman.2006.06.018 PMID:17011771

Chowdhury, A. H., Mohammad, N., Haque, M. R. U., & Hossain, T. (2014). Developing 3Rs (Reduce, Reuse and Recycle) strategy for waste management in the urban areas of Bangladesh: Socioeconomic and climate adoption mitigation option. *IOSR Journal of Environmental Science, Toxicology and Food Technology (IOSR-JESTFT)*, *8*(5), 9-18.

Contreras-Lisperguer, R., Batuecas, E., Mayo, C., Díaz, R., Pérez, F., & Springer, C. (2018). Sustainability assessment of electricity cogeneration from sugarcane bagasse in Jamaica. *Journal of Cleaner Production*, *200*, 390–401. doi:10.1016/j.jclepro.2018.07.322

Corral-Verdugo, V. (1997). Dual 'realities' of conservation behavior: self-reports vs observations of re-use and recycling behavior. *Journal of Environmental Psychology*, *17*(2), 135–145. doi:10.1006/jevp.1997.0048

Daoji, L., & Daler, D. (2004). Ocean pollution from land-based sources: East China Sea, China. *Ambio*, *33*(1), 107–113. doi:10.1579/0044-7447-33.1.107 PMID:15083656

Daskalopoulos, E., Badr, O., & Probert, S. (1998). An integrated approach to municipal solid waste management. *Resources, Conservation and Recycling*, *24*(1), 33–50. doi:10.1016/S0921-3449(98)00031-7

De, S., & Debnath, B. (2016). Prevalence of health hazards associated with solid waste disposal-A case study of kolkata, India. *Procedia Environmental Sciences*, *35*, 201–208. doi:10.1016/j.proenv.2016.07.081

Donoso-Bravo, A., Ortega-Martinez, E., & Ruiz-Filippi, G. (2016). Impact of milling, enzyme addition, and steam explosion on the solid waste biomethanation of an olive oil production plant. *Bioprocess and Biosystems Engineering*, *39*(2), 331–340. doi:10.100700449-015-1519-z PMID:26670779

Dutta, D., & Goel, S. (2017). Electronic Waste (E-Waste) Generation and Management. In *Advances in Solid and Hazardous Waste Management* (pp. 249–266). Springer. doi:10.1007/978-3-319-57076-1_12

El-Gendy, N. S., Hamdy, A., Fatthallah, N. A., & Omran, B. A. (2016). Recycling of some domestic wastes to produce green corrosion inhibitors and biocides for sulfate reducing bacteria. *Energy Sources. Part A, Recovery, Utilization, and Environmental Effects*, *38*(24), 3722–3732. doi:10.1080/15567036.2016.1179364

Emmanuel, E., Perrodin, Y., Keck, G., Blanchard, J.-M., & Vermande, P. (2002). Effects of hospital wastewater on aquatic ecosystem. *Proceedings of the XXVIII Congreso Interamericano de Ingenieria Sanitaria y Ambiental.*

Favas, J., Monteiro, E., & Rouboa, A. (2017). Hydrogen production using plasma gasification with steam injection. *International Journal of Hydrogen Energy*, *42*(16), 10997–11005. doi:10.1016/j.ijhydene.2017.03.109

Garlapati, V. K. (2016). E-waste in India and developed countries: Management, recycling, business and biotechnological initiatives. *Renewable & Sustainable Energy Reviews*, *54*, 874–881. doi:10.1016/j.rser.2015.10.106

Guerrero, M., Re, M., Kazimierski, L. D., Menéndez, Á. N., & Ugarelli, R. (2013). Effect of climate change on navigation channel dredging of the Parana River. *International Journal of River Basin Management, 11*(4), 439-448.

Hassan, M. N., Rahman, R. A., Chong, T. L., Zakaria, Z., & Awang, M. (2000). Waste recycling in Malaysia: Problems and prospects. *Waste Management & Research*, *18*(4), 320–328. doi:10.1034/j.1399-3070.2000.00133.x

Hassan, S. A., Jameel, N. G. M., & Şekeroğlu, B. (2016). Smart solid waste monitoring and collection system. *International Journal (Toronto, Ont.)*, *6*(10).

Hassan, S. H., Abd el Nasser, A. Z., & Kassim, R. M. (2019). Electricity generation from sugarcane molasses using microbial fuel cell technologies. *Energy*, *178*, 538–543. doi:10.1016/j.energy.2019.04.087

Hawash, S. I., Farah, J. Y., & El-Diwani, G. (2017). Pyrolysis of agriculture wastes for bio-oil and char production. *Journal of Analytical and Applied Pyrolysis*, *124*, 369–372. doi:10.1016/j.jaap.2016.12.021

Hohenthal, C., Leon, J., Dobon, A., Kujanpää, M., Meinl, G., Ringman, J., & Forsström, U. (2019). The ISO 14067 approach to open-loop recycling of paper products: Making it operational. *Journal of Cleaner Production, 224*, 264–274. doi:10.1016/j.jclepro.2019.03.179

Hu, J., Huang, Y., Yao, Y., Pan, G., Sun, J., Zeng, X., & Wong, C.-P. (2017). Polymer composite with improved thermal conductivity by constructing a hierarchically ordered three-dimensional interconnected network of BN. *ACS Applied Materials & Interfaces, 9*(15), 13544–13553. doi:10.1021/acsami.7b02410 PMID:28362080

Idris, A., Inanc, B., & Hassan, M. N. (2004). Overview of waste disposal and landfills/dumps in Asian countries. *Journal of Material Cycles and Waste Management, 6*(2), 104–110. doi:10.100710163-004-0117-y

Ihedioha, J. N., Ukoha, P. O., & Ekere, N. R. (2017). Ecological and human health risk assessment of heavy metal contamination in soil of a municipal solid waste dump in Uyo, Nigeria. *Environmental Geochemistry and Health, 39*(3), 497–515. doi:10.100710653-016-9830-4 PMID:27170184

Ilankoon, I. M. S. K., Ghorbani, Y., Chong, M. N., Herath, G., Moyo, T., & Petersen, J. (2018). E-waste in the international context–A review of trade flows, regulations, hazards, waste management strategies and technologies for value recovery. *Waste Management (New York, N.Y.), 82*, 258–275. doi:10.1016/j.wasman.2018.10.018 PMID:30509588

Işıldar, A., Rene, E. R., van Hullebusch, E. D., & Lens, P. N. L. (2018). Electronic waste as a secondary source of critical metals: Management and recovery technologies. *Resources, Conservation and Recycling, 135*, 296–312. doi:10.1016/j.resconrec.2017.07.031

Jiya, A., Ijah, U. J., Galadima, M., & Akpan, U. (2019). Isolation of Microorganisms Associated with Biodegradation of Household Domestic Wastes for Biogas Production in Niger State, Nigeria. *Journal of BioMedical Research and Clinical Practice, 2*(1), 70–76. doi:10.46912/jbrcp.82

Kliopova, I., Staniškis, J. K., Stunžėnas, E., & Jurovickaja, E. (2019). Bio-nutrient recycling with a novel integrated biodegradable waste management system for catering companies. *Journal of Cleaner Production, 209*, 116–125. doi:10.1016/j.jclepro.2018.10.185

Kothari, R., Tyagi, V. V., & Pathak, A. (2010). Waste-to-energy: A way from renewable energy sources to sustainable development. *Renewable & Sustainable Energy Reviews, 14*(9), 3164–3170. doi:10.1016/j.rser.2010.05.005

Kumar, A., Holuszko, M., & Espinosa, D. C. R. (2017). E-waste: An overview on generation, collection, legislation and recycling practices. *Resources, Conservation and Recycling, 122*, 32–42. doi:10.1016/j.resconrec.2017.01.018

Kumar, A., Holuszko, M., & Espinosa, D. C. R. (2017). E-waste: An overview on generation, collection, legislation and recycling practices. *Resources, Conservation and Recycling, 122*, 32–42. doi:10.1016/j.resconrec.2017.01.018

Kumar, A., Samadder, S. R., Kumar, N., & Singh, C. (2018). Estimation of the generation rate of different types of plastic wastes and possible revenue recovery from informal recycling. *Waste Management (New York, N.Y.), 79*, 781–790. doi:10.1016/j.wasman.2018.08.045 PMID:30343811

Lal, R. M., Nagpure, A. S., Luo, L., Tripathi, S. N., Ramaswami, A., Bergin, M. H., & Russell, A. G. (2016). Municipal solid waste and dung cake burning discoloring the Taj Mahal and human health impacts in Agra. *Environmental Research Letters*, *11*(10), 104009. doi:10.1088/1748-9326/11/10/104009

Li, K., Yang, F., Zhang, G., Song, S., Li, Y., Ren, D., & Song, C.-P. (2017). AIK1, a mitogen-activated protein kinase, modulates abscisic acid responses through the MKK5-MPK6 kinase cascade. *Plant Physiology*, *173*(2), 1391–1408. doi:10.1104/pp.16.01386 PMID:27913741

Mena, C., Adenso-Diaz, B., & Yurt, O. (2011). The causes of food waste in the supplier–retailer interface: Evidences from the UK and Spain. *Resources, Conservation and Recycling*, *55*(6), 648–658. doi:10.1016/j.resconrec.2010.09.006

Mendoza, F. J. C., Altabella, J. E., & Izquierdo, A. G. (2017). Application of inert wastes in the construction, operation and closure of landfills: Calculation tool. *Waste Management (New York, N.Y.)*, *59*, 276–285. doi:10.1016/j.wasman.2016.10.041 PMID:27816471

Messerle, V. E., Ustimenko, A. B., & Lavrichshev, O. A. (2016). Comparative study of coal plasma gasification: Simulation and experiment. *Fuel*, *164*, 172–179. doi:10.1016/j.fuel.2015.09.095

Mian, M. M., Zeng, X., Nasry, A. A. N. B., & Al-Hamadani, S. M. (2017). Municipal solid waste management in China: A comparative analysis. *Journal of Material Cycles and Waste Management*, *19*(3), 1127–1135. doi:10.100710163-016-0509-9

Mian, M. M., Zeng, X., Nasry, A. N. B., & Al-Hamadani, S. M. (2017). Municipal solid waste management in China: A comparative analysis. *Journal of Material Cycles and Waste Management*, *19*(3), 1127–1135. doi:10.100710163-016-0509-9

Mohammadshirazi, A., Akram, A., Rafiee, S., & Bagheri Kalhor, E. (2014). Energy and cost analyses of biodiesel production from waste cooking oil. *Renewable & Sustainable Energy Reviews*, *33*, 44–49. doi:10.1016/j.rser.2014.01.067

Mostafa, N. A., Farag, A. A., Abo-dief, H. M., & Tayeb, A. M. (2018). Production of biodegradable plastic from agricultural wastes. *Arabian Journal of Chemistry*, *11*(4), 546–553. doi:10.1016/j.arabjc.2015.04.008

Mustafa, M. F., Liu, Y., Duan, Z., Guo, H., Xu, S., Wang, H., & Lu, W. (2017). Volatile compounds emission and health risk assessment during composting of organic fraction of municipal solid waste. *Journal of Hazardous Materials*, *327*, 35–43. doi:10.1016/j.jhazmat.2016.11.046 PMID:28038430

Narbón-Perpiñá, I., & Prior, D. (2020). Management of Waste Electrical and Electronic Equipment in European Union Countries: A Comparison. In *E-waste Recycling and Management* (pp. 127–141). Springer. doi:10.1007/978-3-030-14184-4_7

Negi, S., Dhar, H., Hussain, A., & Kumar, S. (2018). Biomethanation potential for co-digestion of municipal solid waste and rice straw: A batch study. *Bioresource Technology*, *254*, 139–144. doi:10.1016/j.biortech.2018.01.070 PMID:29413914

Nintendo, E., Nintendo, E., Miyamoto, S., Tezuka, T., & Kondo, K. (2019). *The Legend of Zelda*. Paper presented at the Spin.

Nowicki, M. A., Castro, N. J., Plesniak, M. W., & Zhang, L. G. (2016). 3D printing of novel osteochondral scaffolds with graded microstructure. *Nanotechnology*, *27*(41), 414001. doi:10.1088/0957-4484/27/41/414001 PMID:27606933

Oppelt, E. T. (1987). Incineration of hazardous waste. *JAPCA*, *37*(5), 558–586. doi:10.1080/08940630.1987.10466245 PMID:3328776

Pansuk, J., Junpen, A., & Garivait, S. (2018). Assessment of air pollution from household solid waste open burning in Thailand. *Sustainability*, *10*(7), 2553. doi:10.3390u10072553

Pickin, J. (2008). Representations of environmental concerns in cost–benefit analyses of solid waste recycling. *Resources, Conservation and Recycling*, *53*(1), 79–85. doi:10.1016/j.resconrec.2008.09.003

Polyakova, V., Degaev, E., & El Haddad, P. (2018). *Reduction of Ecological and Economic Risks in Utilization of Solid Domestic Wastes and Construction Waste.* Paper presented at the MATEC Web of Conferences. 10.1051/matecconf/201825106017

Psomopoulos, C. S., Bourka, A., & Themelis, N. J. (2009). Waste-to-energy: A review of the status and benefits in USA. *Waste Management (New York, N.Y.)*, *29*(5), 1718–1724. doi:10.1016/j.wasman.2008.11.020 PMID:19153036

Quintas, P. Y., Dotto, G. L., Da Silva, S. M., & Escudero, L. B. (2019). Removal of thallium from environmental samples using a raw and chemically modified biosorbent derived from domestic wastes. *Environmental Science and Pollution Research International*, *26*(31), 32285–32297. doi:10.100711356-019-06442-x PMID:31598928

Ragaert, K., Delva, L., & Van Geem, K. (2017). Mechanical and chemical recycling of solid plastic waste. *Waste Management (New York, N.Y.)*, *69*, 24–58. doi:10.1016/j.wasman.2017.07.044 PMID:28823699

Roychoudhuri, R., & Debnath, B. (2020). Risk Assessment of E-Waste Recycling with a Focus on ISO 9000 Standard. In *Sustainable Waste Management: Policies and Case Studies* (pp. 431–440). Springer. doi:10.1007/978-981-13-7071-7_38

Salles, A., Wolff, D., & Silveira, G. (2012). Solid wastes drained in an urban river sub-basin. *Urban Water Journal*, *9*(1), 21–28. doi:10.1080/1573062X.2011.633612

Sanlisoy, A., & Carpinlioglu, M. O. (2017). A review on plasma gasification for solid waste disposal. *International Journal of Hydrogen Energy*, *42*(2), 1361–1365. doi:10.1016/j.ijhydene.2016.06.008

Shi, X., Luo, C., Lu, H., & Yu, K. (2019). Primary recycling of anhydride-cured engineering epoxy using alcohol solvent. *Polymer Engineering and Science*, *59*(s2), E111–E119. doi:10.1002/pen.24997

Singh, H., Chauhan, G., Jain, A. K., & Sharma, S. K. (2017). Adsorptive potential of agricultural wastes for removal of dyes from aqueous solutions. *Journal of Environmental Chemical Engineering*, *5*(1), 122–135. doi:10.1016/j.jece.2016.11.030

Singhabhandhu, A., & Tezuka, T. (2010). The waste-to-energy framework for integrated multi-waste utilization: Waste cooking oil, waste lubricating oil, and waste plastics. *Energy*, *35*(6), 2544–2551. doi:10.1016/j.energy.2010.03.001

Srivastava, R., Krishna, V., & Sonkar, I. (2014). Characterization and management of municipal solid waste: A case study of Varanasi city, India. *International Journal of Current Research and Academic Review*, *2*(8), 10–16.

Sujauddin, M., Huda, S., & Hoque, A. R. (2008). Household solid waste characteristics and management in Chittagong, Bangladesh. *Waste Management (New York, N.Y.)*, *28*(9), 1688–1695. doi:10.1016/j.wasman.2007.06.013 PMID:17845843

Tansel, B. (2017). From electronic consumer products to e-wastes: Global outlook, waste quantities, recycling challenges. *Environment International*, *98*, 35–45. doi:10.1016/j.envint.2016.10.002 PMID:27726897

Tanskanen, P. (2013). Management and recycling of electronic waste. *Acta Materialia*, *61*(3), 1001–1011. doi:10.1016/j.actamat.2012.11.005

Tatsuno, H., Nishina, K., Shibayama, Y., Osaka, T., & Hirakawa, M. (2019). *Focus adjustment system for an image display device*. Google Patents.

Tıkız, İ., & Pehlivan, H. (2018). Experimental Investigation for Alternative Use of Agricultural and Domestic Wastes. *Kırklareli Üniversitesi Mühendislik ve Fen Bilimleri Dergisi*, *4*(1), 142–152.

Troschinetz, A. M., & Mihelcic, J. R. (2009). Sustainable recycling of municipal solid waste in developing countries. *Waste Management (New York, N.Y.)*, *29*(2), 915–923. doi:10.1016/j.wasman.2008.04.016 PMID:18657963

Uçkun Kiran, E., Trzcinski, A. P., Ng, W. J., & Liu, Y. (2014). Bioconversion of food waste to energy: A review. *Fuel*, *134*, 389–399. doi:10.1016/j.fuel.2014.05.074

Wang, M., Mao, M., Zhang, M., Wen, G., Yang, Q., Su, B., & Ren, Q. (2019). Highly efficient treatment of textile dyeing sludge by CO2 thermal plasma gasification. *Waste Management (New York, N.Y.)*, *90*, 29–36. doi:10.1016/j.wasman.2019.04.025 PMID:31088671

Wang, W., Zeng, C., Sardans, J., Zeng, D., Wang, C., Bartrons, M., & Penuelas, J. (2018). Industrial and agricultural wastes decreased greenhouse-gas emissions and increased rice grain yield in a subtropical paddy field. *Experimental Agriculture*, *54*(4), 623–640. doi:10.1017/S001447971700031X

Wang, Y., Geng, S., Zhao, P., Du, H., He, Y., & Crittenden, J. (2016). Cost–benefit analysis of GHG emission reduction in waste to energy projects of China under clean development mechanism. *Resources, Conservation and Recycling*, *109*, 90–95. doi:10.1016/j.resconrec.2016.02.010

Wiemes, L., Pawlowsky, U., & Mymrin, V. (2017). Incorporation of industrial wastes as raw materials in brick's formulation. *Journal of Cleaner Production*, *142*, 69–77. doi:10.1016/j.jclepro.2016.06.174

Yan, M., Khoshnevisana, B., Tsapekos, P., Fotidis, I., & Angelidaki, I. (2019). *Continuous biomethanation of organic fraction of municipal solid waste under extreme ammonia levels.* Paper presented at the 16th World Congress on Anaerobic Digestion.

Yilmaz, E., Wzorek, M., & Akçay, S. (2018). Co-pelletization of sewage sludge and agricultural wastes. *Journal of Environmental Management*, *216*, 169–175. doi:10.1016/j.jenvman.2017.09.012 PMID:28943060

Zeng, X., Yang, C., Chiang, J. F., & Li, J. (2017). Innovating e-waste management: From macroscopic to microscopic scales. *The Science of the Total Environment*, *575*, 1–5. doi:10.1016/j.scitotenv.2016.09.078 PMID:27723459

Zhang, W., Yang, J., Wu, X., Hu, Y., Yu, W., Wang, J., Dong, J., Li, M., Liang, S., Hu, J., & Kumar, R. V. (2016). A critical review on secondary lead recycling technology and its prospect. *Renewable & Sustainable Energy Reviews*, *61*, 108–122. doi:10.1016/j.rser.2016.03.046

Zhang, Z., Zhu, Y., Yang, T., Li, L., Zhu, H., & Wang, H. (2017). Conversion of local industrial wastes into greener cement through geopolymer technology: A case study of high-magnesium nickel slag. *Journal of Cleaner Production*, *141*, 463–471. doi:10.1016/j.jclepro.2016.09.147

Zhao, X., Jiang, G., Li, A., & Wang, L. (2016). Economic analysis of waste-to-energy industry in China. *Waste Management (New York, N.Y.)*, *48*, 604–618. doi:10.1016/j.wasman.2015.10.014 PMID:26514312

Zhao, X., Xia, Y., Zhan, L., Xie, B., Gao, B., & Wang, J. (2018). Hydrothermal treatment of e-waste plastics for tertiary recycling: Product slate and decomposition mechanisms. *ACS Sustainable Chemistry & Engineering*, *7*(1), 1464–1473. doi:10.1021/acssuschemeng.8b05147

Zhou, C., Gong, Z., Hu, J., Cao, A., & Liang, H. (2015). A cost-benefit analysis of landfill mining and material recycling in China. *Waste Management (New York, N.Y.)*, *35*, 191–198. doi:10.1016/j.wasman.2014.09.029 PMID:25453315

Zink, T., Geyer, R., & Startz, R. (2018). Toward estimating displaced primary production from recycling: A case study of US aluminum. *Journal of Industrial Ecology*, *22*(2), 314–326. doi:10.1111/jiec.12557

Chapter 4
Assessment of Phytoremediation Efficiency of *Coriandrum sativum* in Metal Polluted Soil and Sludge Samples:
A green approach

Deepshekha Punetha
D. S. B. Campus, Kumaun University, Nainital, India

Geeta Tewari
D. S. B. Campus, Kumaun University, Nainital, India

Chitra Pande
D. S. B. Campus, Kumaun University, Nainital, India

Girish Kharkwal
D. S. B. Campus, Kumaun University, Nainital, India

Sonal Tripathi
NM College of Agriculture, Navsari Agriculture University, India

ABSTRACT

A greenhouse pot experiment was conducted to evaluate the phytotoxic effect of heavy metals (Zn, Cu, Ni, and Cr) on the growth (plant height, plant and seed weight) of Coriandrum sativum and uptake of metals. For this purpose, the polluted soil and sludge samples were collected from three sites of Moradabad: Karula nala (KS), Dhauri nala (DS), and Karula nala sludge (KSL). Metal content in the plant and soil was determined by AAS technique. The plants were also hydrodistilled using Clevenger apparatus and

DOI: 10.4018/978-1-7998-4921-6.ch004

the extracted oil was analyzed by GC and GC/MS. The content of copper in KS and KSL, Zn in KS, Ni in DS and KSL, and Cr in KS was above the permissible limits according to Indian standards. Plant weight was significantly affected in the different soil samples. Among all the collected polluted soil and sludge samples, DS sample showed the highest metal accumulation, while in the KSL sample, the plant could not survive. The study revealed that coriander can be used to remediate the contaminated soil with economic return and metal free final product, essential oil.

INTRODUCTION

One of the major health and environmental problems of our modern society is heavy metal toxicity and the danger of their bioaccumulation in food chain. Fertilizers, pesticides, sewage sludge, burning of fossil fuel, mining and smelting of metallic ferrous ores and municipal waste are the main sources of metal contamination (Peng et al., 2006). Copper (Cu), zinc (Zn), nickel (Ni), chromium (Cr), mercury (Hg) and lead (Pb) are the most common heavy metals (USEPA, 1997). Metals and metalloids having densities greater than 5g cm^{-3} are known as heavy metals and generally associated with pollution toxicity. Although some of these elements (essential metals) are necessary to the organisms in low concentrations (Adriano, 2001). Zinc is used in the metabolism of carbohydrates proteins, auxins, phosphate, in RNA and ribosome formation in plants. It is also the key component in several enzymes (peptidases, dehydrogenases, proteinases) (Mengel and Kirkby, 1982). Copper involves in many physiological processes in plants as respiration, photosynthesis, carbohydrate distribution, nitrogen, cell wall metabolism and seed production along with disease resistance (Kabata and Pendias, 1992).

The utilization of plants and related microbes for environmental cleanup is known as phytoremediation (Rupassara et al., 2002). As this technology is cost effective and eco-friendly, it has a significant application in environmental and ecological research. One main disadvantage of phytoremediation is that it requires a long-term promise, depending on plant growth, bioaccumulation capacity and tolerance to toxicity (Burken et al., 2011). Research and application of phytoremediation for the treatment of petroleum hydrocarbon pollution over the past decades has given much helpful information that can be applied to plan effective remediation systems and for more improvement and innovation (Kamath et al., 2007). The random use of chemicals at production process increase substances over maximum permissible concentrations in air, soil, water and meal. Industrial and agricultural activities in several area destroyed soil and water quality (Wosten, 1997).

The volatile organic components of fragrant plant matter, which can be isolated by physical processes, contribute to both fragrance and flavour (Waterman, 1993). *Coriandrum sativum* L. belonging to the family Apiaceae (Umbelliferae) is also known as dhania, dhanya, coriander, collender, chinese parsley, coriandro, cilantro, cilandrio, cilantro, yuan sui, hu sui, coriandolo, dhanayaka, kusthumbari, koriannon, korion, Koriander, Wanzendill, Schwindelkorn and koendoro. Several medicinal uses of coriander have been reported including anticonvulsant (Karami et al., 2015), effect on memory (Cioanca et al., 2013), neuroprotective (Vekaria et al., 2012), antibacterial and antifungal (Baratta et al., 1998), antimicrobial (Casetti et al., 2012), antioxidant (Wong and Kitts, 2006), anti-inflammatory and analgesic (Hashemi et al., 2003), antidiabetic (Mazhar and Mazumder, 2013), mutagenic and antimutagenic (Reyes et al., 2013) and anticancer effect (Omez et al., 2010).

For production of essential oils, aromatic crops are used and they may be best alternative crops in heavy metal polluted agricultural soils. Accumulation of heavy metals from polluted soil by some aromatic plants was observed by Zheljazkov and Warman (2003). Essential oils are low-volume high-value products that are commonly used as aromatic agents in different non-food industries, such as cosmetics, perfumery and aromatherapy. There is one report on the use of *Coriandrum* species for phytoremediation of metal contaminated sites (Zheljazkov et al., 2008).

As the tested soil and sludge samples had the copper, zinc, nickel and chromium content above the permissible limits and not much work on phytoremedial potential of *Coriandrum* has been explored from India, the main purpose of this work was to evaluate the growth and accumulation of metals (Cu, Zn, Ni and Cr) in the *C. sativum* L. grown in polluted soils.

MATERIALS AND METHODS

Study Area

Moradabad is home to one of the oldest brassware clusters in India, known as Brass City of India and is situated at a distance of 167 km from the national capital, New Delhi (NH 24), on the bank of river Ramganga and located at 28° 83' N 78° 00'E (Diagnostic, 2008). An extensive survey was conducted to identify different polluted areas in the city. Based on this survey, it was found that Karula nala soil (KS), Dhauri nala soil (DS) and Karula nala sludge (KSL) were highly polluted with different metals due to the presence of electroplating industries near these sites around Moradabad city (0-4 km from the city). The composite surface soil (n=20) and sludge (n=10) samples (0-20 cm) were collected from different polluted sites of Moradabad (0-4 Km from the city centre) and labelled properly. Simultaneously, control soil sample was also collected from a site almost 4Km away from the polluted site. The samples brought back to the lab and stored at room temperature.

Pot Experiment

The soils (control, Karula nala and Dhauri nala) and sludge (Karula nala) samples were air dried (1 week), ground and passed through 2mm sieve before use and analysis. Three kilogram of each sample was transferred into free draining 5 kg earthen pots (10 inches in diameter). The pots were prepared in four replicates. To prevent the leaching of metals, each pot was placed in a separate dish. Seeds of *C. sativum* were sown in each pot. After 30 days (first cutting), thinning of plants was done and the plant number was reduced to ten per pot. The plants were watered regularly and the plants were harvested after 90 days of growth for analysis.

Physicochemical Properties of Soil and Sludge Samples

The pH and EC of soil and sludge samples were measured in soil/sludge-water suspension (1:2) using an electric shaker for half an hour (Jackson, 1958). For heavy metal analysis, well mixed 2g of composite soil or sludge samples were digested with 8 mL of aqua regia on a sand bath for 2 hours (Lokeshwari and Chandrappa, 2006). Soil or sludge samples were also extracted for DTPA (Diethylene triamine-pentaacetic acid) extractable metals following the procedure developed by Lindsay and Norvell (1969). Other physicochemical properties such as organic carbon, nitrogen, potassium, phosphorus, heavy metals (Zn, Ni, Cu and Cr) and cation exchange capacity were also measured in soil and sludge samples as per standard methods (Jackson, 1958). Texture of soil samples was determined by pipette method, using hydrogen peroxide (30%) for the removal of organic matter and sodium hexametaphosphate (5%) as dispersing agent (Kilmer and Alexander, 1949).

Heavy Metal Analysis in Soil, Sludge, Plant, Extracted Water and Oil

The plants were carefully separated from the soil at full blooming stage and sequentially washed with tap water, 1N-HCl solution and finally with distilled water to remove all adhered soil particles. Roots and aerial parts were separated and cut into small pieces, air dried for 2 days and finally dried at 60 ± 1°C in a hot air oven for 48 hours. Dry plant tissue was finely ground and wet- ashed using HNO_3:H_2SO_4:$HClO_4$ (10:1:4 v/v) (Jackson, 1958). The metal content in the acid extracts were determined by atomic absorption spectrophotometer (AAS) (Advance Scientific Instrument Service (Model AAS4141). Standard solutions were regularly checked to run the sensitivity of the instrument.

Metal content in each water sample left after distillation and oil was determined using the method given by Zheljazkov and Nielsen (1996).

Extraction of Oil

The fresh plant material after harvesting was hydrodistilled for 3 h using a Clevenger apparatus. The hydrodistilled oil was dried over anhydrous sodium sulphate.

Analysis of Oil and Identification of Components

Gas Chromatographic data of the essential oil samples were obtained by using GLC instrument Nucon 5765 (Figure 1; control plant) and Shimadzu 2010 (Figure 2; DS and Figure 3; KS plants). The conditions of both the instruments were as follows:

In Shimadzu 2010 instrument, the oven temperatures were programmed at 80°C, 210°C and 300°C with hold time of 2 min, 5 min and 15 min respectively. The increment of temperature was programmed at 3°C/ min and 20°C/min respectively. Flame ionization detector was used with 280°C temperature. Carrier gas was N_2 with Rtx5 column (30m× 0.25mm× 0.25μm film thickness). Injection temperature was 270°C with 0.2μL injection volume (manual injection) and 1:20 spilt ratio.

In Nucon GC, the oven temperatures were programmed at 90°C, 210°C and 240°C with hold time of 2 min, 5 min and 15 min respectively. The increment of temperature for was programmed at 4°C/ min and 20°C/min respectively. FID was used with 240°C temperature. Carrier gas was N_2 with Rtx5 column (30m× 0.25mm× 0.25μm film thickness). Injection temperature was 230°C with 0.2μL injection volume (manual injection) and 1:40 spilt ratio.

Gas Chromatographic /Mass Spectrometric analysis of all the essential oil samples was done on 2010 GC, coupled with Shimadzu QP 2010 plus MS. The oven temperatures were programmed at 80°C, 210°C and 300°C with hold time of 2 min, 5 min and 15 min respectively. The temperature for Ramp 1 and Ramp 2 was 3°C/ min and 20°C/min respectively. Ion source temperature was 200°C with EI (70eV) mode and 1 mL/min flow rate. Carrier gas was Helium with Rtx-5 column °C. Injection temperature was 230°C with 0.2μL injection volume and 1:30 spilt ratio. Mass range for MS was 40-600amu.

Sample components present in the essential oils were identified by matching their mass spectra with in WILEY and NIST library. Furthermore, the identification of essential oil components was done by calculation of their retention indices (RI) relative to C_9 to C_{33} *n*-alkane series and comparison with those of the literature data (Adams, 2007). Retention indices were experimentally determined by using the following formula:

$$\mathrm{RI} = 100\,\mathrm{x}\,\mathrm{No} + 100\,\mathrm{x}\,\frac{\left(RT_{substance} - RT_{No}\right)}{\left(RT_{NI} - RT_{No}\right)}$$

(Retention time) $RT_{substance}$ = RT value of the compound to be identified
RT_{No} = n-alkane eluted before unknown peak
RT_{N1} = n-alkane eluted after unknown peak
No = Carbon number from which the standardization is done

Quality Assurance

Appropriate quality assurance procedures and precautions were taken to ensure the reliability of the results. Samples were carefully handled to avoid contamination. The chemicals used in this work were of AR Grade supplied by E. Merck (India), and Lobachemie (India). Glasswares were properly cleaned, and reagents were of analytical grades. All the glasswares used during the course of investigation were of Corning grade manufactured by Borosil India. All the plastic wares and glassware used for the experiments were previously soaked in 14% HNO_3 (v/v), washes with tap water and finally rinsed with distilled water.

Data Analysis

Transfer factor (TF), bioaccumulation factor (BCF) and translocation factors (TLF) have to be considered while examining metal bioavailability, absorption, and translocation and also for evaluating whether a particular plant species is a metal excluder, a metal tolerant, a metal accumulator, or a hyperaccumulator (Ma et al., 2001). The transfer factor (TF) of each metal in the plants was calculated by dividing the metal concentration (mg kg^{-1}) of the harvestable plant material (foliage) by the total metal concentration of the soil (mg kg^{-1}) (Cui et al., 2004). While Bioconcentration factor (BCF) was calculated by dividing metal content in the plant root by metal content in the soil (Yoon et al., 2006). Further, the Translocation Factor (TLF) was calculated by dividing the total metal content in the foliage by total metal content in roots (Wang et al., 2008). All the factors were calculated on dry weight basis.

Bioconcentration factor (BCF) = Metal concentration in roots / Metal concentration in soil
Translocation factor (TLF) = Metal concentration in aerial parts / Metal concentration in roots
Transfer factor (TF) =Metal concentration in plant / Metal concentration in soil

Statistical Analysis

The mean values and standard deviations were calculated for all treatments using MS Excel. The recorded data were subjected to One-way analysis of variance (ANOVA) to determine the significant difference between groups, considering a level of significance of less than 5% and agglomerative hierarchical clustering analysis (Dendrogram) by using SPSS 16.0 software.

RESULTS AND DISCUSSION

Physicochemical Properties of the Experimental Soil and Sludge Samples

The physicochemical properties of studied soil and sludge samples are presented in Table 1. According to the reference values as given in literature, the samples analyzed in this study were characterized as acidic to alkaline (pH=6.5-7.9) having EC (0.1-1.8 dS m^{-1}). Organic carbon varied from 0.9-3.2% for the samples, DS sample showed the highest percentage of organic carbon as well as organic matter (3.2 and 5.5%) respectively. Control soil was sandy loam with 70%, 20% and 10% sand, silt and clay respectively. DS sample was characterized as loamy sand having 78% sand, 16% silt and 6% clay. Sandy clay loam texture was found for KS sample having 52%, 34% and 14% sand silt and clay respectively. The mean content of Cu was above permissible limits in KS (643.2 mg kg^{-1}) and KSL (870.6 mg kg^{-1}), Zn in KS (544.0 mg kg^{-1}), Ni in DS (150.0 mg kg^{-1}) and KSL (184.1 mg kg^{-1}) and Cr in KS (131.0 mg kg^{-1}) (Awasthi, 2000; Cu 270 mg kg^{-1}; Zn 300 mg kg^{-1}; Ni 150 mg kg^{-1}) Cr 100 mg kg^{-1} (European Community Commission). Soil contamination with Pb and Ni was reported by Hashim et al. (2017) during their investigation on the effect of vehicle exhaust on heavy metal accumulation in soil and plants adjacent to Cario-Alexandria agricultural highway of Toukh city. The order of different metals in the collected soil and sludge samples is as follow (Table 1):

For Cu

KSL >KS > DS > Control

For Zn

KS >KSL > DS > Control

For Ni

KSL >DS > KS > Control

For Cr

KS >KSL > DS >Control

Coal burning power stations, petroleum industries, high tension wires and nuclear power stations emits various heavy metals (Zn, B, Cu, Be, Se, Ni, Cd and Cs), which contributes to heavy metal pollution in the soil (Zhu et al., 2016). The present work shows that the concentration of metals was higher in the soil and sludge samples collected near industrial areas. Similar results were shown by Krishna and Govil (2007) and Sharma et al. (2009) as the higher concentration of metals were obtained in the samples collected near the industrial area. Popova, (2016) observed difference in the metal accumulation ability of *Calamagrostis langsdorffii* (Link) Trin. and *Scutellariaga lericulata* L. and noted maximum Zn, Cu, Sr, and Pb accumulation tendency by *Calamagrostis langsdorffii.*

Effect of Metal Contamination in the Growth of *C. Sativum*

Effect of metal contaminated soil on the growth (plant height, fresh plant and seed weight) of coriander is shown in Table 2. It was found that the plant height had a negative effect, while plant and seed weight showed positive response and the highest plant and seed weight was found for DS sample (177.97g and 60.19g) respectively which may be due to the difference in soil properties.

Heavy Metals in Soil and Sludge Samples after Harvesting *C. Sativum*

The concentrations of heavy metals in the samples after harvesting *C. sativum* ranged from 46.0-400.9 mg kg^{-1} for Cu, 29.0-344.8 mg kg^{-1} for Zn, 20.0-75.15 mg kg^{-1} for Ni and 12.0-81.66 mg kg^{-1} for Cr for the first cutting (after 30 days) and ranged from 31.4-245.8 mg kg^{-1} for Cu, 28.0-216.2 mg kg^{-1} for Zn, 10.0-37.6 mg kg^{-1} for Ni, and 8.0-47.6 mg kg^{-1} for Cr for the second cutting (after 90 days) (Table 3; Table 5; Table 7; Table 9).

Heavy Metal Accumulation in *C. Sativum*

Phytoextraction or phytoaccumulation is referred to as the uptake and translocation of metal pollutants in the soil through the roots into the aerial part of plants (Do Nascimento and Xing, 2006).

Copper

The concentration of Cu in different soil and sludge samples are shown in Table 3, which ranged from 4.00-22.57 mg kg^{-1}. The accumulation of Cu was found to be the highest in DS sample as compared to the other samples and the plant cannot survive in the sludge sample. According to Zheljazkov et al. (2008), the concentration of Cu in growth medium increased, plant foliage for basil was also increased. Two cuttings were done after 30 and 90 days and the results showed that accumulation of Cu increased after second cutting in the contaminated soils and accumulation decreased after second cutting in the control soil sample.

Bioconcentration and translocation factors showed that the DS sample had the highest BCF and TLF which suggested that there was higher accumulation of Cu in roots as compared to the aerial parts. Different plant part shows different ability to accumulate various metals. Leaves were found to contain the highest content of Cu followed by roots and stems of *Athyrium esculentum*, *Chromolaena odorata* and *Lantana camara* (Sulaiman and Hamzah, 2018). The order of BCF and TLF in *C. sativum* for Cu was as follows;

0.82 (DS) > 0.11 (Control) > 0.08 (KS) (BCF)

0.63 (DS) > 0.08 (Control) > 0.06 (KS) (TF)

0.78 (DS) > 0.77 (Control) > 0.72 (KS) (TLF)

The concentration of Cu in coriander was found to be below the permissible limit (30 mg kg^{-1}).

Zinc

Coriandrum sativum grown in control and contaminated soil and sludge samples showed variation in the Zn accumulation by plant tissues (Table 5). The accumulation of this metal was found to be higher in the contaminated soil samples as compared to the control soil. Similar results obtained for waste water-irrigated vegetables and were compared to the fresh water-irrigated samples (Otaru et al., 2013). The Zn accumulation was found to be higher after second cutting. Concentration of Zn was found to be below the permissible limit (50 mg kg^{-1}) as given by Awasthi (2000). The accumulation was defined in terms of BCF and TLF factors.

1.02 (DS) > 0.39 (Control) > 0.26 (KS) (BCF)
0.88 (DS) > 0.21 (KS) > 0.07 (Control) (TF)
0.86 (DS) > 0.81 (KS) > 0.19 (Control) (TLF)

It was seen that BCF of Zn was the highest for both the cuttings in DS sample, which showed that the maximum accumulation of Zn occurred in the DS soil sample. The TLF among the soil and sludge samples showed that as the time passed, metal transferred to the aerial parts rather than to remain in the roots for KS sample while the reverse for the DS and control sample. Similar results were also shown by Punetha et al. (2015), who also supported more Zn accumulation in roots than leaves and Sulaiman and Hamzah (2018) observed higher concentration of Cd and Pb in roots as compared to the stems and leaves for *Athyrium esculentum*, *Chromolaena odorata* and *Lantana camara* collected from roadside area of Malaysia.

Nickel

Nickel content in *C. sativum* when grown in polluted soil and sludge sample ranged from 1.24- 45.52 mg kg^{-1}. Accumulation of Ni was found to be higher in contaminated soil than the control soil (Table 7). These results were similar with the results of Prasad et al. (2012), who stated that there was higher accumulation of different metals (Cr, Cd, Ni and Pd) in root and shoot tissues with increasing metal content. The results obtained showed that Ni accumulation was highest in DS sample.

1.52(KS) > 1.27 (DS) > 0.18 (Control) (BCF)
1.21 (DS) > 1.02 (KS) > 0.05 (Control) (TF)
0.96 (DS) > 0.67 (KS) > 0.30 (Control) (TLF)

Bioconcentration and translocation factor showed the ability of plant to accumulate metal in the roots and transfer to the aerial parts of the plant. Both the factors (BCF as well as TLF) were found to be the highest for KS and DS respectively. Whether metals are translocated to the aerial part or remained in roots is a major subject of importance (Singh et al., 2010). Therefore, in the present study, BCF was found to be more as compared to the TLF for coriander (in case of DS and KS), which showed that there was higher accumulation of Ni in roots rather to transfer to the aerial parts of the plant. In contrast, Hashim et al. (2017) found that Ni accumulated in relatively low concentration in roots while sensible concentrations were translocated to aerial parts of the plant especially within the cores.

All the contaminated soil samples had metal concentration above the permissible limit (1.5 mg kg^{-1}) as reported by Awasthi, (2000) except for noncontaminated soil.

Chromium

Chromium accumulation was found to be higher in the contaminated soil samples as compared to the control soil and varied significantly from 1.64-70.00 mg kg^{-1}. The accumulation of Cr was found to be higher in the second cutting rather than the first cutting, which was done after 30 days (Table 9).

Bioconcentration and translocation factor for Cr showed that the Cr content accumulated more in roots as compared to the aerial parts for both the cuttings. DS sample showed the highest accumulation of Cr metal and least by the control sample, while the plant cannot survive in the KSL sample. The results showed that all the plant samples grown in contaminated soils had much higher concentration of Cr than the permissible limit (20 mg kg^{-1}) reported by Awasthi, (2000). The order of BCF and TLF in *C. sativum* for Cr metal is as follows;

4.18 (DS) > 1.60 (KS) > 0.35 (Control) (BCF)

2.96 (DS) > 1.43 (KS) > 0.11 (Control) (TF)

0.89 (KS) > 0.71 (DS) > 0.31 (Control) (TLF)

DS sample had the highest BCF rather than its TLF which showed that there was less transfer of Cr to the aerial part than to accumulate in roots and these results were in good agreement with Peralta-Videa et al. (2002), who showed more accumulation of heavy metals in roots than in shoots of alfalfa plant and Punetha et al. (2015), who also reported the more Cr accumulation in roots than in shoots.

Therefore, the overall order of metal accumulation in *C. sativum* is as follow:

Cr > Ni > Zn > Cu (BCF)

Ni > Cr > Zn > Cu (TLF)

From the above comparison, it may be concluded that the highest accumulation of Cr was observed in the roots whereas the highest accumulation of Ni in the aerial parts of the plant.

Major Correlation Among Metals, Physicochemical Properties of Soil and Growth Parameters

Correlations were also found between different metals and the different plant growth parameters (plant height, plant weight and seed weight) as shown in Table 4, 6, 8 and 10 for Cu, Zn, Ni and Cr respectively.

Copper content in the plant was positively correlated with pH of the soil ($r2=0.912$, $P<0.01$). TF of Cu showed positive correlation with EC ($r2=0.928$, $P<0.01$) and negative with fresh plant and seed weight ($r2= -0.930$, $P<0.01$ and $r2= -0.841$, $P<0.05$) respectively (Table 4). Soil pH influences heavy metal availability to plants (Kabata-Pendias, 2011) and in our results, zinc content in the plant was positively correlated with the pH of soil ($r2=0.929$, $P<0.01$), TF was positively and significantly correlated with EC of the soil ($r2=0.823$, $P<0.05$). Electrical Conductivity of the soil was also negatively correlated with fresh plant weight and seed weight ($r2= -0.969$, $P<0.01$ and $r2= -0.832$, $P<0.05$) respectively (Table 6). Nickel content in the plant showed significant positive correlation with TF ($r2=0.879$, $P<0.05$) and EC ($r2=0.987$, $P<0.01$) and negative correlation with fresh plant weight ($r2= -0.922$, $P<0.01$) (Table 8). Chromium content in the plant was significantly and positively correlated with pH of the soil ($r2=0.814$, $P<0.05$) (Table 10). According to Glavac et al. (2017), Cd, Pb and Zn content in plants were found to be correlated significantly to the respective concentration in soil, while no significance correlation was observed for Cu, Fe and Mn.

Effect of Contaminated Soil and Sludge Samples on Oil Yield

Essential oil yield of *C. sativum* significantly varied from 0.1% - 0.3% (w/w) as shown in Table 2 (Punetha et al., 2014 & 2019). The highest oil yield was noted for the plants grown in control soil sample (0.3% w/w) and as the contamination increased, oil yield showed negative effect. Similar results were shown by Zheljazkov et al. (2008), who reported that the highest Cu treatment of 150 mg L^{-1} reduced the oil yield as compared to the control for basil and peppermint.

Effect of Contaminated Soil and Sludge Samples on Essential Oil Composition

The GC (Figure 1; Figure 2; Figure 3) and GC/MS analysis of *C. sativum* essential oil showed the presence of 35 (control), 35 (DS) and 38 (KS) compounds representing 93.64%, 91.87% and 92.13% respectively of the total oil (Punetha et al., 2014 & 2019). Linalool, 2(*E*)-decenal, caryophyllene oxide, and (*E*)-3-pentadecen-2-ol were the major constituents and had more than 5% contribution with some other minor constituents such as α-pinene, 1-octanol, *cis*-linalool oxide furanoid, fenchone, n-decanal, carvacrol, geranyl acetate, 2-undecanol, undec-2(*E*)-enal, n-dodecanal, *trans*-tetradec-2-enal and myristyl palmitate which contributed more than 1% (ranged from 1.12%-4.61%) and some trace constituents ranged from 0%-0.90% to the total oil. Linalool, the major constituent and responsible for the characteristic aroma in that plant showed a variation from 11.10% to 55.88% (Punetha et al., 2014 & 2019). Control soil sample had the highest concentration of linalool (55.88%) and was the major phytoconstituent from control sample and it was also noted that there was a significant decrease in linalool content as the soil become polluted. In contrast, Ghorbanpour et al. (2016) showed increase in the linalool and methyl chavicol content in sweet basil treated with different levels of Cu and Zn. These results were in agreement with Zheljazkov et al. (2006), who showed decrease in menthol component from heavy metal contaminated sample as compared to the control. The main constituents of coriander leaf essential oil were alcohols and aldehydes while the oil from *C. sativum* seeds showed monoterpenoid linalool as the major content (Padalia et al., 2011). Similarly, linalool was the major component of *C. sativum* grown in metal contaminated soils treated with different manure levels (Punetha et al., 2019). 2(*E*)-Decenal and caryophyllene oxide varied from 3.96%-37.10% and 2.83%-10.27% respectively, while (*E*)-3-pentadecen-2-ol and n-octacosane ranged from 3.96%-23.10% and 2.83%-10.27% respectively and both these component were absent in the control sample. Aliphatic (2*E*)-alkenals and alkanals from the fresh leaves of the *Coriandrum sativum* L. were found to have bactericidal activity against *Salmonella choleraesuis* ssp (Pietro et al., 2004).

Citronellol and *cis*-4-decenal were the minor phytoconstituent which were present only in the control sample and absent in the metal polluted samples (Figure 4). Kilic and Kilic, (2017) also suggested the toxic effect of Cd on essential oil content and showed that essential oil amount was decreased by 22%, 70% and 97% when exposed to 10, 20 and 30 mg/kg of Cd respectively, compared to the control. Some previous reports showed linalool (37.65-75.73%) as the major component in the oil of *C. sativum*, while citronellol, which was a major component in the essential oils from control sample in the present work had been reported to be either absent or present in traces (Zoubiri and Baaliouamer, 2010; Anwar et al., 2011). Some phytoconstituents were absent in the samples from different soils as shown in Table 11. Essential oil of *C. sativum* is reported to show many beneficial effects. Aliphatic (2*E*)-alkenals and alkanals extracted from the leaves of the *C. sativum* were found to have bactericidal activity against *Salmonella choleraesuis* ssp. (Kubo et al., 2004; Roy et al., 2010). The oil from this plant also has a potential antimicrobial compound against *C. jejuni* in food (Rattanachaikunsopon and Phumkhachorn, 2010).

Figure 1. Gas Chromatogram of essential oil of Coriandrum sativum cultivated in control soil

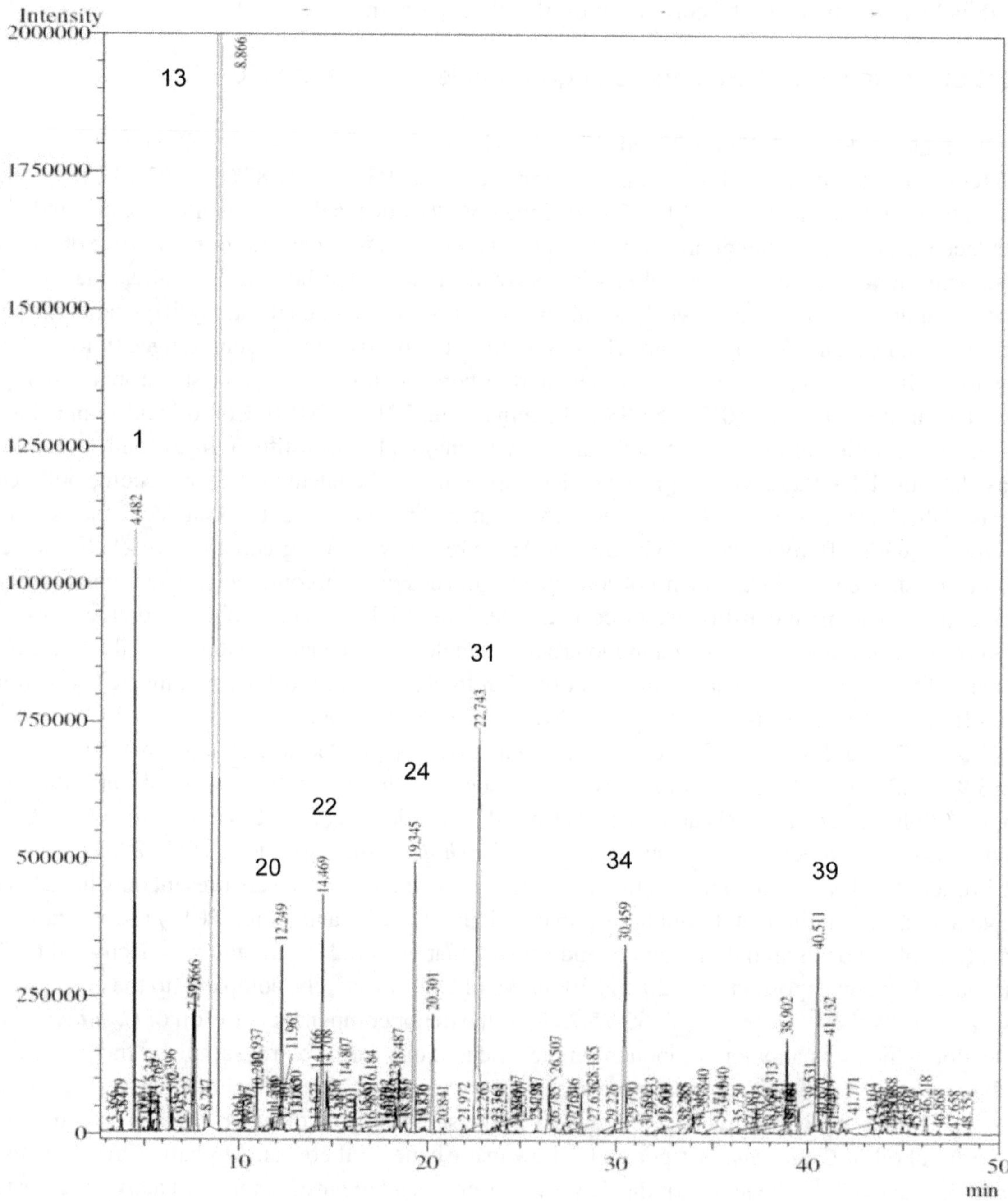

Correlation Among Physicochemical Propertied of Soil, Growth Parameters and Essential Oil Constituents

Oil yield was significantly and positively correlated with linalool, while negatively correlated with (*E*)-3-pentadecen-2-ol, pH and caryophyllene oxide. Fresh seeds and plant weight were negatively correlated

Figure 2. Gas Chromatogram of essential oil of Coriandrum sativum cultivated in Karula nala soil

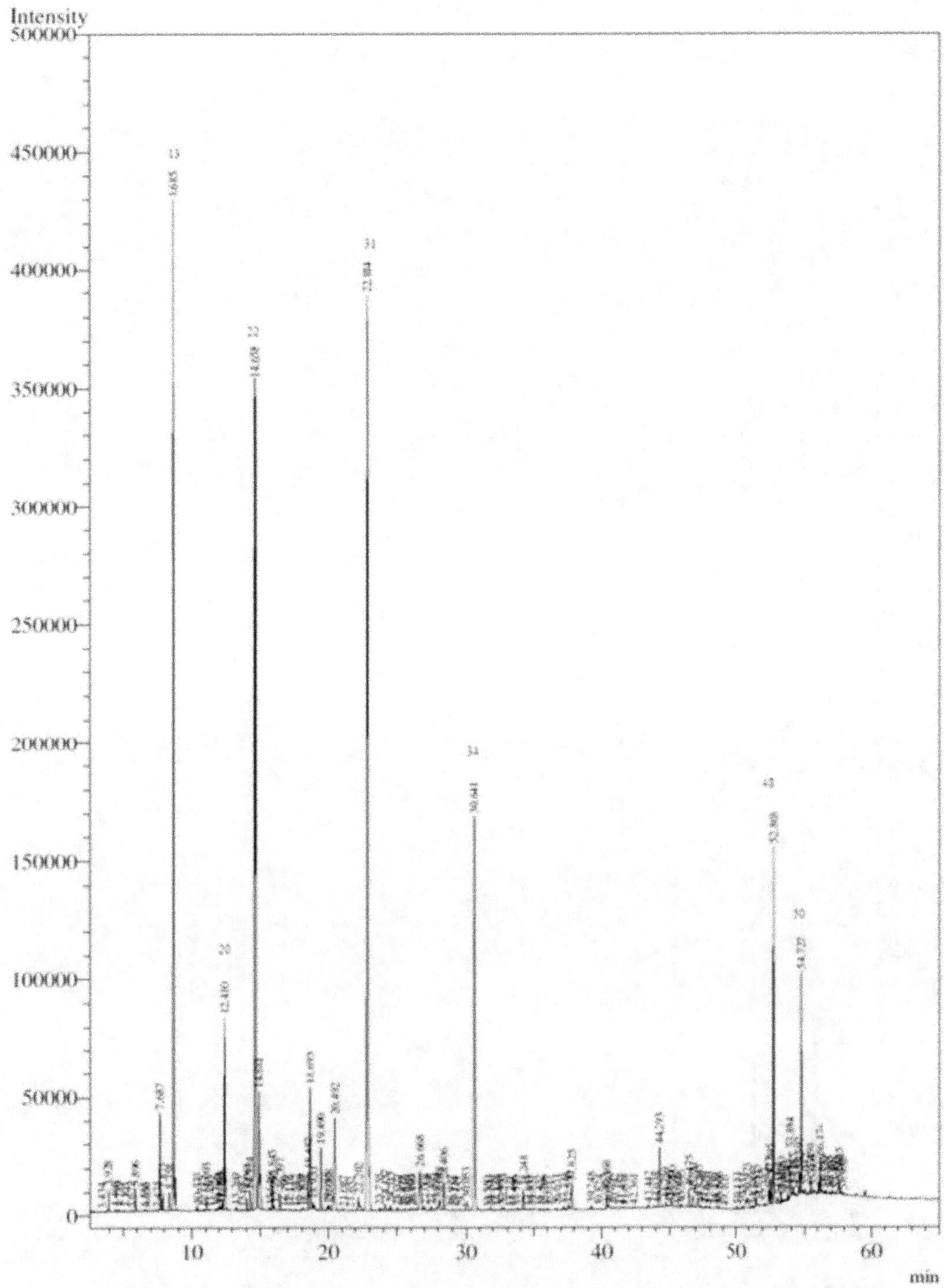

Figure 3. Gas Chromatograms of essential oil of Coriandrum sativum cultivated in Dhauri nala soil

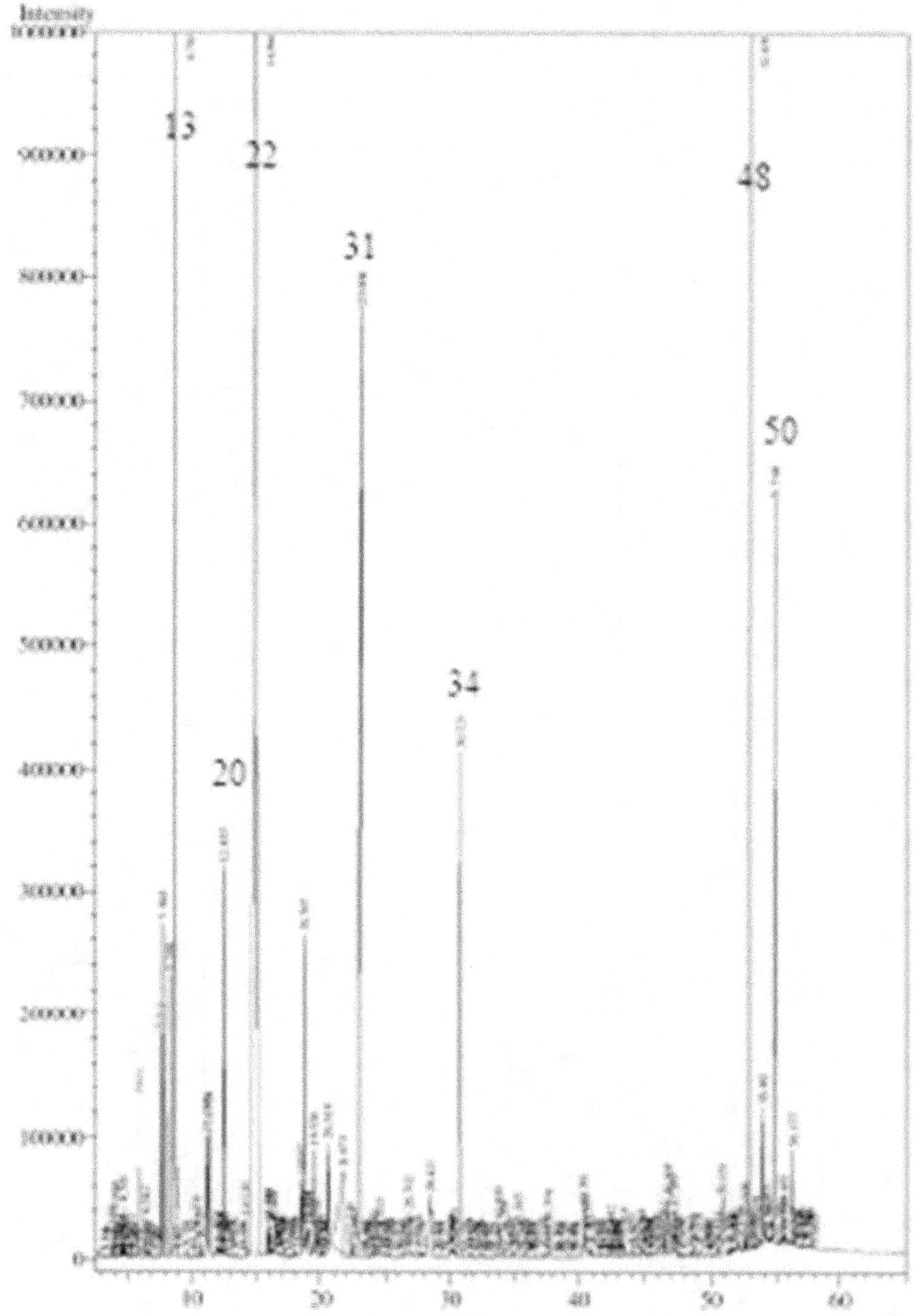

Figure 4. Effect of metal contamination on the essential oil composition of C. sativum

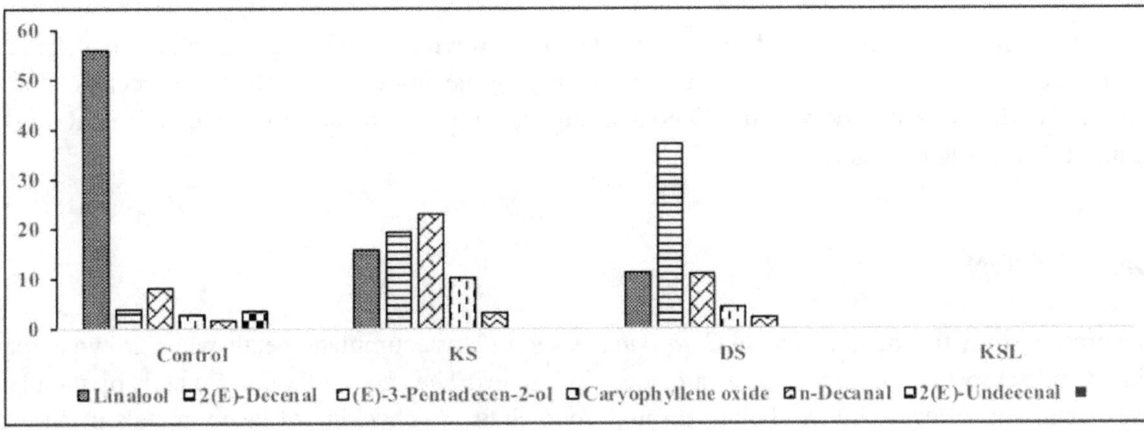

with pH. Linalool was negatively correlated with EC and (E)-3-pentadecen-2-ol as well as caryophyllene oxide were positively correlated with pH (Table 12).

Cluster Analysis

The essential oils of *C. sativum* grown in different soil type were analyzed by cluster analysis as shown in Figure 5. This analysis grouped these essential oils into two clusters depending on the differences in their chemical constituents and allowing them to be characterized into two different groups.

Group I- KS and DS sample (Linalool, 2(*E*)-decenal and (*E*)-3-pentadecen-2-ol)

Group II- Control sample (Linalool, caryophyllene oxide)

Figure 5. Agglomerative hierarchical clustering analysis (Dendrogram) of essential oil of C. sativum grown in different soil type

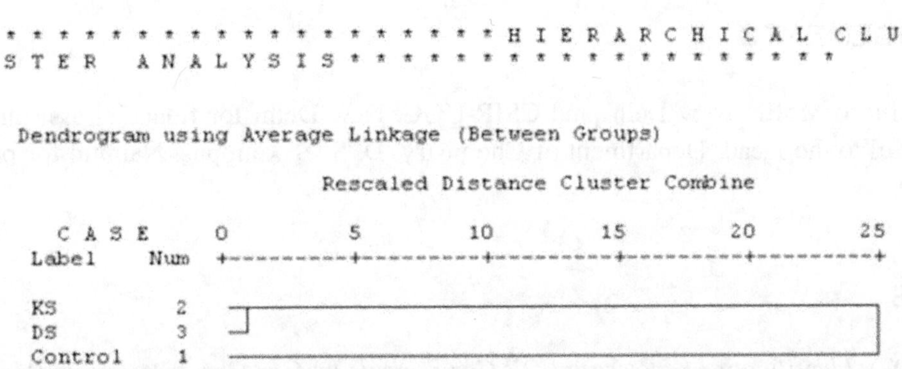

Metal Content in Oil and Extracted Water

C. sativum is an aromatic compound and this plant can be grown in metal polluted soil as an alternative crop and the data showed that there was very low or negligible amount of all the tested metals (Zn, Ni, Cu and Cr) in the oil. Zheljazkov et al. (2008) also suggested that there was no transfer of metal into the essential oil and water extracted.

CONCLUSION

In the present study the medicinal plant *C. sativum* was found to accumulate metals when grown in metal polluted soil samples collected from the area near industries. The accumulation of metals by the plants causes hazardous effect to human beings through food chain, so checking of heavy metals in different medicinal plants as well as in different vegetables must be needed to prevent excessive increase of these metals in human food chain.

It may be concluded that coriander can also be applicable to remediate the different metal (Zn Ni, Cu and Cr) contaminated soil samples. For control sample, more Zn was accumulated in roots and Cu in aerial parts, whereas DS sample showed highest content of Cr in roots and Ni in aerial parts and in case of KS sample Cr showed highest value of BCF and TLF among all the tested metals but as BCF was greater than TLF, so Cr prefers to remain in roots. The experimental plant (*C. sativum*) grew in both control and polluted soil samples except Karula nala sludge. Ni showed the highest TLF suggesting maximum transfer of these metals to the aerial parts. Cr showed the highest BCF suggesting maximum metal remains into the roots. Ni content was negatively correlated with plant and seed weight While Cr content in the soil was positively correlated with seed weight.

Control soil sample showed linalool (55.88%) as the major component, which was 253.4% and 403.42% more than the contaminated samples (KS and DS) respectively. KS had the (*E*)-3-pentadecen-2-ol (23.10%) as a major phytoconstituent and it was 178.6% and 110.38% more compared to the control and DS sample respectively. DS had 2(*E*)-decenal (37.10%) as a major component, which was 836.87% more than the control sample. The coriander plant could not survive in KSL sample which had the highest concentration of Cu.

ACKNOWLEDGEMENTS

Authors are grateful to MoEF, New Delhi and CSIR-UGC, New Delhi for financial assistance. The authors are thankful to the Head, Department of Chemistry, D. S. B. Campus, Nainital for providing necessary facilities.

REFERENCES

Adams, R. P. (2007). *Identification of Essential Oil Components by Gas Chromatography/Mass Spectrometry* (4th ed.). Allured Publishing.

Adriano, D. C. (2001). *Trace Elements in Terrestrial Environments; Biochemistry, Bioavailability and Risks of Metals*. Springer-Verlag. doi:10.1007/978-0-387-21510-5

Anwar, F., Sulman, M., Hussain, A. I., Saari, N., Iqbal, C., & Rashid, U. (2011). Physicochemical composition of hydro-distilled essential oil from coriander (*Coriandrum sativum* L.) seeds cultivated in Pakistan. *Journal of Medicinal Plants Research*, *5*(15), 3537–3544.

Awasthi, S. K. (2000). Prevention of Food Adulteration Act no 37 of 1954. Central and State Rules as Amended for 1999, 303-327.

Baratta, M. T., Dorman, H. J. D., Deans, S. G., Biondi, D. M., & Ruberto, G. (1998). Chemical composition, antimicrobial and antioxidative activity of laurel, sage, rosemary, oregano and coriander essential oils. *The Journal of Essential Oil Research*, *10*(6), 618–627. doi:10.1080/10412905.1998.9700989

Burken, J., Vroblesky, D., & Balouet, J. C. (2011). Phytoforensics, dendrochemistry and phytoscreening: New green tools for delineating contaminants from past and present. *Environmental Science & Technology*, *45*(15), 6218–6226. doi:10.1021/es2005286 PMID:21749088

Casetti, F., Bartelke, S., Biehler, K., Augustin, M., Schempp, C. M., & Frank, U. (2012). Antimicrobial activity against bacteria with dermatological relevance and skin tolerance of the essential oil from *Coriandrum sativum* L. fruits. *Phytotherapy Research*, *26*(3), 420–424. doi:10.1002/ptr.3571 PMID:21815228

Cioanca, O., Hritcu, L., Mihasan, M., & Hancianu, M. (2013). Cognitive-enhancing and antioxidant activities of inhaled coriander volatile oil in amyloid β (1-42) rat model of Alzheimer's disease. *Physiology & Behavior*, *120*, 193–202. doi:10.1016/j.physbeh.2013.08.006 PMID:23958472

Cui, Y. J., Zhu, Y. G., Zhai, R. H., Chen, D. Y., Huang, Y. Z., Qui, Y., & Liang, J. Z. (2004). Transfer of metals from near a smelter in Nanning, China. *Environment International*, *30*, 785–791. doi:10.1016/j.envint.2004.01.003 PMID:15120196

Do Nascimento, C. W. A., & Xing, B. S. (2006). Phytoextraction: A review on enhanced metal availability and plant accumulation. *Scientia Agrícola*, *63*(3), 299–311. doi:10.1590/S0103-90162006000300014

Ghorbanpour, M., Asgari Lajayer, H., & Hadian, J. (2016). Influence of copper and zinc on growth, metal accumulation and chemical composition of essential oils in sweet Basil (*Ocimum basilicum* L.). *Faslnamah-i Giyahan-i Daruyi*, *15*(59), 132–144.

Glavac, N. K., Djogo, S., Razic, S., Kreft, S., & Veber, M. (2017). Accumulation of heavy metals from soil in medicinal plants. *Arhiv za Higijenu Rada i Toksikologiju*, *68*(3), 236–244. doi:10.1515/aiht-2017-68-2990 PMID:28976884

Hashemi, V. H., Ghanadi, A., & Sharif, B. (2003). Anti-inflammatory and analgesic effects of *Coriandrum sativum* L. in animal models. *Shahrekord University of Medical Sciences Journal*, *5*(2), 8–15.

Hashim, T. A., Abbas, H. H., Farid, I. M., El-Husseiny, O. H. M., & Abbas, M. H. H. (2017). Accumulation of some heavy metals in plants and soils adjacent to Cario-Alexandria agricultural highway. *Egyptian*. *Journal of Soil Science*, *57*(2), 215–232.

Jackson, M. L. (1958). *Soil Chemical Analysis*. Prentice Hall.

Kabata-Pendias, A. (2011). *Trace Element in Soils and Plants* (4th ed.). Boka Raton, Taylor and Francis Group.

Kabata-Pendias, A., & Pandias, H. (1992). *Trace Element in Soils and Plants* (2nd ed.). CRC Press.

Kamath, R., Rentz, J. A., Schnoor, J. L., & Alvarez, P. J. J. (2007). *Phytoremediation of Hydrocarbon-contaminated Soils; Principles and Applications*. USA University of Iowa.

Karami, R., Hosseini, M., Mohammadpour, T., Ghorbani, A., Sadeghnia, H. R., Rakhshandeh, H., Vafaee, F., & Esmaeilizadeh, M. (2015). Effects of hydroalcoholic extract of *Coriandrum sativum* on oxidative damage in pentylenetetrazole-induced seizures in rats. *Iranian Journal of Neurology, 14*(2), 59–66. PMID:26056549

Kilic, S., & Kilic, M. (2017). Effects of cadmium-induced stress on essential oil production, morphology and physiology of Lemon balm (Melissa officinalis L., Lamiaceae). *Applied Ecology and Environmental Research, 15*(3), 1653–1669. doi:10.15666/aeer/1503_16531669

Kilmer, V. J., & Alexander, L. T. (1949). Soil texture methods of making mechanical analyses of soils. *Soil Science, 68*(1), 15–24. doi:10.1097/00010694-194907000-00003

Krishna, A. K., & Govil, P. K. (2007). Soil contamination due to heavy metals from an industrial area of Surat, Gujarat, Western India. *Environmental Monitoring and Assessment, 124*(1-3), 263–275. doi:10.100710661-006-9224-7 PMID:17058016

Kubo, I., Fujita, K. I., Kubo, A., Nihei, K. I., & Ogura, T. (2004). Antibacterial activity of coriander volatile compounds against *Salmonella choleraesuis. Journal of Agricultural and Food Chemistry, 52*(11), 3329–3332. doi:10.1021/jf0354186 PMID:15161192

Lindsay, W. L., & Norvell, W. A. (1978). Development of a DTPA soil test for zinc, iron, manganese and copper. *Soil Science Society of America Journal, 42*(3), 421–428. doi:10.2136ssaj1978.03615995 004200030009x

Lokeshwari, H., & Chandrappa, G. T. (2006). Impact of heavy metal contamination of Bellandur Lake on soil and cultivated vegetation. *Current Science, 91*(5), 622–627.

Ma, L. Q., Komar, K. M., Tu, C., Zhang, W. A., Cai, Y., & Kennelley, E. D. (2001). A fern that hyperaccumulates arsenic. *Nature, 409*(6820), 579. doi:10.1038/35054664 PMID:11214308

Mazhar, J., & Mazumder, A. (2013). Evaluation of antidiabetic activity of methanolic leaf extract of *Coriandrum sativum* in alloxan induced diabetic rats. *Research Journal of Pharmaceutical, Biological and Chemical Sciences, 3*(4), 500–507.

Mengel, K., & Kirkby, E. A. (1982). *Principles of Plant Nutrition*. International Potash Institute, Bern.

Omez-Flores, R., Hernández-Martínez, H., Tamez-Guerra, P., Tamez-Guerra, R., & Quintanilla-Licea, R., Monreal- Cuevas, R., & Rodríguez-Padilla, C. (2010). Antitumor and immunomodulating potential of *Coriandrum sativum, Piper nigrum* and *Cinnamomum zeylanicum. Journal of Natural Products, 3*, 54–63.

Otaru, A. J., Ameh, C. U., Okafor, J. O., Odigure, J. O., Abdulkareem, A. S., & Ibrahim, S. (2013). Study on the effectiveness of phytoremediation in the removal of heavy metals from soil using Corn. *International Journal of Computational Engineering Research, 3*(4), 87–93.

Padalia, R. C., Karki, N., Sah, A. N., & Verma, R. S. (2011). Volatile constituents of leaf and seed essential oil of *Coriandrum sativum* L. *Journal of Essential Oil-Bearing Plants, 14*(5), 610–616. doi:10.1080/0972060X.2011.10643979

Peng, K., Li, X., Luo, C., & Shen, Z. (2006). Vegetation composition and heavy metal uptake by wild plants at three contaminated sites in Xiangxi area, China. *Journal of Environmental Science and Health. Part A, Environmental Science and Engineering & Toxic and Hazardous Substance Control, 40*, 65–76. PMID:16401571

Peralta-Videa, J. R., Gardea-Torresdey, J. L., Gomez, E., Tiemann, K. J., Parsons, J. G., Rosa, G. D., & Carrillo, G. (2002). Potential of alfalfa plant to phytoremediate individually contaminated montmorillonite-soils with cadmium (ii), chromium (vi), copper (ii), nickel (ii), and zinc (ii). *Bulletin of Environmental Contamination and Toxicology, 69*(1), 74–81. doi:10.100700128-002-0012-y PMID:12053260

Pietro, L. C., Nicola, S., Lacobellis, A., Adriana, D. M., Francesco, C., & Felice, S. (2004). Antibacterial activity of *Coriandrum sativum* L. and *Foeniculum vulgare* Miller Var. *vulgare* (Miller) essential oils. *Journal of Agricultural and Food Chemistry, 52*(26), 7862–7866. doi:10.1021/jf0493122 PMID:15612768

Popova, E. (2016). Accumulation of heavy metals in the "soil-plant" system. *AIP Conference Proceedings, 1772*, 050006. doi:10.1063/1.4964576

Prasad, A., Chand, S., Kumar, S., Chattopadhyay, A., & Patra, D. D. (2012). Heavy metals affect yield, essential oil compound and Rhizosphere Microflora of Vetiver (*Vetiverial zizanioides* L. Nash) grass. *Communications in Soil Science and Plant Analysis, 45*(11), 1511–1522. doi:10.1080/00103624.2014.904334

Punetha, D., Pande, C., Tewari, G., & Tewari, K. (2014). Qualitative variation in Coriandrum sativum L. seed oil with manure applications. *Indian Perfumer, 58*(2), 27–31.

Punetha, D., Tewari, G., Pande, C., Kharkwal, G. C., & Tewari, K. (2015). Investigation on heavy metal content in common grown vegetables from polluted sites of Moradabad district, India. *Journal of the Indian Chemical Society, 92*, 97–103.

Punetha, D., Tewari, G., Pande, C., Rana, L., & Tripathi, S. (2019). *Effect of manure on the metal efficiency of Coriandrum sativum. In Zero Waste: Management Practices for Environmental Sustainability*. Taylor & Francis.

Rattanachaikunsopon, P., & Phumkhachorn, P. (2010). Potential of coriander (*Coriandrum sativum*) oil as a natural antimicrobial compound in controlling *Campylobacter jejuni* in raw meat. *Bioscience, Biotechnology, and Biochemistry, 74*(1), 31–35. doi:10.1271/bbb.90409 PMID:20057155

Reyes, M. R., Reyes-Esparza, J., Angeles, O. T., & Rodríguez-Fragoso, L. (2010). Mutagenicity and safety evaluation of water extract of *Coriandrum sativum* leaves. *Journal of Food Science, 75*(1), T6–T12. doi:10.1111/j.1750-3841.2009.01403.x PMID:20492211

Roy, S. D., Thakur, S., Negi, A., Kumari, M., Sutar, N. Z., & Jana, G. K. (2010). *In vitro* antibiotic activity of volatile oils of *Carum carvi & Coriandrum sativim. International Journal of Chemical and Analytical Science*, *1*(7), 149–150.

Rupassara, S. I., Larson, R. A. A., Sims, G. K., & Marley, K. A. (2002). Degradation of Atrazine by hornwort in aquatic systems. *Bioremediation Journal*, *6*(3), 217–224. doi:10.1080/10889860290777576

Sharma, R. K., Agrawal, M., & Marshall, F. M. (2009). Heavy metals in vegetables collected from production and market sites of a tropical urban area of India. *Food and Chemical Toxicology*, *47*(3), 583–591. doi:10.1016/j.fct.2008.12.016 PMID:19138719

Singh, A., Agrawal, M., & Marshall, F. M. (2010). The role of organic vs. inorganic fertilizers in reducing phytoavailability of heavy metals in a wastewater-irrigated area. *Ecological Engineering*, *36*(12), 1733–1740. doi:10.1016/j.ecoleng.2010.07.021

Sulaiman, F. R., & Hamzah, H. A. (2018). Heavy metals accumulation in suburban roadside plants of a tropical area (Jengka, Malaysia). *Ecological Processes*, *7*(28), 1–11. doi:10.118613717-018-0139-3

USEPA. (1997). *Cleaning Up the Nation's Waste Sites: Markets and Technology Trends, United States Environmental Protection agency EPA/542/R-96/005*. Office of Solid Waste and Emergency Response.

Vekaria, R. H., Patel, M. N., Bhalodiya, P. N., Patel, V., Desai, T. R., & Tirgar, P. R. (2012). Evaluation of neuroprotective effect of *Coriandrum sativum* Linn. against ischemic - reperfusion insult in brain. *International Journal of Phytopharmacology*, *3*(2), 186–193.

Wang, X., Liu, Y., Zeng, G., Chai, L., Xiao, X., Song, X., & Min, Z. (2008). Pedological characteristics of Mn mine tailings and metal accumulation by native plants. *Chemosphere*, *72*(9), 1260–1266. doi:10.1016/j.chemosphere.2008.05.001 PMID:18555510

Waterman, P. G. (1993). The chemistry of volatile oils. In R. K. M. Hay & P. G. Waterman (Eds.), *Volatile oil crops* (pp. 47–62). Logman Scientific & Technical.

Wong, P. Y., & Kitts, D. D. (2006). Studies on the dual antioxidant and antibacterial properties of parsley (*Petroselinum crispum*) and cilantro (*Coriandrum sativum*) extracts. *Food Chemistry*, *97*(3), 505–515. doi:10.1016/j.foodchem.2005.05.031

Wosten, J. H. M. (1997). Pedotransfer functions to evaluation soil quality. In E. G. Gregorich & M. R. Carter (Eds.), *Soil quality for crop production and ecosystem health* (pp. 221–245). Elsevier science Ltd. doi:10.1016/S0166-2481(97)80037-2

Yoon, J., Cao, X., Zhou, Q., & Ma, Q. L. (2006). Accumulation of Pb, Cu and Zn in native plants growing on a contaminated Florida site. *The Science of the Total Environment*, *368*(2-3), 456–464. doi:10.1016/j.scitotenv.2006.01.016 PMID:16600337

Zheljazkov, V., & Nielsen, N. E. (1996). Studies on the effect of heavy metals (Cd, Pb, Cu, Mn, Zn and Fe) upon the growth, productivity and quality of lavender (*Lavandula angustifolia* Mill) production. *The Journal of Essential Oil Research*, *8*(3), 259–274. doi:10.1080/10412905.1996.9700612

Zheljazkov, V. D., Craker, L. E., & Xing, B. (2006). Effect of Cu, Pb and Cu on growth and essential oil contents in dill, peppermint and basil. *Environmental and Experimental Botany*, *58*(1-3), 9–16. doi:10.1016/j.envexpbot.2005.06.008

Zheljazkov, V. D., Craker, L. E., Xing, B., Nielson, N. E., & Wilcox, A. (2008). Aromatic plant production on metal contaminated soils. *The Science of the Total Environment*, *395*(2-3), 51–62. doi:10.1016/j.scitotenv.2008.01.041 PMID:18353428

Zheljazkov, V. D., & Warman, P. R. (2003). Application of high Cu compost to Swiss chard and basil. *The Science of the Total Environment*, *302*(1-3), 13–26. doi:10.1016/S0048-9697(02)00390-X PMID:12526895

Zhu, C., Tian, H., Cheng, K., Liu, K., Wang, K., Hua, S., Gao, J., & Zhou, J. (2016). Potentials of whole process control of heavy metals emissions from coal-fired power plants in China. *Journal of Cleaner Production*, *114*, 343–351. doi:10.1016/j.jclepro.2015.05.008

Zoubiri, S., & Baaliouamer, A. (2010). Essential oil composition of *Coriandrum sativum* seed cultivated in Algeria as food grains protectant. *Food Chemistry*, *122*(4), 1226–1228. doi:10.1016/j.foodchem.2010.03.119

APPENDIX

Table 1. Physicochemical properties of collected soil and sludge samples

Samples	pH (Mean ± SD)	EC (dS m^{-1})	Organic carbon (%) ±SD	Organic matter (%) ±SD	Texture	Total Nitrogen (%)	% P_2O_5	% K_2O	Cu (mg kg^{-1})	Zn (mg kg^{-1})	Ni (mg kg^{-1})	Cr (mg kg^{-1})
Control soil	7.1±0.0	0.1±0.0	2.3±0.3	4.0±0.8	Sandy loam	0.302±0.001	0.0049	0.0089	52.8±0.6	91.6±1.3	33.7±2.0	12.5±1.5
KS	7.9±0.2	0.4±0.1	0.9±0.1	1.5±0.4	Sandy clay loam	0.224±0.007	0.0047	0.0075	643.2±1.3	544.0±4.6	35.3±0.2	131.0±
DS	7.1±0.1	0.6±0.3	3.2±0.2	5.5±0.4	Loamy sand	0.129±0.010	0.0040	0.0281	117.7±3.9	150.0±0.5	150±0.3	50.0±
KSL	6.5±0.1	1.8±0.2	1.7±0.2	2.9±0.4	-	0.396±0.044	0.0016	0.2130	870.6±1.1	277.3±0.3	184.1±0.1	97.01±0.

DS = Dhaurinala soil; KS = Karulanala soil; KSL = Karulanala sludge

Table 2. Effect of metal contamination on the yield of C. sativum

Samples	% Oil yield (w/w)	Plant Height (I month)	Plant Height (III month)	Plant weight (g)	Seed weight (g)
Control soil	0.3^{a}	10^{a}±2	39^{a}±4	156.39^{a}±5.54	43.29^{a}±1.75
KS	0.1^{b}	9^{a}±3	38^{a}±3	78.79^{b}±2.51	27.64^{a}±5.79
DS	0.2^{c}	8^{a}±1	35^{a}±3	177.97^{c}±2.76	60.19^{a}±3.20
KSL	-	-	-	-	-

The mean values followed by same alphabets (a-c) at superscript within a column are not significantly different at $P \leq 0.05$ according to Duncan's test.

Table 3. Copper content in soil and C. sativum (Permissible limit: 30 mg kg^{-1})

	Sample	Soil (mg kg^{-1})		Aerial parts (mg kg^{-1})		Roots (mg kg^{-1})	Transfer factor (TF) (Plant/Soil)		Bioconcentration Factor (BCF) (Root/Soil)	Translocation factor (TLF) (Shoot/Root)
		I cutting	II cutting	I cutting	II cutting		I cutting	II cutting		
Cu	CS Co	46.00^{a}±1.09	48.00^{a}±1.59	4.98^{a}±0.87	4.00^{a}±1.41	5.16^{a}±1.67	0.11^{a}±0.03	0.08^{a}±0.01	0.11^{a}±0.21	0.77^{a}±0.07
	DS Co	60.81^{b}±1.85	31.42^{b}±1.78	22.57^{b}±2.07	19.98^{b}±1.48	25.64^{b}±2.27	0.37^{b}±0.03	0.63^{b}±0.10	0.82^{b}±0.05	0.78^{a}±0.06
	KS Co	400.97^{c}±4.73	245.83^{c}±3.12	16.22^{c}±1.53	14.76^{c}±2.03	20.46^{c}±1.75	0.04^{c}±0.01	0.06^{a}±0.02	0.08^{a}±0.10	0.72^{a}±0.08
	KSL Co	-	-	-	-	-	-	-	-	-

The mean values followed by same alphabets (a-c) at superscript within a column are not significantly different at $P<0.05$ according to Duncan's test

Table 4. Correlation between Cu with growth parameters of C. sativum and physicochemical properties of soil

	Cu content in soil	Cu content in plant	Cu content (DTPA)	TF	pH	EC	Plant height	Plant weight	Seed weight
Cu content in soil	1	.189	1.00**	-.542	.517	-.439	-.289	.633	.829*
Cu content in plant		1	.177	.615	.912*	.784	-.162	-.609	-.316
Cu content (DTPA)			1	-.545	.509	-.449	-.280	.641	.835*
BCF				1	.342	.928**	.218	-.930**	-.841*
pH					1	.486	-.070	-.254	.081
EC						1	.013	-.969**	-.832*
Plant height							1	-.034	-.059
Plant weight								1	.944**
Seed weight									1

*Correlation coefficient significant at $p<0.05$;**Correlation coefficient significant at $p<0.01$

Table 5. Zinc content in soil and C. sativum (Permissible limit: 50 mg kg^{-1})

	Sample	Soil (mg kg^{-1})		Aerial parts (mg kg^{-1})		Roots (mg kg^{-1})	Transfer factor (TF) (Plant/Soil)		Bioconcentration Factor (BCF) (Root/Soil)	Translocation factor (TLF) (Shoot/Root)
		I cutting	II cutting	I cutting	II cutting		I cutting	II cutting		
Zn	CS Co	29.00^a±2.65	28.05^a±3.83	3.60^a±1.98	2.03^a±0.31	10.89^a±1.75	0.12^a±0.04	0.07^a±0.02	0.39^a±0.03	0.19^a±0.05
	DS Co	73.95^b±2.14	36.46^b±2.64	26.00^b±2.69	32.00^b±2.50	37.25^b±2.18	0.35^b±0.05	0.88^b±0.09	1.02^b±0.12	0.86^b±0.27
	KS Co	344.80^c±3.15	216.16^c±3.02	49.25^c±3.46	45.00^c±2.30	55.29^c±2.56	0.14^a±0.04	0.21^c±0.06	0.26^c±0.05	0.81^b±0.11
	KSL Co	-	-	-	-	-	-	-	-	-

Co = *C. sativum*; DS = Dhaurinala soil; KS = Karulanala soil; KSL = Karulanala sludge; The mean values followed by same alphabets (a-c) at superscript within a column are not significantly different at $P<0.05$ according to Duncan's test.

Table 6. Correlation between Zn and growth parameters of C. sativum and physicochemical properties of soil

	Zn content in soil	Zn content in plant	Zn content (DTPA)	TF	pH	EC	Plant height	Plant weight	Seed weight
Zn content in soil	1	.822*	1.00**	-.292	.594	-.355	-.306	.562	.782
Zn content in plant		1	.818*	.250	.929**	.145	-.079	.103	.424
Zn content (DTPA)			1	-.293	.584	-.369	-.293	.574	.792
BCF				1	.473	.823*	.339	-.777	-.639
pH					1	.486	-.070	-.254	.081
EC						1	.013	-.969**	-.832*
Plant height							1	-.034	-.059
Plant weight								1	.944**
Seed weight									1

*Correlation coefficient significant at $p<0.05$; **Correlation coefficient significant at $p<0.01$

Table 7. Nickel content in soil and C. sativum (Permissible limit: 1.5 mg kg^{-1})

	Sample	Soil (mg kg^{-1})		Aerial parts (mg kg^{-1})		Roots (mg kg^{-1})	Transfer factor (TF) (Plant/Soil)		Bioconcentration Factor (BCF) (Root/Soil)	Translocation factor (TLF) (Shoot/Root)
		I cutting	II cutting	I cutting	II cutting		I cutting	II cutting		
Ni	CS Co	$30.05^{a}\pm2.43$	$28.05^{a}\pm3.83$	$1.24^{a}\pm0.75$	$1.50^{a}\pm0.57$	$4.96^{a}\pm1.18$	$0.04^{a}\pm0.03$	$0.05^{a}\pm0.02$	$0.18^{a}\pm0.21$	$0.30^{a}\pm0.22$
	DS Co	$75.15^{b}\pm2.82$	$37.58^{b}\pm2.43$	$40.24^{b}\pm3.02$	$45.52^{b}\pm2.74$	$47.59^{b}\pm2.70$	$0.53^{b}\pm0.11$	$1.21^{b}\pm0.22$	$1.27^{b}\pm0.20$	$0.96^{b}\pm0.25$
	KS Co	$20.00^{c}\pm3.08$	$10.00^{c}\pm2.26$	$8.15^{c}\pm2.41$	$10.21^{c}\pm2.10$	$15.24^{c}\pm2.75$	$0.41^{b}\pm0.10$	$1.02^{b}\pm0.24$	$1.52^{c}\pm0.18$	$0.67^{a,b}\pm0.06$
	KSL Co	-	-	-	-	-	-	-	-	-

The mean values followed by same alphabets (a-c) at superscript within a column are not significantly different at $P<0.05$ according to Duncan's test

Table 8. Correlation between Ni with growth parameters of C. sativum and physicochemical properties of soil

	Ni content in soil	Ni content in plant	Ni content (DTPA)	TF	pH	EC	Plant height	Plant weight	Seed weight
Ni content in soil	1	.675	.998**	.257	.104	.776	-.367	-.829*	-.819*
Ni content in plant		1	.683	.879*	.597	.987**	.071	-.922**	-.746
Ni content (DTPA)			1	.287	.073	.787	-.348	-.850*	-.851*
BCF				1	.749	.817	.316	-.734	-.535
pH					1	.486	-.070	-.254	.081
EC						1	.013	-.969**	-.832*
Plant height							1	-.034	-.059
Plant weight								1	.944**
Seed weight									1

*Correlation coefficient significant at $p<0.05$; **Correlation coefficient significant at $p<0.01$

Table 9. Chromium content in soil and C. sativum (Permissible limit: 20 mg kg^{-1})

	Sample	Soil (mg kg^{-1})		Aerial parts (mg kg^{-1})		Roots (mg kg^{-1})	Transfer factor (TF) (Plant/Soil)		Bioconcentration Factor (BCF) (Root/Soil)	Translocation factor (TLF) (Shoot/Root)
		I cutting	II cutting	I cutting	II cutting		I cutting	II cutting		
Cr	CS Co	$12.00^a \pm 1.39$	$15.00^a \pm 1.18$	$3.00^a \pm 0.69$	$1.64^a \pm 0.54$	$5.26^a \pm 1.01$	$0.25^a \pm 0.09$	$0.11^a \pm 0.01$	$0.35^a \pm 0.02$	$0.31^a \pm 0.07$
	DS Co	$20.00^b \pm 3.71$	$8.00^b \pm 0.77$	$30.00^b \pm 1.50$	$23.68^b \pm 1.49$	$33.47^b \pm 2.43$	$1.50^b \pm 0.57$	$2.96^b \pm 0.76$	$4.18^b \pm 0.03$	$0.71^b \pm 0.11$
	KS Co	$81.66^c \pm 1.94$	$47.63^c \pm 1.37$	$70.00^c \pm 1.40$	$68.05^c \pm 1.40$	$76.39^c \pm 1.71$	$0.86^{a,b} \pm 0.15$	$1.43^c \pm 0.33$	$1.60^c \pm 0.01$	$0.89^b \pm 0.10$
	KSL Co	-	-	-	-	-	-	-	-	-

The mean values followed by same alphabets (a-c) at superscript within a column are not significantly different at $P<0.05$ according to Duncan's test

Table 10. Correlation of Cr metal with growth parameter of C. sativum and physicochemical properties of soil

	Cr content in soil	Cr content in plant	Cr content (DTPA)	TF	pH	EC	Plant height	Plant weight	Seed weight
Cr content in soil	1	.871*	1.00**	-.188	.511	-.418	-.340	.607	.800
Cr content in plant		1	.867*	.237	.814*	-.109	-.146	.351	.641
Cr content (DTPA)			1	-.192	.499	-.435	-.324	.623	.812*
BCF				1	.727	.870	.287	-.802	-.620
pH					1	.486	-.070	-.254	.081
EC						1	.013	-.969**	-.832*
Plant height							1	-.034	-.059
Plant weight								1	.944**
Seed weight									1

*Correlation coefficient significant at $p \leq 0.05$; **Correlation coefficient significant at $p \leq 0.01$

Table 11. Essential oil composition of C. sativum grown in different soil and sludge samples

S. No.	Compounds	RI (Calculated)	RI (Adams, 2007)	Percentage of oil (Control) (CS)	Percentage of oil Karula nala soil (KS) (Punetha et al., 2019)	Percentage of oil Dhauri nala soil (DS) (Punetha et al., 2014)	Percentage of oil Karula sludge (KSL)	Mode of identification
	Nonane	906	900	3.28	-	-	-	A, B
	α-pinene	934	932	0.31	0.02	0.07	-	A, B
	Camphene	954	946	-	0.01	0.01	-	A, B
	Sabinene	975	969	0.01	-	-	-	A,B
	Myrcene	990	988	0.25	0.31	0.65	-	A, B
	n-Octanal	1003	998	-	0.02	0.01	-	A, B
	p-Cymene	1024	1020	0.02	-	-	-	A,B
	Limonene	1029	1024	0.44	0.02	0.15	-	A, B
	ϒ-Terpinene	1059	1054	0.01	-	-	-	A, B
	1-octanol	1068	1063	0.01	-	-	-	A, B
	Cis-linalool oxide furanoid	1072	1067	1.44	1.51	3.64	-	A, B
	Fenchone	1086	1083	1.81	1.51	1.29	-	A, B
	Linalool	1096	1095	$55.88^{a}\pm0.8$	$15.81^{b}\pm0.5$	$11.10^{c}\pm0.7$	-	A, B, C
	n-Nonanal	1160	1165	0.25	-	0.04	-	A, B
	Camphor	1146	1141	0.57	-	-	-	A, B
	Cis-linalool oxide pyranoid	1174	1170	0.04	0.26	0.78	-	A, B
	Trans-linalool oxide pyranoid	1176	1173	0.21	0.02	0.66	-	A, B
	α-Terpineol	1188	1186	0.01	-	-	-	A, B
	Cis-4-decenal	1194	1193	0.02	-	-	-	A, B
	n-Decanal	1205	1201	$1.80^{a}\pm0.0$	$3.23^{b}\pm0.6$	$2.22^{b}\pm0.6$	-	A, B
	Citronellol	1225	1223	0.01	-	-	-	A, B
	2(*E*)-Decenal	1263	1260	$3.96^{a}\pm1.2$	$19.24^{b}\pm1.0$	$37.10^{c}\pm1.2$	-	A, B, C
	2(*E*)-Decen-1-ol	1271	1268	0.35	0.34	0.20	-	A, B
	Carvacrol	1300	1298	0.12	2.90	1.99	-	A, B
	8(*Z*)-Undecenal	1303	1300	-	-	0.73	-	A, B
	2-Undecenal	1305	1301	0.01	-	-	-	A, B
	2(*E*)-Undecenal	1360	1357	3.57	-	-	-	A, B
	Geranyl acetate	1381	1379	1.33	-	0.65	-	A, B
	n-Dodecanal	1412	1408	0.11	1.29	2.96	-	A, B
	n-Dodecanal	1412	1408	0.11	1.29	2.96	-	A, B
	(*E*)-3-Pentadecen-2-ol	1444	-	$8.29^{a}\pm0.4$	$23.10^{b}\pm0.7$	$10.98^{c}\pm0.3$	-	A, B, C
	2(*E*)-Dodecenal	1466	1464	0.84	0.31	0.22	-	A, B
	2(*E*)-Tridecen-1-al	1570	1567	0.45	0.78	0.01	-	A, B
	Caryophyllene oxide	1586	1582	$2.83^{a}\pm1.3$	$10.27^{b}\pm1.4$	$4.31^{c}\pm0.8$	-	A, B
	Myristaldehyde	1622	-	-	0.04	0.01	-	A, B
	Trans-2-tetradecenal	1669	-	0.1	0.01	-	-	A, B
	14-Methy-8(*Z*)-hexadecenal	1762	-	-	0.08	0.04	-	A, B
	n-Hexadecanol	1870	1874	1.22	0.71	0.09	-	A, B
	Diidobutyl phthalate	1913	-	2.84	0.29	0.27	-	A, B
	Phytol	1940	1942	1.25	0.03	0.31	-	A, B
	Isopropyl palmitate	2035		-	0.15	0.01	-	A, B
	n-Nonadecanol	2170	-	-	0.64	0.11	-	A, B
	Heptyl decanoate	2226	-	-	0.07	0.15	-	A, B
	Octyl laurate	2259	-	-	-	0.63	-	A, B
	Octadecanoic acid, methyl ester	2327	-	-	0.53	0.90	-	A, B
	Octyl-octanoate	2600	-	-	0.28	0.03	-	A, B
	Octyl-octanoate	2600	-	-	0.28	0.03	-	A, B
	n-Octacosane	2810	2800	-	4.61	6.49	-	A, B
	Myristic acid tetradecyl ester	3000	-	-	0.74	0.56	-	A, B
	Myristyl palmitate	3185	-	-	2.25	2.52	-	A, B
	n-Pentatriacontane	3515	3500	-	0.31	0.22	-	A, B
	Dodecanoic acid, phenyl methyl ester	3598	-	-	0.18	0.02	-	A, B
Total				93.64%	91.87%	92.13%	-	

continued on following page

Table 11. Continued

S. No.	Compounds	RI (Calculated)	RI (Adams, 2007)	Percentage of oil (Control) (CS)	Percentage of oil Karula nala soil (KS) (Punetha et al., 2019)	Percentage of oil Dhauri nala soil (DS) (Punetha et al., 2014)	Percentage of oil Karula sludge (KSL)	Mode of identification
Aliphatic Hydrocarbon				3.53	4.92	6.71	-	
Oxygenated Monoterpene				56.00	20.22	14.38	-	
Aliphatic Aldehyde				7.26	25.00	43.34	-	
Aliphatic Alcohol				9.86	24.79	11.38	-	
Oxygenated Sesquiterpene				2.83	10.27	4.31	-	

Mode of identification: A= retention index (RI) relative to n-alkane homologous series (C_8-C_{32}) on Rtx-5 column; B= NIST, WILLEY library and MS data and literature (Adams, 2007); Mean values followed by different alphabets (a-c) at their superscript are significantly differet at a significant level of 0.05.

Table 12. Correlation between physicochemical properties of soil, different growth parameters and major constituents of C. sativum

	Linalool	2(*E*)-Decenal	(*E*)-3-Pentadecen-2-ol	Caryophylleneoxide	pH	EC	OM	Oil yield	Plant height	Plant weight	Seed weight	Zn content	Cu content	Ni content	Cr content
Linalool	1	-.890**	-.564	-.579	-.415	-.951**	.047	.814**	.759*	.218	-.118	-.979**	-.912**	-.578	-.814**
2(*E*)- Decenal		1	.126	.144	-.045	.987**	.412	-.461	-.972**	.251	.557	.964**	.626	.143	.989**
(*E*)-3-Pentadecen-2-ol			1	1.000**	.985**	.282	-.852**	-.939**	.109	-.929**	-.754*	.385	.853**	1.000**	-.020
Caryophylleneoxide				1	.982**	.300	-.842**	-.945**	.091	-.922**	-.742*	.402	.862**	1.000**	-.002
pH					1	.115	-.929**	-.866**	.277	-.978**	-.855**	.222	.751*	.982**	-.191
EC						1	.262	-.596	-.923**	.093	.418	.994**	.742*	.298	.953**
OM							1	.619	-.614	.985**	.986**	.155	-.452	-.843**	.541
Oil yield								1	.240	.744*	.481	-.680*	-.981**	-.944**	-.326
Plant height									1	-.470	-.736*	-.875**	-.426	.093	-.996**
Plant weight										1	.944**	-.016	-.598	-.922**	.390
Seed weight											1	.316	-.299	-.743*	.672*
Zn content												1	.811**	.401	.915**
Cu content													1	.861**	.505
Ni content														1	-.004
Cr content															1

**. Correlation is significant at the 0.01 level (2-tailed).

*. Correlation is significant at the 0.05 level (2-tailed).

Chapter 5
Previous, Contemporary, and Prospects of E-Waste and Its Management

Vara Saritha
https://orcid.org/0000-0003-0473-993X
Department of Environmental Science, GITAM Institute of Science, GITAM University (Deemed), India

Madhavi Konni
https://orcid.org/0000-0002-2022-5528
Department of Chemistry, GITAM Institute of Science, GITAM University (Deemed), India

Bhavya Kavitha Dwarapureddi
Department of Environmental Science, GITAM Institute of Science, GITAM University (Deemed), India

R. S. S. Srikanth Vemuri
Department of Basic Science and Humanities, Vignan's Institute of Engineering for Women, India

Manoj Kumar Karnena
https://orcid.org/0000-0001-9241-2990
Department of Environmental Science, GITAM Institute of Science, GITAM University (Deemed), India

ABSTRACT

The contemporary world is driven by electronic gadgets without which the survival of mankind is perceived to be incomplete. The uncontrolled dependence of mankind on electronic gadgets has resulted in enhanced production of these gadgets leading to the accumulation of e-waste. Both technological innovation and market expansion have played an important role in electronic waste (e-waste). Owing to hazardous material composition, electronic waste causes environmental problems during the waste management phase if not properly pre-treated. Growing attention is being given to the impacts of these

DOI: 10.4018/978-1-7998-4921-6.ch005

hazardous components from e-waste on the environment. Many countries have drafted legislation to improve the reuse, recycling, and other forms of recovery of such wastes so as to reduce disposal problems. The purpose of this chapter is to present an overview of electronic waste, the current status of management of electronic waste, and recycling technologies for the recovery of metals from end-of-life electronic equipment.

INTRODUCTION

The electronic industry has been raised as one of the world's leading and inventive sector as one of its kind. This is attributed to the profound privilege of the contemporary generation, which enjoys their entire social and economic life surrounded by electronics. This has led to an exceptional expansion of electronics, making them cheaper and more available. Unfortunately, the waste management system has not improved at a similar pace (Masud et al., 2019). All these reasons and others have contributed to the unprecedented accumulation of electronic waste. Other reasons for e-waste present the anthropocentric view of humans since most of the reasons do not have a requirement to dispose of e-waste, and some are merely for personal satisfaction. These include very less lifespan of electronic items since the consumer would like to upgrade technologically (even when the old or previous item is in good condition) and technological advancements that are being released into the market every day. For instance, in 2008, the users of computers are increased and surpassed 1 billion, and these goods are obsoleted in the next five years. For these and many other petty known reasons, e-waste ends up in dump yards resulting in environmental pollution. Components of e-waste yet times enter the food chain and affect humans also.

Pollution from e-waste is a global problem. According to the UN, global e-waste is projected to exceed 40 million tons per year. Recycling end of product life results in enhanced pollution, economically unviable, and is unregulated in most countries. Severe health concerns are caused by e-waste in millions of people, precisely in developing countries like Europe, Africa, and Asia. It is reported that more than 200 million people around the world are at threat from exposure to toxic waste. Recycling of valuable elements in e-waste like copper and gold has been a source of income typically in the informal sector from emerging industrialized countries. However, some of the recycling techniques employed like cable burning for retention of inherent copper leads to hazardous substances exposure to both adult and child workers and their families (Lakshmi and Raj, 2017; Li et al., 2020).

The objective of this chapter is to present an overview of electronic waste, the current status of electronic waste management, and recycling technologies for the recovery of metals from end-of-life electronic equipment.

Definition of E-Waste

Some known definitions of E-waste are as follows:

- According to the Sinha-Khetriwal, 2002, the power-driven electronic appliances that have reached its End of life (EOL) is called E-Waste.
- Puckett et al., 2002, described that Electronic devices were ranging from individual household items to personal computers that have been discarded after EOL.
- The APME, 1995, defined as it is a combination of Ferrous, Non-ferrous, Ceramic, and Plastic Materials.

The term E-Waste and WEE (Waste Electronic Equipment's) is repeatedly used in the present scenario as the usage of electronic devices in our day to day life are increasing. Discarded electronic appliances belong to E-waste Category, whereas WEE additionally refers to EOL appliances. The European Commission Defined WEE as Wide-Ranging EOL is the raw material that can be utilized to produce, measure, and transfer electromagnetic current (EU RoHS Directive 2002/95/EC, 2003).

Classes of E-Waste Based on their Application and Size

Since much difference is not made between E-waste and WEEE, Indian legislation has considered e-waste for E-waste management (E-Waste Rules, 2011; Rodrigues et al., 2020). In this chapter, we use the term "E-waste" to refer to all the discarded/broken/surplus/obsolete electrical and electronic devices, as stated by Pathak et al., 2017. More specifically, E-waste can be divided into six categories Balde et al., 2015, as shown in **Table 1.**

Illustrations	Examples	Illustrations	Examples
Equipment's that Exchange heat and temperature comprises heat pumps, Air Conditioners, Refrigerators.		Large size equipment, Comprises Xerox machines, Washing Machines, Electronic Stoves.	
Computers, televisions, laptops, notebooks, and tablets		Electric lamps comprise Fluorescent lamps & LED Bulbs	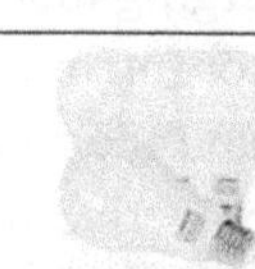
Small size equipment Comprises household items such as toasters, microwaves, electric kettles, vacuum cleaner, calculators, electrical toys, video cameras, and small medical devices		Small ICT equipment Comprises landline telephones, mobile phones, and routers, printers, GPS (global positioning systems)	

CHARACTERISTICS OF ELECTRONIC WASTE

Hazardous Components in e-Waste

E-waste consists of components like ferrous, non-ferrous, ceramic, and plastic in large quantities with different size and shapes; some of these materials contain hazardous components which need separate and precise treatment for their removal shown in Table 2 (Cooper, 2000).

Composition of E-Waste

Table 2. Hazardous components in WEE

Materials	Elucidation
Batteries	Batteries consist of heavy metals like Pb (Lead), Hg (Mercury), and Cd (Cadmium).
CRTs (Cathode ray tubes)	Lead is present in the cone glass and Fluorescent coatings inside the glass panel
Switches and its components containing mercury	Thermostats, sensors, and relay Switches generally consist of mercury. In addition, it is also used in mobile phones, data transmission devices, and medical equipment.
Asbestos left-over	Separate treatment is required for treating these left-overs
Cartridge tonners and colors	Toners and Colours consist of Toxic Chemical and Hazardous Substances, and these are abstemiously toxic if acute exposure occurs.
Printed Circuit Boards	In these boards, cadmium and lead are common and present in solder; SMD chip resistors, Semiconductors, and Infrared detectors contain cadmium.
Capacitors containing PCBs (Polychlorinated biphenyl)	Capacitors containing PCBs have to be removed for Safe Obliteration.
LCDs (Liquid Crystal Displays)	LCDs with older versions and Designs consist of Lead and Mercury in larger quantities, But the Surface area greater than 100 Cm^2 is removed from WEE.
Plastics with halogenated flame retardants.	Toxic substances will be released during the combustion and incineration of these types of plastics.
Equipment containing CFC HCFC or HFCs.	The CFC, HFC, or CFCs present in the foam and the refrigerating circuit must be properly extracted and destroyed or recycled.
Gas discharge lamps	Mercury has to be removed before discarding.

With reference to material composition, electronic waste can be defined as a mixture of various metals, particularly copper, aluminum, and steel, which are either attached to or covered with or mixed with various types of plastics and ceramics (Hoffmann, 1992). Precious metals have wide application in the manufacture of electronic appliances (Table 3), serving as contact materials due to their high chemical stability and their good conducting properties. Platinum group metals are used, among other things, in switching contacts (relays, switches) or as sensors to ascertain the electrical measure and as a function of the temperature (Teller, 2006).

Table 3. Composition of metals of different electronic scrap

E- Waste (Scrap)	Mass (%)					Mass (mg kg^{-1})			Reference
	Fe	Cu	Al	Pb	Ni	Ag	Au	Pd	Hageluken,2006
TV boards	28	10	10	1.0	0.3	280	20	10	Hageluken,2006
PC boards	7	20	5	1.5	1	1000	250	110	Hageluken,2006
Mobile phone	5	13	1	0.3	0.1	1380	350	210	Hageluken,2006
Portable audio	23	21	1	0.14	0.03	150	10	4	Hageluken,2006
DVD player	62	5	2	0.3	0.05	115	15	4	Hageluken,2006
Calculator	4	3	5	0.1	0.5	260	50	5	Hageluken,2006
PC mainboard	4.5	14.3	2.8	2.2	1.1	639	566	124	Legarth et al., 1995
Printed circuit boards	12	10	7	1.2	0.85	280	110	-	Zang and Forssberg, 1997
TV scrap (CRTs removed)	-	3.4	1.2	0.2	0.038	20	<10	<10	Cui & Forssberg, 2007
Electronic	8.3	8.5	0.71	3.15	2.0	29	12	-	Ilyas et al., 2007
PC	20	7	14	6	0.85	189	16	3	Patil and Sharma, 2017
Typical electronic	8	20	2	2	2	2000	1000	50	Sum, 1991
E-scrap sample 1	37.4	18.2	19	1.6	-	6	12	-	Busselle et al., 1991
E-scrap sample 2	27.3	16.4	11.0	1.4	-	210	150	20	Busselle et al., 1991
Printed circuit boards	5.3	26.8	1.9	-	0.47	3300	80	-	Theo, 1998
E-scrap (1972 sample)	26.2	18.6	-	-	-	1800	220	30	Bhuie et al., 2004
Nokia Cellular phones	8	19	9	0.9	1	9000	-	-	Bhuie et al., 2004

Note that "-" denotes not reported.

GLOBAL GENERATION OF E-WASTE

The Global development of electronic items has overstated modes of communication, transportation, entertainment, edification, and healthcare resulting in a mammoth amount of E-waste, which is discarded periodically and frequently. The global E-waste Monitor Report 2017 revealed that 44.7 Million Metric Tons of electronic waste had been produced in 2016 (Balde et al., 2017), which might surge beyond 50 MMT in the years to come. Fuelling is the major concern of sustainable management of E-waste, and it contributes a share of 5% of the entire solid waste generated in the world-wide (Kiddee et al., 2013). Developed countries like the United States and the leading countries of Europe are considered to be the main manufacturers and generators of the E-Waste (Robinson, 2009).

E-waste generation statistics change rapidly in the present scenario. Nations like China are leading in the generation of E-Waste (7.2 MMT/year), leaving the United States in second place. The share of another fast-developing country, India, is also increasing and expected to overtake the japan by 2020 for the third rank in global E-waste generation. The top 10 countries generating E-waste against total amount of generation and per capita consumption is as follows: (Figure 1)

During the year 2016, 44.7 MMT of E-waste was generated, out of which, small types of equipment has a major share with 16.8 MMT, large types of equipment with 9.1 MMT, temperature exchange equip-

Figure 1. The amount of E-waste generation: the top 10 countries in the year 2016

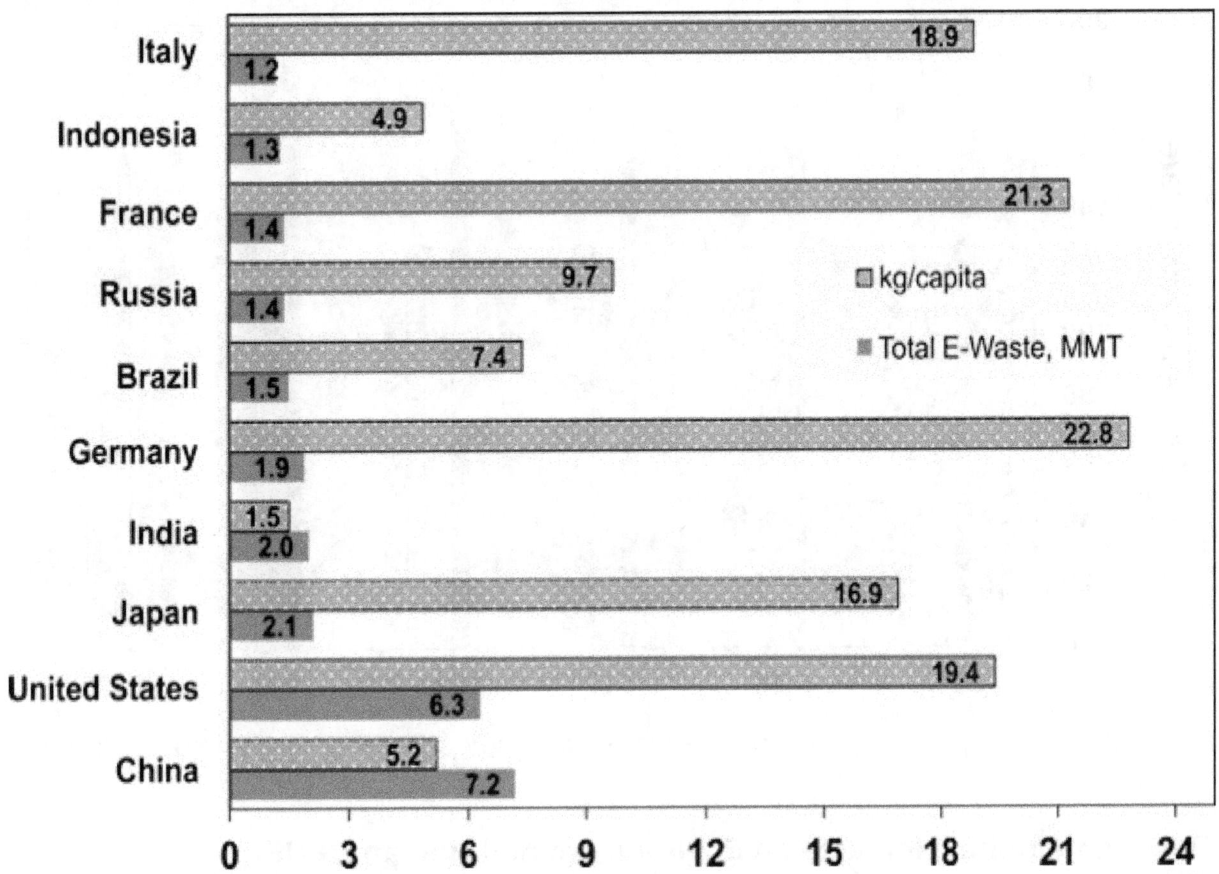

ment like freezing and cooling contributing 7.6 MMT, screens and monitors with 6.6 MMT, small ICT with 3.9 MMT and electric lamps with 0.7 MMT (E-Waste Recycling Facts and Figures, 2018). It was understood that the annual growth rate generated (E-waste) per category differs, but the overall amount will be increased in the coming years (Kumar et al., 2020).

CHALLENGES TO THE ENVIRONMENT

E-Waste is a protuberant environmental management problem that is alarmingly increasing with the pace of the global population and consumption. It has become an environmental curse to be battled. Owing to the stringent rules on the disposal of E-waste, the need for effective management has become mandatory, thus attracting researchers towards finding solutions. Apart from being the biggest contributors to E-waste, nations like China and India are also acting as dumping grounds for world-wide E-waste trade (Pathak et al., 2017; Zeng et al., 2017). Major reasons governing E-waste movement into these countries are deficiency of treatment facilities, valuable secondary materials recycling, and inexpensive operating costs for the transfer of waste (Figure 2 and 3).

Figure 2. Categorical growth rate of E-waste world-wide

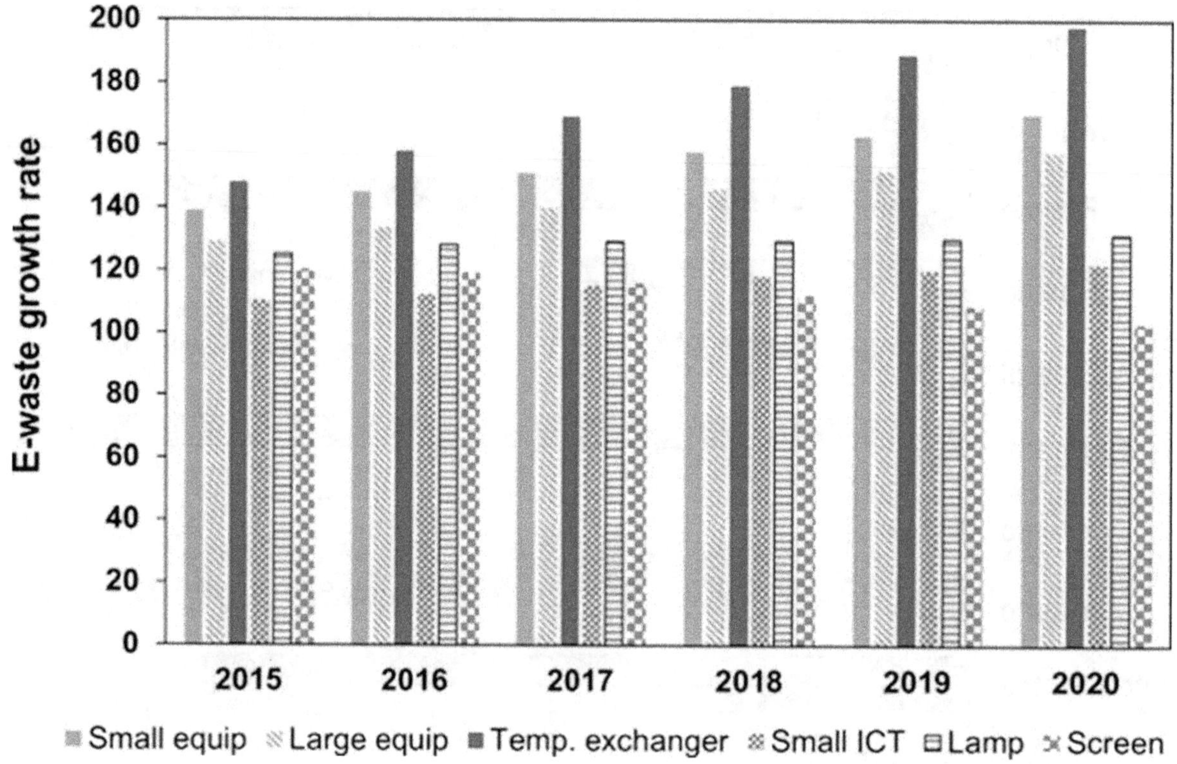

Direct and Indirect Impacts on Environment and Human Health

Human health and the environment are adversely affected by the improper management of E-waste, either directly or indirectly. In order to reduce the cost of recycling, methods that are unsafe and informal are being used for recycling E-waste; these methods being unscientific, have a direct impact on humans and the environment. To illustrate these direct impacts, some cases in the different parts of India where the informal recycling taking place are accounted for here.

A study conducted by Pathak et al., 2017; Pradhan et al., 2017 reported severe environmental problems causing serious threats to human health at Mandoli industrial area in New Delhi. Kumar, 2014 analyzed soil and water around E-waste recycling areas revealed that heavy metal concentrations and hazardous materials are beyond the acceptable levels (Cadmium – 1.29mg/kg; Copper – 115.50mg/kg; Arsenic – 17.08mg/kg; Zinc – 776.84mg/kg; Selenium – 12.67mg/kg and Lead – 2,645.31mg/kg) which is attributed to the leaching of metals from E-waste into soil and groundwater. Additionally, Ha et al., 2009, found that the soils near the recycling areas consist of hazardous substances like organochlorine compounds (7mg/kg soil).

In Moradabad, Uttar Pradesh, severe pollution was observed due to the burning of PCBs (Printed Circuit Board) containing dioxins and furan and practicing improper disposable methods like dumping the burnt ashes in a subsidiary of Ganga, i.e., Ramganga river created severe acute water pollution as these ashes generally contain a high amount of heavy metals and poly-carbonaceous materials (Centre for Science and Environment, 2015).

Figure 3. The generation of (A) computers and (B) mobile phones in India as predicted by the mathematical modeling (with T5% extrapolation)

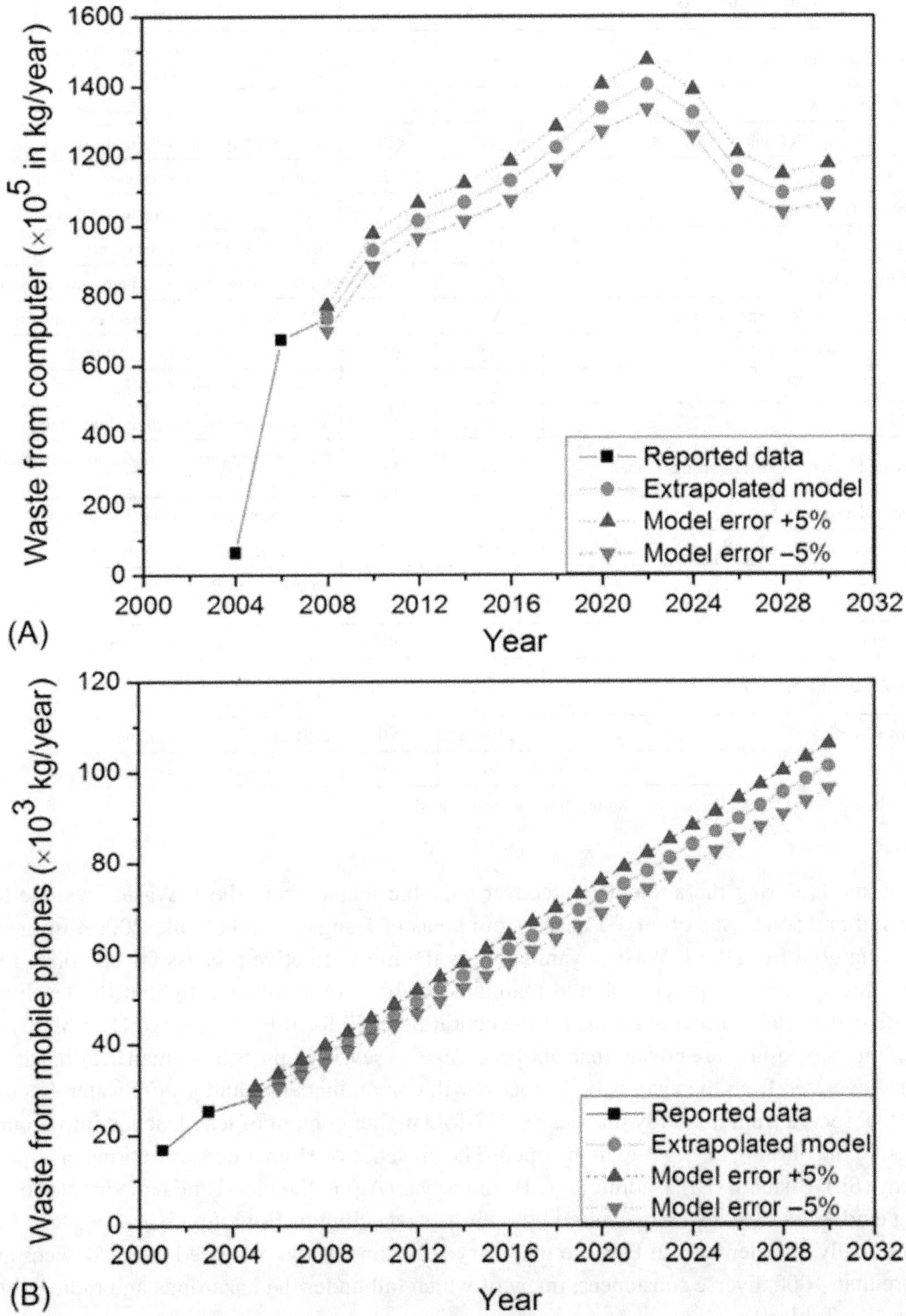

Table 4. Hazardous and radioactive waste category

E-waste Components	Toxic Metals	Limit (ppm)	Disease Caused by the Exposure to Above Permissible Limit
Batteries and Capacitors made up of Ceramics	Ag^x	5.0	It Causes blue pigmentation on the body, damages the liver, kidney, and brain.
Light-emitting devices	As^y	5.0	Skin disease, cancer, and impaired nerve signaling
A fluorescent lamp, CRT gun	Ba^y	<100	brain swelling, weakness in the muscle
Motherboards	Be^y	0.75	Causes lung cancer, skin disease, berylliosis.
PVC Cables, PCB casings	Br^y	0.1	hormonal issues, hearing loss, DNA damages.
Printer ink, toners	Cd^y	1.0	Irreversible effects on the kidney.
Printed circuit boards	CN^y	< 0.5	Cyanide poisoning, >2.5 ppm cause to coma
Computer hosing, cabling, hard discs.	Cr $(VI)^y$	5.0	DNA damage and permanent eye impairment
LCD, switches, backlight bulbs	Hg^y	0.2	Damages brain, kidney, and fetuses
Mobile batteries	Li^x	< 10^a	Diarrhea, vomiting, drowsiness,
CRT, PCB	Ni^x	20.0	allergic reaction, bronchitis, lung cancers
LED lead-acid battery, solder, fluorescent tubes	Pb^z	5.0	Damages the nervous system and reproductive system,
CRT glass and a solder alloy	Sb^y	< 0.5	Carcinogen, diarrhea, and stomach ulcer
A fax machine, photoelectric cells	Se^y	1.0	High concentration causes selenosis
CRT, batteries	Sr^y	1.5	Somatic as well the genetic changes due to this cancer in the bone
Luminous substances	Zn^y	250.0	Nausea, vomiting, pain, cramps, and diarrhea
Cooling units and insulation foam	$CFCs^y$	<1.0	Depletes the ozone layer and leads to the incidence of skin cancer
Transformer, capacitor	$PCBs^y$	5.0	PCB causes cancer in animals
Computer input units	PVC^y	0.03	Hazardous and toxic air contaminants

^{x}Critical, ^{y} hazardous and toxic, ^{z} radioactive waste, ^{a} limit in serum/blood.

Furthermore, during the attempts to recover valuable metals from the E-Waste, cyanide leaching into the soil and water was observed in the slum areas of Bangalore (Ha et al., 2009). In the process of extracting gold from the E-Waste, cyanidation is the most effective process for complete recovery; however, this cyanidation process should maintain alkaline conditions throughout the leaching as its pKa value is 9.5. Maintaining other than these conditions will form HCN gas, which is highly toxic in nature; if these exposures are greater than 500ppb, it causes severe impacts on human health. Discharging effluents generated from the cyanidation process will contaminate soil and groundwater. Sepulveda et al., 2010, observed from his study that nearly 247-fold higher content of lead leached into groundwater attributed to the burning of E-waste in the open. The presence of Higher concentrations of copper (Cu), Antimony (Sb), Bismuth (Bi), Cadmium (Cd), and Silver (Ag) in the blood and hair samples of workers is strong evidence of the adversity caused by improper recycling of E-waste.

Additionally, the elements in E-waste may vary according to the produced goods/devices and contain more than 1000 diverse substances, many of which fall under the hazardous and radioactive waste category (see Table 4).

Figure 4. Impacts of E-Waste on human health and environment

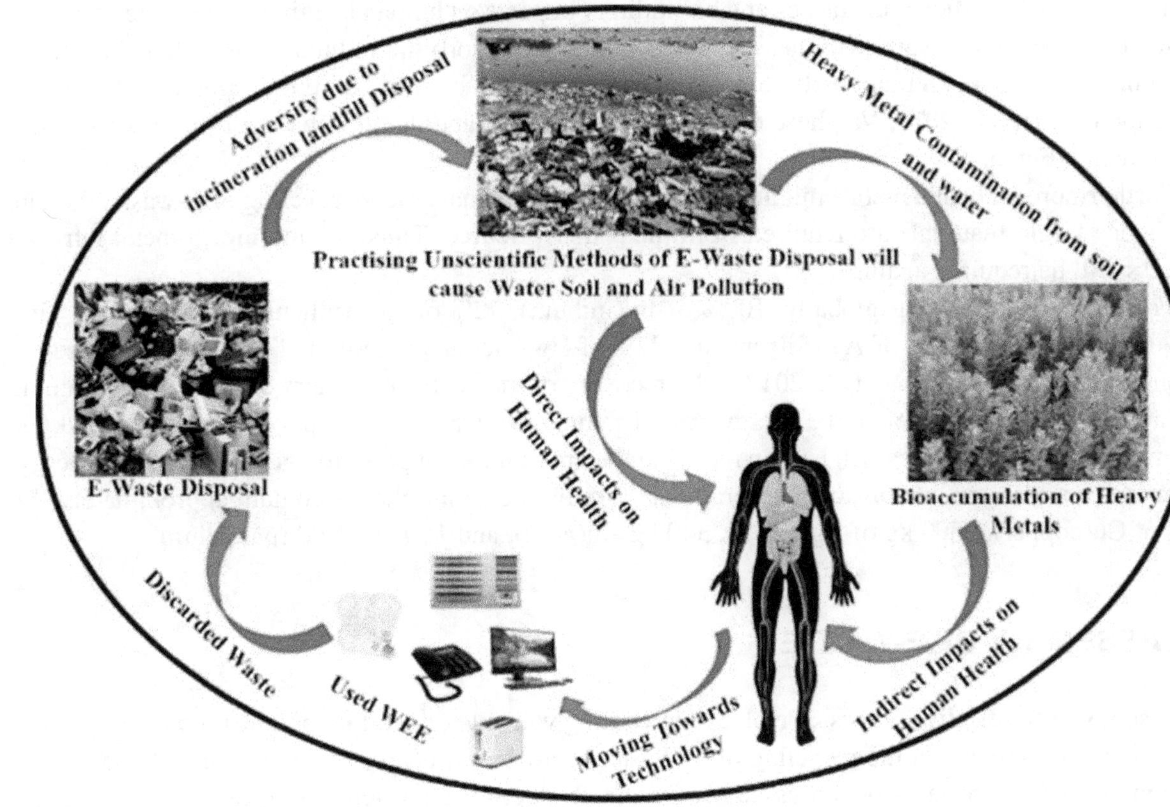

The penalties due to these substances are primarily physical health outcomes, for example, dysfunction of lungs, thyroid, the nervous system, and fertility in mammals (Agrawal et al., 2012; Sepulveda et al., 2010). It has been observed that local refurbishes in India are not practicing eco-friendly disposable techniques as these methods are expensive, and locals follow the rudimentary techniques to reduce the cost of disposing of to gain more profits. The E-waste cycle (Negative Impacts of improper handling of E-waste) shown in Figure 4: highlights their sustainable management. Dumping of E-Waste in landfilling sites releases a huge amount of heavy metals and toxic substances that can eventually drain to pollute the groundwater (Kiddee et al., 2013).

Up to 600 m^3 of the water can be polluted by the cadmium content present in one mobile phone. Awasthi et al., 2017 and Wang et al., 2005 stated that electronic appliances containing PCB and other plastic components are highly inflammable in nature, releasing acute toxic gases like dioxins and furans on burning and these cause severe threat to the ecosystem. The Physico-chemical parameters of the soil are altered by the heavy metals after leaching of these substances from the landfill. During the process of dismantling and shredding, the suspended particles released into the atmosphere causes atmospheric pollution. The atmospheric pollution increases the particulate matter in the ambient air and causes severe adverse respiratory and non-respiratory problems like the enhanced concentration of particulate matter in the ambient air causing breathing difficulties, eye and respiratory irritation, coughing, choking, pneumonitis, tremors, and neuropsychiatric problems for humans.

Air pollution is caused by practicing unscientific incineration practices for E-waste, and this results in the release of hazardous substances and has indirect impacts on human health. For instance, the crops grown in contaminated soils with heavy metals can easily absorb the pollutants. Post-harvesting, the contaminants enter the human body through an indirect route, but they have the same effect as direct pollution (Vara et al., 2019, Varghese et al., 2016). The direct and direct impacts of the E-waste cycle are shown in Fig. 5.

Furthermore, valuable metals might be lost by improper management/recycling of E-waste, therein, and hence virgin materials are required from the primary source. Thus, the mining-to-metal retrieval process will be required again.

For instance, every year globally, 100% of In (indium), 72% of Ru (ruthenium), 50% of Sn (tin), 44% of Cu (Copper), 34% of Ag (Silver), and 22% of Hg (mercury) are mined resulting in severe environmental pollution (Zeng et al., 2017). The recovery of metals from primary sources during mining can approximately save 41% of the energy cost of mining and water consumption up to 391.5 m^3/kg in ore milling, and this process will help in reducing the environmental pollution generated by the energy-intensive process. Remarkably, on an average of 10 million cellular phones contain approximately 160 tons of Cu(copper), 3500 kg of Ag(silver), 340 kg Au(gold), and 150 kg of Pd (palladium).

DISASSEMBLY OF E-WASTE

Disassembly is a selective process, involves segregating valuable or singling of hazardous components, and it is a vital process in the recycling of E-Waste. Moreover, it is a systematic approach that allows the removal of components or sub-assembly from a product for a particular purpose. Researchers are focussed on innovation in disassembly facilities and the disassembly planning process (DPP). The main motto behind the DPP is to advance software and procedures in disassembly strategies.

Gupta et al., 1996 proposed the following phases for the DPP plan:

- In the phase of input and output product analysis, valuable, reusable, and hazardous components are defined after the identification of optimal disassembly and preliminary cost analysis.
- In the following phase (Assembly analysis), component hierarchy, joining elements, and former assembly sequences are analyzed.
- Uncertainty issues analysis: originates from defective parts or joints in the products while upgrading or downgrading the quality of the product for consumer satisfaction.
- Determination of dismantling strategy: this a final phase for deciding whether to use destructive or non-destructive disassembly. Highly flexible tools are required for a process plan and implementation of good disassembly. Even though the electronic equipment used for assembly is well advanced, the automation for disassembly of E-waste is too complicated and not developed.

The automated disassembly is being prevented from becoming commercially successful due to the following reasons:

(1) Similar type of products; (2) the number of products that are similar is too small; (3) un-friendly product design for general disassembly; (4) problem in logistics during returns; and (5) dissimilarities in giving back products that to be disassembled.

Dismantling of electronic equipment in the recycling process in indispensable because; (1) the first priority is to reuse the components; (2) Hazardous components dismantling is essential; (3) High valuable and high-grade materials used in printed circuit boards, cables, and engineering plastics are common and simple to recover. In general, the manual dismantling process is practiced by recycling plants. Ragn-Sells Elektronika°tervinning AB in Sweden is a typical electronics-recycling corporation. Figure 5, present the contemporary disassembly process being used (Cui & Forssberg, 2003). Wide – a variety of tools are involved in the process of dismantling for the removal of hazardous components and for the recovery of valuable and reusable materials.

RECYCLING TECHNOLOGY OF ELECTRONIC WASTE

Magnetic Separation

Figure 5. The recycling process developed by RagneSells Elektronika°tervinning AB in Sweden

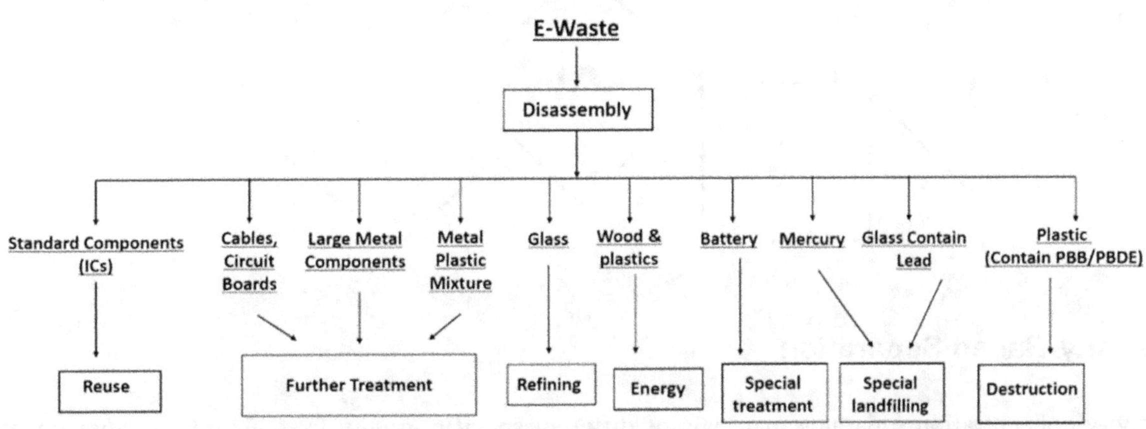

Magnetic separators, i.e., low-intensity drum separators, are mostly used for separating the ferromagnetic metals from non-ferrous metals. The high-intensity magnetic separators design has been modified after the introduction of rare earth alloy permanent magnets capable of providing very high-field strengths and gradients.

Eddy Current Separation

Introducing eddy current separators in recycling industries is one of the most significant development. Figure 6 illustrates the principle of eddy current. Eddy currents can be actuated in an electrically conductive particle by a time-dependent attractive magnetic field (magnet rotor in the figure). Further, the eddy currents will, in return, result in magnetic fields that oppose the inducing fields, giving rise to a so-called Lorentz force, a repulsive force.

Figure 6. Illustration of Eddy current separation: magnet rotor (left) produces a time-dependent magnetic field B, including Eddy currents in a particle (right) resulting in a particle magnetic moment M

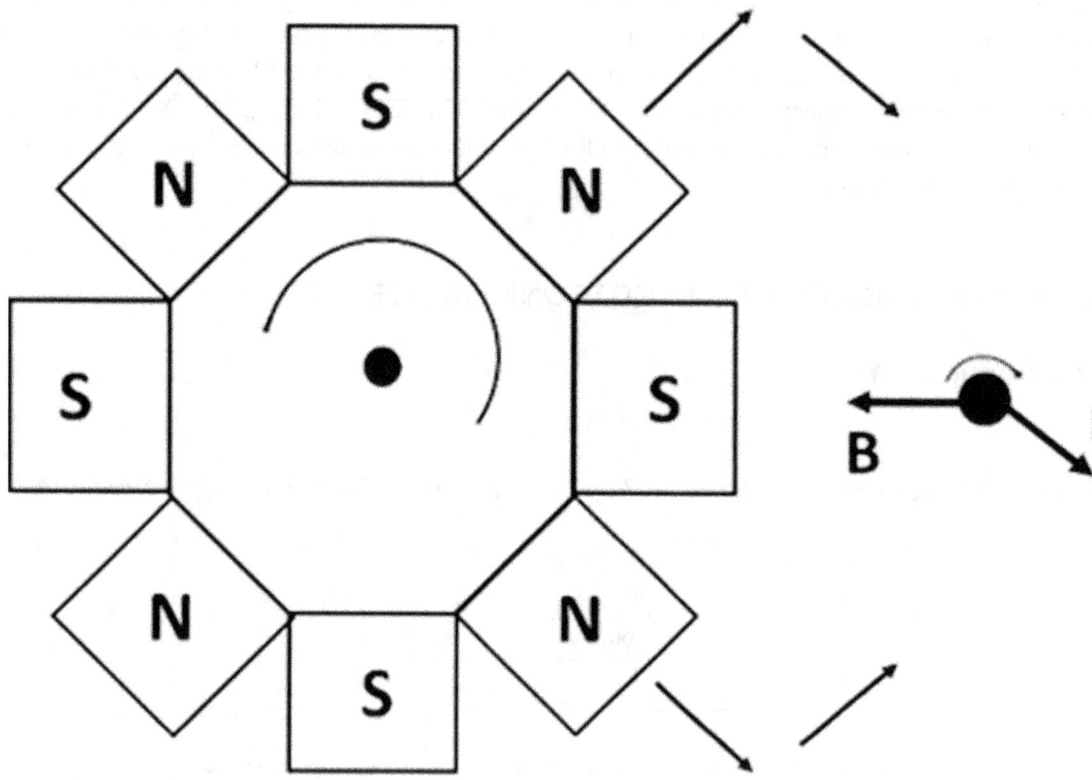

Density-Based Separation

Gravity concentration separates materials of different specific gravity by their relative movement in response to the force of gravity and one or more other forces, the latter often being the resistance to motion offered by a fluid, such as water or air (Wills & Finch, 2015). The motion of a particle in a fluid is dependent not only on the density of the particle but also on its size and shape; large particles will be affected more than smaller ones. In practice, close size control of feeds to gravity processes is required to reduce the effect on the size and make the relative motion of the particle specific gravity-dependent.

Optical Sorting

With the fast development of the Charge- Coupled Device (CCD) sensor, computing, and software technology, optical sorting processes have been developed in both the recycling and mineral processing industry. In addition, recording more and better data with sensors improves the separation performance of automated sorting equipment. The measuring of particle properties such as color, texture, morphology, conductivity, and others allows high-quality sorting of mixed materials into almost pure fractions. Hence, multisensory systems involving the use of two or more different sensors have been developed over the past few years.

Table 5. Metals recovery from various process

Techniques	Metals recovered	Main process features	Main results	References
Noranda process	Cu, Au, Ag, Pt, Pd, Se, Te, Ni	Copper smelter is feeding with 14% of copper for total throughput; Converter upgrading; Metal recovery by electrorefining	Copper and precious metals are recovered in huge amounts	Veldhuizen & Sippel, 1994
Boliden Ronnskar Smelter	Cu, Ag, Au, Pd, Ni, Se, Zn, Pb	Kaldo reactor is feeding with high concentrations of lead; Coper converter up-gradation and refining process	Copper and precious metals are recovered in huge amounts	Theo, 1998; Lehner,2003
Test at Ronnskar Smelter	Copper and precious metals	Computer component scrap feeding to Zinc fuming process in 1:1 ratio with revert slag; Plastics were used as fuel and reducing agents; Copper smelter recovers the precious metals and Copper that are following the copper collector	Complete recovery of precious metals and copper in the Zinc fuming process	Mark & Lehner, 1999
Umicore's Precious metal refining process	Base metals, Precious metals, Platinum group metals and Selenium, Tellurium, Indium	Electrowinning, IsaSmelt copper leaching and precious metal recovery for precious metal operation; E-waste cover up to 10% of the feed; Plastics substitute coke, which acts a reducing and fuel to the IsaSmelt; Off gas emission control system is installed at the IsaSmelt furnace	Special metals like Sb, Bi, Sn, Se, Te, In are recovered along with base metals and precious metals.	Hagelüken, 2006; Hageluken & Art, 2007
Full-scale trial at Umicore's smelter	Metals in electronic scrap	Coke is replaced by plastic rich materials from WEEE, and these substances act as a reducing agent as well as an energy source for the IsaSmelt	Using WEEE (6%) and coke (1%) doesn't negatively affect the metal recovery process (instead of using only coke (4.5%))	Brusselaers et al., 2005
Dunn's patent for gold refining	Gold	Gold scrap subjected to chlorine at 300ºC to 700ºC; Impurity-Metal chlorides are washed with Hydrochloric acid; Silver chloride is dissolved by ammonium hydroxide and nitric acid washing; Samples should contain more than 80% of gold	Gold with 99.9% purity was recovered from gold scraps	Dunn et al., 1991
Day's patent for refractory ceramic precious metals scraps	platinum and palladium	In the plasma arc furnace, scrap was subjected to heat (at least 1400 C); Precious metals in molten metallic phase and collector metals were produced; Ceramic residues went into a slag phase;	Recoveries of 80.3% and 94.2% were obtained for platinum and palladium, respectively	Day, 1984
Aleksandrovich's Patent for recovery of platinum group metal and gold from electronic scraps	Platinum group metal and gold	Fusing together scraps based on chalcogenides of base metals with carbon reducer; After the settling and cooling of melted materials, solidification and separation of solidified products are facilitated by formed phase boundaries	Platinum group metal and gold was recovered	Aleksandrovich et al., 1998

Pyrometallurgical Process

Pyrometallurgical processing is a traditional method for recovering precious and non-ferrous materials from electronic waste. The pyrometallurgical process includes incineration, smelting in the blast furnace, sintering, melting, dross formation, and reactions in a gas phase at high temperatures. In this process, the scrap that is crushed was subjected to heating in a furnace or in a molten bath to remove plastics and the refractory oxides from the slag phase together with some metals.

The above table gives a summary of typical pyrometallurgical methods for the recovery of metals from electronic waste. The traditional technology of pyrometallurgy has been used for the recovery of copper and precious metals from electronic waste equipment for years. However, most methods involving the pyrometallurgical processing of electronic waste has the following limitations (Hageluken, 2006; Sum, 1991):

- Integrated smelters cannot recover aluminum and iron as metals; they are transferred into the slag component. Unfortunately, aluminum has an influence on the slag properties that are undesirable in most cases.
- The presence of halogenated flame retardants in the smelter feed can lead to the formation of dioxins unless special installations and measures are present. Traditional smelters designed for the treatment of mining concentrates or simple copper scrap encounter some challenges for electronic waste treatment. However, state-of-the-art smelters are highly dependent on investments.
- Ceramic components and glass in the e-waste increase the amount of slag from blast furnaces, which results in a loss of precious metals and base metals from the scrap.
- Energy recovery and the use of organic constituents as reducing agents are beginning to be used only now.
- Only partial separation of metals can be achieved using pyrometallurgy, resulting in a limited upgrading of the metal value. Subsequent hydrometallurgical techniques and/or electrochemical processing are therefore necessary.
- Precious metals take a long time to separate in pyrometallurgical processes and are thus only obtained at the very end of the process.

Hydrometallurgical Process

Precious metals contribute the highest value to electronic scrap. From an economic point of view, the recovery of precious metals from e-waste is most attractive. In the past two decades, the most active research area on the recovery of metals from electronic scraps is recovering precious metals by hydrometallurgical techniques (Quinet et al., 2005; Macaskie et al., 2007; Ogata & Nakano, 2005). Compared with pyrometallurgical processing, the hydrometallurgical method is more exact, more predictable, and more easily controlled. The main steps in hydrometallurgical processing consist of a series of acid or caustic leaches of solid material. The solutions are then subjected to separation and purification procedures such as precipitation of impurities, solvent extraction, adsorption, and ion-exchange to isolate and concentrate the metals of interest. Consequently, the solutions are treated by electrorefining process, chemical reduction, or crystallization for metal recovery (Safarzadeh et al., 2007; Ritcey, 2006; Tavlarides et al., 1987).

A summary of new developments in the recovery of metals from electronic waste by hydrometallurgical techniques is listed in Table 6. It can be seen that most of the hydrometallurgical techniques for the

Table 6. Recovery of Metals from Electronic Waste by Hydrometallurgical Techniques (most practiced)

Metals recovered	Main process features	Main product	References
Au	Computer chips were treated in nitric acid to dissolve base metals; the residue was leached with aqua regia; ferrous sulfate precipitation was used for gold recovery	Gold flakes	Sheng & Etsell, 2007
Au and Ag	E-waste with size - 0.5 mm was treated with a combination of KI and I2 or NaCl and bleaching powder; solvent extraction was used for gold and silver recovery	Au and Ag	Shibata & Matsumoto, 1999
Ni	Leaching of nickel from waste multi-layer ceramic capacitors was performed by using 1M HNO3 at 90_C, 90-min reaction time, and 5 g/l pulp density	Ni in solution	Kim et al., 2007
Au (98%), Pd (96%), Pt (92%), Ag (84%)	H2SO4 and MgCl2 for dissolution of base metals; HCl and bromide ions were used to dissolve precious metals; cementation of gold by zinc powder	Au and platinum group metals powders	Kogan, 2006
Cu (98%)	Copper was dissolved by sulphuric acid or aqua regia; electrowinning was performed for copper recovery	Cu	Veit et al., 2006
Cu, Ag (93%), Pd (99%), Au (95%)	Sulphuric acid leaching of copper, chloride leaching of palladium, thiourea or cyanide leaching of gold and silver, activated carbon adsorption of gold, silver, and palladium	AgCl, Cu, Pd, Au	Quinet et al., 2005
92% for Au, Ag, Pd	HCl or H2SO4 for dissolution of base metals, leaching of silver by HNO3, leaching of gold and palladium by HCl and NaClO3; precipitation of Au by FeCl2	Gold sponge	Zhou et al., 2005
Au	E-scrap was treated with a leaching solution based on NaCl, CuCO3, and HCl	Gold residue	Olper et al., 2004
Sn, Pb	The solder was dissolved with a solution comprising Ti(IV) and an acid. Tin and lead were recovered by electrowinning	Sn and Pb	Gibson et al., 2004
Cu, Pb, and Sn	HNO3 leaching of PCBs, electrodeposition recovery of base metals	Cu, Pb, and Sn	Mecucci & Scott, 2002
Au	Thermal treatment, HNO3 leaching, and aqua regia leaching for gold dissolution, solvent extraction of gold by diethyl malonate, ferrous sulfate solution was used for gold precipitation	Metallic gold	Chmielewski et al., 1997
Au and Ni	Leaching of base metals by sulphuric acid and oxidant (ferric sulfate) and aqua regia leaching of precious metals	Ni and Au in solution	Zakrewski et al., 1992

recovery of metals are involved in acid leaching and/or halide leaching. This is because acid leaching is the most feasible approach for the removal of base metals resulting in the exposure of precious metals surfaces. However, to develop an environment-friendly technique for the recovery of precious metals from e-waste, more attention should be paid to evaluate the environmental impact of the various techniques.

REGULATIONS DEALING WITH ELECTRONIC WASTE

It is a well-known fact that the environmental quality is dwindling at a very fast pace due to such informal, unplanned, and untrained treatment of E-wastes. But one significant cause for such environmental degradation due to E-wastes is the inability of the law to deter violators. In many countries, there are now legislations and statutes in place specifically to deter such crimes.[37] In India also many environmental legislations and the rules are there, which are dealing with E-waste either directly or indirectly. Such legislations have, inopportunely, not resulted in preventing environmental dilapidation as penal provisions in environmental laws dealing with electronic wastes are weak, lenient, hard to impose, and unlikely to affect 'business as usual.' The situation is aggravated by other problems like slow justice delivery system, poor monitoring and enforcement capacity of regulators, and lack of comprehensive databases to evidence violations, among others.

There are various legislations/rules dealing with E-wastes in environmental laws either directly and indirectly, to name a few are:

1. Electronic waste (management) Amendment rules, 2018.
2. Hazardous and Other Wastes (Management and Transboundary Movement) Rules, 2016.
3. The Environment (Protection) Act, 1986
4. The Air (Prevention and Control of Pollution) Act, 1981
5. The Water (Prevention and Control of Pollution) Act, 1974

Out of these legislations and legal provisions, only a few legislations provide for penal provisions to deal with the menace produced by improper disposal of electronic waste. Improper disposal like burning and segregating electronic waste in open areas and not in a manner in which environmentally sound causes leakage in the environment, which results in pollution to the air, water, and soil.

STRATEGIES TOOK FOR THE MANAGEMENT OF E-WASTE

Inventory Management: Proper control over the materials used in the manufacturing process is an important way to reduce waste generation. By reducing both the number of hazardous materials used in the process, and the amount of excess raw materials in stock, the quantity of waste generated can be reduced. This can be done in two ways, i.e., establishing material-purchase review and control procedures and inventory tracking system. Developing review procedures for all material purchased is the first step in establishing an inventory management program. Procedures should require that all materials be approved prior to purchase. In the approval process, all production materials are evaluated to examine if they contain hazardous constituents and whether alternative

Non-hazardous materials are available. Another inventory management procedure for waste reduction is to ensure that only the needed quantity of material is ordered. This will require the establishment of a strict inventory tracking system. Purchase procedures must be implemented which ensure that materials are ordered only on an as-needed basis and that only the amount needed for a specific period of time is ordered (Zhang et al., 2020).

Production-Process Modification: Changes can be made in the production process, which will reduce waste generation. This reduction can be accomplished by changing the materials used to make the product or by the more efficient use of input materials in the production process or both.

Waste reduction can be accomplished by the more efficient use of input 0020wq

1. Material change and

2. Process-equipment modification.
    a. **Material Change:** Hazardous material used in the production process may be replaced by the less hazardous material. For example, a circuit board manufacturer can replace the solvent-based product with water-based flux and simultaneously replace the solvent vapor degreaser with the detergent part washer.
    b. **Process Equipment Modification:** Modifying existing equipment to take advantage of better production techniques can significantly reduce waste generation. For example, in many electronic manufacturing operations, which involve porting a product, such as electroplating or painting, the chemical is used to strip off coating from rejected products so that they can be recoated.

Improvements in the operation and maintenance of process equipment can result in significant waste reduction. This can be accomplished by reviewing current operational procedures and examination of the production process for ways to improve its efficiency. Instituting standard operation procedures can optimize the use of raw materials in the production process and reduce the potential for materials to be lost through leaks and spills. A strict maintenance program, which stresses corrective maintenance, can reduce waste generation caused by equipment failure. An employee-training program is a key element of any waste reduction program. Training should include correct operating and handling procedures, proper equipment uses, recommended maintenance and inspection schedules, correct process control specifications, and proper management of waste materials (Esenduran et al., 2019).

Hazardous materials used in either a product formulation or a production process may be replaced with less hazardous or non-hazardous material. Implementation of this waste reduction technique may require only some minor process adjustments, or it may require extensive new process equipment (Esenduran et al., 2019).

For example, a circuit board manufacturer can replace the solvent-based product with water-based flux and simultaneously replace the solvent vapor degreaser with a detergent parts washer. Installing more efficient process equipment or modifying existing equipment to take advantage of better production techniques can significantly reduce waste generation. New or updated equipment can use process materials more efficiently, producing less waste.

Modifying existing process equipment can be a very cost-effective method of reducing waste generation. In many cases, the modification can just be relatively simple changes in the way the materials are handled within the process to ensure that they are not wasted. For example, in many electronic manufacturing operations, which involve coating a product, such as electroplating or painting, chemicals are used to strip off coating from rejected products so that they can be recoated. These chemicals, which can include acids, caustics, cyanides, etc., are often hazardous waste and must be properly managed. By reducing the number of parts that have to be reworked, the quantity of waste can be significantly reduced.

- **Volume Reduction:** Volume reduction includes those techniques that remove the hazardous portion of waste from a non-hazardous portion. These techniques are used to reduce the volume, and thus the cost of disposing of waste material. The techniques that can be used to reduce waste-stream volume can be divided into two general categories:

Source segregation and Waste concentration.

- **Segregation of waste** is, in many cases, a simple and economical technique for waste reduction. Wastes containing different types of metals can be treated separately so that the metal value in the sludge can be recovered.
- **The concentration of a waste** stream may increase the likelihood that the material can be recycled or reused. Methods include gravity and filtration, ultra-filtration, reverse osmosis, freeze vaporization, etc.
- **Recovery and Reuse:** The technique could eliminate waste disposal costs, reduce raw material costs, and provide income from a saleable waste. Waste can be recovered on-site, or at an off-site recovery facility, or through the inter-industry exchange. Several physical and chemical techniques are available to reclaim a waste material such as reverse osmosis, electrolysis, condensation, electrolytic recovery, filtration, centrifugation, etc. for example, a printed circuit board manufacturer can use electrolytic recovery to reclaim metals from copper and tin-lead plating bath (Islam & Huda, 2020; Song et al., 2019).

However, recycling of hazardous products has little environmental benefit if it simply moves the hazards into secondary products that eventually have to be disposed of. Unless the goal is to redesign the product to use non-hazardous materials, such recycling is a false solution (Figure 7).

Figure 7. Sustainable life cycle of E-waste

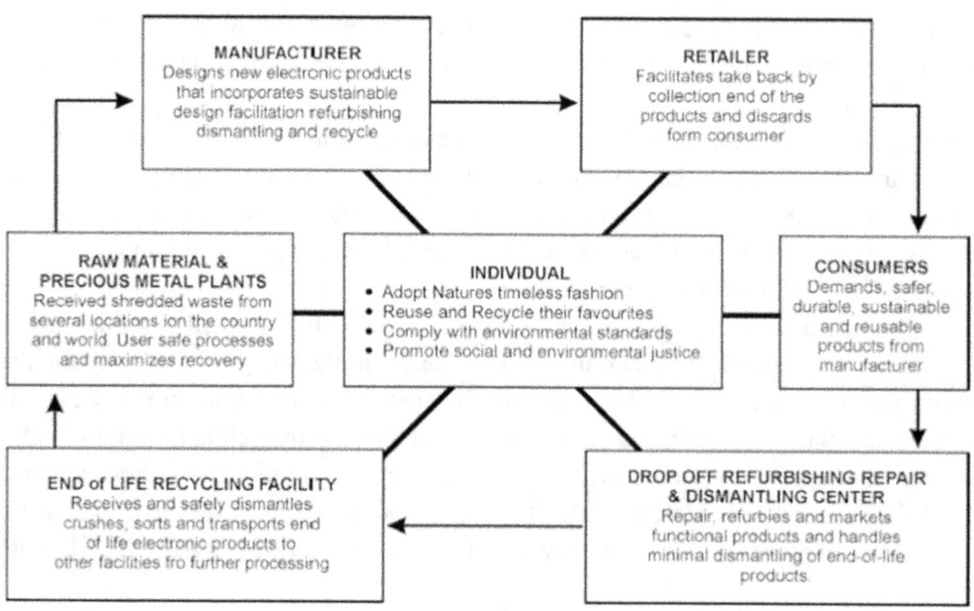

CONCLUSION

It is evident from the above discussion, pollution from improper handling of E-waste cannot be ignored owing to the magnitude of health and environmental impacts. To escape from hazardous effects, it is vital to appropriately manage E-waste. Inadequate infrastructure and insufficient budget are two issues that restrict E-waste management. This is high time to follow scientific and safe methods for recycling E-waste. Sustainable E-waste management methods discussed in the article, along with their challenges, might prove to be useful in implementing strategies. Integrating these solutions with existing waste management systems might be a possible way to solve problems related to E-waste management.

REFERENCES

Agrawal, A., Pathak, P., Mishra, D., & Sahu, K. K. (2012). Solvent mediated interactions for the selective recovery of Cadmium from Ni–Cd battery waste. *Journal of Molecular Liquids*, *173*, 77–84. doi:10.1016/j.molliq.2012.06.017

Aleksandrovich, S., Nicolaevich, E., & Ivanovich, E. (1998). *Method of processing of products based on ahalcogenides of base metals containing metals of platinum group and gold*. Russian Patent, RU2112064, C22B, 11(02).

Association of Plastics Manufacturers in Europe (APME). (2004). 1995. Plastics—a material of choice for the electrical and electronic industry—plastics consumption and recovery. In *Western Europe* (Vol. 1). APME.

Awasthi, A. K., Zeng, X., & Li, J. (2016). Environmental pollution of electronic waste recycling in India: A critical review. *Environmental Pollution*, *211*, 259–270. doi:10.1016/j.envpol.2015.11.027 PMID:26774773

Balde, C. P., Forti, V., Gray, V., Kuehr, R., & Stegmann, P. (2017). *The Global E-Waste Monitor—2017. United Nations University (UNU), International Telecommunication Union (ITU) & International Solid Waste Association*. ISWA.

Balde, C. P., Kuehr, R., Blumenthal, K., Fondeur Gill, S., Kern, M., Micheli, P., . . . Huisman, J. (2015). E-waste statistics-Guidelines on classification, reporting and indicators. United Nations University, IAS-SCYCLE.

Bhuie, A. K., Ogunseitan, O. A., Saphores, J. D., & Shapiro, A. A. (2004, May). Environmental and economic trade-offs in consumer electronic products recycling: a case study of cell phones and computers. In *IEEE International Symposium on Electronics and the Environment, 2004. Conference Record. 2004* (pp. 74-79). IEEE. 10.1109/ISEE.2004.1299691

Bisschop, L. (2016). How e-waste challenges environmental governance. In *Hazardous Waste and Pollution* (pp. 27–43). Springer. doi:10.1007/978-3-319-18081-6_3

Brusselaers, J., Hageluken, C., Mark, F., Mayne, N., & Tange, L. (2005). An eco-efficient solution for plastics-metals-mixtures from electronic waste: the integrated metals smelter. *5th Identiplast*.

Busselle, L. D., Moore, T. A., Shoemaker, J. M., & Allred, R. E. (1999, May). Separation processes and economic evaluation of tertiary recycling of electronic scrap. In *Proceedings of the 1999 IEEE International Symposium on Electronics and the Environment* (Cat. No. 99CH36357) (pp. 192-197). IEEE. 10.1109/ISEE.1999.765874

Centre for Science and Environment. (2015). *Recommendations to Address the Issues of Informal Sector Involved in E-Waste Handling*. http://www.cseindia.org/ userfiles/moradabad-e-waste.pdf

Chmielewski, A. G., Urbañski, T. S., & Migdał, W. (1997). Separation technologies for metals recovery from industrial wastes. *Hydrometallurgy*, *45*(3), 333–344. doi:10.1016/S0304-386X(96)00090-4

Cooper, T. (2000). WEEE, WEEE, WEEE, WEEE, all the way home? An evaluation of proposed electrical and electronic waste legislation. *European Environment*, *10*(3), 121–130. doi:10.1002/1099-0976(200005/06)10:3<121::AID-EET226>3.0.CO;2-N

Cui, J., & Forssberg, E. (2003). Mechanical recycling of waste electric and electronic equipment: A review. *Journal of Hazardous Materials*, *99*(3), 243–263. doi:10.1016/S0304-3894(03)00061-X PMID:12758010

Cui, J., & Forssberg, E. (2007). Characterization of shredded television scrap and implications for materials recovery. *Waste Management (New York, N.Y.)*, *27*(3), 415–424. doi:10.1016/j.wasman.2006.02.003 PMID:16624540

Day, J. G. (1984). *U.S. Patent No. 4,427,442*. Washington, DC: U.S. Patent and Trademark Office.

Dunn, W. E., Jr., Carda, D. D., & Storbeck, T. A. (1991). U.*S. Patent No. 5,004,500*. Washington, DC: U.S. Patent and Trademark Office.

E-Waste Rajyasabha Report. (2011). https://rajyasabha.nic.in/rsnew/publication_electronic/EWaste_in_india.pdf/

E-Waste Recycling Facts and Figures. (2018). https://www.thebalance.com/e-waste-recycling-factsand-figures-2878189

Esenduran, G., Atasu, A., & Van Wassenhove, L. N. (2019). Valuable e-waste: Implications for extended producer responsibility. *IISE Transactions*, *51*(4), 382–396. doi:10.1080/24725854.2018.1515515

EU RoHS Directive 2002/95/EC. (2003). https://eur-lex.europa.eu/homepage.html

Gibson, R. W., Goodman, P. D., Holt, L., Dalrymple, I. M., & Fray, D. J. (2003). *U.S. Patent No. 6,641,712*. Washington, DC: U.S. Patent and Trademark Office.

Gupta, S. M., & McLean, C. R. (1996). Disassembly of products. *Computers & Industrial Engineering*, *31*(1-2), 225–228. doi:10.1016/0360-8352(96)00146-5

Ha, N. N., Agusa, T., Ramu, K., Tu, N. P. C., Murata, S., Bulbule, K. A., Parthasaraty, P., Takahashi, S., Subramanian, A., & Tanabe, S. (2009). Contamination by trace elements at e-waste recycling sites in Bangalore, India. *Chemosphere*, *76*(1), 9–15. doi:10.1016/j.chemosphere.2009.02.056 PMID:19345395

Hageluken, C. (2006, May). Improving metal returns and eco-efficiency in electronics recycling-a holistic approach for interface optimisation between pre-processing and integrated metals smelting and refining. In *Proceedings of the 2006 IEEE International Symposium on Electronics and the Environment*, 2006 (pp. 218-223). IEEE. 10.1109/ISEE.2006.1650064

Hageluken, C., & Art, S. (2007). *Recycling of e-scrap in a global environment: opportunities and challenges. Tackling e-waste towards efficient management techniques*. TERI Press.

Hoffmann, J. E. (1992). Recovering precious metals from electronic scrap. *JOM*, *44*(7), 43–48. doi:10.1007/BF03222275

Ilyas, S., Anwar, M. A., Niazi, S. B., & Ghauri, M. A. (2007). Bioleaching of metals from electronic scrap by moderately thermophilic acidophilic bacteria. *Hydrometallurgy*, *88*(1-4), 180–188. doi:10.1016/j.hydromet.2007.04.007

Islam, M. T., & Huda, N. (2020). Assessing the recycling potential of "unregulated" e-waste in Australia. *Resources, Conservation and Recycling*, *152*, 104526. doi:10.1016/j.resconrec.2019.104526

Kiddee, P., Naidu, R., & Wong, M. H. (2013). Electronic waste management approaches: An overview. *Waste Management, 33*(5), 1237-1250.

Kim, E. Y., Lee, J. C., Kim, B. S., Kim, M. S., & Jeong, J. (2007). Leaching behavior of nickel from waste multi-layer ceramic capacitors. *Hydrometallurgy*, *86*(1-2), 89–95. doi:10.1016/j.hydromet.2006.11.007

Kogan, V. (2006). *Process for the recovery of precious metals from electronic scrap by hydrometallurgical technique*. International Patent, WO/2006/013568 (C22B 11/00), WIP Organization.

Kumar, A. (2019). Exploring young adults'e-waste recycling behaviour using an extended theory of planned behaviour model: A cross-cultural study. *Resources, Conservation and Recycling*, *141*, 378–389. doi:10.1016/j.resconrec.2018.10.013

Lakshmi, S., & Raj, A. (2017). A Review Study of E-Waste Management in India. *Asian Journal of Applied Science and Technology*, *1*, 33–36.

Legarth, J. B., Alting, L., & Baldo, G. L. (1995, May). Sustainability issues in circuit board recycling. In *Proceedings of the 1995 IEEE International Symposium on Electronics and the Environment ISEE* (Cat. No. 95CH35718) (pp. 126-131). IEEE. 10.1109/ISEE.1995.514963

Lehner, T. (2003, May). E&HS aspects on metal recovery from electronic scrap. In *IEEE International Symposium on Electronics and the Environment*, 2003 (pp. 318-322). IEEE. 10.1109/ISEE.2003.1208097

Li, H., Oraby, E., & Eksteen, J. (2020). Extraction of copper and the co-leaching behaviour of other metals from waste printed circuit boards using alkaline glycine solutions. *Resources, Conservation and Recycling*, *154*, 104624. doi:10.1016/j.resconrec.2019.104624

Macaskie, L. E., Creamer, N. J., Essa, A. M. M., & Brown, N. L. (2007). A new approach for the recovery of precious metals from solution and from leachates derived from electronic scrap. *Biotechnology and Bioengineering*, *96*(4), 631–639. doi:10.1002/bit.21108 PMID:16917944

Mark, F. E., & Lehner, T. (1999). *Plastics recovery from waste electrical and electronic equipment in non-ferrous processes*. Academic Press.

Masud, M. H., Akram, W., Ahmed, A., Ananno, A. A., Mourshed, M., Hasan, M., & Joardder, M. U. H. (2019). Towards the effective E-waste management in Bangladesh: A review. *Environmental Science and Pollution Research International*, *26*(2), 1250–1276. doi:10.100711356-018-3626-2 PMID:30456610

Mecucci, A., & Scott, K. (2002). Leaching and electrochemical recovery of copper, lead and tin from scrap printed circuit boards. Journal of Chemical Technology & Biotechnology: International Research in Process. *Environmental & Clean Technology*, *77*(4), 449–457.

Ogata, T., & Nakano, Y. (2005). Mechanisms of gold recovery from aqueous solutions using a novel tannin gel adsorbent synthesized from natural condensed tannin. *Water Research*, *39*(18), 4281–4286. doi:10.1016/j.watres.2005.06.036 PMID:16221485

Olper, M., Maccagni, M., & Cossali, S. (2004). *U.S. Patent Application No. 10/798,143*. Washington, DC: US Patent Office.

Pathak, P., Srivastava, R. R., & Ojasvi. (2017). Assessment of legislation and practices for the sustainable management of waste electrical and electronic equipment in India. *Renewable & Sustainable Energy Reviews*, *78*, 220–232. doi:10.1016/j.rser.2017.04.062

Patil, S. A., & Sharma, N. M. (2015). Electronic Waste-A Literature Review. *International Journal of Science and Research*.

Pradhan, J. K., & Kumar, S. (2014). Informal e-waste recycling: Environmental risk assessment of heavy metal contamination in Mandoli industrial area, Delhi, India. *Environmental Science and Pollution Research International*, *21*(13), 7913–7928. doi:10.100711356-014-2713-2 PMID:24652574

Puckett, J., Byster, L., & Westervelt, S. (2002). *Exporting Harm: The High-Tech Trashing of Asia, The Basel Action Network (BAN) and Silicon Valley Toxics Coalition (SVTC), 2002*. https://ban. org/E-waste/technotrashfinalcomp.pdf

Quinet, P., Proost, J., & Van Lierde, A. (2005). Recovery of precious metals from electronic scrap by hydrometallurgical processing routes. *Mining, Metallurgy & Exploration*, *22*(1), 17–22. doi:10.1007/BF03403191

Ritcey, G. M. (2006). Solvent extraction in hydrometallurgy: Present and future. *Tsinghua Science and Technology*, *11*(2), 137–152. doi:10.1016/S1007-0214(06)70168-7

Robinson, B. H. (2009). E-waste: An assessment of global production and environmental impacts. *The Science of the Total Environment*, *408*(2), 183–191. doi:10.1016/j.scitotenv.2009.09.044 PMID:19846207

Rodrigues, A. C., Boscov, M. E., & Günther, W. M. (2020). Domestic flow of e-waste in São Paulo, Brazil: Characterization to support public policies. *Waste Management (New York, N.Y.)*, *102*, 474–485. doi:10.1016/j.wasman.2019.10.052 PMID:31751920

Safarzadeh, M. S., Bafghi, M. S., Moradkhani, D., & Ilkhchi, M. O. (2007). A review on hydrometallurgical extraction and recovery of cadmium from various resources. *Minerals Engineering*, *20*(3), 211–220. doi:10.1016/j.mineng.2006.07.001

Sepulveda, A., Schluep, M., Renaud, F. G., Streicher, M., Kuehr, R., Christian, H., & Gerecke, A. C. (2010). A review of the environmental fate and effects of hazardous substances released from electrical and electronic equipment during recycling: Examples from China and India. *Environmental Impact Assessment Review*, *30*(1), 28–41. doi:10.1016/j.eiar.2009.04.001

Sheng, P. P., & Etsell, T. H. (2007). Recovery of gold from computer circuit board scrap using aqua regia. *Waste Management & Research*, *25*(4), 380–383. doi:10.1177/0734242X07076946 PMID:17874665

Shibata, J., & Matsumoto, S. (1999). *Development of environmentally friendly leaching and recovery process of gold and silver from wasted electronic parts.* Available from: http://www. environmentalcenter. com/articles/article320/article320. htm

Sinha-Khetriwal, D. (2002). *The management of electronic waste: A comparative study on India and Switzerland* (MS thesis). St. Gallen, Switzerland: University of St. Gallen.

Song, S., Duan, Y., Zhang, T., Zhang, B., Zhao, Z., Bai, X., Xie, L., He, Y., Ouyang, J., Huang, X., & Sun, H. (2019). Serum concentrations of bisphenol A and its alternatives in elderly population living around e-waste recycling facilities in China: Associations with fasting blood glucose. *Ecotoxicology and Environmental Safety*, *169*, 822–828. doi:10.1016/j.ecoenv.2018.11.101 PMID:30597781

Sum, E. Y. (1991). The recovery of metals from electronic scrap. *JOM*, *43*(4), 53–61. doi:10.1007/BF03220549

Tavlarides, L. L., Bae, J. H., & Lee, C. K. (1987). Solvent extraction, membranes, and ion exchange in hydrometallurgical dilute metals separation. *Separation Science and Technology*, *22*(2-3), 581–617. doi:10.1080/01496398708068970

Teller, M. (2006). Recycling of electronic waste material. In *Sustainable metals management* (pp. 563–576). Springer. doi:10.1007/1-4020-4539-5_23

Theo, L. (1998, May). Integrated recycling of non-ferrous metals at Boliden Ltd. In *Proceedings of the 1998 IEEE International Symposium on Electronics and the Environment. ISEE-1998* (Cat. No. 98CH36145) (pp. 42-47). IEEE.

Vara, S., Karnena, M. K., Dwarapureddi, B. K., & Chintalapudi, B. (2019). Will Single Use Products Lead to Sustainability? *International Journal of Social Ecology and Sustainable Development*, *10*(2), 37–52. doi:10.4018/IJSESD.2019040104

Varghese, J., Saritha, V., Bhavya Kavitha, D., Yasaswitha, O., & Karnena, M. K. (2016). Journal of Global Biosciences. *Journal of Global Biosciences*, *5*(3), 3776–3784.

Veit, H. M., Bernardes, A. M., Ferreira, J. Z., Tenório, J. A. S., & de Fraga Malfatti, C. (2006). Recovery of copper from printed circuit boards scraps by mechanical processing and electrometallurgy. *Journal of Hazardous Materials*, *137*(3), 1704–1709. doi:10.1016/j.jhazmat.2006.05.010 PMID:16757116

Veldhuizen, H., & Sippel, B. (1994). Mining discarded electronics. *Ind. Environ.(Switzerland)*, *17*(3), 7–11.

Wang, D., Cai, Z., Jiang, G., Leung, A., Wong, M. H., & Wong, W. K. (2005). Determination of polybrominated diphenyl ethers in soil and sediment from an electronic waste recycling facility. *Chemosphere*, *60*(6), 810–816. doi:10.1016/j.chemosphere.2005.04.025 PMID:15946727

Wills, B. A., & Finch, J. (2015). *Wills' mineral processing technology: an introduction to the practical aspects of ore treatment and mineral recovery*. Butterworth-Heinemann.

Zakrewski, J., Chamer, R., & Koscielniak, A. (1992). *Method for recovering gold and nickel from electronic scrap and waste*. Polish Patent, PL158889B (C22B7/00).

Zeng, X., Duan, H., Wang, F., & Li, J. (2017). Examining environmental management of e-waste: China's experience and lessons. *Renewable & Sustainable Energy Reviews*, *72*, 1076–1082. doi:10.1016/j.rser.2016.10.015

Zhang, B., He, Y., Zhu, H., Huang, X., Bai, X., Kannan, K., & Zhang, T. (2020). Concentrations of bisphenol A and its alternatives in paired maternal–fetal urine, serum and amniotic fluid from an e-waste dismantling area in China. *Environment International*, *136*, 105407. doi:10.1016/j.envint.2019.105407 PMID:31955035

Zhang, S., & Forssberg, E. (1997). Electronic scrap characterization for materials recycling. *Journal of Waste Management and Resource Recovery*, *3*, 157–168.

Zhou, P., Zheng, Z., & Tie, J. (2005). *Technological process for extracting gold, silver and palladium from electronic industry waste*. Chinese Patent, CN1603432A (C22B 11/00).

Chapter 6
Design and Development of Small-Scale Oil Palm Waste Shredder for Achieving Environmental Sustainability

Man Djun Lee
https://orcid.org/0000-0002-1250-9459
University College of Technology, Sarawak, Malaysia

Pui San Lee
https://orcid.org/0000-0001-9631-0983
University College of Technology, Sarawak, Malaysia

ABSTRACT

This chapter presents the design and development of an economically and technically feasible small-scale oil palm waste shredder to contribuite to oil palm waste management. Most available oil palm waste shredders are unaffordable due to high initial investment cost, large, heavy and unportable, which have been the drawbacks for small and medium enterprises to participate in the industry that relates with oil palm waste management. To address this issue, a low-cost and small-scale oil palm waste shredder were designed and produced. The central design concept deduced is to compromise acceptable efficiency with an affordable price by integrating all mandatory components of oil palm waste shredder into a comparatively small and portable support structure. The innovations made in the design are size reduction and adoption of double-shafts cutters with seven jaws technology. Preliminary calculations supported the design. The cutting force produced is up to 19.45kN with safety factor of 4.5. This design can handle other oil palm wastes as oil palm frond is harder than other oil palm wastes.

DOI: 10.4018/978-1-7998-4921-6.ch006

BACKGROUND

Oil palm tree, with species name given as *Elaeis Guineenis Jacq.* emanated from West Africa and has been well-known as one of the most precious gifts from nature in agricultural world. The reason behind is that the palm oil, extracted from its fruits is the most popular edible plant oil used in food preparation which its consumption can be traced back to 5,000 years ago and nowadays, palm oil is consumed by more than three billions people over 150 nations in the world (Azhar, 2009). Since Malaysia is located near the equator, the strategic geographical factor has made the climate in the nation excellent for cultivation of oil palm. Malaysian palm oil industry has been the economic pillar of agricultural sector since the launching of the New Economic Policy (NEP) in 1970s which aimed to eradicate rural poverty by diversifying commercial crop rather than solely dependent on rubber (Otieno et al., 2016). On top of that, the palm oil industry has experienced significant growth under the implementation of three consecutive Industry Master Plans (IMP1~IMP3) until the nation being the first-runner-up of palm oil producer in global today. Despite the prosperity encompassed by the palm oil production, the rapid development of oil palm plantations and palm oil mills also come with a drawback, which a vast scale of waste is generated from both plantation and milling activities (Sukiran et al., 2017). All plantation and milling activities which result in the production of different types of solid waste are presented in Figure 1.

All of the solid wastes produced are organic waste which own high potential for harnessing biomass energy because they are made up of lignocelluloses (Hassan & Abd-aziz, 2011). These solid wastes are generated from both plantation and milling activities which are known as oil palm frond (OPF), oil palm trunk (OPT), empty fruit bunch (EFB), mesocarp fiber (MF), palm kernel shell (PKS) and palm kernel cake. These solid wastes are generated in a substantial amount annually as their production comprises of two stages. During the first stage, OPF and OPT are produced massively in pruning and harvesting, as well as replanting activities respectively. During the second stage, palm oil mill activities such as oil extraction, FFB pressing and nut cracking etc. produce EFB, MF and PKS abundantly (Hassan and Abd-aziz, 2011). Availability of each type of solid waste together with corresponding characteristics are tabulated in Table 1.

Table 1. Production quantity of each solid waste (Adapted from Goh et al., 2010)

Types	Quantity (ktonnes)	Moisture Content (wt%)	Dry Weight (ktonnes)
EFB	18,022	65.0	6,308
OPF	46,837	60.0	18,735
OPT	10,827	75.9	2,609
MF	11,059	42.0	6,414
PKS	4,506	7.0	4,190

From Table 1, it is prominent that EFB and OPF have a lion share in overall solid waste production. Since OPF and EFB are the main oil palm solid wastes, the management or treatment methods of these wastes are essential to be implemented properly to ensure the sustainability of palm oil industry.

PROBLEM STATEMENT

It was reported that both oil palm plantations and palm oil mills are currently facing three main issues:

1. Environmental problems/Pollutions due to abundant oil palm waste
2. Lack of sustainable and yet affordable oil palm waste management technologies
3. Inactive participation of small and medium enterprises (SMEs) in waste management

Uprising Environmental Issues due to Abundant Oil Palm Waste

Inevitably, the drawback of development and booming of any industry is the environmental issues created from the waste production. Traditionally, EFB are handled by three approaches, namely composition, incineration and dumped (land-filled) (Otti, Ifeanyichukwu, Nwaorum, & Ogbuagu, 2014). Incineration is the most widely used method due to its lowest cost and undoubtedly, the consequence is the arising of air pollution issue because incineration of EFB emits white smoke with high ash content (Loh, 2016). Direct dumping of EFB with residual oils results in oil spills, which not only contaminates the soil but also poses as a wastage (Wang, 2016). For OPF's management, the most widely used method is natural composting whereby the fall-off OPF after pruning and harvesting are left to be decomposed naturally on land for soil fertilization. Even though natural composting of OPF does not cause any environmental problem, it is considered as a huge wastage because OPF consist of high lignocellulose content which has great potential in other utilization after being treated properly and correctly (Sukiran, Abnisa, Wan Daud, Abu Bakar, & Loh, 2017).

Lack of Sustainable and yet Affordable Oil Palm Waste Management Technologies

In palm oil industry, solving environmental issues is not easy in Malaysia because the production of oil palm waste is of enormous quantity per year, which has resulted in the underutilisation and inadequate energy recovery of oil palm waste as to compromise with economical prospect (Nurhayati Abdullah & Town, 2013). Although modern oil palm waste treatment technologies are available and advocated under the enforcement of Environmental Quality Act (EQA), it is difficult for palm oil mills to fully meet the requirements of environmentally friendly biomass energy conversion because the abundance of waste production surpluses the availability of conversion technologies (Lim Meng Hon, 2010). The crisis of waste treatment technologies' application is shown in the following Figure 2.

From Figure 1, the readers can observe that the application count of new technologies or patents in oil palm waste management is lower as compared to palm oil production, even though more patent families are found in waste treatment. This phenomenon is relevant to current scenario of oil palm waste treatment in Malaysia, in which palm oil industry has to juggle between environmental issues and economic reasons. Hence, abundant oil palm wastes are managed by combustion or incineration and eventually causing environmental issues, which is common route for most palm oil mills to compromise with economical reason (Lim Meng Hon, 2010).

Figure 1. Patents and applications count in oil palm waste treatment from 1995 to 2015 (Wang, 2016)

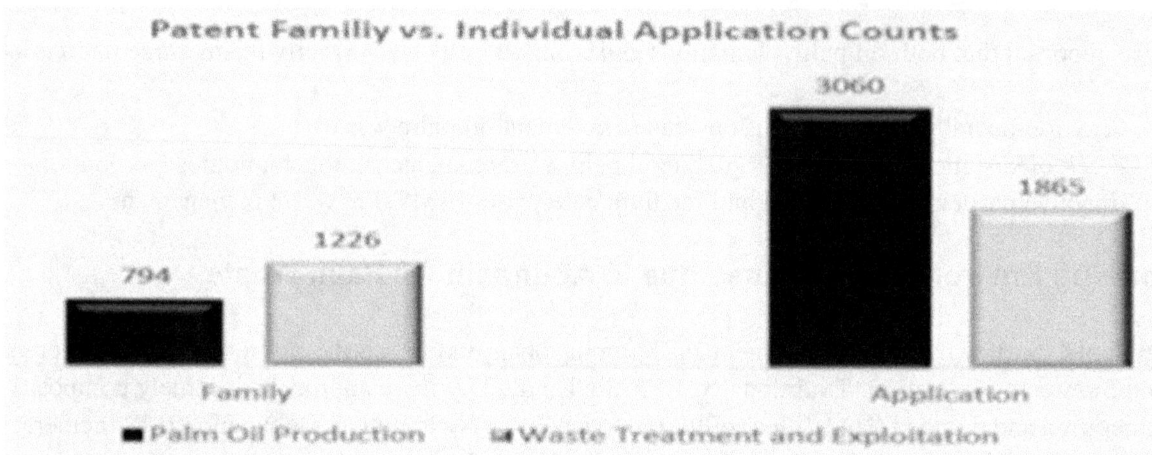

Inactive Engagement of Small and Medium Enterprises (SMEs) in Sustainable Oil Palm Waste Management

It was found that most of the biomass waste is unutilised though Malaysia is abundant with biomass energy resource (annual generation of 168 million tonnes of green waste), which is mainly composed of oil palm waste. Involvement of SMEs in sustainable oil palm waste management is inadequate to convert the waste into high value-added biomass products, despite high participation of SMEs in oil palm plantation and palm oil industry (Abd Rahman, 2013). Engagement of SMEs in oil palm waste management is essential because statistics from Small and Medium Corporation Malaysia (SME Corp) shows that proportion of SMEs in overall Malaysian's business or enterprises constitute 97.3% (Zainul,

Figure 2. Percentage share of SMEs' GDP and Malaysia's GDP in 2016 (Department of Statistics Malaysia, 2017)

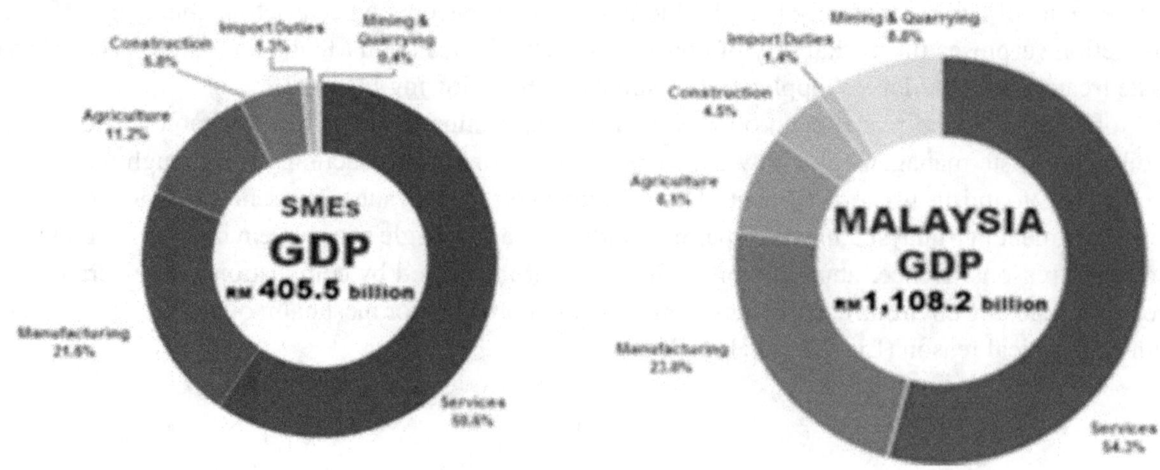

2013). Among them, a total of 645,000 SMEs are involved in oil palm sector (Zainul, 2013). Contribution of SMEs to GDP of Malaysia is shown in the following Figure 2.

Based on Figure 2, agriculture sector's contribution to GDP of SMEs accounted for 11.2%, which was the third largest contributor in 2016. On top of that, report from The Star Online (2017) has stated:

'SMEs GDP led the agriculture sector with a share of 50.7%, up 1.7% from 2015 due to the rubber, oil palm, livestock and other agriculture sub-sector especially in oil palm segment' (The Star, 2017).

Therefore, it is undeniable that SMEs will contribute significantly to the growth of oil palm waste management with their active participation in prospect. However, study by Malaysian Industry-Government Group for High Technology (MIGHT) in 2013 has revealed the prime factor causing the engagement in oil palm waste management to be skewed towards big private organisations instead of SMEs: lack of technologies with local patents in oil palm waste management (Abd Rahman, 2013). SMEs are facing financial burden to purchase costly imported technologies or machinery with foreign patents. This explains why most SMEs are inactive in oil palm waste management.

LITERATURE REVIEW

Characterization of Oil Palm Solid Wastes

Over the last forty years, consumer (dis)satisfaction has taken a prominent position in the marketing research literature (e.g. Anderson, 1973; Anderson & Sullivan, 1993; Cardozo, 1965; Churchill & Surprenant, 1982). This attention is justified since consumer (dis)satisfaction (directly or indirectly) impacts upon repurchase intention (Szymanski & Henard, 2001), consumer retention (Anderson, 1994; Mittal and Kamakura, 2001) and eventually upon firm performance (Anderson et al., 1994). Consumer (dis) satisfaction is a summarizing response that results from a consumer's post-consumption cognitive and affective evaluation of a product or service performance given pre-purchase expectations (Anderson and Sullivan, 1993; Oliver, 1993; Tse & Wilton, 1988). EFB and OPF are the most abundant solid waste generated from palm oil mill and oil palm plantation respectively. In order to develop or improvise treatment methods of EFB and OPF, their crucial characteristics need to be studied and acknowledged as these parameters own the potential to affect the design considerations of waste treatment machine or methodology in advance.

The size of EFB is quite large as it can be 50cm in maximum diameter whereas its weight can go up to 20kg (Malaysian-German Chamber of Commerce and Industry (MGCC), 2017). In term of moisture content (MC), EFB usually contain a high proportion of water (>60%) (N. Abdullah, Sulaiman, & Gerhauser, 2011). High MC has made treatment of EFB difficult because it limits the processing duration of EFB. The processing of EFB need to be completed within three days after production. Otherwise, they will start rotting due to high MC (Malaysian-German Chamber of Commerce and Industry (MGCC), 2017). This explains why treatment of EFB is difficult sometimes due to logistics problem to transport the EFB to the treatment plant in time (Malaysian-German Chamber of Commerce and Industry (MGCC), 2017). Their biomass values tend to deteriorate with time due to degradation issue. MC and other impurities constituents of EFB in contrast to other solid wastes are shown in the following Table 2.

Table 2. Proximate analysis of various oil palm solid wastes (Sukiran et al., 2017)

Type of Solid Wastes	Moisture Content (wt.%)	Volatile Matter (wt.%)	Fixed Carbon (wt.%)	Ash Content (wt.%)
EFB	66.0-69.0	86.5-87.7	10.8-14.5	3.7-5.3
OPF	62.0-77.0	83.6-88.3	3.2-14.8	3.2-3.8
OPT	67.0-81.0	86.3-88.3	4.9-7.8	2.9-3.7
PKS	11.0-13.0	82.7-84.4	13.2-20.4	1.4-4.8
MF	35.0-48.0	84.0-85.6	7.6-17.4	5.0-7.4

From Table 2, it can be observed that EFB, OPF and OPT show similarity in high MC. Biomass wastes with high MC are inappropriate to be used as biomass fuel for any thermal combustion process due to low calorific value of a wet fuel. Calorific value is defined as the amount of energy released during combustion (Loh, 2016). Therefore, calorific value of a biomass waste is inversely proportional to its MC. This means that EFB, OPF and OPT are more preferable to undergo other bioprocessing because of other high constituents, which are depicted in Figure 3 below.

Figure 3. Chemical composition of various oil palm solid wastes (Loh, 2016)

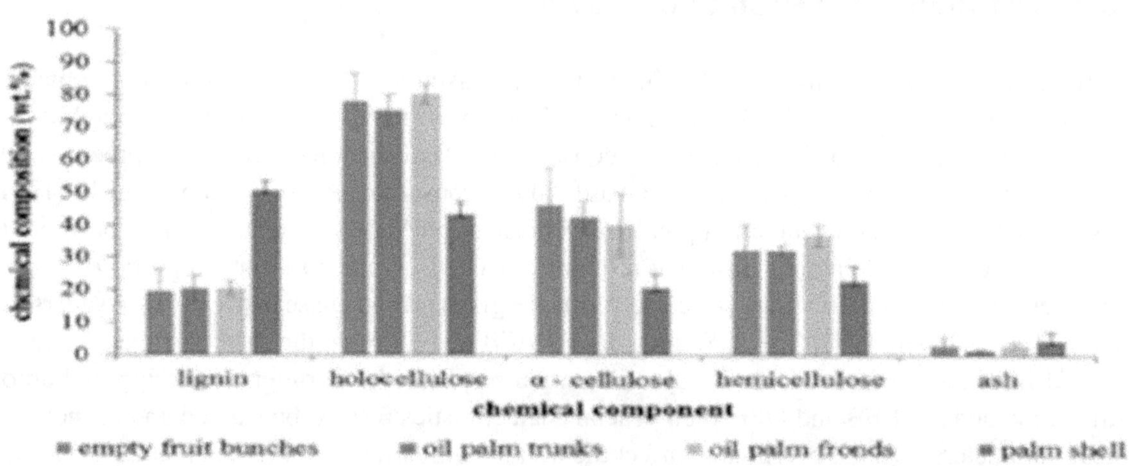

From Figure 3, it can be deduced that oil palm wastes are made up of lignocellulosic materials which contain a mixture of various celluloses and lignin. According to Sukiran (2017), cellulose is defined as 'linear homopolysaccharide that contains crystalline and amorphous structures' whereas lignin is 'a highly cross-linked polyphenolic polymer without any ordered repeating units; it is amorphous and is a high molecular weight phenolic compound' (Sukiran et al., 2017). In other words, cellulose is sugar and hemicellulose is a mixture of various sugars in complex form. Hence, high cellulose and hemicellulose content simply means the waste is more readily to undergo bioconversion, which are depicted by EFB, OPF and OPT. On the other hand, lignin is much harder to decompose and this means waste with high lignin needs to undergo process with high thermal stability conditions (Sukiran et al., 2017). Again, this shows that PKS is more suitable to be used as direct biomass fuel.

Similarity of EFB, OPF and OPT is shown again in their chemical composition. They have been identified as the wastes which need to be shredded for other bioconversion processes later. However, the availability of OPT is not as high as EFB and OPF because OPT is produced only during replanting activity in oil palm plantations (Hassan & Abd-aziz, 2011). Meanwhile, the size and strength of OPT are much higher than EFB and OPF which cause the necessity of OPT to undergo pre-treatment process such as chopping before shredding. Due to these drawbacks, OPT should not be the main target but it could pose as a challenge or limitation test for mini shredder in this research.

Despite OPT, OPF should be the main solid waste which will be shredded in this research apart from EFB. However, shredding OPF would not be as easy as EFB because of its structural complexity and also different chemical compositions according to different parts of OPF (Bulan, Mandang, & Hermawan, 2015). Hence, it is essential to review and study the specific characteristics of OPF in this research. OPF is comprised of four major parts, namely rachis, leaflet, stem and petiole, which is shown in the following Figure 4.

Figure 4. Components of OPF (Oslan, Ahmad and Mohd, 2013)

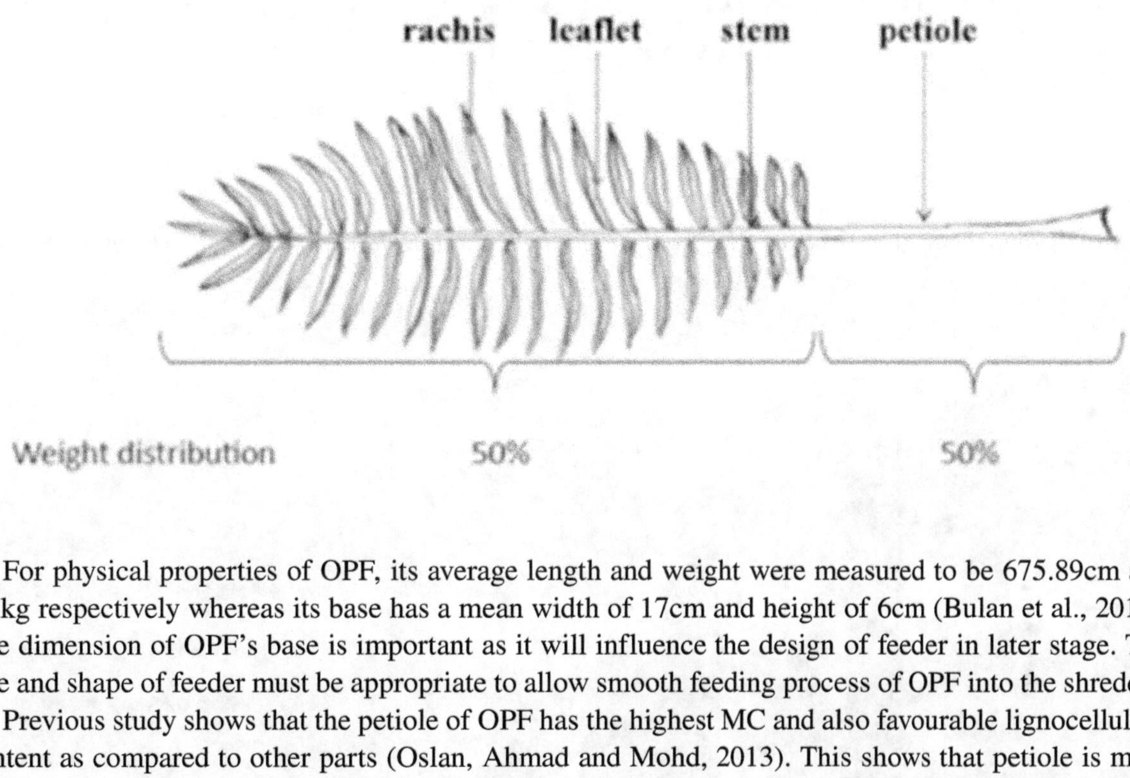

For physical properties of OPF, its average length and weight were measured to be 675.89cm and 9.5kg respectively whereas its base has a mean width of 17cm and height of 6cm (Bulan et al., 2015). The dimension of OPF's base is important as it will influence the design of feeder in later stage. The size and shape of feeder must be appropriate to allow smooth feeding process of OPF into the shredder.

Previous study shows that the petiole of OPF has the highest MC and also favourable lignocellulose content as compared to other parts (Oslan, Ahmad and Mohd, 2013). This shows that petiole is more suitable to be converted into other high value products through bioconversion process. Another supportive reason is that petiole only makes up 34% of total nutrient of OPF, which means its nutrient to be returned back to soil through natural decomposition in plantation is comparatively low (Oslan et al., 2013). For leaflet and rachis which are rich in crude protein, starch and pectin, they should be decomposed in the plantation to enhance the recovery of nutrient in soil. Stem is actually also suitable for bioconversion. However, study from Oslan (2013) shows that 'Leaving leaflets in the plantation facilitates the return of nutrients to soil, while stem and rachis provide structural strength for aeration and space' (Oslan et al.,

Figure 5. Compression test result of OPF from 5 years old plant (Bulan, Mandang and Hermawan, 2015)

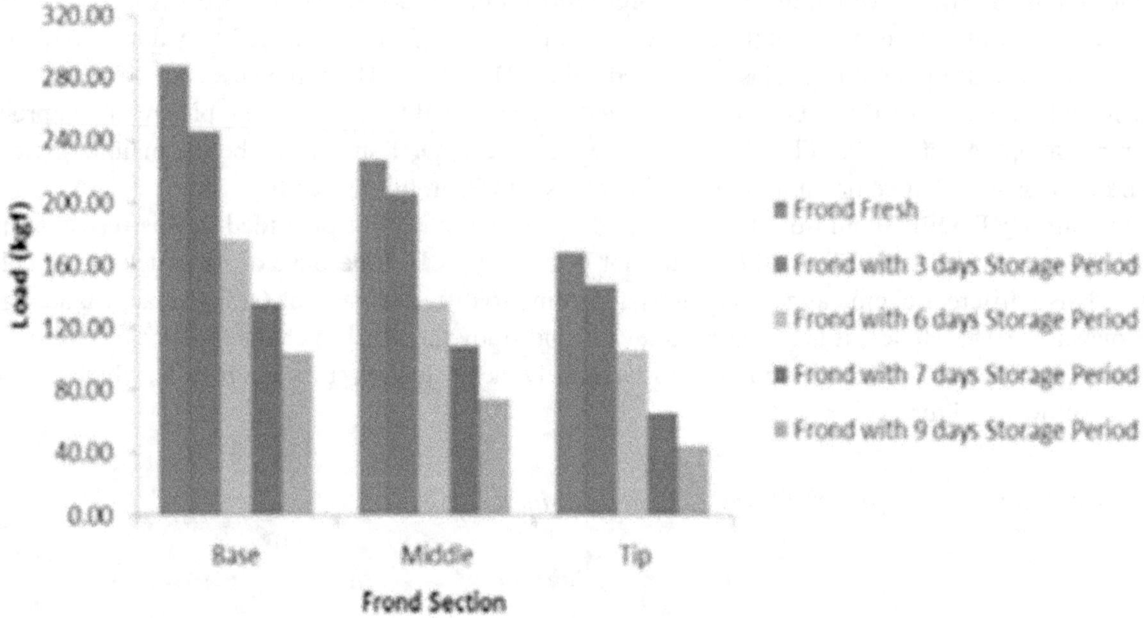

Figure 6. Compression test result of OPF from 20 years old plant (Bulan, Mandang and Hermawan, 2015)

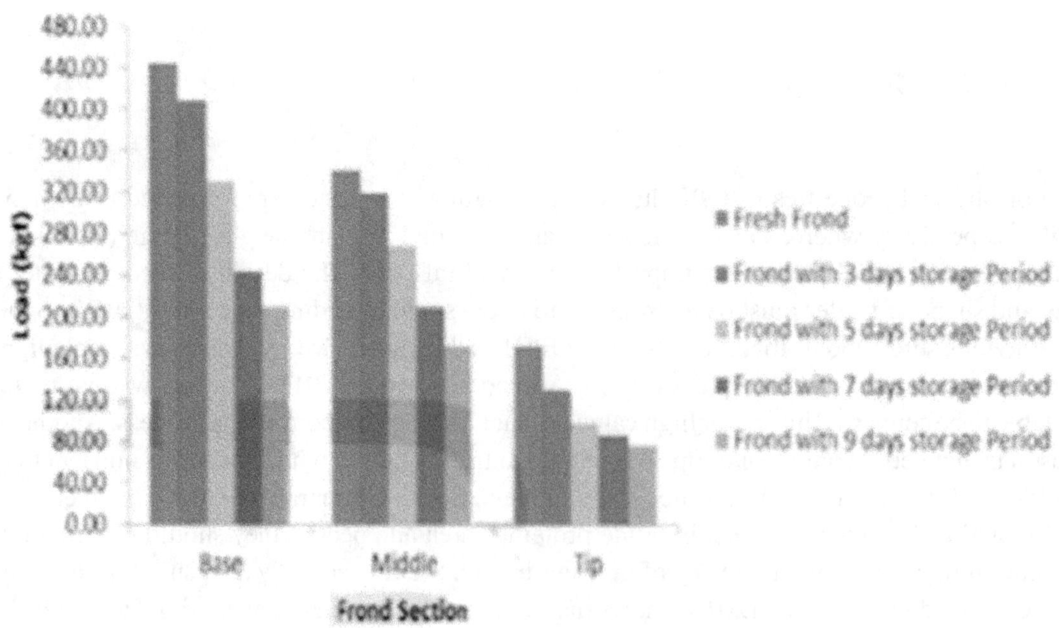

2013). By this, the main component of OPF which requires shredding has been identified, which is its petiole. Meanwhile, this also means that OPF needs pre-treatment process before being shredded, which its petiole has to be chopped and transported out while other parts remain in plantation.

Apart from identifying the right component of OPF to be shredded, it was also found that the age and storage period of OPF would affect its MC and thus the required compression force in order to shred its petiole (Bulan et al., 2015). Compression test result of OPF with different ages and storage periods is depicted in the following Figures 5 and 6 accordingly to give a clear comparison.

Based on Figures 5 and 6, it was observed that the compression load required to shred OPF decreased in longer storage period. Meanwhile, the base part (petiole) of OPF was found to have the highest compression strength as petiole contains the most compact fibre structure. In term of age, OPF from 20 years old plant was found to have higher compression strength than OPF from 5 years old plant. From the perspective of compression strength, it can be deduced that compression strength of 5 years old and 20 years old petiole falls between the range of 287.56 kgf (2.820kN) and 443.78 kgf (4.351kN) respectively (Bulan et al., 2015). Thus, the torque of the shredder must be high enough to generate cutting force above 4.351kN in order to shred the petiole of OPF.

Potential Applications of Shredded Oil Palm Fibres

The selling price of oil palm fibre falls within RM400-RM600/tonne which is quite lucrative in term of raw fibre. Moreover, the value of final products made from the fibre will be enhanced by multiple folds. It was found that non-hazardous and biodegradable oil palm fibre has a plenty of usages which can be summarised as follow:

Figure 7. Distribution of patent family among oil palm fibre exploitation technologies

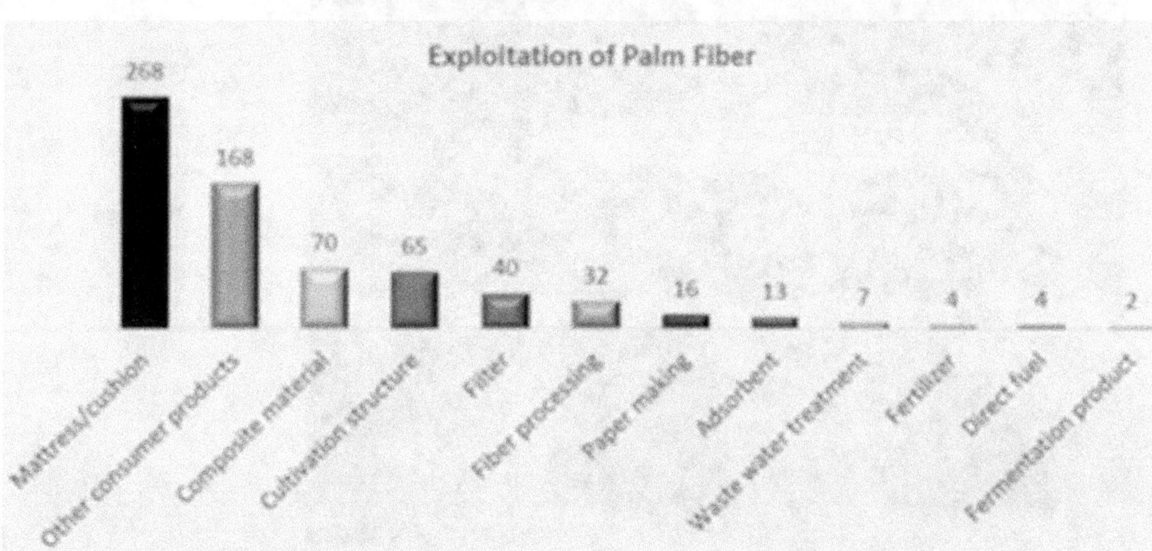

Mattress and Cushion

In recent years, oil palm fibre has been well-known as a superior alternative material to coconut fibre for producing mattress and cushion because oil palm fibre has higher strength due to its strong inter-molecular bond (Teoh, 2018). Mattress and cushion made from oil palm fibre has successfully gain favourable market demand in China, Japan and Korea (Wang, 2016). The manufacturing technology is led by Guizhou Daziran Technology Co., Ltd from China which is famous and capable of manufacturing ergonomic mattress from oil palm fibre, as described in patent CN102861755 (Wang, 2016).

Composite Materials

Various new composite materials or products can be formed with the combination of oil palm fibre and other materials like polymers and concrete (Aziz, 2015). One of the most successful example is sound dampening sheet which has been commercialised and utilized in various automotive production, such as Proton (ESX-62223-6), Hyundai (MS-731-06) and Nissan (NES-M-9001), as shown in the following figure 8.

Cultivation Structures for Other Plants

Oil palm fibre is ideal structural material for constructing growing bed of other plants such as ornamental plant and oyster mushrooms (Trisakti *et al.*, 2018; Zakil *et al.*, 2008). This is achieved by composting

Figure 8. Installation of sound dampening sheet in Automotive (Aziz, 2015)

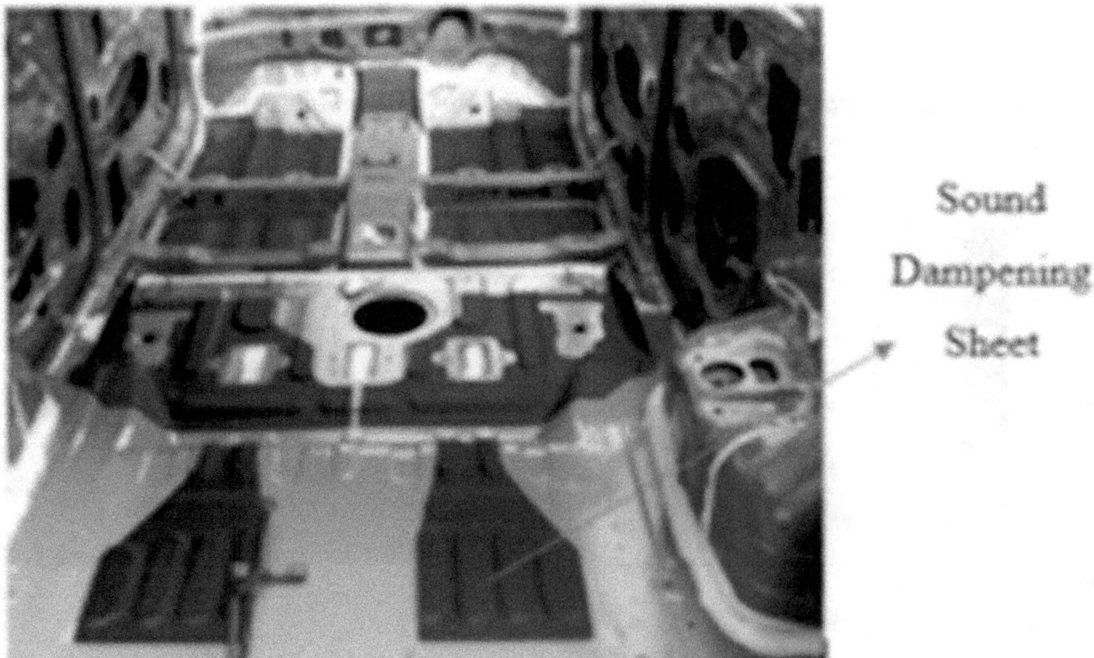

oil palm fibre together with other organic wastes. For cultivation of ornamental plant in study from Trisakti (2018), active liquid organic fertilizer (ALOF) is mixed with oil palm fibre to produce compost material with high nutrients and microbial sources. Procedures quoted by Trisakti (2018) are: 'Composting process was conducted by shredding and cutting EFB to size of 1-3 cm, measuring the weight of EFB total, and putting into the composter. Next, ALOF was poured and mixed until moisture content of compost reaches 55 - 65% and maintained at that value range during the composting proceed. During the process, compost pile was turned every three days in order to reduce and equalize the moisture content' (Trisakti et al., 2018). Similarity of starting procedure is shown as well in cultivation of oyster mushroom in study conducted by Zakil (2008), which initiated the composting process with 'The fresh empty fruit bunch are cut into smaller pieces, dried and then shredded using a shredder' (Zakil et al., 2008). These two studies show the significance of shredding as preliminary step in producing compost material for cultivation of other plants.

Solid Biomass Fuel

Raw EFB is not suitable to be the fuel for boiler to generate electricity. However, shredded EFB fibre has posed as another better feedstock fuel material after processing the EFB fibre into biomass briquettes and pellets (Aziz, 2015). Biomass briquettes are manufactured from oil palm fibre by using piston press technology or screw extrusion technology. Meanwhile, pelletizing of oil palm fibre serves as an improvised method to solve the durability and degradation problem of oil palm fibre to enhance the capability of oil palm fibre in supporting combustion (Bakar et al., 2015). The conversion of oil palm fibre into solid biomass fuel is represented in the following Figure 9.

Figure 9. Various solid biomass fuel made from oil palm fibre (Aziz, 2015)

Fermentation Product (Biofuel)

Ordinary fossil fuels such as diesel and petroleum are non-renewable and hazardous to environment as burning of fossil fuel causes severe air pollution. Therefore, biofuel has been discovered as diversification to reduce the dependence on fossil fuel under the National Bio-fuel Policy implemented by Malaysian government in March 2006 (Lim Meng Hon, 2010).

As oil palm biomass is mainly composed of lignocellulose materials, fermentation of oil palm biomass can produce biofuel such as bioethanol and biodiesel by extracting sugars (cellulose and hemicellulose) from them. The main materials for bioprocessing to be converted to biofuel are cellulose and hemicellulose (Lim Meng Hon, 2010). In order to extract cellulose and hemicellulose, oil palm fibre needs to undergo pre-treatment process, which is named saccharification. In this process, the oil palm fibre is added with acids and enzymes for glucose/sucrose conversion (Lim Meng Hon, 2010). The overall process for producing biofuel is shown in Figure 10 below.

Figure 10. Production of biofuel from oil palm biomass (Hassan & Abd-aziz, 2011)

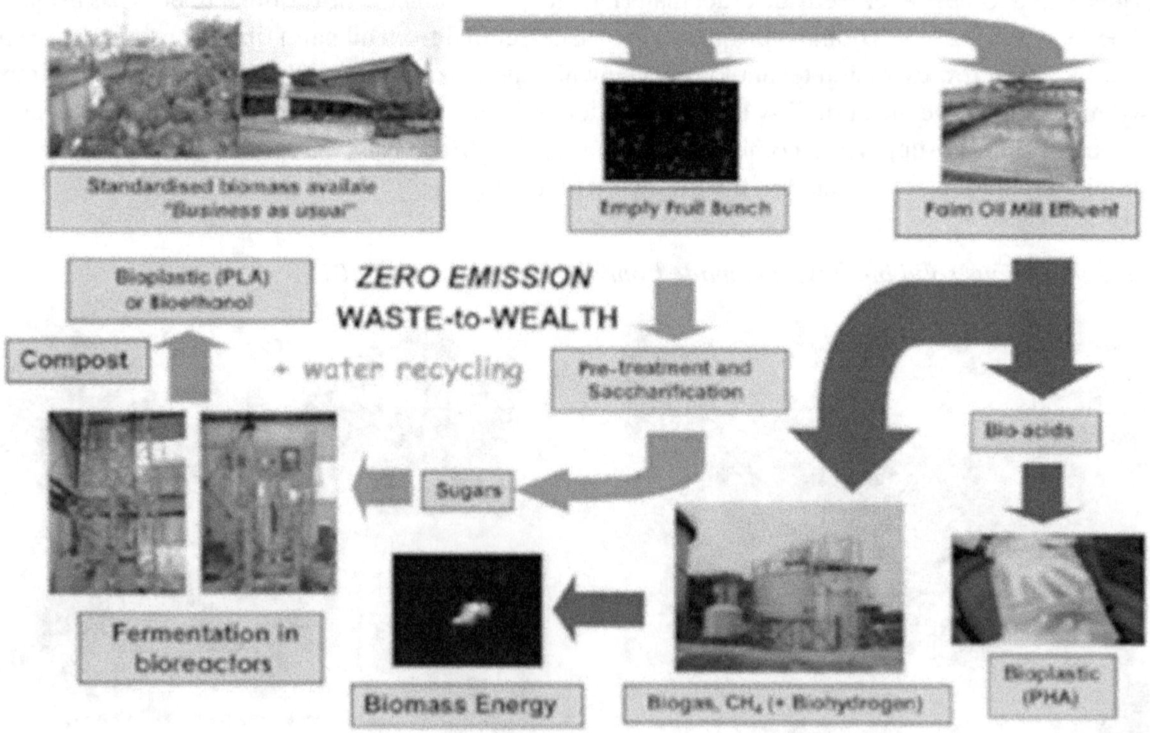

Besides, it was also found that sugars extracted from OPF juice, which is the waste product from shredding OPF can be used to make bioethanol as well by extracting sugars from it. Study conducted by Oslan, Ahmad and Mohd (2013) has proved that 'Sugars from saccharification and OPF juice can be used for bioconversion into bioethanol, bioplastics and other biopolymers, while biomass residue from these processes can be used as a source for biofertilizers as well as bio-composites' (Oslan et al., 2013).

This means that OPF juice should be collected instead of disposal after fibre is obtained from shredding due to its high capability of sugar recovery by saccharification process.

Paper

The global consumption of paper is enormous, which is estimated to be 300 million tonnes annually. Consequently, forest resource is diminishing as the ordinary raw material for making paper is wood (Rosli, Daud, & Law, 2011). However, it was found that oil palm fibre has exhibited great potential as alternative papermaking material which is non-wood-based. This has resulted in the booming demand of oil palm fibre in papermaking industry worldwide.

To produce paper with high quality, oil palm solid wastes like OPF and EFB need to be shredded to generate sludge waste and undergo chemical pulping treatment process. It was reported that the oil palm fibre is favourable in pulp production because the cell wall of oil palm fibre is much thicker than other wood fibre, which gives the pulp s significantly higher rigidity index (Rosli et al., 2011). In short, oil palm fibre has been discovered as one of the most superior substitutes to wood-based papermaking material.

Other Consumer Products

It was found that the capability of oil palm fibre is also shown in various consumer daily products such as shoes, brushes, fabrics and even polylactic acid (PLA) bioplastic (Wang, 2016). This is because more and more new technologies have been emerging in recent years to utilize the cellulosic fibre of oil palm waste as substitute material for manufacturing comparatively high value product. This is because oil palm fibre is 100% natural, biodegradable and non-hazardous. Meanwhile, oil palm fibre is of high availability due to abundant oil palm solid wastes production especially in Malaysia and Indonesia, which ranked second and first in worldwide CPO production. The value of oil palm fibre is of high prospect to increase from time to time as more potential utilizations are discovered and commercialised.

EXISTING WASTE SHREDDING TECHNOLOGIES AND DEDUCTIONS

Manual Chaff Cutter Machine

Figure 11 depicts the manual chaff cutter machine invented by P.B. Khope and J.P. Modak (2013) which functions to reduce the forage size by chopping and to be used as animal feedstock. This machine increases the efficiency of chopping forage, output uniformity and also enhances the safety as compared to traditional chopping by using axe.

The primary innovation of this machine is that it utilizes human-powered flywheel motor concept in transmission of kinetic energy from human muscles to rotational kinetic energy of flywheel by paddling, which is similar to a bicycle mechanism. This is a smart way of reducing the energy consumption in strenuous traditional chopping process by using leg muscles, which are stronger and able to generate more force than arm muscles (Khope & Modak, 2013). Another advantage is that the device is of 100% environmentally friendly and convenience because no electric power supply or diesel generator is needed to operate it. This is due to the consideration of huge scarcity of electric power supply scenario in India (Khope & Modak, 2013).

Figure 11. Manual chaff cutter machine developed in India (Khope and Modak, 2013)

However, this machine is not applicable in Malaysia as Malaysia does not have shortage of electricity supply. The efficiency of using electric driven motor in shredding is much higher than human powered bicycle mechanism. Efficiency is the main concern in design of oil palm waste shredder in Malaysia because of the abundance of oil palm wastes generated.

Single Shaft Low-Speed Waste Granulator

It was found that a Malaysian company named Ant Spirits Sdn Bhd has developed a low-cost waste granulator as shown in Figure 13 above. According to definition quoted by Oxford Dictionary 6th Edition (2004), granulator is a machine which is able to reduce the dimension of a substance into grain size. This means only fibre grain can be produced if oil palm solid wastes such as EFB and OPF is handled by the machine. Despite low customization of output fibre length, efficiency of the single shaft granulator is comparatively much lower, as compared to double shaft shredder (Sakthivel, Rajeshkannan, Naveen-kumar, & Muralimanokar, 2017).

However, there are certain advantages which we can adopt from the machine. The first advantage is the utilisation of electric motor as main power source, which has increased the efficiency greatly as

Figure 12. Low-speed mini waste granulator from Ant Spirits Sdn Bhd (Ant Spirits Sdn Bhd, 2018)

compared to manual cutting mechanism. The second innovative feature of the granulator is that it is small in size and portable with four small wheels installed on its base.

Mini Double Shaft Agricultural Waste Shredder

Figure 13 above depicts a mini double shaft shredder developed by I.M. Sanjay Kumar (2015) from Sri Siddhartha Institute of Technology in India. The main focus of the shredder is agricultural waste such as coconut leaves and arecanut leaves. The efficiency of this machine is comparatively higher as it utilizes an electric motor to drive double shafts with 4 cutters each in shredding process. The production rate

Figure 13. Mini double shaft shredder developed in India (Kumar, 2015)

is high, which was tested to be 80-90kg of coconut powder per hour (Kumar, 2015). In addition, the machine is quite affordable, which it only costs about 17000 INR/- (RM 1,010/-).

However, it was reported that the belt-driven gears and pulley mechanism deployed in this machine has one crucial weakness, which is the incapability of reducing the input rotational speed of motor to generate sufficient cutting force to the cutters. This explains why the output rotational speed of shafts (550rpm) is comparatively high and is only sufficient to shred organic waste with low strength such as coconut leaves. In fact, a reducing gearbox transmission system should replace the belt-driven pulley system as reducing gearbox is more effective and consistent in reducing the input rotational speed of motor to desired level to increase the torque generated (Pavankumar, Sachin, Shankar, Thyagaraja, & Madhusudhan, 2018). Increasing torque means more cutting force can be gained by the cutter to shred harder materials such as OPF.

Snipper Press Machine

Figure 14. Snipper press machine developed by Hur Far Engineering Works Sdn. Bhd in Malaysia (RAMLI et al., 2013)

Based on observations from its working process, it was found that the machine has high efficiency in production of oil palm fibre, which is relevant to the data provided by RAMLI (2013), 7 to 8 tonnes per hour. The machine is very powerful as it is equipped with electric motor of 110 kW connected to an effective reducing gearbox, named Sumitomo (538, 2013). Apart from efficiency, the machine has two outstanding advantages. The first outstanding achievement is that the machine is capable of handling three types of oil palm solid wastes, EFB, OPF and OPT. Besides, the machine is installed with juice channelling system which enables the juice to be collected for other bioconversion processes later. The most crucial drawback of this machine is its high cost, which constitutes RM 218,000 in total (538, 2013). Another weakness is the lack of collecting tank for fibre produced, which causes inconvenience to collect the fibre manually and the spill of fibre also affects the hygiene of the workplace. Besides, the machine is not portable.

Figure 15. Heavy Duty EFB Shredder developed by YKL Engineering Sdn. Bhd. from Malaysia (YKL Engineering Sdn. Bhd., 2002)

KH-77 Heavy Duty EFB Shredder Machine

As the name implies, KH-77 heavy duty shredder is extremely efficient with two electric motors (22 kW and 37 kW) and reducing gearboxes (SEW Model of MC3PLSF04 and MC3PLSF06). The production rate of oil palm fibre is 7 to 8 Mt per hour. Besides, this shredder is also installed with juice collecting system just like the snipper press machine discussed before. In fact, this machine shows similar pros and cons with snipper press machine with the only difference in efficiency and motor power. Portability, high cost (about RM 19,9400) and huge size are the weaknesses of this machine. More installation area is needed due to huge size.

SHREDDER DESIGN CONSIDERATIONS

Overall shredder design considerations are enlisted as follow:

1. Primary design consideration of shredder machine is efficiency as to handle the abundant oil palm waste in Malaysia. However, efficiency must compromise with costing as SMEs are the main market target.
2. The significance of shredder tends to increase with plenty applications of oil palm fibre.
3. Juice produced from shredding must be collected due to its potential for sugar extraction.
4. The shredder should be portable and of appropriate size for convenience and also reduction of investment modal as smaller space is required for machine operation.

DESIGN CALCULATION AND CONFIGURATION

Motor Power and Reducing Gear Train Value

It was found that conventional configuration of motor power in low-cost shredders ranges from 1hp (0.7457 kW) to 2hp (1.4914 kW) (Kumar, 2015). Based on QFD result, the most important parameters that would affect the efficiency of machine are motor power and reducing gear train value. Therefore, as to increase the efficiency of the mini shredder, the motor is configured to be three-phase inducted with power of 5.5 kW and 1440 rpm. Meanwhile, the reducing gear train value is set to be 1/48 to reduce the output rotational speed of the shaft connected to the shredder.

Calculation of output rotational speed, ω_{out} is as follow:

$$\omega_{out} = \omega_{in} \times TV$$

Where ω_{in} is input rotational speed from the motor

TV is reducing gear train value.

$$\omega_{out} = 1440 \times \frac{1}{48} = 30rpm$$

Shredder Main Body Case

As reduction in price and size is second design consideration, the shredder main body case design needs to be small and compact. Besides these three major components (shredding mechanism, motor and reducing gearbox), it was found that other auxiliary components such as feed hopper, base with caster wheels and juice collecting tank are also needed for safety, portability and cleanliness respectively. On top of that, overall dimension after assembly should not be more than $2m^3$ to be considered mini in size.

Therefore, the internal dimension of shredder main body case is set to be:

Cutter

Figure 17 shows the design of a cutter. There are several parameters of the cutter which are needed to calculated and configured as follow:

1. Number of cutter jaws

The most conventional number of cutter jaws are 3, 5, 7 and 9 whereby the output fibre length decreases accordingly with the increasing jaw number. Based on QFD result, short fibre is more favourable in market. Thus, 7 jaws are chosen for optimum fibre length production.

2. The thickness of cutter, t

Figure 16. Internal dimensions of shredder main body case

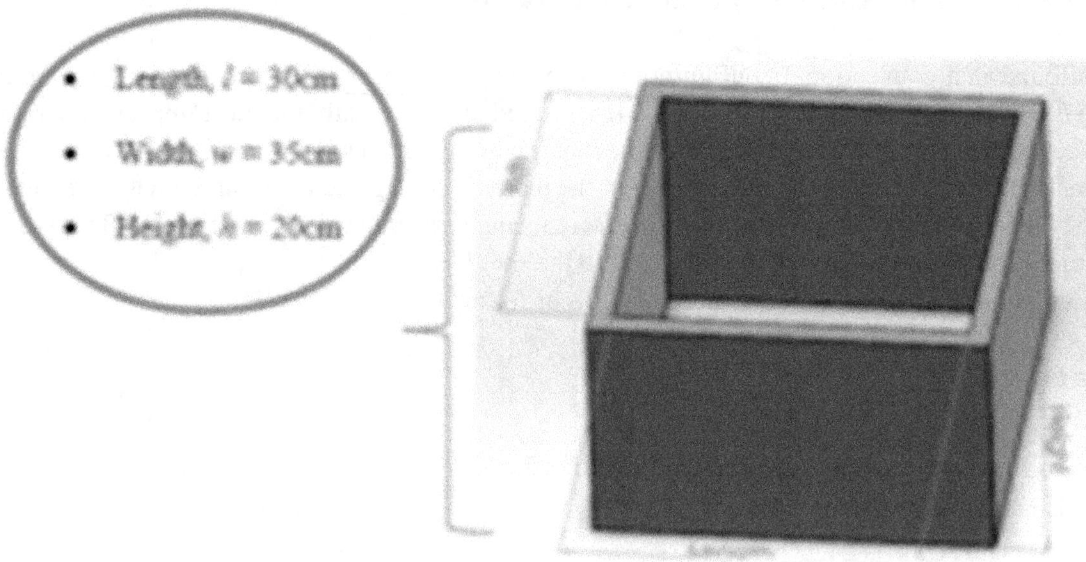

The thickness of the cutter would affect the number of cutters installing within the small shredder case. In order to install 20 cutters, the thickness is to calculated as follow:

$$t = \frac{Maximum\ shredder\ internal\ length}{Number\ of\ cutters}$$

$$t = \frac{30}{20} = 1.5cm$$

3. The outer diameter of the cutter, D

As the height of the shredder case is limited to be 20cm, the outer diameter of cutter has to be less than 20cm. It was found that 18cm will be appropriate from the perspective of compacity and dimensional tolerance.

4. Cutting Force, Fc

$$F_c = \frac{Torque, \tau}{Radius\ of\ rotation, r}$$

Figure 17. Schematic diagram of cutter

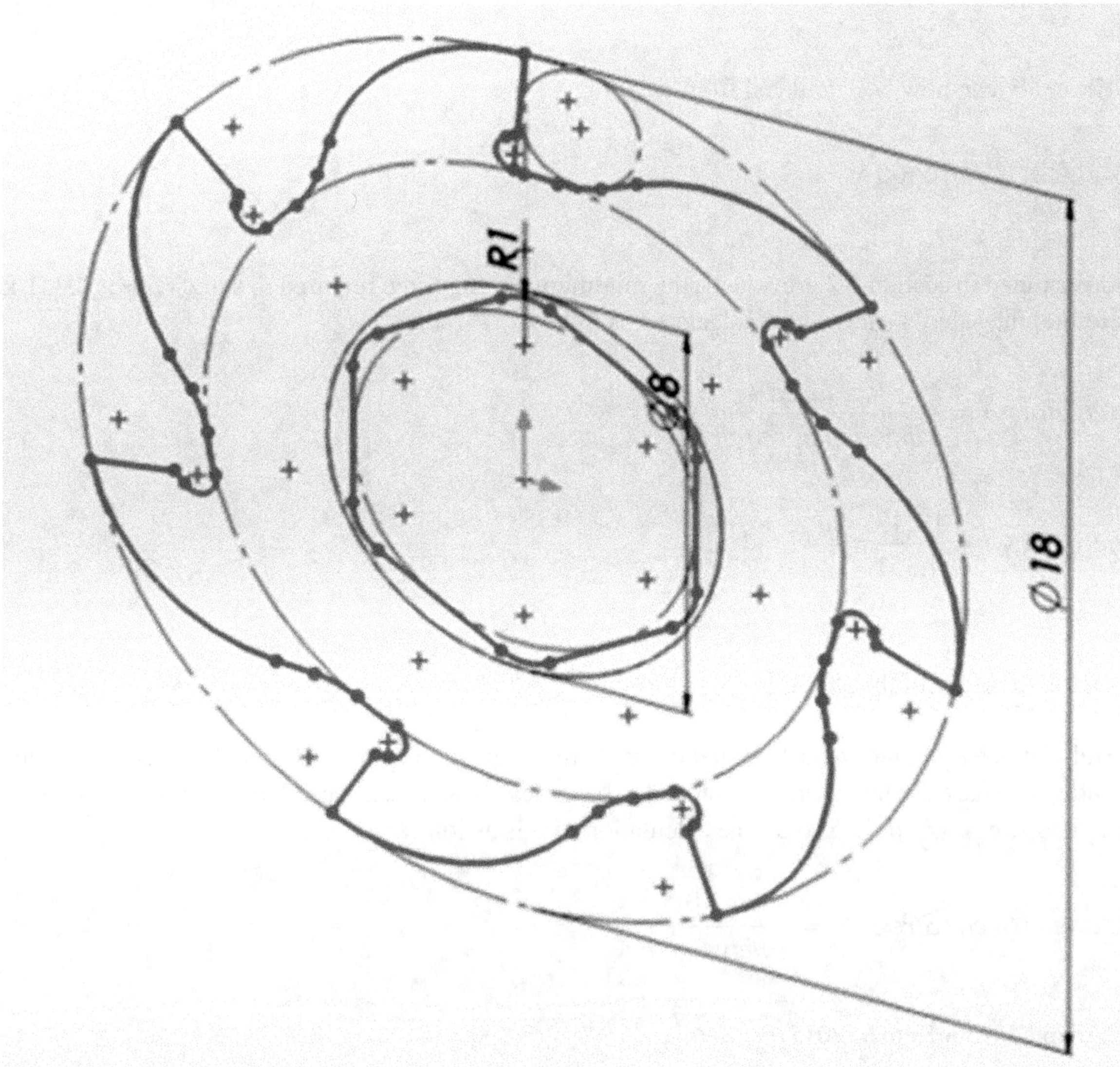

$$\tau = \frac{Motor\ Power, P(W)}{Angular\ Speed\ of\ Shaft, \omega_{shaft}\left(\frac{rad}{s}\right)}$$

$$\tau = \frac{5.5 \times 10^3}{30 \times \frac{2\pi}{60}} = 1750.70 N \cdot m$$

$$r = \frac{D}{2} = \frac{18}{2} = 9cm = 0.09m$$

Hence, F_c can now be calculated from τ *and* r.

$$F_c = \frac{1750.70}{0.09} = 19.45kN$$

As determined in section 2.2 previously, the minimum cutting force required to shred OPF is 4.351 kN. Therefore, the safety factor can be calculated as follow:

$$Safety\ Factor = \frac{Produced\ F_c}{Minimum\ F_c\ required}$$

$$Safety\ Factor = \frac{19.45}{4.351} = 4.47 \approx 4.5$$

5. <u>Hub diameter of the cutter, *d*</u>

Hub diameter of cutter must be equal to the diameter of rotating shaft. To ease and secure the cutter mounting on shaft, the hub shape is designed to be helical. The material used for shaft is NiCr steel with high yield stress, σ_y of 275MPa. The calculation of *d* is as follow:

$$Maximum\ Shear\ Stress, S_s = \frac{\sigma_y \times 0.5}{Safety\ Factor}$$

(Adapted from Kumar, 2015)

$$S_s = \frac{275 \times 10^6 \times 0.5}{4.5} = 30.56MN / m^2$$

$$d = \sqrt[3]{\frac{16\tau}{\pi S_s}}$$

(Adapted from Augusto, 2017)

$$d = \sqrt[3]{\frac{16 \times 1750.70}{\pi \times 30.56 \times 10^6}} = 0.0663m = 6.63cm$$

Thus, the standard rotating shaft with $d = 8cm$ (>6.63cm) is selected.

6. 6. Transmission Ratio and Teeth of Spur Gears

Velocity Ratio, $VR = \frac{Teeth\, of\; Driven\, Gear, N_2}{Teeth\, of\; Pinion\, Gear, N_1}$

To ensure the rotational speed of both rotating shafts to be consistent, gear transmission ratio or velocity ratio, *VR* is chosen to be 1:1

Meanwhile, interference checking is performed to determine the number of gear teeth to ensure a smooth gear meshing.

$$Diametral\ Pitch, P_d = \frac{Number\ of\ teeth, N}{Pitch\, diameter, d}$$

In this case, N_1 or N_2 and *d* are pre-configured as 16 and 10.5cm respectively.

$$P_d = \frac{16}{10.5} = 1.52$$

Interference checking formula is given by:

$$N_2 < \frac{\{N_1^2 \sin^2 \varnothing - 4k^2\}}{4k - 2N_1 \sin^2 \varnothing}$$

(Adapted from Cleghorn, 2016)Where

$N_1\ and\ N_2\ are\, teeth\, number\ of\ pinion\, and\ driven\ gears\ respectively$

$\varnothing\ is\ standard\ pressure\, angle\, of\ gear\left(14.5°, 20°\ or\ 25°\right); 20°\ is\ chosen$

$k = addendum\, of\ gear, a \times P_d\ which\, always\ equals\, 1\ for\, \varnothing = 20°$

Hence,

$$\frac{\{N_1^2 \sin^2 \varnothing - 4k^2\}}{4k - 2N_1 \sin^2 \varnothing} = \frac{\{16^2 \sin^2 20 - 4(1)^2\}}{4(1) - 2(16)\sin^2 20}$$

$$\frac{\left\{N_1^2 \sin^2 \varnothing - 4k^2\right\}}{4k - 2N_1 \sin^2 \varnothing} = 101.07$$

Since $N_2 < \frac{\left\{N_1^2 \sin^2 \varnothing - 4k^2\right\}}{4k - 2N_1 \sin^2 \varnothing}$ (16<101.07), N_1 and N_2 =16 is valid for VR of 1:1 in gear transmission. No interference will occur.

DEVELOPMENT OF THE SHREDDER

CAD/3D-Modelling

3D drawings of product design generated by using Solidworks 2016 as the CAD tool.

Figure 18. 3D design layout of small-scale oil palm waste shredder

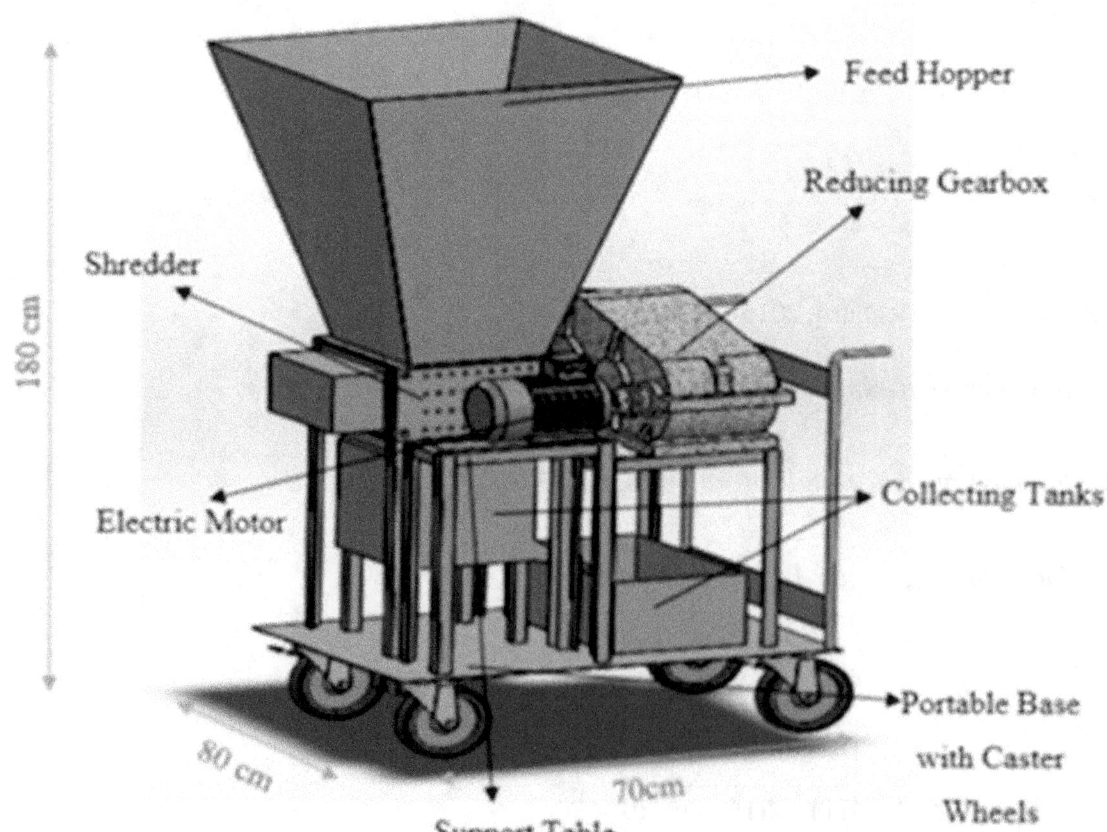

Manufacturing of Product

Fabricated Model

During the fabrication process of machine (both main and auxiliary components), revision has been done to the original design and minor improvement is made on several features to further enhance the feasibility of the machine. In this context, major improvement is performed on the auxiliary components whereas the main components are only improvised slightly.

EXPERIMENTAL VALIDATION OF THE SHREDDER

The developed oil palm waste management machine is operated to carry out shredding test on a few fresh and dry OPF samples and the parameters of output fiber (length and diameter) are measured by scanning electron microscope (SEM). Meanwhile, the moisture content of the OPF fiber is also determined based on oven dry weight basis.

FINDINGS AND SCENARIO ANALYSIS

The developed machine is capable of shredding OPF into pieces and producing fiber within several seconds. Hence, it can be deduced that the machine is efficient in oil palm solid waste management since the hard and long OPF can be easily shredded, as shown in 21.

Table 3. General Product Design Specification (PDS)

S.No.	Components	Quantity	Material
1	Cutters with seven jaws	20	CrSi Steel Alloy
2	Rotating Shaft	2	NiCr Steel Alloy
3	Shaft Collar	20	NiCr Steel Alloy
4	Spacer	20	Aluminium 1060 Alloy
5	Shredder Frame	4	ASTM A36 Steel
6	Spur Gear with 20 teeth	2	20MnCr5 Stainless Steel
7	Bearing Housing	4	ASTM A36 Steel
8	Gear Housing	1	AISI 1020 Low-carbon Steel
9	Feed Hopper	1	Aluminium 1060 Alloy
10	Collecting Tank	1	Aluminium 1060 Alloy
11	Support Table	1	ASTM A36 Steel
12	Machine Base	1	ASTM A36 Steel
13	Handle	2	AISI 1020 Low-carbon Steel
14	Collecting Tank with Flow Channel	1	Aluminium 1060 Alloy

Most of the fiber's cross-sections are in oval or irregular shape. To simplify the analysis, the shapes are assumed to be circular and the diameter is calculated from the average of maximum horizontal and vertical distances of the shapes.

Based on Figure 24, generally the output fiber is short (between 30mm and 80mm). The length of output fiber decreases significantly with the increase of shredding number undergone by OPF samples. On the other hand, the number of shredding does not affect significantly the diameter of the output fiber. Generally, the output fiber is very thin, with a mean value range of 0.53-0.58mm.

Table 4. Manufacturing specification of various components

S.No.	Components	Manufacturing Technology	Assembling Technology	Production/Assembly Machine
1	Cutters with seven jaws	1) Powder Metallurgy (PM) 2) Vacuum Hardening 3) Nitrocarburizing 4) CNC Surface Grinding	Heat Shrink Assembly	1) PM Machine 2) Vacuum Hardening Furnace 3) Nitrocarburizing furnace 4) CNC Milling Machine
2	Rotating Shaft	1) Metal Extrusion 2) Turning	Heat Shrink Assembly	1) Direct extrusion press machine 2) CNC Lathe Machine
3	Shaft Collar	PM	Heat Shrink Assembly	PM Machine
4	Spacer	1) Die Casting 2) CNC Surface Grinding 3) CNC Tapping	Mechanical Fastening	1) Electrical Discharge Machining (EDM) Die-sinking Machine 2) Cold Chamber Die-casting Machine 3) CNC Milling Machine 4) CNC Robotic Arm
5	Shredder Case	1) EDM Cutting 2) CNC Milling 3) CNC Tapping	Mechanical Fastening	1) EDM Machine 2) CNC Milling Machine 3) CNC Robotic Arm
6	Spur Gear with 20 teeth	None (Purchased)	Heat Shrink Assembly	CNC Gear-shaft Hobbing Machine
7	Radial Ball Bearing	None (Purchased)	Heat Shrink Assembly	CNC Gear-shaft Hobbing Machine
8	Bearing Housing	1) Die Casting 2) CNC Surface Grinding 3) CNC Tapping	Mechanical Fastening	1) EDM Die-sinking Machine 2) Cold Chamber Die-casting Machine 3) CNC Milling Machine 4) CNC Robotic Arm
9	Gear Housing	1) Guerin Process 2) CNC Tapping	Mechanical Fastening	1) Punch and Die Device 2) CNC Robotic Arm
10	Feed Hopper	1) Spinning Process 2) CNC Tapping	Mechanical Fastening	1) CNC Milling Machine 2) CNC Robotic Arm
11	Collecting Tank	Ironing Process	None	Punch and Die Machine
12	Support Table	1) Cutting 2) Welding	Fusion Welding	1) Electric Cutter 2) CNC Robotic Arm
13	Machine Base	Cutting	Mechanical Fastening	1) Electric Cutter 2) CNC Robotic Arm

Figure 19. Fabricated model of small-scale oil palm waste shredder machine

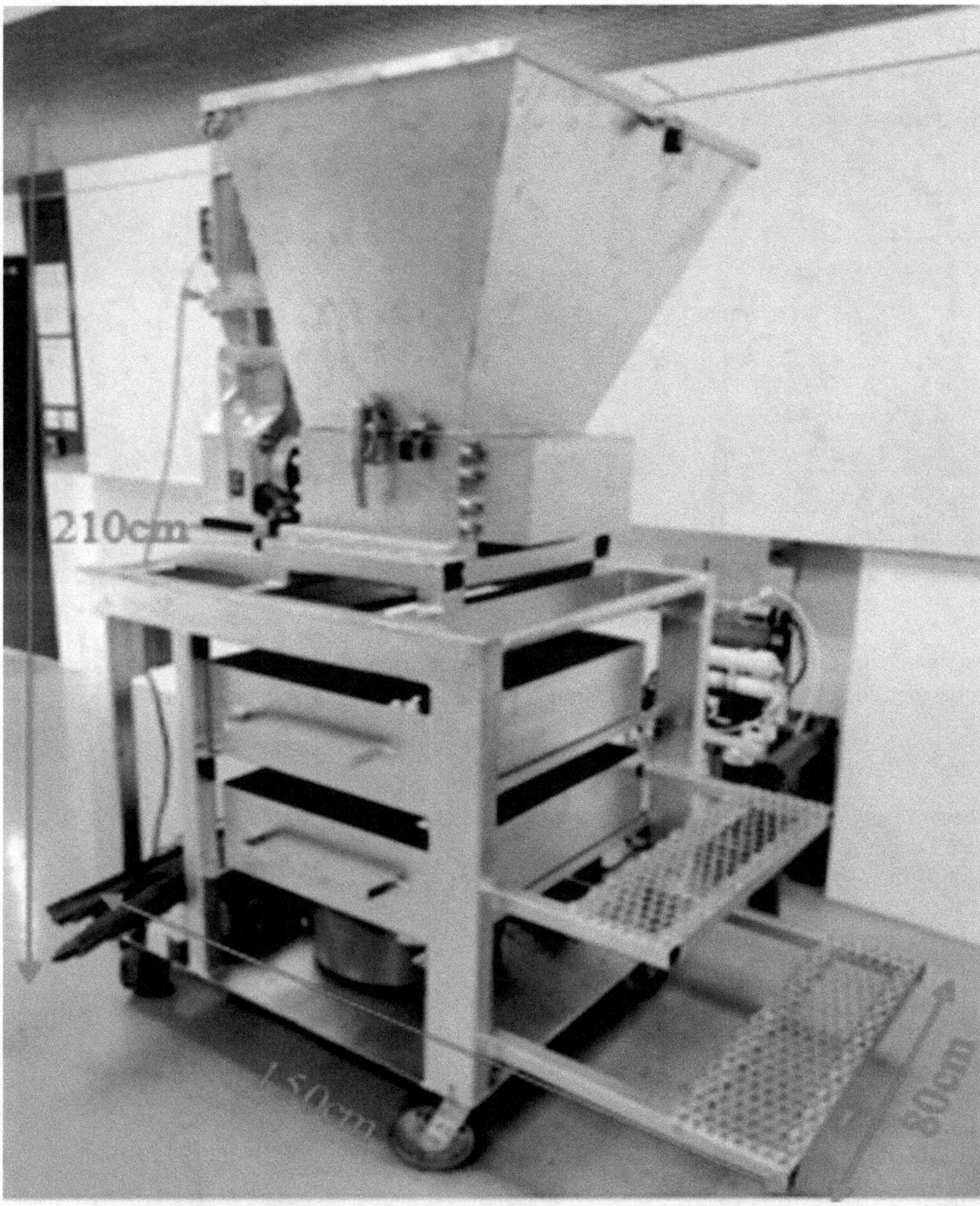

Based on Figure 25, moisture content of fresh OPF fiber is approximately double the dry OPF fiber (64.85% versus 32.88%). It was found that the storage period and condition of the OPF affect the moisture content of output OPF fiber significantly, likewise in the case above (fresh OPF sample is stored within

Figure 20. Output of OPF fiber after shredding

7 days while dry OPF sample is stored more than 1 month). Hence, the moisture content of output fiber can be controlled by the storage period of OPF depending on the exploitation of the fiber. Typically,

Figure 21. Parameter test of output OPF fiber by SEM

Figure 22. Measuring the diameter of fiber in SEM software

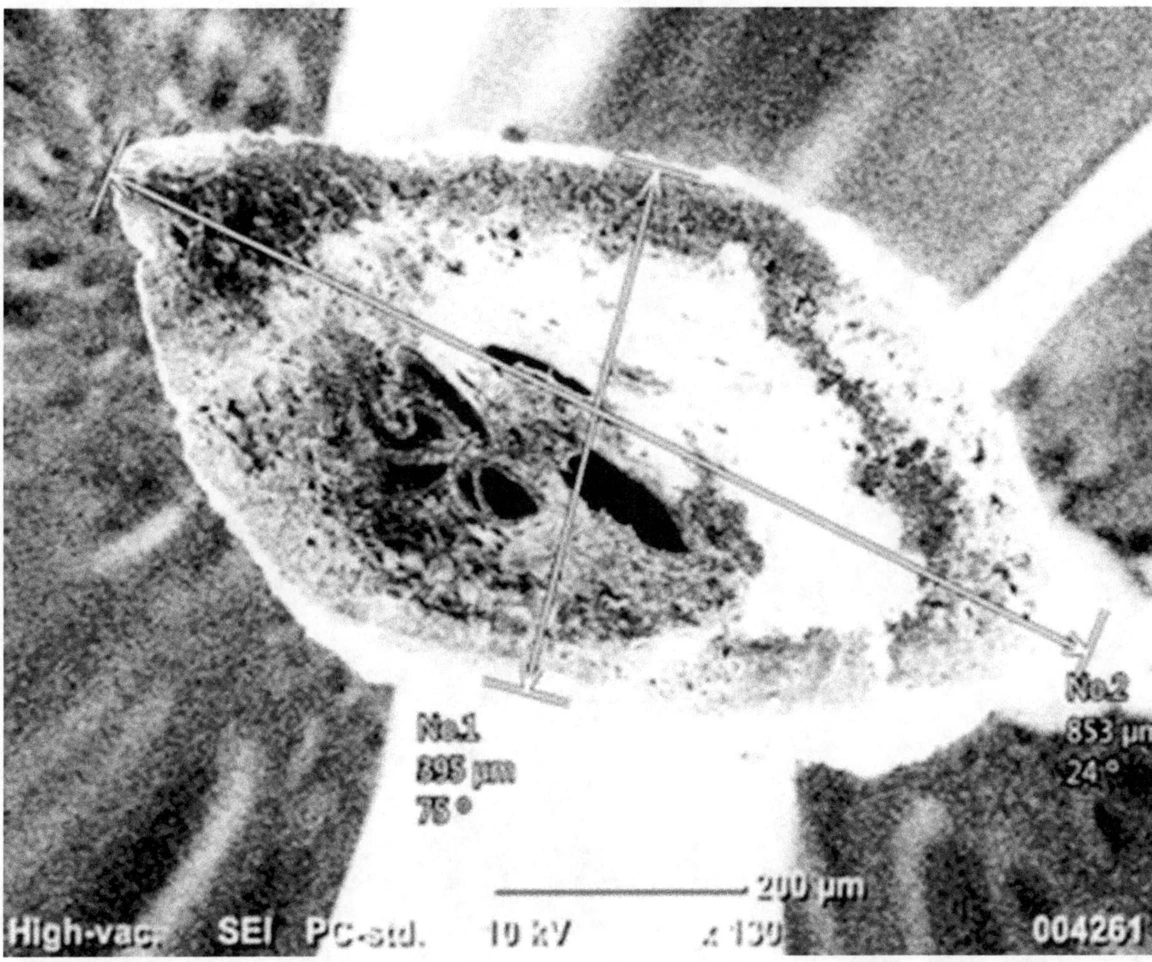

palm fiber with high moisture content is ideal raw material for producing bio-fertilizers whereas dry palm fiber is ideal to be feedstock for industrial boilers. Different applications of output oil palm fiber from the developed machine is summarised as follow:

CONCLUDING REMARKS

The fabricated small-scale oil palm waste shredder in this project is technically feasible from the perspective of cost, portability and efficiency. This machine poses as potential to resolve the major drawbacks of most existing industrial oil palm waste shredders which are unaffordable, heavy, large (requires a huge installation area) and unportable. To rectify the problems, the design concept adopted to develop this machine is to compromise acceptable efficiency with affordable price by integrating all mandatory components of oil palm waste shredder into a comparatively small and portable support structure.

The innovations made in the modified design are: 2 to 3 multiples of size reduction, adoption of double-shafts cutters with 7 jaws technology, installation of oil palm fibre and waste juice collecting

Figure 23. Result of output OPF fiber's length

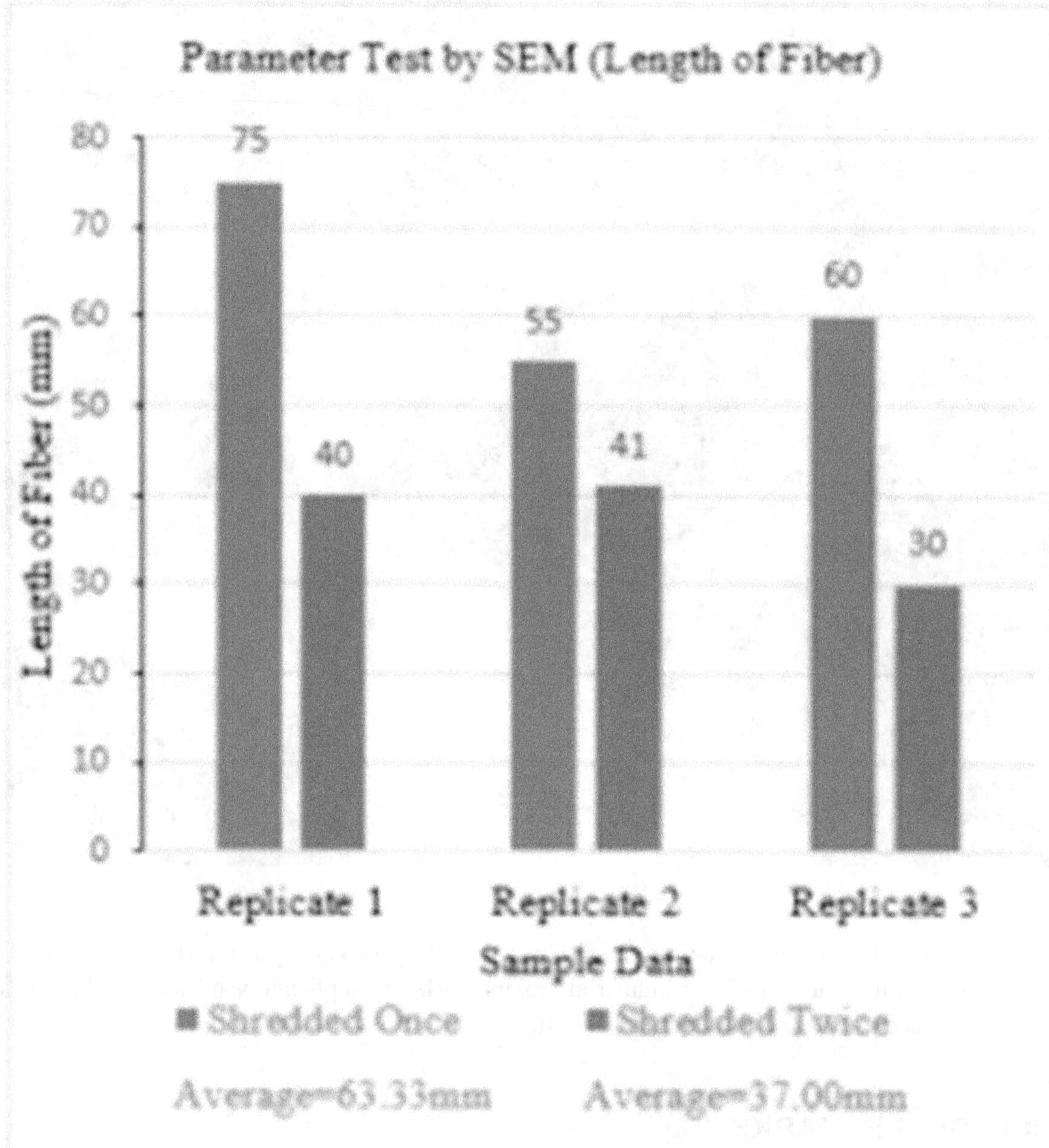

system and deployment of a portable machinery base equipped with caster wheels. Preliminary calculations supported the conceptual design which consists of torque generated from 5.5kW electric motor connected to a ZQ500 reducing gearbox. The cutting force produced is expected up to 19.45kN, which has a safety factor of 4.5 as compared to the maximum compression strength of oil palm fronds (OPFs) which is only 4.351kN. This conceptual design is expected to be able to handle other oil palm wastes like empty fruit bunches (EFBs) as oil palm fronds (OPF) are generally harder than other oil palm wastes.

Figure 24. Result of output OPF fiber's diameter

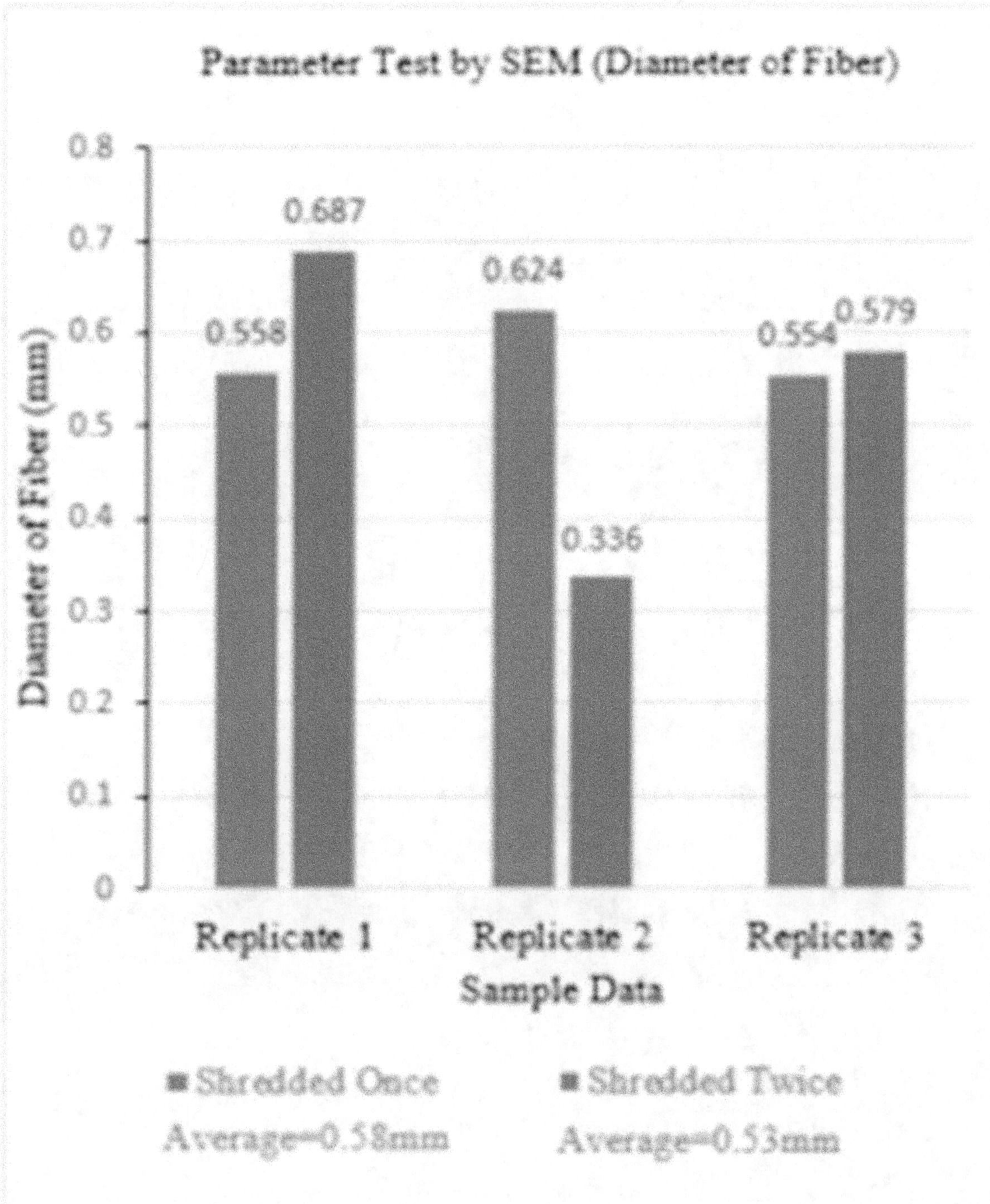

Figure 25. Result of output OPF fiber's moisture content

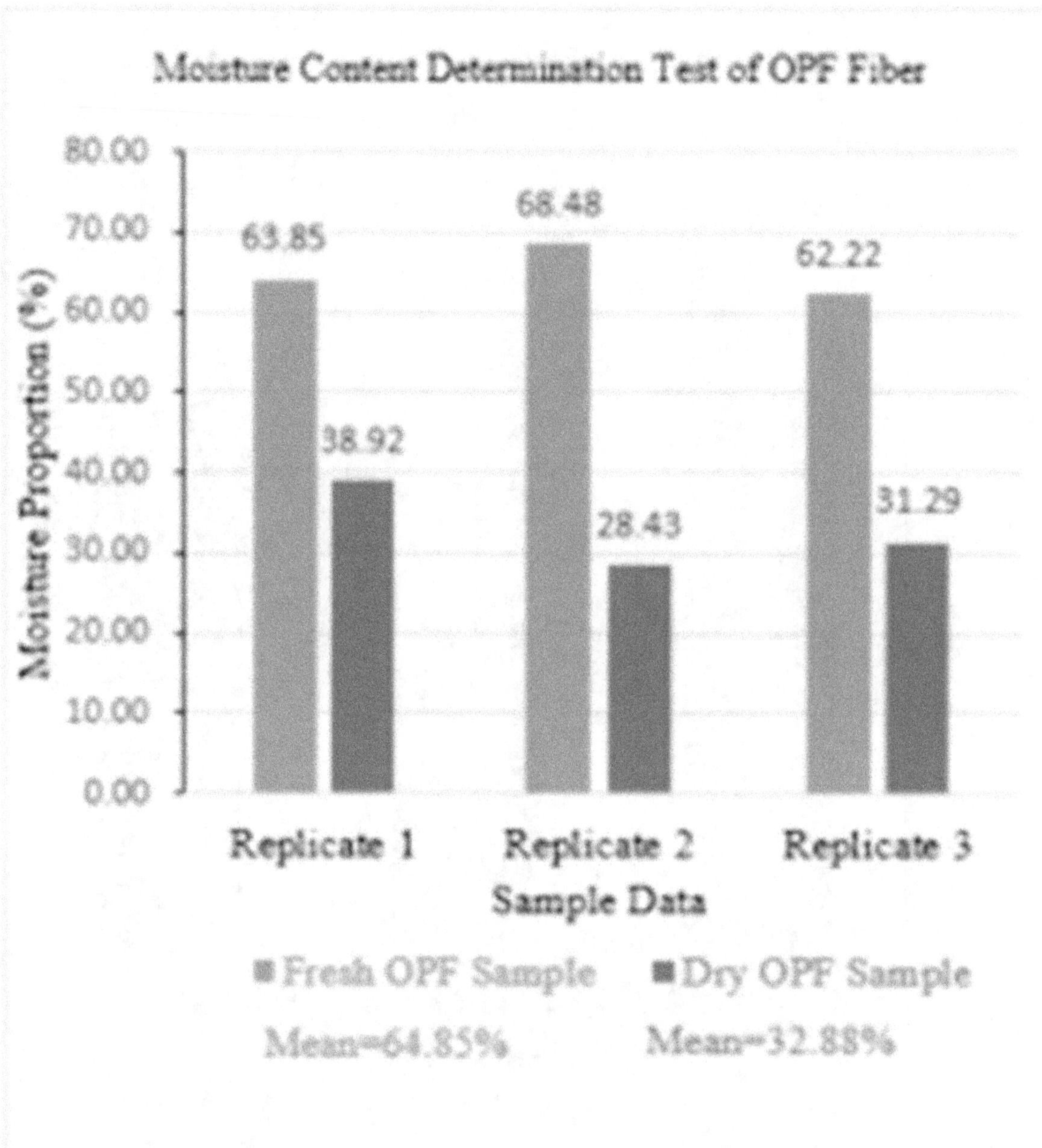

Figure 26. Moisture content of output fiber from different categories

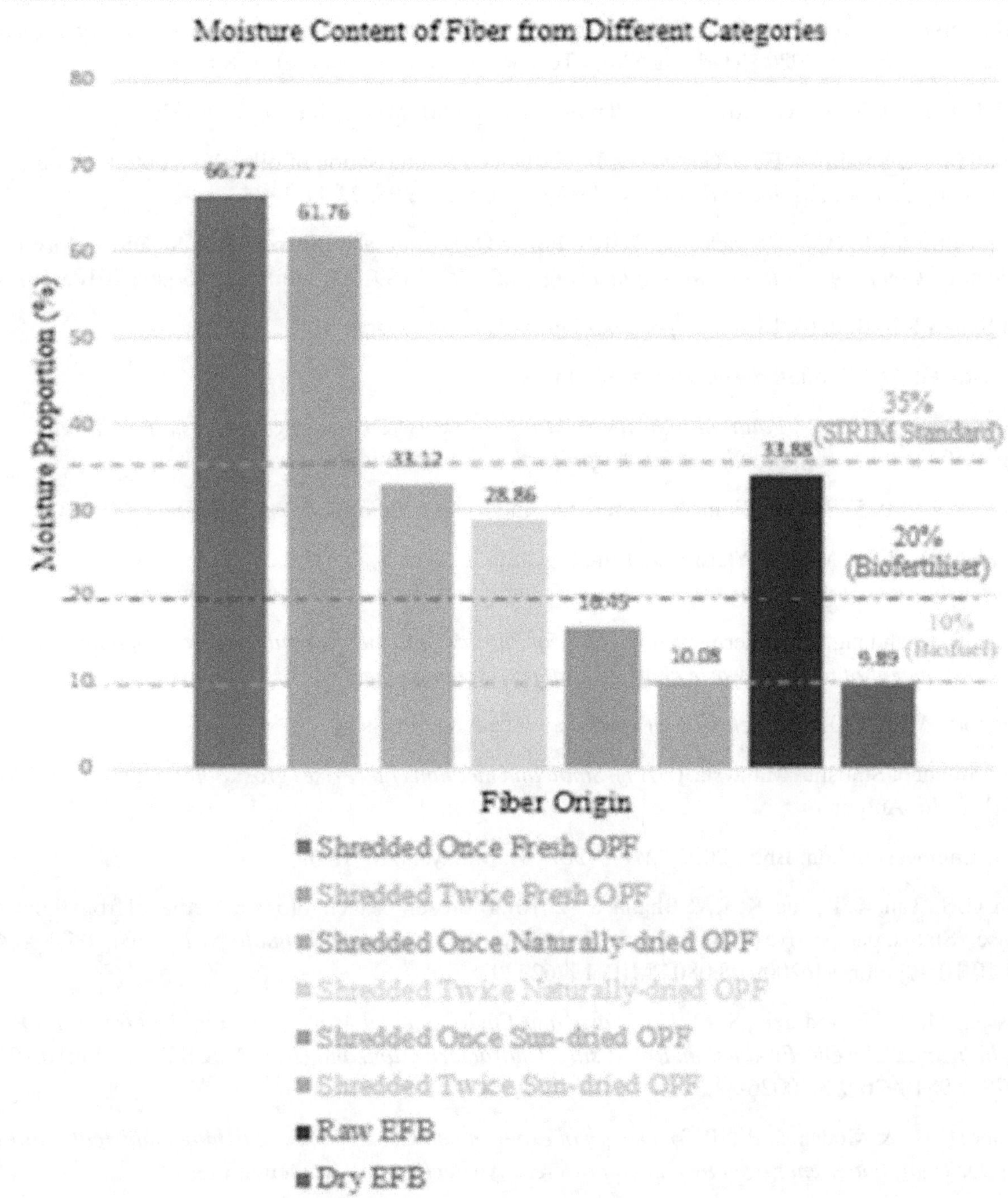

REFERENCES

Abd Rahman, N. I.-G. G. for H. T. (2013). Driving SMEs towards sustainable future. *Malaysian Biomass Industry Action Plan 2020, 1*(Driving SMEs Towards Sustainable Future), 1–80.

Abdullah, N., & Town, G. (2013). *The oil palm wastes in Malaysia*. doi:10.5772/55302

Abdullah, N., Sulaiman, F., & Gerhauser, H. (2011). Characterisation of oil palm empty fruit bunches for fuel application. *The Journal of Physiological Sciences; JPS*, *22*(1), 1–24.

Akkawuttiwanich, P., & Yenradee, P. (2018). Fuzzy QFD approach for managing SCOR performance indicators. *Computers & Industrial Engineering*, *122*(May), 189–201. doi:10.1016/j.cie.2018.05.044

Ant Spirits Sdn Bhd. (2018). *Low speed mini granulator*. Author.

Augusto, G. (2017). *Shaft design*. Academic Press.

Azhar, D. D. P. (2009). *Palm oil industry in Malaysia skills & knowledge for sustained development origin of palm oil palm oil value chain & applications*. Academic Press.

Aziz, D. A. (2015). MIDA Biomass Conference. In *Driving National Biomass Agenda*. MPOB.

Bakar, N. A. B. U., May, C. Y., Lim, J., Lim, S., Chin, E., Loh, S. O. H. K., & Soon, L. I. M. W. (2015). *Improved process for the production of low-ash empty fruit bunch pellet*. Academic Press.

Bulan, R., Mandang, T., & Hermawan, W. (2015). *Physical and mechanical properties of palm frond for the development of palm oil waste chopper and pressing machine design*. Academic Press.

Cleghorn, W. L. (2016). *Mechanics of machines*. Academic Press.

Department of Statistics Malaysia. (2017). *Small and medium enterprises gross domestic product (SMEs GDP) 2016*. Author.

YKL Engineering Sdn. Bhd. (2002). *MY-142874-A*. Malaysia: MyIPO.

Goh, C. S., Tan, K. T., Lee, K. T., & Bhatia, S. (2010). Bioresource technology bio-ethanol from lignocellulose : Status, perspectives and challenges in Malaysia. *Bioresource Technology*, *101*(13), 4834–4841. doi:10.1016/j.biortech.2009.08.080 PMID:19762229

Hassan, M. A., & Abd-aziz, S. (2011). *Waste and Environmental Management in the Malaysian Palm Oil Industry. Palm Oil: Production, Processing, Characterization, and Uses*. AOCS Press. doi:10.1016/B978-0-9818936-9-3.50026-5

Khope, P. B., & Modak, J. P. (2013). *Design of experimental set-up for establishing empirical relationship for Chaff cutter energized by human powered flywheel motor*. Academic Press.

Kumar, I. M. S. (2015). *Design and development Of agricultural waste shredder machine*. Academic Press.

Lim Meng Hon, J. (2010). *A case study on palm empty fruit bunch as energy feedstock*. Academic Press.

Loh, S. K. (2016). The potential of the Malaysian oil palm biomass as a renewable energy source. *Energy Conversion and Management*. Advance online publication. doi:10.1016/j.enconman.2016.08.081

Malaysian-German Chamber of Commerce and Industry (MGCC). (2017). *Oil palm biomass & biogas in Malaysia,2017*. Retrieved from www.malaysia.ahk.de

Oslan, A. M. R., Ahmad, M., & Mohd, K. (2013). *Investigation of oil palm frond properties for use as biomaterials and biofuels investigation of oil palm frond properties for use as biomaterials and biofuels*. Academic Press.

Otieno, N. E., Dai, X., Barba, D. De, Bahman, A., Smedbol, E., Rajeb, M., & Jaton, L. (2016). *Palm oil production in Malaysia : An analytical systems model for balancing economic prosperity, forest conservation and social welfare*. Academic Press.

Otti, V. I., Ifeanyichukwu, H. I., Nwaorum, F. C., & Ogbuagu, F. U. (2014). *Sustainable oil palm waste management in engineering development*. Academic Press.

Pavankumar, S. B., Sachin, K. R., Shankar, R., Thyagaraja, B., & Madhusudhan, T. (2018). *Design and fabrication of organic waste shredding machine*. Academic Press.

Rosli, W., Daud, W., & Law, K. (2011). *Oil palm fibers as papermaking material: potentials and challenges*. Academic Press.

Sakthivel, M., Rajeshkannan, G., Naveenkumar, M., & Muralimanokar, D. M. (2017). Design and analysis of twin shaft shreader using Pro-E and Hyperworks software. *International Journal of Advanced Research in Basic Engineering Sciences and Technology (IJARBEST)*, *3*(24), 836–858.

Sukiran, M. A., Abnisa, F., Wan Daud, W. M. A., Abu Bakar, N., & Loh, S. K. (2017). A review of torrefaction of oil palm solid wastes for biofuel production. *Energy Conversion and Management*, *149*, 101–120. doi:10.1016/j.enconman.2017.07.011

Teoh, J. (2018). *Oil palm fibre*. Academic Press.

The Star. (2017). *SMEs growth outpaces country's GDP growth*. Star Media Group Berhad.

Trisakti, B., Mhardela, P., Trisakti, B., Lubis, J., Hanum, H., & Tantawi, A. R. (2018). *Production of oil palm empty fruit bunch compost for ornamental plant cultivation Production of oil palm empty fruit bunch compost for ornamental plant cultivation*. doi:10.1088/1757-899X/309/1/012094

Wang, D. H. (2016). *Patent landscape report : palm oil production and waste treatment technologies*. Academic Press.

Zainul, I. F. (2013). *SMEs in palm oil urged to venture into refined downstream activities*. Star Media Group Berhad.

Zakil, F. A., Ali, N., Hassan, O., & Pretreatment, A. (2008). *The usage of empty fruit bunch (EFB) and palm pressed fibre (PPF) as substrates for the cultivation of Pleurotus ostreatus*. Academic Press.

ADDITIONAL READING

Ben-David, A. (1995). Monotonicity maintenance in information-theoretic machine learning algorithms. *Machine Learning*, *19*(1), 29–43. doi:10.1007/BF00994659

Breiman, L. (2001). Random Forests. *Machine Learning*, *45*(1), 5–32. doi:10.1023/A:1010933404324

Cardoso, J. S., & da Costa, J. E. P. (2007). Learning to classify ordinal data: The data replication method. *Journal of Machine Learning Research*, *8*, 1393–1429.

Demšar, J. (2006). Statistical comparison of classifiers over multiple datasets. *Journal of Machine Learning Research*, *7*, 1–30.

Guyon, I., & Elisseeff, A. (2003). An introduction to variable and feature selection. *Journal of Machine Learning Research*, *3*, 1157–1182.

KEY TERMS AND DEFINITIONS

Attribute Evaluation: A data mining procedure which estimates the utility of attributes for a given task (usually prediction). Attribute evaluation is used in many data mining tasks, for example in feature subset selection, feature weighting, feature ranking, feature construction, decision and regression tree building, data discretization, visualization, and comprehension.

Context of Attributes: In a given problem, the related attributes, which interact in the description of the problem. Only together, these attributes contain sufficient information for classification of instances. The relevant context may not be the same for all the instances in a given problem.

Evaluation of Ordered Attributes: For ordered attributes, the evaluation procedure should take into account their dual nature: they are nominal, but also behave as numeric attributes. So each value may have its distinct behaviour, but values are also ordered and may have an increasing impact.

Feature Subset Selection: Procedure for reduction of data dimensionality to select the most relevant set of features for a given task trying not to sacrifice the performance.

Feature Weighting: Under the assumption that not all attributes (dimensions) are equally important feature weighting assigns different weights to them and thereby transforms the problem space. It used in data mining tasks where the distance between instances explicitly taken into account.

Non-Myopic Attribute Evaluation: An attribute evaluation procedure which does not assume conditional independence of attributes but takes context into account. It allows proper evaluation of attributes which take part in strong interactions.

Ordered Attribute: An attribute with nominal, but ordered values, for example, increasing levels of satisfaction: low, medium, and high.

Chapter 7
Performance of Chitosan and Polyglutamic Acid in Palm Oil Mill Effluent Treatment

Man Djun Lee
https://orcid.org/0000-0002-1250-9459
University College of Technology, Sarawak, Malaysia

Pui San Lee
https://orcid.org/0000-0001-9631-0983
University College of Technology, Sarawak, Malaysia

ABSTRACT

This chapter presents the study on pollutant removal efficiency in palm oil mill effluent using chitosan and polyglutamic acid (PGA). Up until today, palm oil mill effluent (POME) has been considered one of the significant sources of environmental pollution. The characteristics of POME include contaminating the source of drinking water, which is also harmful to the aquatic ecosystem by creating a highly acidic environment or causing eutrophication. With increasing public awareness of environmental pollution, it creates the need to address this issue. Both chitosan and PGA are non-polluting food-based anionic and biodegradable biopolymers that are environmentally friendly in wastewater treatment. The critical parameter to determine the effectiveness of pollutants removal is chemical oxygen demand, colour, and total suspended solids. In this aspect, this chapter also discussed some of the significant findings done in previous studies to provide proper understandings and implications on this topic.

BACKGROUND

Palm oil industry is a significant industry sector and plays a significant role in Malaysia's economy as one of the largest palm oil producers in the entire world (Liew, Kassim, Muda, Loh, & Affam, 2014). The palm oil industry in Malaysia contributes about 39% of the world palm oil production and also 44% of palm oil world export (Bello & Abdul Raman, 2017). Due to this importance, a large area of land

DOI: 10.4018/978-1-7998-4921-6.ch007

has been converted into oil palm plantation estate. At the same time, more and more palm oil mill has been built to process the increasing amount of oil palm fresh fruit bunch (FFB) into crude palm oil. In Malaysia, there are 400 hectares land planted with oil palm in the year 1920, subsequently with 54,000 hectares in the year 1960, 2,692,286 hectares in the year 1996 and 5,229,739 hectares in the year 2013. At the year 2011, a record of total 426 palm oil mills is in operation in Malaysia with 250 mills in peninsular Malaysia and 176 mills in Sabah and Sarawak (Liew et al., 2014). The growth of the industry at the same time indicates the increase of wastewater or palm oil mill effluent (POME) produced and released into the watercourse, which will bring harm to the environment.

The process of extracting crude palm oil from the fresh fruit bunch consumes much water and therefore produces a large volume of wastewater. In Malaysia, a record of 0.67 cubic meters of POME generated in order to process one ton of FFB (Liew et al., 2014). Approximately 5-7.5 tons of water is required to produce one ton of crude palm oil. Eventually, more than 50% of these water would become POME which is shown in Figure 1 (Bello & Abdul Raman, 2017).

It is approximately 48-72 million tons, and 49-74 million tons of POME was generated in the year 2013 and 2014, respectively. In the year 2014, it estimated 19.66 million tons of crude palm oil produced with roughly 44 million cubic meters of POME generated (Parthasarathy, Mohammed, Fong, Gomes, & Manickam, 2016). In POME generated by processing 1 ton of FFB, it contains about 29-30kg of 30°C, 3-days Biochemical Oxygen Demand (BOD_3) (Liew et al., 2014). From the data of POME produced in the year 2014, if the raw POME discharged into the environment without any further treatment, the BOD discharged is equal to the waste generated by 75 million people which is the 2.5 times of current Malaysia's population (Parthasarathy et al., 2016). POME is also said to be 100 times polluting than domestic sewage (Liew et al., 2014). According to the Department of Environment (DOE) practice, there are two ways of discharging treated POME, which are into water course or land. For the discharge into the watercourse, there are seven contaminants contained in the POME regulated. The regulated parameters are BOD_3, suspended solids (SS), oil and grease (O&G), ammoniacal nitrogen (AN), total nitrogen (TN), pH and temperature. For the discharge onto the land, the only parameter is BOD_3 which set at 5000mg/L. Table 1 shows the characteristic of raw and treated POME obtained from the discharge point of the local palm oil mill in Malaysia and DOE discharge limit (Bello & Abdul Raman, 2017).

The most popular method to treat the POME in Malaysia is the ponding system due to low equipment cost and the system is easy to operate. In Malaysia, there are more than 85% of palm oil mills that are currently adopting this method to reduce the BOD of POME into an acceptable limit which is less than 100mg/L in West Malaysia and 50mg/L in East Malaysia. In the ponding system, the POME undergoes biological treatments which include anaerobic digestion process followed by aerobic ponding with the hydraulic retention time of 40 days or above. However, the ponding system also causes some drawbacks which are long hydraulic retention time (HRT), vast land needed and the release of greenhouse gases (methane). There are also many palm oil mills which are unable to achieve the discharge limit only by using the ponding system (Parthasarathy et al., 2016).

If untreated POME discharge into the watercourse, it will undergo biodegradation process and consume dissolved oxygen in the water which eventually will kill the marine animals, especially fish in the river. The untreated POME, which is acidic, will cause the watercourse to turn acidic and affect the aquatic life. Moreover, the oil content in untreated POME tends to form a thick layer on the water surface that will prevent the absorption of oxygen. The dark brown colour and unpleasant smell of POME will turn the stream into brownish and unacceptable for public consumption (Bello & Abdul Raman, 2017). Apart from that, the high concentration of suspended solids will remain at the bottom of the river and

Figure 1. Palm oil mill effluent

undergo biodegradation, which will produce sludge oxygen demand (SOD) and deplete the dissolved oxygen [4]. In order to protect the environment, DOE Malaysia establishes a standard where the final discharge of treated POME that came out from the mill must be less than 100mg/L of COD. Hence, for POME to have the minimum or no impact on the environment when discharging and to comply with the discharge limits, the palm oil mill must have an effective POME wastewater treatment system. The cost of maintenance and operation of the POME wastewater treatment system, availability of land and location of mill greatly influencing the choice and selection of POME wastewater treatment systems in Malaysia. In return, it will stress the industry players, especially small and medium scale palm oil mill financially. Therefore, the central idea of this chapter is to provide an inexpensive and uncomplicated method for small and medium scale palm oil industries to process POME before discharging to the

Table 1. Characteristic of Raw POME and DOE Discharge Limit (Hassan, Yacob, Shirai, & Hung, 2006)

Parameters	Raw POME	DOE Discharge Limit
Temperature (°C)	85	45
pH	4.2	5.0-9.0
Oil & Grease (mg/L)	6000	50
BOD (mg/L)	25000	100
COD (mg/L)	51000	-
TS (mg/L)	40000	1500
TSS (mg/L)	18000	400
TVS (mg/L)	34000	-
TN (mg/L)	750	200
Color (ADMI)	Above 500	200

watercourse. This chapter provides insights into utilising chitosan and polyglutamic acid in the POME treatment process to remove pollutants that contribute to high COD, colour, and TSS of POME.

PALM OIL MILL EFFLUENT

Source of Palm Oil Effluent

The most common way in extracting palm oil from fresh fruit bunches (FFB) is the wet palm oil milling process. Several stages of wet palm oil milling process required a tremendous amount of water and steam for washing and sterilizing. As a result, this generates a considerable quantity of wastewater or better known as POME from palm oil mill. Figure 2 shows a simplified process flow diagram to produce palm oil.

In a palm oil mill operation using a conventional manufacturing process, there are three primary processing operations responsible for producing the POME. These three primary processes are sterilization of FFB, clarification of the extracted crude palm oil and hydro-cyclone separation of the cracked mixture of kernel and shell. Sterilization process customarily carried out in horizontal cylindrical autoclaves known as sterilizers where the FFBs are cooked with steam at the pressure about 3 atm for 1 to 1.5 hours. The sterilization process aims to inactivate the natural enzymes in the fruits (lipases) and inhibit the splitting of fat into free fatty acid (FFA) and cause oil loss. Besides, the steam sterilization process loosens the fruits from the bunch and soften the mesocarp to ease the oil extraction. This station contributes approximately 36% of total POME (Liew et al., 2014). The clarification process is to separate the oil produced from the press station, which is mixed with water and solid from the bunch fibre. The oil usually is separated from the mixture in the clarifier tank by using gravity, de-sander and also decanter. This station contributes the majority part of the POME, which is 60% (Ta Yeong Wu, Mohammad, Jahim, & Anuar, 2010). The nuts from the nut silo will be cracked by nutcracker in the ripple mill. These cracked kernel and shell mixture are separated in air columns and by a water bath in hydrocyclone. This station only produces around 4% of POME. The POME generated from sterilizer condensate, clarifica-

Figure 2. Typical palm oil process flowchart (Lam & Lee, 2011)

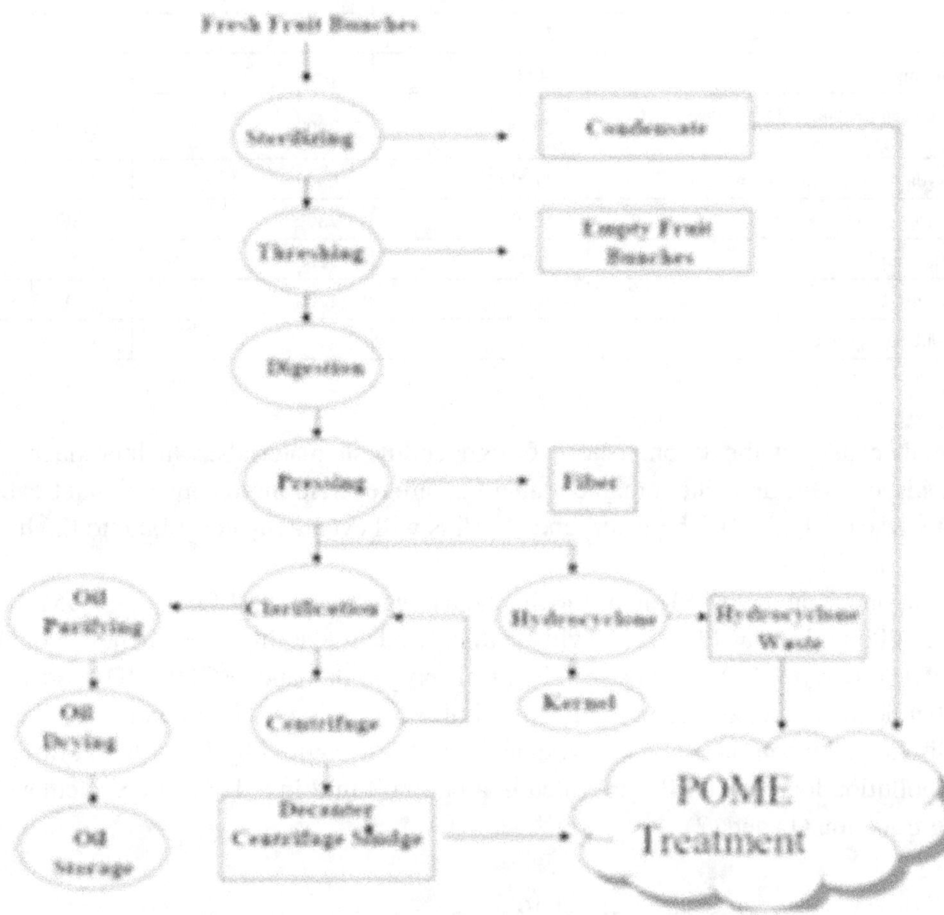

tion of oil and hydro-cyclone is in the ratio of 9:15:1 (36%: 60%: 4%). Table 2 shows the characteristics of different source of wastewater in palm oil mill that combined to produce POME (Liew et al., 2014).

Characteristics of POME

The POME from different mills would have different characteristics due to different oil extraction technique, FFB quality, climate, condition of palm oil processing and mill requirement on POME discharge limit (Ta Yeong Wu et al., 2010). POME is a mixture of water (up to 95%), oil and fine suspended solids (Manickam et al., 2014). The suspended solid (TSS) is the vegetative matter such as cell walls, organelles, short fibres, water-soluble carbohydrates (glucose, reducing sugar and pectin), nitrogenous compound (protein and amino acid), free organic acid, lipids and also combined small organic and mineral constituents (Edewor, 2000). POME is considered as non-toxic waste as the palm oil mills usually do not use any harmful chemical in the entire milling process (Liew et al., 2014). The dark colour of

Table 2. Characteristic of Different Sources of Wastewater

Parameter	Sterilizer condensate	Clarification wastewater	Hydro-cyclone wastewater
pH	5.0	4.5	-
Oil and grease; mg/L	4000	7000	300
Biochemical oxygen demand (BOD) 3 days, 30°C; mg/L	23000	29000	5000
Chemical oxygen demand (COD) ; mg/L	47000	64000	15000
Suspended solids; mg/L	5000	23000	7000
Dissolved solids; mg/L	34000	22000	100
Total nitrogen; mg/L	600	1200	100
Ammoniacal-nitrogen; mg/L	20	40	-

POME is usually caused by the decomposition of lignocellulosic materials; which produces lignin, tannin, humic acids, carotene and other organic matter that are recalcitrant to conventional treatment (Tan, Goh, Lai, Lau, & Ismail, 2014). These suspended solids will eventually contribute to the high BOD of POME (Liew et al., 2014).

In term of organic content, based on the biochemical oxygen demand (BOD), raw POME has an average BOD of 25,000mg/L. Raw POME is highly acidic. Biodegradability of effluent can be determined from the BOD/COD ratio. COD stands for chemical oxygen demand. BOD/COD ratio indicates the fraction of chemically oxidized organics which is eligible for biological degradation. In East Malaysia, the POME discharged when the BOD is less than 50mg/L as required by Department of Environment (DOE). The pollution load of POME generated in a palm oil mill in a day can be calculated based on the following Equation (1) and (2):

$$\textbf{Pollution Load,}\frac{\textbf{kg}}{\textbf{d}}=\frac{\text{POME Flow Rate,}\frac{m^3}{d}\text{ x Concentration, mg / L}}{1000} \quad (1)$$

$$\textbf{POME Flow Rate,}\frac{\textbf{m}^3}{\textbf{d}}=\text{Process capacity}\frac{\text{ton FFB}}{\text{h}}\text{ x Process Efficiency, \%}$$

$$\text{x Operating Hours,}\frac{h}{d}\text{ x POME generated,}\frac{m^3}{\text{ton}} \quad (2)$$

Biochemical Oxygen Demand (BOD)

Biochemical oxygen demand (BOD) is the measure of the amount of oxygen that bacteria will consume during the decomposition of organic matter content under aerobic conditions. BOD test should be carried out according to APHA Standard Method 5510B (American Public Health Association, 2001). BOD is determined by incubating a sealed sample of water for five days and measuring the loss of oxygen from the beginning to the end of the test. The samples are usually diluted before the incubation because the bacteria could deplete all of the oxygen in the bottle before the test is complete (Brown and Caldwell, 2001). It is essential to determine the sample size and dilution ratio before the BOD test, as this will ensure valid BOD results. The pH value of the samples should be in the range of 6.0-8.0, as alkalinity or acidity of samples can prevent bacteria from growing during the BOD test. pH can adjust by adding sodium hydroxide (NaOH) and sulfuric acid (H_2SO_4) (American Public Health Association, 2001). When the test carries out in this way, the BOD usually abbreviated as BOD_5. BOD is a severe problem in natural waters because the dissolved oxygen (DO) of the water can be stressed by BOD oxidation (Doran, Dee, & Techknowledge, 2008).

Chemical Oxygen Demand (COD)

Chemical oxygen demand is a measure of the amount of oxygen required to oxidize all organic material into carbon dioxide and water. COD values usually are higher than BOD values, but COD measurements can be obtained in a few hours while BOD measurements will take around five days (Brown and Caldwell, 2001). Samples heated for 2 hours with sulfuric acid and strong oxidizing agent potassium dichromate. The reduction reaction is shown in Equation (3).

$$6Cl^{-} + Cr_2O_7^{2-} + 14H^{+} \rightarrow 3Cl_2 + 2Cr^{3+} + 7H_2O \quad (3)$$

The amount of Cr^{3+} produced is measured at wavelengths and reflected in mg/L of COD.

Total Suspended Solids (TSS)

Total suspended solids are a measure of suspended matter contained in the wastewater. Suspended solids contain BOD and can impair water quality by adding turbidity and reducing aesthetics. Discharges of SS also caused deposits that developed at the bottom of waterways. In the laboratory, standard filtration and drying method used to measure SS, where the increase of weight of a container/filter is measured, for a known volume of wastewater filtered (Doran et al., 2008). The TSS before and after the experiment measured according to Standard Methods Section 2540 D, and total solids dried from 103°C to 105°C. The treated and the untreated POME samples were evaporated in a weighed dish and dried to a constant weight in an oven from 103°C to 105°C. The increase in weight over the empty dish represents the total solids. TSS calculation is shown in Equation (4).

$$TSS\left(\frac{mg}{L}\right) = \frac{\left(Weight\ of\ dried\ residue + dish - weight\ of\ dish\right) mg\ x1000}{sample\ volume, ml} \quad (4)$$

Conventional Method in POME Treatment

The natural chemical properties of the POME make it easily treated by a biological approach. Currently, there are three biological processes employed in the palm oil industry which are anaerobic, facultative anaerobic, and aerobic treatments. The anaerobic treatment is the major part which is removing pollutant (BOD). It can remove BOD up to 95% (Hassan, Yacob, Shirai, & Hung, 2005). There are four main stages which are hydrolytic, acidogenic, acetogenesis and methanogenic. The hydrolysis process begins with bacteria of insoluble organic polymers (carbohydrate) and complex organic compound (protein and lipid) to make them available for other bacteria. Hydrolytic microorganisms will secrete extracellular enzymes for hydrolysis. This process will convert organics into simpler molecule such as amino acids, glycerol, triglycerides, sugar and fatty acids. Meanwhile, in acidogenesis process, the hydrolyzed or soluble products from the first stage are further broken down by acidogen into simpler organic compound such as volatile fatty acid (VFA), ammonia, carbon dioxide, hydrogen and hydrogen sulfide. In the acetogenesis process, the simple molecule from the previous stage is further digested by acetogens to produce carbon dioxide, hydrogen and acetic acid. For the methanogenesis process, the acetic acid, hydrogen and VFA are converted to methane, carbon dioxide and water by methanogens.

The ponding system is a combination of a series anaerobic, facultative, and algae (aerobic) ponds, as shown in Figure 3 and 4. Ponding system primarily anaerobic and facultative ponds require less energy as it does not need mechanical mixing, operation control or monitoring. The major drawback of the ponding system is a large area of land is needed to accommodate a series of ponds to achieve the discharge limit (Hassan et al., 2005). In constructing the ponds, depth is the primary consideration for different types of ponds. However, the optimum depth for the anaerobic pond is 5-7m, the facultative anaerobic pond is 1-1.5m and aerobic pond is 0.5-1m. The sufficient hydraulic retention time (HRT) of anaerobic, facultative anaerobic and aerobic ponds are 45, 20 and 14 days, respectively (Hassan et al., 2005). The problems arise from the ponding system is the formation of scum. Scum form when the bubbles rise to the surface together with the fine suspended solids. It is caused by the presence of oil and grease in the POME. Another drawback of the ponding system is the solid sludge accumulates at the bottom of the ponds. It will affect the effectiveness of the pond as it decreases the volumetric capacity and hydraulics retention time (HRT) (Hassan et al., 2005). Therefore, de-sludging is required when the sludge is more than one-third of the pond. About 85% of the palm oil mills that POME in Malaysia adopted ponding system because it is inexpensive, low capital, simple and easy to handle (Zainal, Jalani, Mamat, & Astimar, 2017). The palm oil industry is widely favoured to the ponding system as only clay lining of ponds is needed and can be constructed easily by excavating hence low marginal cost (Hassan et al., 2006).

The combination of open digester and ponding system is another type of conventional POME treatment system that combines an open digester tank with a series of ponds. Figure 5 shows a typical open digester tank. Digester tank may build with various volumetric capacities ranging from 600 until 3600m^3. In this treatment method, digester has the same function as the anaerobic pond. It carries out the anaerobic digestion. The output of the POME from the digester will then enter facultative anaerobic ponds and then algae (aerobic) ponds. The digester can decrease the BOD in a shorter time than the pond. The HRT for digester is only 20-25 days which is a lot shorter than anaerobic ponding system. Although it is proven that the digester is more effective than anaerobic ponds, it brings some drawback to the operator. The disadvantages of digester include scum formation on the top, sludge accumulation at the bottom and the corrosion of the steel structure of the digesters due to prolonged exposure to hydrogen sulfide. There are incidents which reported that the digester burst and collapsed (Hassan et al., 2005).

Figure 3. Series of Ponds for POME treatment

Extended aeration is to complement the previous conventional treatment system, which shown in Figure 6. In this treatment method, mechanical surface aerators are introduced at the aerobic ponds to supply more oxygen to the ponds. It can reduce the BOD in POME effectively by aerobic processes. Usually, the surface aerators are installed at the end of the ponds before discharging the POME. This

Figure 4. Typical configuration for ponding treatment system for POME (Zainal et al., 2017)

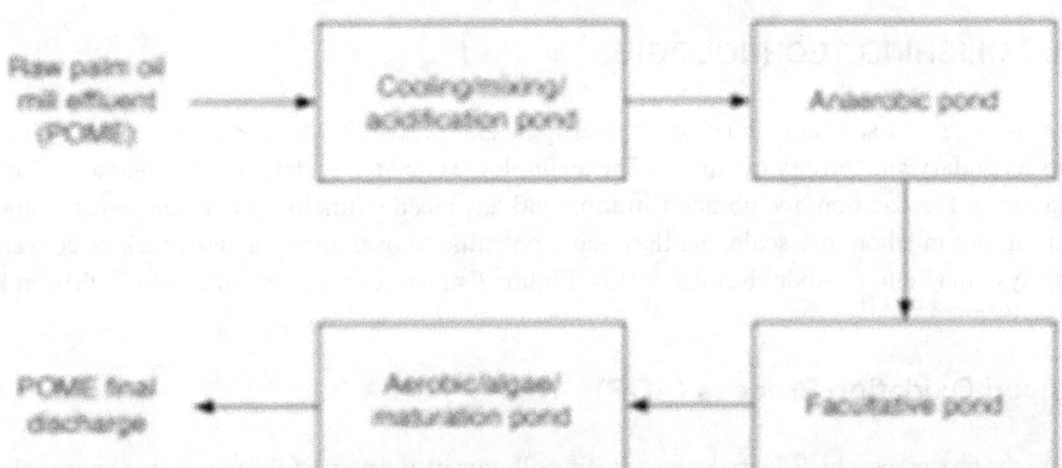

Figure 5. Typical open digester tank

treatment method is useful only used when the land area is a constraint and does not permit extensive wastewater treatment (Hassan et al., 2005).

POME POLISHING TECHNOLOGIES

In recent years, many studies conducted to investigate alternative POME treatment technologies, especially in secondary and tertiary treatment. The technologies that are widely investigated are adsorption, coagulation or flocculation, membrane filtration and advanced oxidation processes. Most of these investigations are in laboratory scale, but they show potential to overcome the drawback of conventional ponding system (Bello & Abdul Raman, 2017). Figure 7 shows an overview of recent POME polishing technologies.

Advanced Oxidation Process (AOP)

AOPs are the processes which degrade the organic pollutant by the powerful and reactive hydroxyl radical (OH•). Hydroxyl radical (OH•) generated would have an oxidation potential of 2.8eV OH• can generate

Figure 6. Surface aerator for POME ponds

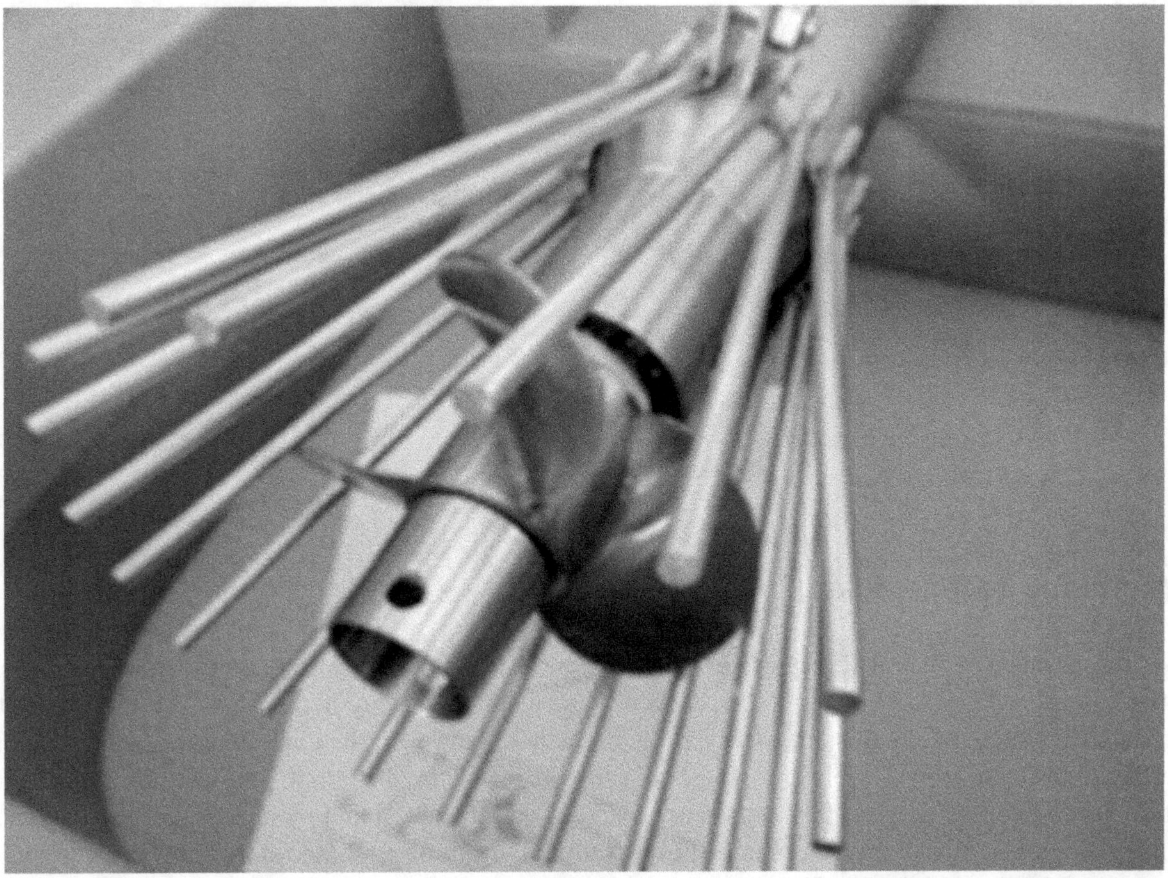

through either one or a combined of chemical oxidation by using H_2O_2, ozone, ultrasound and radiation assisted source (ultraviolet) (Soon & Hameed, 2011). During the treatment of wastewater, OH• will attack the organic pollutants and convert them to CO_2, H_2O and inorganic salt (Neyens & Baeyens, 2003). AOPs can effectively degrade the pollutants and have its advantages of non-selectively, mineralisation of pollutants and ease of operation compared to the conventional methods. The most popular AOPs are Fenton oxidation, photocatalysis, ultrasound cavitation and ozonation. Fenton oxidation uses the reaction between Fe^{2+} and H_2O_2 to produce OH• (Gar, Tawfik, & Ookawara, 2015). Photocatalysis applies metal oxide (such as TiO_2) in the presence of irradiation (UV and Vis) to produce OH• (Dong, Zhang, He, Dong, & Wang, 2014). Ozonation uses the ozone, which is a powerful oxidant with high thermodynamic oxidation potential (Aparicio, Eiroa, Kennes, & Veiga, 2007). Ultrasound (US) cavitation uses ultrasound to oxidize the pollutants. AOPs are more effective when combined two or more AOPs due to more OH• is generated, lower catalyst consumption and shorter process time (Buthiyappan, Raman, Aziz, Mohd, & Wan, 2016). AOPs have successfully adopted as tertiary treatment of wastewater such as olive oil mill wastewater (OOWW), agrochemicals, pulp and paper, textile wastewater and pharmaceutical (Bello & Abdul Raman, 2017).

Figure 7. Overview of Recent POME Polishing Technologies

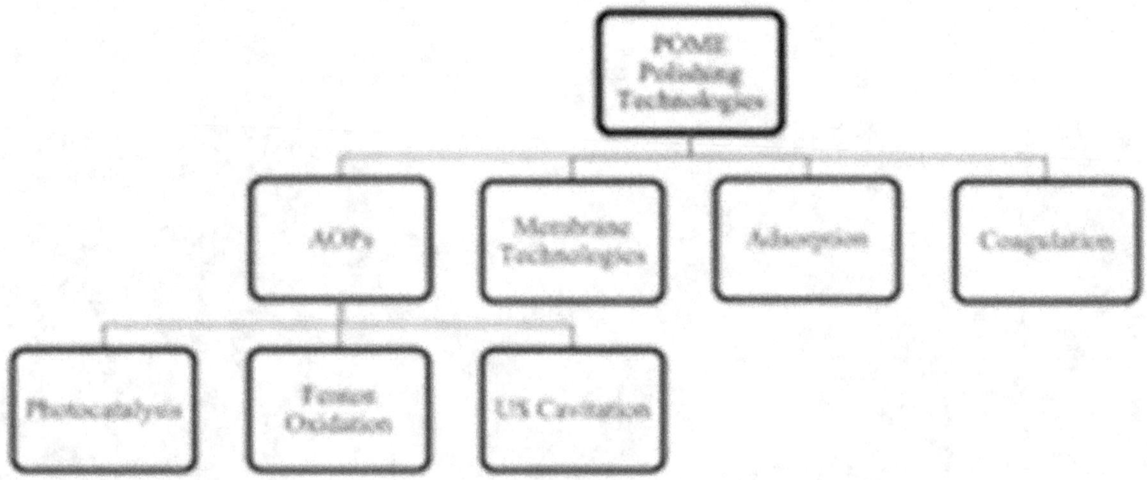

Membrane Technologies

Besides advance oxidation processes (AOPs), membrane technology is also one of the popular polishing methods of POME. The advantages of membrane technologies are high removal rate, modularity, and ability to integrate with other water treatment method. However, the main drawback of membrane technologies is that the membrane fouling will cause significant reduction in permeate flow due to the surface and pore-blocking of the membrane. The high initial capital and maintenance costs have also limited the application of the membrane. The most commonly used membrane in membrane technologies are microfiltration (MF), ultrafiltration (UF), nano-filtration (NF) and reverse osmosis (RO) (Bello & Abdul Raman, 2017). There is an argument about membrane technologies in removing COD from POME compared to other technology. Higher pressure might have provided higher treatment efficiency but at the same time also contributes to the increasing rate of membrane fouling. The effectiveness of membrane technologies in POME treatment depends on the properties of the membrane. Nano-filtration performs better than ultrafiltration, but it has a higher level of fouling compared to ultrafiltration. Membrane technologies can be combined with other technology such as coagulation and flocculation to increase their treatment effectiveness (Bello & Abdul Raman, 2017). Table 3 shows some of the application of membrane technologies used in POME polishing.

Adsorption Technologies

For adsorption technologies, it is a physicochemical separation process involving inter-phase transfer between an adsorbent and a solution. The pollutant in the solution (adsorbate) absorb onto the surface of the adsorbent. Adsorption can be a reversible process which offers the possibility of adsorbent regeneration through desorption (Fu & Wang, 2011). Adsorption mechanism mainly influenced by the physical forces (physisorption) or chemical interactions (chemisorption) between the adsorbent and adsorbate. The adsorption is also influenced by characteristics of the adsorbent such as specific surface area, porosity and surface charge. Chemical structure of the adsorbate and environmental condition such as temperature,

Table 3. The Application of Membrane Technologies for POME Polishing

Process	Pressure (bar)	pH	Time (hr)	COD removal (%)	BOD removal (%)	Solids removal (%)	Color Removal (%)	Ref.
Membrane NF+UF	1-5	-	4	-	-	-	97.9	(Amat et al., 2015)
Membrane UF	5	9.05	4	88	-	80	-	(Said, Ahmad, Wahab, Tusirin, & Nor, 2014)
Membrane UF	2	8	1.5	90	90	-	-	(N. S. Azmi & Yunos, 2014)
Membrane UF	-	6.6-7	-	95	-	79	-	(N. A. Azmi, Yunos, & Zakaria, 2012)
Membrane UF	0.2-0.8	-	4.5	57	-	97.7	-	(T Y Wu, Mohammad, Jahim, & Anuar, 2007)
Membrane UF	1-7	1-14	-	60	60	-	97	(Ahmad, Ismail, & Bhatia, 2005)
Membrane UF+RO	UF: 0-7 RO: 0-60	5.5-6 0.5	-	98.8	99.4	-	-	(Latifahmad, Ismail, & Bhatia, 2003)

pH, solubility and ionic strength will influence the adsorption performance (Kyriakopoulos, Doulia, & Hourdakis, n.d.). Table 4 shows some of the application of adsorption process in POME polishing.

Coagulation and Flocculation Technologies

Coagulation process commonly used to remove the organic matter and suspended solids (SS) from the wastewater. During the coagulation process, the chemical is added into the wastewater to enhance the flocculation and sedimentation. It will help in removing dissolved solids and suspended solids from the wastewater. Aluminium and iron-based compounds coagulant are often used in water treatment, as they are simple, easy to handle, cheap, and have excellent removal efficiency (Keeley, Smith, Judd, & Jarvis, 2014).

Nevertheless, the residual aluminium and iron concentrations may inhibit the biological treatment process in wastewater due to the reduction of microorganism respiration rate and low organic matter elimination (Lees, Noble, Hewitt, & Parsons, 2001). The drawbacks of these chemicals are high cost, non-biodegradable and possible adverse effect of the chemical. Recently, interests have shift to natural and biodegradable coagulants such as PGA, cotton, chitosan, natural seed gum, *Jatropha curcas* seeds, and *Moringa oleifera* (Ang, Mohammad, Benamor, & Hilal, 2016; Bhatia, Othman, & Ahmad, 2007; Nourani, Baghdadi, Javan, & Bidhendi, 2016; Pui, Shak, & Wu, 2014). It is because chemical coagulants are non-biodegradable, costly and not environmental-friendly. The coagulation technologies can also combine with other polishing technologies such as adsorption, membrane filtration and AOPs to achieve better pollutant removal. Table 5 shows the application of coagulation-flocculation for POME polishing.

CHITOSAN

Chitosan is a biopolymer coagulant which is non-toxic, biodegradable, renewable and environmental friendly (Crini & Badot, 2008). Chitosan is a type of marine polymer which has widely used in practical fields such as wastewater management, pharmacology, biochemistry and biomedical. Chitosan is a cellulose-like polyelectrolyte biopolymer which derived from de-acetylation of chitin, as shown in

Table 4. Application of Adsorption for POME Polishing

Process/Adsorbent dosage	pH	Time (min)	COD removal (%)	Colour removal (%)	TSS removal (%)	Ref.
BP/Natural clay: 5g/L	7	90	95	95	95	(Said, Hasan, Tusirin, & Nor, 2015)
BP/GAC: 150g/L	4	87.9	-	90	-	(Alkhatib, Mamun, & Akbar, 2014)
BP/Fly ash: 90g/L	4	60	89	97	96	(Bello & Abdul Raman, 2017)
FB/Resin: 0.3mBH	9.28	-	88	98	-	(Bello & Abdul Raman, 2017)
BP/AC: 10g/L	8.5	30	98	-	99	(Ridzuan et al., 2013)
BP/Banana peel: 300g/L	2	1800	100	95.96	100	(Fong, 2014)
BP/AC: 200g/L	-	120	98.99	79.3	98.45	(Mohammed, Ketabachi, & Mckay, 2014)
FB/Resin: 0.3mBH	3	-	72	64	-	(Belo et al., 2013)
BP/Zeolite: 10g/L	3	50	-	-	-	(Shavandi, Haddadian, Ismail, Abdullah, & Abidin, 2012)

Figure 8. Chitin can easily found in marine nature, and it is occurring in the insects, yeast, fungi and exoskeletons of crustaceans (Ahmad et al., 2006). Chitosan contains a high amount of amino functions that provide novel binding properties for heavy metals in wastewater (Ravi Kumar, 2000). Chitosan can coagulate effectively at pH less than 4.5 as strong acidic condition exaggerates POME to form unstable flocs (Ahmad, Sumathi, & Hameed, 2005). The mechanisms involved in the coagulation can divide into two main categories which are charge neutralization or electrostatic interaction and sweep coagulation/co-precipitation (Ng et al., 2013). The chitosan coagulation process is charge neutralization while synthetic coagulant such as ferric chloride ($FeCl_3$) is sweep coagulation as shown in Figure 9 (Ang, Mohammad, Teow, Benamor, & Hilal, 2015). The flocs formed by charge neutralization are smaller than the flocs formed through sweep coagulation (Jarvis, Jefferson, & Parsons, 2006). The smaller sized flocs could bring fouling risk to the membrane if membrane technologies are used.

Performance of Chitosan in POME Treatment

The optimum condition for coagulation treatment with chitosan as a coagulant is about 100ppm(mg/L) of POME at pH 4.5. The removal percentage for the COD, the colour and the TSS is 15.39%, 85.79 and 97%, respectively. The results are shown in Figure 10. The graph in Figure 10 showed that any further increase in dosage does not increase the colour and the TSS removal significantly. However, a further increase in dosage causes the COD to increase. The negative result of the COD removal observed with the addition of chitosan, which is a natural biopolymer coagulant (an impurity) that causes the COD to increase when the dosage exceeds its saturation point. In essence, the low COD removal was due to

Table 5. Application of Coagulation-flocculation in POME Treatment

Coagulant type/dosage	pH	Mixing rate (rpm)	Time (min)	COD removal (%)	TSS removal (%)	Oil and Grease removal (%)	Ref.
Calcium lactate: 50mg/L	-	258	23	58	58	-	(Zahrim, Nasimah, & Hilal, 2014)
Alum: 2124mg/L	6	-	20	59	-	-	(Taylor, Malakahmad, & Chuan, 2013)
Chitosan: 0.5g/L	4	100	15	-	95	95	(Ahmad, Sumathi, & Hameed, 2006)
Mango Pit: 50g/L Fly ash: 90g/L	4	200	60	89	96	-	(Bello & Abdul Raman, 2017)
PAC: 0.6g/L + AC: 10g/L	8.5	50	30	98	99	-	(Ridzuan et al., 2013)
PAC: 2g/L	-	150	5	93	-	-	(Poh, Ong, Lau, & Chong, 2014)

natural properties of POME as chitosan is not effective in removing dissolved solid (Musonge, 2014). Typical raw POME has a total solid of 40,000 mg/L, while 34,000 mg/L of it is dissolved solid (Bello & Abdul Raman, 2017). Furthermore, TSS removal is very effective at low chitosan dosage (Ahmad et al., 2006). However, chitosan is effective in removing suspended solids that contributes to the COD but not total dissolved solids (TDS).

Performance of Chitosan Paired with other Method in POME Treatment

This study was done by combining ultrasound (US) cavitation, chitosan and ferric chloride ($FeCl_3$) in different ways to determine the best combination and order of treatment. Every treatment method is conducted by following the result of optimum condition obtained from previous studies. The result is shown in Figure 11.

From the graph in Figure 11, the COD removal for the combination of ultrasound (US) cavitation, followed by ferric chloride coagulation treatment, is the highest, at 56.26%. Besides, the colour removal for this combination is the highest among other combinations, at 92.41%.

Furthermore, among all these combinations where chitosan is present, the COD removal percentage is less substantially, ranging from 35.1% to 40.12%, which can be observe at the combination of US- FeC13 (ultrasound and ferric chloride) that shows the COD removal at 56.26%. However, when chitosan added after it, the COD removal percentage drops to 38.61%, due to chitosan being a natural coagulant that is biodegradable (Crini & Badot, 2008). Chitosan is not very useful in coagulating the organic pollutant (COD), which dissolved in the POME (Musonge, 2014). Therefore, chitosan will become the pollutant, contributing to COD and causing the COD removal percentage to decrease. Even though ferric chloride performs better when paired with other polishing methods, the dosage usage of chitosan in POME treatment is lesser and hence more superior in terms of cost-effectiveness and environmentally friendly method for palm oil mills in dealing with wastewaters.

Figure 8. Derivation of chitosan from chitin (Crini & Badot, 2008)

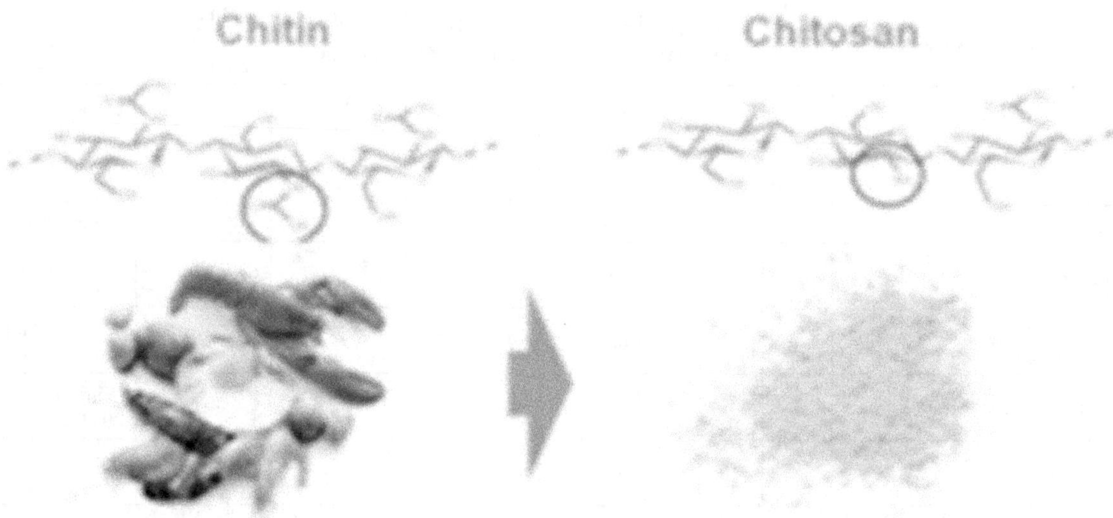

POLYGLUTAMIC ACID (PGA)

PGA is a water-soluble, biodegradable polymer that is produced by microbial fermentation. PGA is produced by several Bacillus species as an extracellular polymer and is completely biodegradable and non-toxic to humans. Figure 12 shows the derivation of PGA from Japanese food source named natto. Due to its non-toxicity, PGA has appeared in many medical applications such as drug delivery, biological glues (Richard & Margaritis, n.d.). PGA' multi-functionality, biodegradability, nontoxicity, compatibility, and edibility have made it a promising biopolymer for application as health food, thickener, osteoporosis-preventing factor, and stabilizer in the food industry; as moisturiser in cosmetics; as chelating agent in waste-water treatment; as hydrogel (especially superabsorbent polymer, SAP) for environmental, agricultural, and biomedical product applications; as biodegradable packing material, and in many other possible applications including Liquid Crystal Displays (LCDs) and conductive display material, drug deliverer, gene vector, curative biological adhesive, dispersant, and enzyme-immobilizing material (Sung et al., 2005). PGA has numerous applications and the potential can be seen in water treatment as natural coagulant that would not contribute any harmful byproducts to the environment after treatment process.

Performance of PGA in POME Treatment

The effect of PGA dosage on the removed percentage of COD under the medium of various pH is shown in Figure 13. The removal of COD was optimum with a dosage of 800ppm at pH 3, where 45.16% of COD removed. As POME is produced in the pH range between 3 to 5, PGA is indeed useful in POME treatment without modifying its pH. This result further clarifies that PGA in wastewater treatment is dosage-dependent, and pH is not the critical parameter affecting its effectiveness (Carvajal-zarrabal et al., 2012).

Figure 9. Chitosan Charge Neutralization and FeCl3 Sweep Coagulation

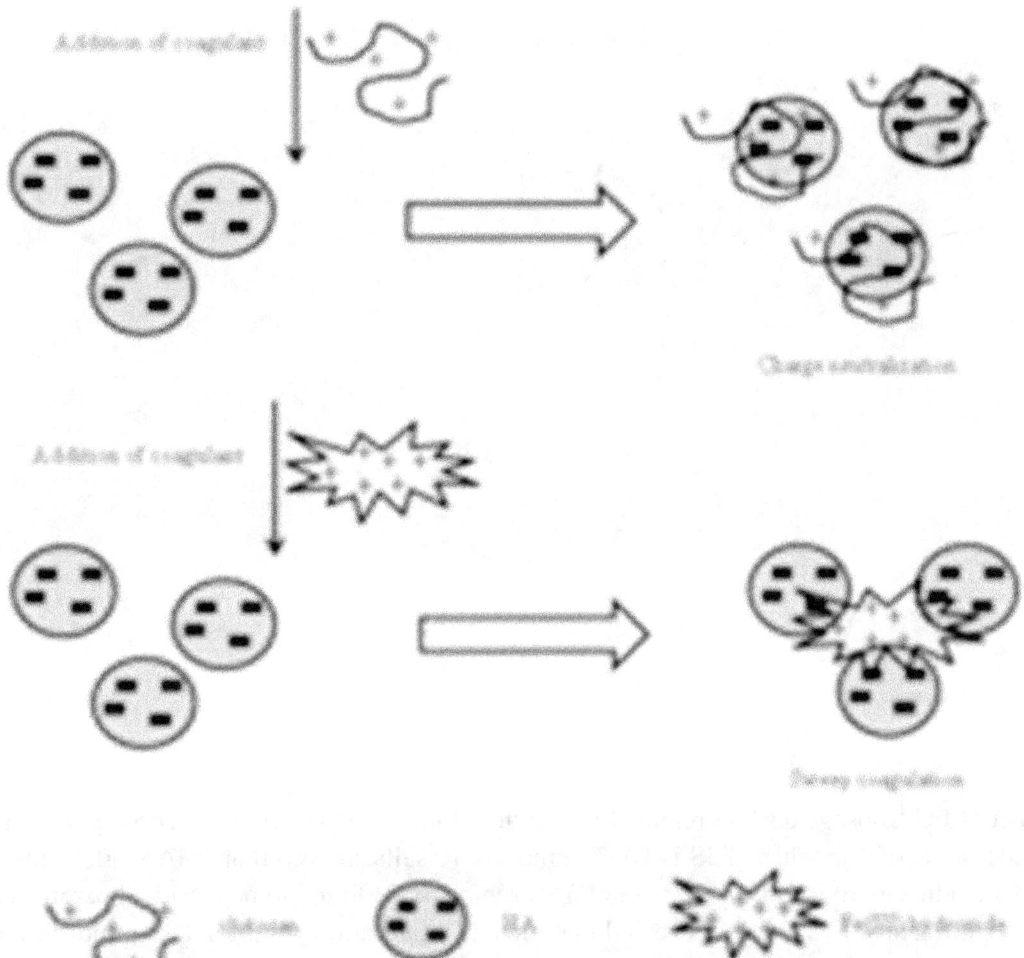

The effect of PGA dosage on colour removal percentage in POME treatment shown in Figure 14. The removal of colour was the highest with the dosage of 1600ppm and pH of 5, where 92.68% of colour removed. It is apparent from the figure that the effectiveness of PGA on colour removal suggest better colour removal with every increasing in PGA dosage. It is because the colour removal of polyglutamic acid is dosage-dependent (Pan, Shi, & Zhang, 2014). However, in industrial practice, increasing dosage means an increase in total production cost. Overall, the result suggests that the optimum dosage of PGA with 800ppm at 77.6% of colour removal as any increase in dosage will not yield any significant results but would only contribute significantly to the total cost. There are similarities between the result from this study with a study conducted by Inbaraj [10] in which PGA (9.9 x 105 Da) is useful in the elimination of basic dyes from aqueous solution. On top of that, up to 98% of dye adsorbed on PGA could recover at pH 1, which promotes the reuse of spent PGA. By recycling the PGA, it would contribute to environmental sustainability.

Figure 10. Effect of Chitosan dosage in pollutant removal (COD, Colour and TSS)

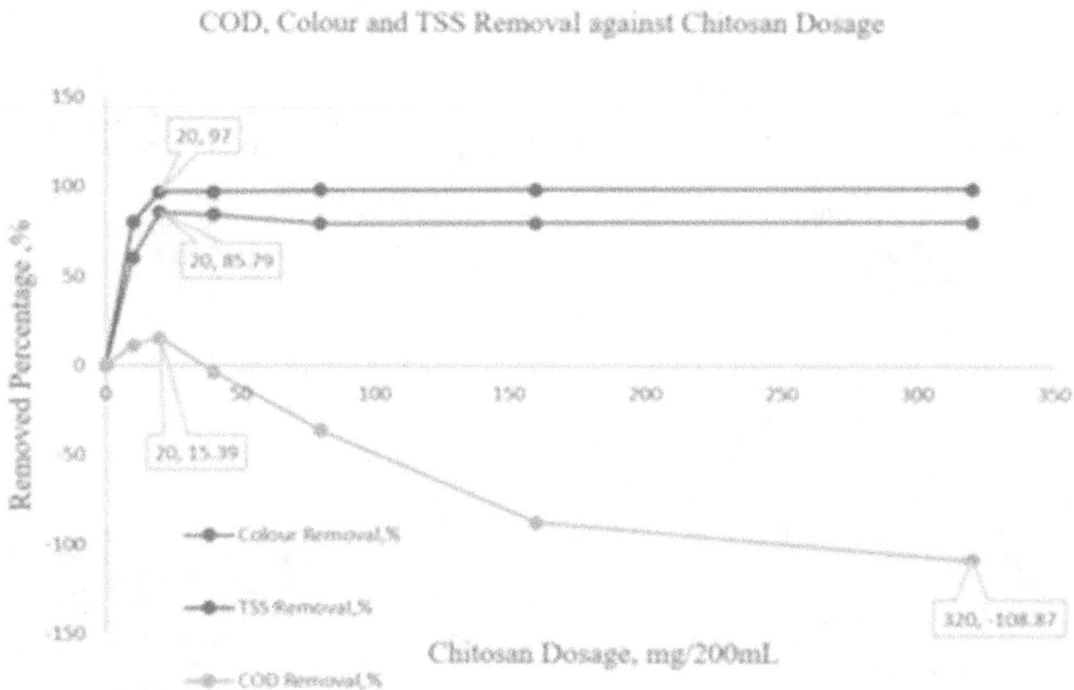

The effect of PGA dosage on TSS removal is shown in Figure 15. The optimum dosage found to be 800ppm under a pH of 3 in which TSS is 100% removed. Results showed that PGA works better with POME under acidic environment. In the case of TSS removal, pH found to be a critical parameter that contributes significantly. Results also revealed that there is a saturation point for TSS removal in every pH. Surprisingly, any increase in pH after the saturation point will not increase TSS removal but contributes directly to the increase in TSS. This is due to the fact that PGA itself become a pollutant after saturation level reached.

The summary of all the result under various pH with the dosage of 800ppm recorded in Figure 16. The best performance for 800ppm of PGA works best under pH of 3 with the highest removal of TSS (100%), colour (77.6%) and COD (45.16%). The findings also show that as pH increases, the pollutant removal efficiency reduces. Since most POME discharged at pH between 3 to 5, PGA is significant in pollutant removal for POME without the need for pH modification. It also contributes to environmental sustainability as no chemicals needed in modifying the pH of the POME. The use of PGA contributes to save time as the pollutant removal process only take minutes and more superior compared to ponding method (bioprocess) could take days to weeks to effectively remove pollutants that contribute to COD, colour and TSS.

Figure 11. COD and colour removal percentage with the different treatment combination

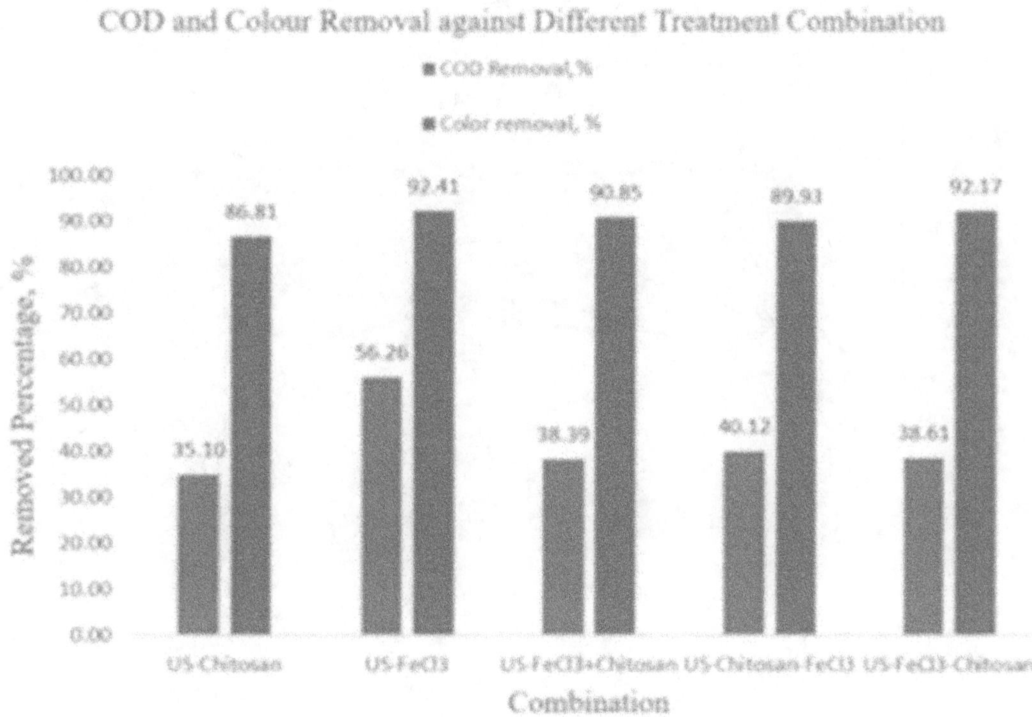

Figure 12. Derivation of polyglutamic acid from natto

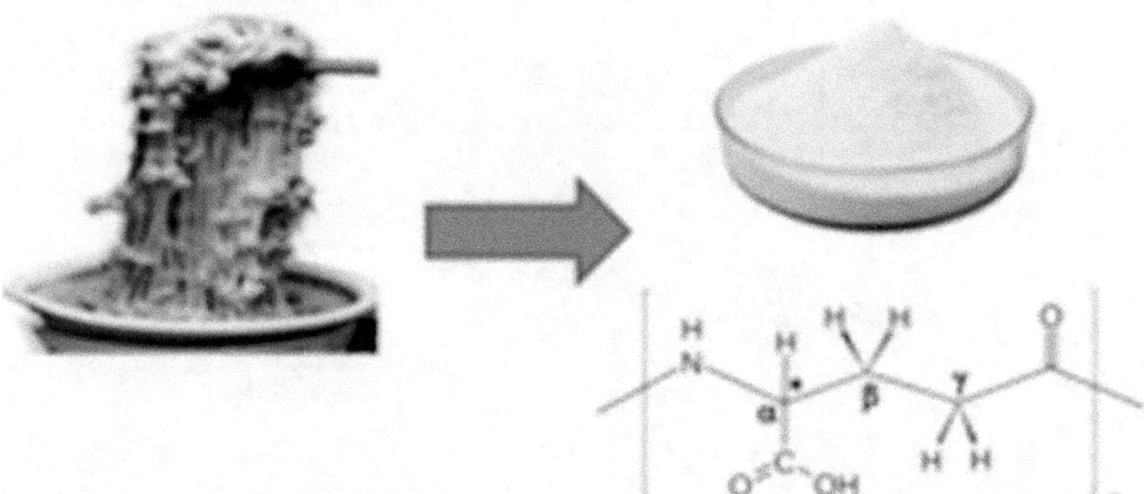

Figure 13. COD removal percentage vs PGA under various pH

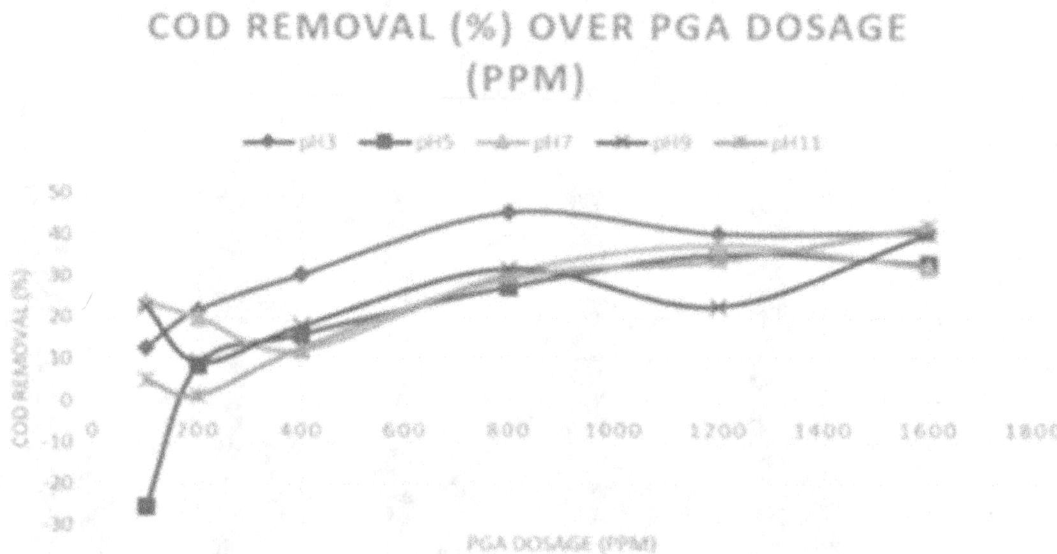

Figure 14. Colour removal percentage vs PGA dosage under various pH

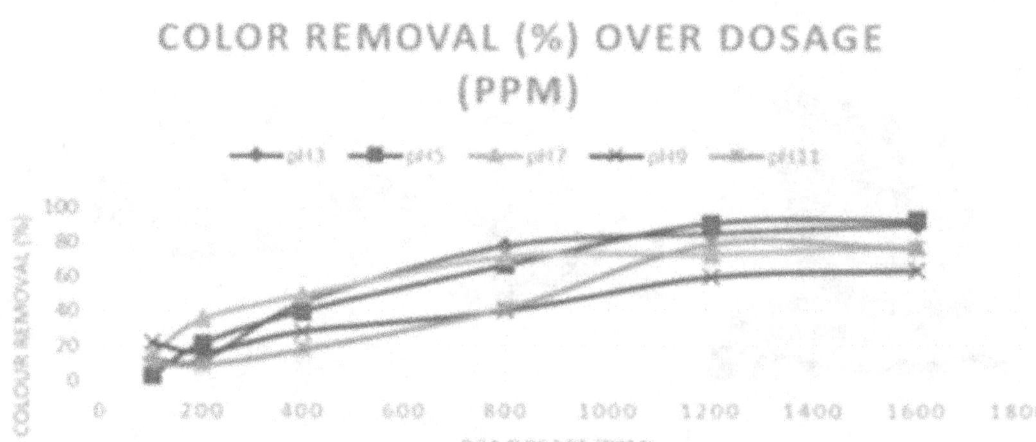

Figure 15. TSS removal percentage vs PGA dosage under various pH

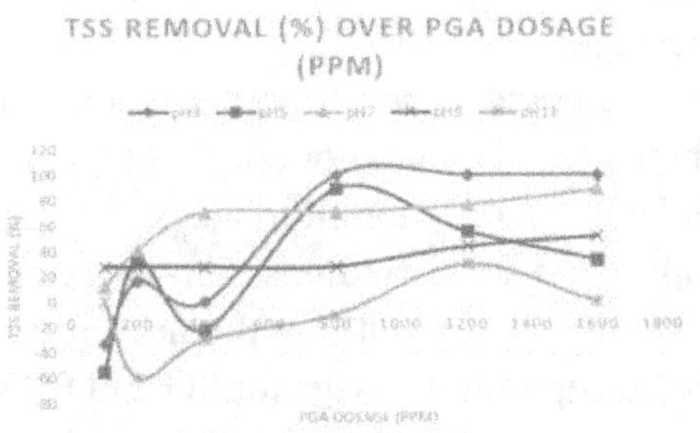

CONCLUDING REMARKS

This chapter presents the treatment performance of palm oil mill effluent by utilizing chitosan and PGA. Both chitosan and PGA are natural, food-based and environmentally friendly biopolymer which has enormous potential to be used to treat POME before discharged to watercourse. Some of the methods for combinations, as suggested in the chapter. The pollutant removal performance measure in terms of COD, colour, and TSS removal percentage. The main contribution of this chapter is to provide a low cost and simple method to small and medium oil palm processing industry in processing their wastewater before discharge to the environment. For chitosan, the main advantage is that low dosage would contribute

Figure 16. COD, Colour, TSS removal percentage under various pH

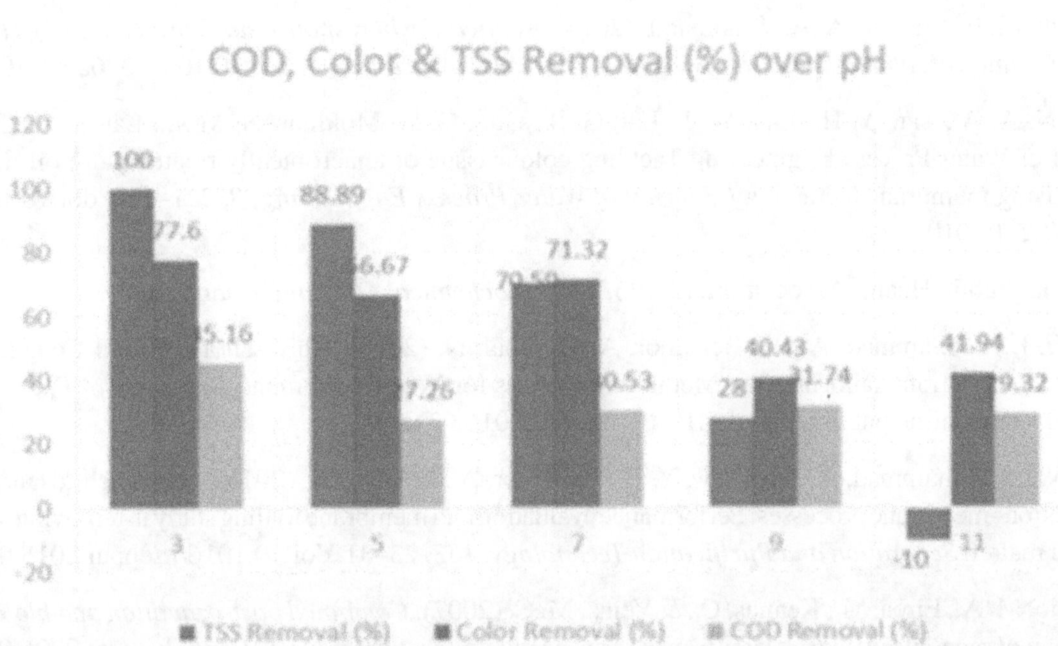

to high removal of suspended solids in POME. However, the disadvantage is that if paired with other methods such as ultrasound cavitation and ferric chloride, it would not have significant improvement in terms of pollutant removal percentage.

On the other hand, ferric chloride could work with other methods to improve pollutant removal significantly. Nevertheless, utilizing chitosan would not contribute to significant increment in the overall treatment cost, which would encourage palm oil mill to adapt this method in treating their wastewater. On the other hand, previous studies and some findings demonstrated that PGA work best in an acidic environment which fits POME characteristics without the need for pH and temperature modification. This chapter also highlighted that the optimum dosage for PGA in POME treatment found to be 800ppm with a pH of 3. For future studies, it recommended pairing PGA with other wastewater polishing methods such as ultrasound cavitation to increase pollutant removal.

REFERENCES

Ahmad, A. L., Ismail, S., & Bhatia, S. (2005). Membrane treatment for palm oil mill effluent: Effect of transmembrane pressure and crossflow velocity. *Desalination, 179*(1-3), 245–255. doi:10.1016/j.desal.2004.11.071

Ahmad, A. L., Sumathi, S., & Hameed, B. H. (2005). Adsorption of residue oil from palm oil mill effluent using powder and flake chitosan: Equilibrium and kinetic studies. *Water Research, 39*(12), 2483–2494. doi:10.1016/j.watres.2005.03.035 PMID:15985277

Ahmad, A. L., Sumathi, S., & Hameed, B. H. (2006). Coagulation of residue oil and suspended solid in palm oil mill effluent by chitosan, alum and PAC. *Chemical Engineering Journal, 118*(1–2), 99–105. doi:10.1016/j.cej.2006.02.001

Alkhatib, M. F., Mamun, A. A., & Akbar, I. (2014). *Application of response surface methodology (RSM) for optimization of color removal from POME by granular activated carbon*. doi:10.100713762-014-0504-4

Amat, N. A. A., Tan, Y. H., Lau, W. J., Lai, G. S., Ong, C. S., Mokhtar, N. M., ... Lai, S. O. (2015). Journal of Water Process Engineering Tackling colour issue of anaerobically-treated palm oil mill effluent using membrane technology. *Journal of Water Process Engineering, 8*, 221–226. doi:10.1016/j.jwpe.2015.10.010

American Public Health Association. (2001). *5210 Biochemical Oxygen Demand*. BOD.

Ang, W. L., Mohammad, A. W., Benamor, A., & Hilal, N. (2016). Chitosan as Natural Coagulant in Hybrid Coagulation-Nanofiltration Membrane Process for Water Treatment. *Biochemical Pharmacology*. Advance online publication. doi:10.1016/j.jece.2016.03.029

Ang, W. L., Mohammad, A. W., Teow, Y. H., Benamor, A., & Hilal, N. (2015). Hybrid chitosan/FeCl3 coagulation–membrane processes: Performance evaluation and membrane fouling study in removing natural organic matter. *Separation and Purification Technology, 152*, 23–31. doi:10.1016/j.seppur.2015.07.053

Aparicio, M. A., Eiroa, M., Kennes, C., & Veiga, M. C. (2007). *Combined post-ozonation and biological treatment of recalcitrant wastewater from a resin-producing factory*. doi:10.1016/j.jhazmat.2006.09.025

Azmi, N. A., Yunos, K. F. M., & Zakaria, R. (2012). Application ofsandwich membrane for the treatment of palm oil mill effluent (POME) for water reuse. Procedia Engineering, 44, 1980–1981. doi:10.1016/j.proeng.2012.09.014

Azmi, N. S., & Yunos, K. F. M. (2014). Wastewater Treatment of Palm Oil Mill Effluent (POME) by Ultrafiltration Membrane Separation Technique Coupled with Adsorption Treatment as Pre-treatment. *Agriculture and Agricultural Science Procedia*, *2*, 257–264. doi:10.1016/j.aaspro.2014.11.037

Bello, M. M., & Abdul Raman, A. A. (2017). Trend and current practices of palm oil mill effluent polishing: Application of advanced oxidation processes and their future perspectives. *Journal of Environmental Management*, *198*, 170–182. doi:10.1016/j.jenvman.2017.04.050 PMID:28460324

Bhatia, S., Othman, Z., & Ahmad, A. L. (2007). *Coagulation –flocculation process for POME treatment using Moringa oleifera seeds extract : Optimization studies*. doi:10.1016/j.cej.2007.01.034

Brown & Caldwell. (2001). Description of Commonly Considered Water Quality Constituents. *Watershed Protection Plan Development Guidebook*, 1–11.

Buthiyappan, A., Raman, A., Aziz, A., Mohd, W., & Wan, A. (2016). *Recent advances and prospects of catalytic advanced oxidation process in treating textile effluents*. doi:10.1515/revce-2015-0034

Carvajal-zarrabal, O., Nolasco-hipólito, C., Barradas-dermitz, D. M., Hayward-jones, P. M., Aguilar-uscanga, M. G., & Bujang, K. (2012). Treatment of vinasse from tequila production using polyglutamic acid. *Journal of Environmental Management*, *95*, S66–S70. doi:10.1016/j.jenvman.2011.05.001 PMID:21600690

Crini, G., & Badot, P. M. (2008). Application of chitosan, a natural aminopolysaccharide, for dye removal from aqueous solutions by adsorption processes using batch studies: A review of recent literature. *Progress in Polymer Science (Oxford)*, *33*(4), 399–447. doi:10.1016/j.progpolymsci.2007.11.001

Dong, S., Zhang, X., He, F., Dong, S., & Wang, B. (2014). *Visible-light photocatalytic degradation of methyl orange over spherical activated carbon-supported and Er 3 + : YAlO 3 -doped TiO 2 in a fluidized bed*. doi:10.1002/jctb.4391

Doran, M. D., Dee, P. E., & Techknowledge, M. D. (2008). *Wastewater and Wastewater Treatment Very Basics*. Academic Press.

Edewor, J. O. (2000). *A Comparison of Treatment Methods for Palm Oil Mill Effluent (POME) Wastes*. Academic Press.

Fong, M. (2014). *Treatment and decolorization of biologically treated Palm Oil Mill Eff luent (POME) using banana peel as novel biosorbent*. doi:10.1016/j.jenvman.2013.11.031

Fu, F., & Wang, Q. (2011). Removal of heavy metal ions from wastewaters : A review. *Journal of Environmental Management*, *92*(3), 407–418. doi:10.1016/j.jenvman.2010.11.011 PMID:21138785

Gar, M., Tawfik, A., & Ookawara, S. (2015). Journal of Water Process Engineering Comparison of solar TiO 2 photocatalysis and solar photo-Fenton for treatment of pesticides industry wastewater : Operational conditions, kinetics, and costs. *Journal of Water Process Engineering*, *8*, 55–63. doi:10.1016/j.jwpe.2015.09.007

Hassan, M. A., Yacob, S., Shirai, Y., & Hung, Y. (2005). Treatment of Palm Oil Wastewaters. *Waste Treatment in the Food Processing Industry*, 101–117. doi:10.1201/9781420037128.ch4

Hassan, M. A., Yacob, S., Shirai, Y., & Hung, Y.-T. (2006). Treatment of Palm Oil Wastewaters. *The Science of the Total Environment*, *366*(1), 187–196. doi:10.1201/9781420037128.ch4 PMID:16125215

Jarvis, P., Jefferson, B., & Parsons, S. A. (2006). Floc structural characteristics using conventional coagulation for a high doc, low alkalinity surface water source. *Water Research*, *40*(14), 2727–2737. doi:10.1016/j.watres.2006.04.024 PMID:16765408

Keeley, J., Smith, A. D., Judd, S. J., & Jarvis, P. (2014). Reuse of recovered coagulants in water treatment : An investigation on the effect coagulant purity has on treatment performance. *Separation and Purification Technology*, *131*, 69–78. doi:10.1016/j.seppur.2014.04.033

Kyriakopoulos, G., Doulia, D., & Hourdakis, A. (n.d.). *International Journal of Environmental Effect of ionic strength and pH on the adsorption of selected herbicides on Amberlite*. doi:10.1080/03067310500247678

Lam, M. K., & Lee, K. T. (2011). Renewable and sustainable bioenergies production from palm oil mill effluent (POME): Win-win strategies toward better environmental protection. *Biotechnology Advances*, *29*(1), 124–141. doi:10.1016/j.biotechadv.2010.10.001 PMID:20940036

Latifahmad, A., Ismail, S., & Bhatia, S. (2003). *Water recycling from palm oil mill effluent (POME) using membrane technology*. Academic Press.

Lees, E. J., Noble, B., Hewitt, R., & Parsons, S. A. (2001). *Respiration rate and sludge characteristics of an activated microbial biomass*. Academic Press.

Liew, W. L., Kassim, M. A., Muda, K., Loh, S. K., & Affam, A. C. (2014). Conventional methods and emerging wastewater polishing technologies for palm oil mill effluent treatment: A review. *Journal of Environmental Management*, *149*, 222–235. doi:10.1016/j.jenvman.2014.10.016 PMID:25463585

Manickam, S., Zainal Abidin, N. B., Parthasarathy, S., Alzorqi, I., Ng, E. H., Tiong, T. J., Gomes, R. L., & Ali, A. (2014). Role of H2O2in the fluctuating patterns of COD (chemical oxygen demand) during the treatment of palm oil mill effluent (POME) using pilot scale triple frequency ultrasound cavitation reactor. *Ultrasonics Sonochemistry*, *21*(4), 1519–1526. doi:10.1016/j.ultsonch.2014.01.002 PMID:24485395

Mohammed, R. R., Ketabachi, M. R., & Mckay, G. (2014). Combined magnetic field and adsorption process for treatment of biologically treated palm oil mill effluent (pome). *Chemical Engineering Journal*, *243*, 31–42. Advance online publication. doi:10.1016/j.cej.2013.12.084

Musonge, P. (2014). *Influence of Effluent Type on the Performance of Chitosan as a Coagulant*. Academic Press.

Neyens, E., & Baeyens, J. (2003). A review of classic Fenton's peroxidation as an advanced oxidation technique. *Journal of Hazardous Materials*, *98*(1–3), 33–50. doi:10.1016/S0304-3894(02)00282-0 PMID:12628776

Ng, M., Liu, S., Chow, C. W. K., Drikas, M., Amal, R., & Lim, M. (2013). Understanding effects of water characteristics on natural organic matter treatability by PACl and a novel PACl-chitosan coagulants. *Journal of Hazardous Materials*, *263*, 718–725. doi:10.1016/j.jhazmat.2013.10.036 PMID:24220196

Nourani, M., Baghdadi, M., Javan, M., & Bidhendi, G. N. (2016). Production of a biodegradable flocculant from cotton and evaluation of its performance in coagulation-flocculation of kaolin clay suspension : Optimization through response surface methodology (RSM). *Biochemical Pharmacology*. Advance online publication. doi:10.1016/j.jece.2016.03.028

Pan, Y., Shi, B., & Zhang, Y. (2014). *Research on Flocculation Property of Bioflocculant PG*. doi:10.5539/mas.v3n6p106

Parthasarathy, S., Mohammed, R. R., Fong, C. M., Gomes, R. L., & Manickam, S. (2016). A novel hybrid approach of activated carbon and ultrasound cavitation for the intensification of palm oil mill effluent (POME) polishing. *Journal of Cleaner Production*, *112*, 1218–1226. doi:10.1016/j.jclepro.2015.05.125

Poh, P. E., Ong, W. Y. J., Lau, E. V, & Chong, M. N. (2014). *Journal of Environmental Chemical Engineering Investigation on micro-bubble flotation and coagulation for the treatment of anaerobically treated palm oil mill effluent (POME)*. Elsevier. doi:10.1016/j.jece.2014.04.018

Pui, K., Shak, Y., & Wu, T. Y. (2014). Coagulation – flocculation treatment of high-strength agro-industrial wastewater using natural Cassia obtusifolia seed gum : Treatment efficiencies and flocs characterization. *Chemical Engineering Journal*, *256*, 293–305. doi:10.1016/j.cej.2014.06.093

Ravi Kumar, M. N. (2000). A review of chitin and chitosan applications. *Reactive & Functional Polymers*, *46*(1), 1–27. doi:10.1016/S1381-5148(00)00038-9

Richard, A., & Margaritis, A. (n.d.). *Poly (glutamic Acid) for Biomedical Applications*. Academic Press.

Ridzuan, M., Ali, M., Shirai, Y., Samsu, A., Amiruddin, A., Ali, M., & Idris, J. (2013). Treatment of effluents from palm oil mill process to achieve river water quality for reuse as recycled water in a zero emission system. *Journal of Cleaner Production*, *10–13*. Advance online publication. doi:10.1016/j.jclepro.2013.12.004

Said, M., Ahmad, A., Wahab, A., Tusirin, M., & Nor, M. (2014). Journal of Industrial and Engineering Chemistry Blocking mechanism of PES membrane during ultrafiltration of POME. *Journal of Industrial and Engineering Chemistry*, 1–7. doi:10.1016/j.jiec.2014.02.023

Said, M., Hasan, H. A., Tusirin, M., & Nor, M. (2015). *Removal of COD, TSS and colour from palm oil mill effluent (POME) using montmorillonite*. doi:10.1080/19443994.2015.1036778

Shavandi, M. A., Haddadian, Z., Ismail, M. H. S., Abdullah, N., & Abidin, Z. Z. (2012). *Removal of Residual Oils from Palm Oil Mill Effluent by Adsorption on Natural Zeolite*. doi:10.100711270-012-1169-6

Soon, A. N., & Hameed, B. H. (2011). Heterogeneous catalytic treatment of synthetic dyes in aqueous media using Fenton and photo-assisted Fenton process. *DES*, *269*(1–3), 1–16. doi:10.1016/j.desal.2010.11.002

Sung, M. H., Park, C., Kim, C. J., Poo, H., Soda, K., & Ashiuchi, M. (2005). Natural and edible biopolymer poly-γ-glutamic acid: Synthesis, production, and applications. *Chemical Record (New York, N.Y.)*, *5*(6), 352–366. doi:10.1002/tcr.20061 PMID:16278834

Tan, Y. H., Goh, P. S., Lai, G. S., Lau, W. J., & Ismail, A. F. (2014). *Jurnal Teknologi Full paper Treatment of Aerobic Treated Palm Oil Mill Effluent (AT-POME) by using TiO 2 Photocatalytic Process*. Academic Press.

Taylor, P., Malakahmad, A., & Chuan, S. Y. (2013). *Desalination and Water Treatment Application of response surface methodology to optimize coagulation – flocculation treatment of anaerobically digested palm oil mill effluent using alum*. doi:10.1080/19443994.2013.791778

Wu, T Y, Mohammad, A. W., Jahim, J., & Anuar, N. (2007). *Palm oil mill effluent (POME) treatment and bioresources recovery using ultrafiltration membrane : Effect of pressure on membrane fouling*. doi:10.1016/j.bej.2007.01.029

Wu, T. Y., Mohammad, A. W., Jahim, J. M., & Anuar, N. (2010). Pollution control technologies for the treatment of palm oil mill effluent (POME) through end-of-pipe processes. *Journal of Environmental Management*, *91*(7), 1467–1490. doi:10.1016/j.jenvman.2010.02.008 PMID:20231054

Zahrim, A. Y., Nasimah, A., & Hilal, N. (2014). Journal of Water Process Engineering Pollutants analysis during conventional palm oil mill effluent (POME) ponding system and decolourisation of anaerobically treated POME via calcium lactate-polyacrylamide. *Journal of Water Process Engineering*, *4*, 159–165. doi:10.1016/j.jwpe.2014.09.005

Zainal, N. H., Jalani, N. F., Mamat, R., & Astimar, A. A. (2017). A review on the development of palm oil mill effluent (POME) final discharge polishing treatments. *Journal of Oil Palm Research*, *29*(4), 528–540. doi:10.21894/jopr.2017.00012

KEY TERMS AND DEFINITIONS

Advanced Oxidation Process (AOP): The process which degrades the organic pollutant by the powerful and reactive hydroxyl radical (OH•). Hydroxyl radical (OH•) generated would have an oxidation potential of 2.8eV OH• can generate through either one or a combined of chemical oxidation by using H_2O_2, ozone, ultrasound and radiation assisted source (ultraviolet).

Biochemical Oxygen Demand (BOD): The measure of the amount of oxygen that bacteria will consume when decomposing organic matter content under aerobic conditions. BOD is determined by incubating a sealed sample of water for five days and measuring the loss of oxygen from the beginning to the end of the test.

Chemical Oxygen Demand (COD): The measure of the amount of oxygen required to oxidize all organic material into carbon dioxide and water. COD values usually are higher than BOD values, but COD measurements can be obtained in a few hours while BOD measurements take five days.

Coagulation-Flocculation: Commonly used wastewater treatment method by removing colour, suspended solids and organic matter from wastewater. Coagulation-flocculation uses chemical agents to assist flocculating, sedimentation and elimination of dissolved and suspended solids from wastewater.

Hydraulic Retention Time (HRT): A measure of the average length of time that a soluble compound remains in a constructed bioreactor. The volume of the aeration tank divided by the influent flowrate is τ (tau), the hydraulic retention time.

Parts per Million (PPM): A way of expressing very dilute concentrations of substances. It use to describe the concentration of something in water or soil. One ppm is equivalent to 1 milligram of something per litre of water (mg/l) or 1 milligram of something per kilogram soil (mg/kg).

Total Suspended Solids (TSS): The measure of suspended matter contained in the wastewater. Suspended solids contain BOD and can impair water quality by adding turbidity and reducing aesthetics. In the laboratory, standard filtration and drying method used to measure SS, where the increase of weight of a container/filter is measured, for a known volume of wastewater filtered.

Chapter 8
Microalgal Photobioreactors as an Integrated Approach for Simultaneous Wastewater Treatment, Carbon Sequestration, and Recovery of Valuable Resources

Priya Banerjee
https://orcid.org/0000-0001-6695-5588
Department of Environmental Studies, DDE, Rabindra Bharati University, India

Aniruddha Mukhopadhyay
Department of Environmental Science, University of Calcutta, India

ABSTRACT

Unplanned and unmonitored growth of civilization and industries recorded in the previous decades have resulted in the introduction of several pollutants of emerging concern like pharmaceuticals and antibiotic drugs, endocrine disrupters, personal care products, etc. in adjacent aquatic environments. Moreover, the primary considerations of recent wastewater treatment strategies include energy neutrality and efficient recovery of valuable products as end/byproducts of the concerned process. The present study reviews such recent advances in designs, development, and implementation of algae based photobioreactors for simultaneous CO2 capture and treatment of effluents containing various types of emerging contaminants with negligible energy input. It also discusses the subsequent usage of algal biomass for production of value-added products like bioethanol, alginate, etc. This study has primarily addressed the potential and limitations of such photobioreactors for simultaneous achievement of wastewater remediation, CO2 sequestration, and bioenergy production with a zero-discharge concept.

DOI: 10.4018/978-1-7998-4921-6.ch008

INTRODUCTION

Microalgae are widely used in present times as feed for animals, extraction of valuable products like polyunsaturated fatty acids (PUFAs) and pigments that are used in human nutrition products, cosmetics, etc. (Spolaore et al., 2006). Microalgae have also been recommended for commercial scale production of different commodities and is also the sole substitute for ensuring sustainable biofuel production (Acién et al., 2017). Large scale cultivation of microalgae may be carried out but utilizing nutrients from different streams of effluents (Acién et al., 2017). Half of the total energy consumed by conventional wastewater treatment plants are spent in proving O_2 for growth of bacteria which in turn release CO_2 and N_2 into the atmosphere by oxidizing organic carbon present in the effluents (Acién et al., 2017). Application of microalgae may be considered as a better alternative to bacteria as it is capable of both reducing energy consumption and CO_2 emission by utilizing the same for photosynthesis (Craggs et al. 2011). Besides, microalgae are also capable of retaining N_2 in its biomass in the form of protein-nitrogen (Acién et al., 2017).

However, commercial-scale cultivation of microalgae is limited by the availability of appropriate photobioreactors (PBRs) (Ting et al., 2017). PBRs are open, closed or partially closed containers prepared from transparent and water-resistant materials and are capable of facilitating ideal environment for growth of photosynthetic microorganisms (Ting et al., 2017). Various types of PBRs have been reported till date. Many of them have achieved success on commercial scales (Acién et al., 2017; Gupta et al., 2015; Sing and Sharma, 2012). This study discusses different types of widely used PBRs, parameters guiding their efficiency and criteria for selecting PBRs for different studies. This chapter also reviews PBR based integrated waste management strategies and recovery of different valuable resources from the same.

TYPES OF PBR

Sufficient availability of light, selection of viable species of microalgae, appropriate environmental conditions, minimum usage of land and microalgae production over the entire year are considered as significant parameters for designing and development of PBRs (Ting et al., 2017). Successful cultivation of microalgae on commercial scale requires maximum possible reduction of cost incurred, enhancement of biomass productivity and prevention of contamination of microalgal cultures from bacteria or fungi (Ting et al., 2017). According to previous reports, PBRs are usually classified into two systems: open and closed. Open systems are usually installed outdoor and make maximum utilization of sunlight. Closed systems may be installed both outdoor and indoor and are supported by natural or artificially controlled environment respectively. Both systems have been described as follows.

Open System PBRs

Three primary types of open system PBRs are described as follows.

1. **Unstirred Open System:** These are mostly natural water bodies like lakes, lagoons, ponds, etc. Absence of stirring often results in poor mixing but incurs low cost for commercial scale units. These water bodies are usually shallow (0.5m in depth) and ensure penetration of light throughout

its depth. According to a previous study, these water bodies may be covered by plastic sheets for better control of temperature.

2. **Circular Pond:** This type of water body usually has a diameter of 40-50 m and a depth of 20-30 cm. A long rotating arm located at the center of the pond acts as a paddlewheel for mixing culture media and algal cells. Though mixing provides better efficiency in comparison to unstirred systems, contamination is inevitable as a result of exposure to the surroundings.
3. **Raceway Pond:** These ponds may be made up of concrete or plastic and is usually 15-50 cm in depth. The culture media used in such ponds may be pure or wastewater. This system is also capable of integrating CO_2 capture facilities in its design. Presence of paddlewheel, baffle and channels avoids sedimentation and ensures that the algal cells are suspended in the medium. This in turn ensures that all algal cells receive appropriate sunlight and CO_2 required for their growth. This type of pond is more efficient than circular ponds or unstirred systems and is therefore preferred for commercial-scale outdoor cultivation of microalgae (Tredici, 2004). Schematic representation of the same has been shown in Figure 1.

Inspite of the cost effectiveness and sustainability offered by the open system PBRs over the previous decades, these are faced by limitations like unsuitability for production of valuable resources like natural

Figure 1. Schematic representation of different PBRs for pure cultivation of microalgae (Ting et al., 2017)

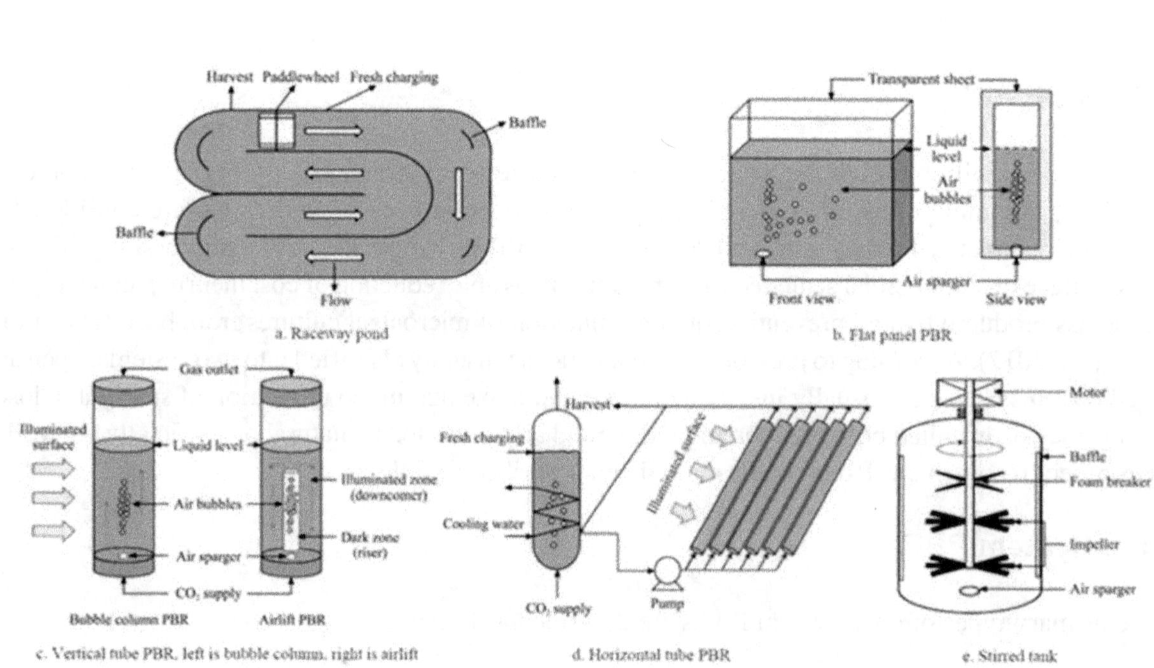

pigments and chemicals, contamination of microalgal cultures from bacteria and fungi, etc. Therefore, though these systems are still considered as primary bioreactors for commercial scale cultivation of microalgae, they require further investigations for improving penetration of light and mixing as well as reducing evaporation of water.

Closed Systems

Closed system PBRs were developed for overcoming the limitations faced by open systems. Closed systems prevent direct interactions between the environment and the culture media and thereby facilitate a controllable environment (Vasumathi et al., 2012; Wang et al., 2014). Air tightness, controlling of process parameters and supplementary approaches for mass transfer offered by these systems increase the cost of the same. Proper illumination necessitates large amounts of energy and renders these systems uneconomical. These limitations hinder the pilot or commercial scale application of these closed systems. However, these closed systems facilitate the production of high value-added products and fine chemicals which are further used for the production of bio-pharmaceuticals, human health supplements, cosmetics and biofuels (Ting et al., 2017). Commonly used configurations of closed system PBRs have been described as follows.

1. **Flat panel:** Schematic representation of this type of PBR has been shown in Figure 1b. it can be constructed at both indoor and outdoor location. The flat panels are prepared from transparent or partially-transparent materials like plexiglass, polycarbonates, glass, plastic bags, etc.). Mixing of algal cells and culture media occurs via air bubbles generated from an air sparger supplemented by an air pump. Emission of the exhaust gas occurs at the interface of the gas and the liquid. In case of reactors installed outdoor, the reactors are inclined for utilizing the maximum intensity of incident light.
2. **Vertical tube:** This type of PBR has two configurations namely bubble column and airlift (as shown in figure 1c, left and right respectively). Both types of reactors have an air sparger fitted to their bottoms that release the spared gas as tiny bubbles into the culture media for ensuring suspension of algal cells, enhancement of mass transfer, CO_2 sequestration and release of O_2. The airlift reactor maybe considered as a modification of the bubble column reactor.
3. **Horizontal tube:** This type of PBR (shown in figure 1d) offers the highest surface to volume ratio and therefore ensure maximum exposure of algal cells to incident light. However, this type of PBR is limited by uneven mass transfer taking place radially, high requirement of land and high maintenance and operational cost incurred and is hence considered unsuitable for commercial-scale applications.
4. **Stirred tank:** This type of indoor reactor (shown in figure 1e) are facilitated with an external source of illumination. Mixing is carried out by an electric motor. Therefore, this type of reactor provides optimum temperature, mass transfer and agitation. CO_2 is provided via an air sparger fitted at the bottom. 20% of free headspace facilitates exchange of gases with the culture medium. However, a low surface to volume ratio limits the penetration of light in turn reducing photosynthetic activity of the microalgal cells.

Hence, it may be concluded that, convenient control of environment makes closed systems more advantageous than open ones. However, the closed systems are difficult to clean and upscale and are not cost effective for commercial scale applications. These limitations should be addressed in future investigations.

Comparison of the Performance of Different PBRs

Few factors considered crucial for designing and development of PBRs are penetration and distribution of light, transfer of mass, shear stress, physiology of the selected microalgae and scalability of the concerned process (Singh and Sharma, 2012). Any single PBR is often unable to cater to the conditions of cultivation required by different strains of microalgae or the environmental conditions essential for the recovery of various valuable target products (Chang et al., 2017). However, hybrid PBRs, developed by combining the components of different PBRs, have been found to yield promising results for mass production of microalgae in comparison to single PBRs (Singh and Sharma, 2012). In such hybrid designs, components of one PBR often compensates for the limitations posed by another (Singh and Sharma, 2012). A comparative analysis of different PBRs have been shown in Table 1.

MAJOR FACTORS AFFECTING PHOTOBIOREACTOR EFFICIENCY

Many studies have reported that proper optimization of environmental conditions will ensure maximum growth rate and biomass concentration of the selected microalgae (Chang et al., 2017). Besides appropriate process optimization will help in decreasing overall production expenses. A proper understanding of both microalgal physiology and concepts of engineering underlining the PBR design are essential for ensuring the economic feasibility of involving phototrophic microorganisms in downstream applications. Few of the significant factors influencing PBR performance efficiency is described below. Any fluctuation in the optimum condition of these parameters severely effect cellular functions and in turn reducing PBR efficiency.

Utilization of Light

Exposure to illumination having optimum intensity and spectral quality is important for achieving maximum photosynthetic efficiency. Microalgal cell growth may be significantly affected by limitation, saturation or inhibition of light (Ho et al., 2012). Increase in intensity of light reportedly results in a rise in photosynthetic efficiency. Response to intensity of light is strain specific. However, it is difficult to maintain a uniform supply of light inside the PBR. Efficient agitation of algal cells ensure that cells are alternatively subjected to light-dark cycles and in turn increase biomass productivity and efficiency of the PBR (Posten, 2009). Besides, photosynthetic microorganisms only utilize the spectrum of natural light between 400-700 nm as their source of energy. This portion of the spectrum is referred to as *photosynthetically active radiation* (PAR) (Razzak et al., 2013). Hence, providing illumination in the range of PAR enhances growth of microalgal cells. Optimum utilization of PAR may be ensured by increasing the surface-to-volume ratio or decreasing the path length of light penetration in a PBR in order to achieve greater surface area for illumination of microalgal cells (Christenson and Sims, 2011).

Table 1. Comparison of the performance of different types of photobioreactors in microalgal cultivation (Chang et al., 2017)

PBR configuration	Types of PBR	Advantages	Disadvantages
Open systems	· Circular ponds · Raceway ponds	· Cost effective · Convenient for cleaning · Convenient maintenance · Suitable for mass cultivation · Low energy requirement	· Limited control over culture parameters · Inefficient mixing · Improper provisions of light and CO_2 · Difficult to sustain growth of microalgae over long periods · Less productivity · Suitable for few microalgal species · Prone to contamination · Small surface area for illumination · Constant water loss · Requirement of large land area · Upscaling results in decrease of surface area for illumination
Closed systems: Tubular		· Cost effective · Suitable for outdoor installation · High yield of biomass · Production of valuable resources · Large area for illumination	· Fluctuating pH gradients · Accumulation of O_2 · Effected by fouling · Needs large area of land · Experiences photo-inhibition
	· Stirred tank	· Production of valuable resources · Large illumination area · Specific control over growth conditions	· Supports cultivation of heterotrophic microalgae only
	Vertical tubular: · Bubble column · Air-lift	· High rates of mass transfer · Efficient mixing · Homogeneous shear · Energy efficient · Conveniently sterilized, reduced fouling · Immense potential for upscaling · Reduced chances of photo-inhibition and photo-oxidation · Suitable for microalgae immobilization · Great photosynthetic efficiency · Maximum exposure to light-dark cycles · Requires minimum land area · Separate zone of disengagement for liquid and gaseous phases	· Small surface area for illumination · Sophisticated materials for construction · Cultured cells subjected to shear stress · Further reduction in surface area receiving illumination upon upscaling · Incurs high cost · Modest scalability
Closed systems: Flat-plate		· Relatively cost effective · Convenient for cleaning · Readily conditioned · Large surface area for illumination · Suitable for outdoor installations · Efficient illumination · High biomass production · Good for microalgae immobilization · Reduced accumulation of O_2 · Shortest oxygen path · Low energy requirement	· Inconvenient for upscaling · Inefficient control over environmental parameters · Costly construction materials · Cells experience aeration induced shear damage · Reduced photosynthetic efficiency · Prone to light inhibition · Susceptible to fouling
Closed systems: Internally illuminated		· Allows internal illumination · Rise in average irradiance · Efficient light penetration · Can be sterilized under heat and pressure · Improved homogeneous distribution of light · Decrease number of day-night cycles · Light intensity varies diurnally	· High capital expenses · Inconvenient for cleaning · Light energy transmission lost through optical fibers
Closed systems: Fermenter-type		· Offers open exchange of gases · Primarily used for process optimization · Facilitates precise monitoring and regulation of every operating parameter	· Permits limited upscaling due to low area to volume ratio

pH of the Culture Medium

pH of the culture medium is a significant factor influencing cultivation of microalgae in terms of cell growth and metabolism as well as cellular ion uptake (Khalil et al., 2010). Moreover, pH of culture medium influences the solubility and availability of essential nutrients and CO_2, intracellular and cell wall secreted enzyme activities as well as production of different economically important commodities. Increase in acidity changes rate of nutrient uptake, affects vital processes of the cell and promotes metal induced toxicity. Increase in basicity reduces the algal affinity for CO_2, improves cell flexibility and lengthens the completion of cell cycle (Juneja et al., 2013). Extreme variations in pH of the culture medium may also influence internal pH of the microalgal cells. Most of the phototrophic microorganisms grow comfortably within pH 6-10 (Rodolfi et al., 2009). Microalgal strains thriving in fresh water are found to demonstrate more tolerance to fluctuations in pH conditions in comparison to marine microalgae (Chang et al., 2017). However, in integrated wastewater treatment processes, flue gas (usually consisting of SO_x, NO_x and CO_2) frequently used as a source of carbon for microalgae may cause alterations in pH of the culture mediums during the process of cultivation (Yen et al., 2015). Therefore, it may be assumed that changes in pH of the culture medium inhibits cell growth. This may be prevented by usage of buffered mediums or by integrating pH-stat systems within PBR configurations (Chang et al., 2017).

Ambient Temperature

Microalgal growth rate is significantly affected by conditions of ambient temperature (Ho et al., 2014). Microalgal growth rate reportedly increases till optimum temperatures are reached and experiences a rapid decline on exceeding the same (Ho et al., 2013). Fluctuation of temperature from optimal values effect cytoplasm viscosity and impair usage of carbon and nitrogen. Temperatures below the optimal value may increase formation of unsaturated fatty acids and weaken the rigidity of cell walls. Temperatures exceeding optimal values may impair protein synthesis and cellular metabolism as well as cause photo-oxidative damage. Outdoor PBRs are subjected to seasonal and diurnal variations in temperature. Moreover, few studies have reported the effect of flue gas induced heat stress on microalgal cells as temperatures of flue gas emitted by steel, power or coal plants may be as high as 120-150 °C (Kao et al., 2014). Implementation of thermophilic strains in PBRs having heat removal systems are found to be more cost effective than heat exchanger systems (Yen et al., 2015). Excess of irradiance may also cause a rise in temperature in PBRs.

Assimilation of Carbon Dioxide

Levels of atmospheric CO_2 (0.03-0.04%) are insufficient for efficient photosynthesis (Ho et al., 2011). Flue gas is a cost effective and concentrated source of CO_2 that can be utilized for large scale cultivation of microalgae (Cheah et al., 2015). However, most species of microalgae have low CO_2-binding capability and have developed carbon concentrating mechanisms (CCMs) for ensuring efficient accumulation of CO_2 within cells (Raven, 2010). The rapid rate of growth of microalgae facilitates higher rates of CO_2 sequestration than terrestrial plants (Fulke et al., 2010). Most efficient carbon sequestration can be achieved by selecting the most appropriate specie of microalgae and optimizing process parameters like availability of nutrients, intensity of light, pH of culture medium and ambient temperature (Kargupta et al., 2015). Different strains of microalgae have demonstrated a CO_2 capture efficiency of 28-53% from flue gas (Kargupta et al., 2015). *Chlorella* and *Scenedesmus* sp. are two microalgal specie reported widely for their immense potential of CO_2 sequestration (Tang et al., 2011; Toledo-Cervantes et al., 2013). However, toxic compounds like CO, NOx and SOx present in flue gas often acidify the culture medium (Van Den Hende et al., 2012) and in turn inhibit the growth and productivity of microalgal biomass (Kumar et al., 2014). Therefore, flue gas essentially requires to undergo expensive pretreatments before being applied for cultivation of microalgae. Hence, development of cost-effective pretreatment strategies is essential for facilitating CO2 capture and usage by microalgae (Yadav et al., 2015).

CO_2 sequestration is most commonly carried out through gas-liquid exchange in case of large-scale microalgae cultivations. Utilization of flue gas (consisting of 6-20% CO_2) as an alternative source of carbon compensates for the low atmospheric CO_2 concentrations and also reduces emissions of greenhouse gases into the atmosphere. However, high CO_2 content of flue gas may cause a reduction of pH thereby resulting in inhibition of cell growth. Rate of flow of inlet gas also plays a significant role in quantifying the total inorganic carbon content of the liquid phase (Van Den Hende et al., 2012). Hence, it is essential to increase the rate of CO_2 transfer from gaseous to liquid phases for ensuring improved concentrations of "dissolved CO_2" (Van Den Hende et al., 2012). Rate of CO_2 transfer may be increase by improving cell mixing using gas recirculation or gas injections (Jacob-Lopes et al., 2010). An ideal PBR should consist of a large illuminated area and advanced CO_2 diffuser facilitating an extended period of CO_2 retention in order to improve penetration of light and dissolution of inorganic carbons (Yen et al., 2015). Appropriate design and implementation of PBRs ensuring high surface to volume ratio, parallel flow of gas and liquid with optimal rates and an influent CO_2 concentration below 10% reportedly facilitates mass transfer of CO2 with simultaneous enhancement of microalgal biomass productivity (Kargupta et al., 2015).

Removal of Oxygen

In most species of microalgae, rubisco enzymes exhibit higher O_2 binding activity than CO_2. Excessive binding with O_2 in comparison to CO_2 results in rates of photorespiration exceeding that of photosynthesis and a parallel decline in efficiency of PBRs (Christenson and Sims, 2011). However, excessive exposure to O_2 and light result in the formation of reactive oxygen species which in turn cause oxidative damage in cells (Chang et al., 2017). However, appropriate removal of excessive O_2 from closed PBR systems remains a challenge (Christenson and Sims, 2011). In closed PBRs O_2 accumulation is found to rise with a corresponding increase in the working volume of the reactor (Kumar et al, 2011). Integration of a separate system for aeration or degassing within the configuration of commercial scale closed PBRs may prevent excessive accumulation of O_2 within the system.

Mixing/ Agitation of Microalgal Cells

Agitation or mixing of microalgal cells reportedly facilitate high yield of biomass (Chang et al., 2017). Mixing distributes nutrients uniformly to all algal cells, ensures heat transfer and efficient utilization of light, facilitates gas exchange and prevents sedimentation of cells. Mixing also improves mass transfer of CO2 and prevents excessive accumulation of heat and O_2. However, mixing can induce hydrodynamic stress, cause cell breakage and inhibit metabolic activity. Besides, Tolerance to turbulence is species specific for microalgae (Mata et al, 2010). Agitation of microalgal cells may be carried out mechanically (as in raceway ponds and stirred tanks) or non-mechanically (as in tubular, air-lift and bubble column PBRs) (Kumar et al, 2011).

CRITERIA FOR THE SELECTION OF PHOTOBIOREACTORS

Form and Quality of the Desired Product

Each PBR has its own specific features, advantages and limitations. Selection of PBR design varies with the type and quality of the target product. If the desired product is biofuel, then biomass having high lipid or carbohydrate content is preferred along with minimum usage of resources and cost-effective processes (like open or raceway ponds). Only when produced under these conditions, biofuel will be able to compete with fossil fuels on a commercial scale. However, more expensive PBRs of closed type are preferred for more valuable target products like health supplements such as docosahexaenoic acid (Ren et al., 2010), EPA (Chen et al., 2013a), phycocyanin (Chen et al., 2013b), etc. as well as hormones and cosmetics (Chang et al., 2017; Demirbas and Demirbas, 2011). Closed systems are selected for yielding valuable products as these systems facilitate precise control over culture conditions enabling synthesis of specific biomolecules by the concerned microalgae. Closed PBRs also have minimum chances of contamination and ensure a biomass yield of better quality in comparison to open systems (Mata et al, 2010).

Tolerance of Selected Microalgal Specie

Two important criteria for identifying the most efficient microalgal strain include the tolerance and the adaptability demonstrated by the selected strain in response to fluctuations in environmental conditions in open systems, shear forces as well as O_2 accumulation in PBR setups (Chang et al., 2017). When cultivated in open systems, selected strains have to compete for resources with fungal and bacterial contaminations. In case of contamination with other strains of microalgae, the selected strain has to grow rapidly to out compete the other strains of microalgae. On the other hand, in closed PBRs the selected microalgal strain has to be able to tolerate stress induced by strong shearing forces resulting from air sparging or pumping (Chang et al., 2017). Therefore, efficiency of the selected PBR configuration also depends upon the adaptability and the tolerance of the microalgal strain selected for the concerned process to fluctuations in its environmental conditions.

Process Performance and Cost

The maximum yield of biomass depends upon the type of PBR selected. Closed systems facilitate the growth of biomass having significantly better quality than those cultivated in open systems. Size of the selected PBR varies with the target rate of production of microalgae. Open systems can be upscaled conveniently and cost-effectively for facilitating larger biomass production rates. Nevertheless, open systems offer poor CO_2 transfer efficiency and productivity of biomass, high risk of contamination and uncontrolled evaporation of growth medium (Chang et al., 2017). In comparison to open systems, closed systems facilitate efficient mixing and improved CO_2 mass transfer thereby yielding a biomass having a larger content of desired biomolecules. However, upscaling of closed PBRs is difficult and cost incurring. Moreover, commercial scale PBRs often damage microalgal cells by exposing them to high shear stress arising from mixing and high levels of O_2 accumulated within the setup (Brennan and Owende, 2010). Hence, it is essential to determine which type of PBR is most suitable for providing maximum biomass yield with minimum cost for most efficient production of any specific target resource.

Resource Availability

Selection of PBR configuration and size is significantly dependent upon availability of water, CO_2, light and land. Open systems require large flat lands exposed to sufficient sunlight for maintaining a shallow growth medium for ensuring maximum photosynthetic efficiency and optimum mixing. Geological positioning is also a more significant parameter for open systems in comparison to closed ones. Closed PBRs may be expanded vertically and hence requires lesser land area in comparison to open systems. However, most of the cost incurred by cultivation of microalgae on commercial scale accounts for CO_2 and water (Slade and Bauen, 2013). Integrated waste management systems use waste resources like flue gas, wastewater, seawater, etc. for cultivation of microalgae. This drastically reduces cost of production and is hence highly desirable from an economical aspect. Weather also influences microalgal growth rate significantly. For example, open systems cannot be operated during winters or monsoon. Efficient monitoring and maintenance of PBRs also necessitate professional labor having knowledge regarding microalgal growth and skills for operating PBRs. However, concerned personnel may require further training if sophisticated PBRs and other associated equipment are used.

ROLE OF PBR IN INTEGRATED WASTEWATER TREATMENT

Microalgal treatment of wastewater was first investigated in the 1970s. Different streams of wastewater have been treated using microalgae as well as used for growing microalgae as its composition is similar to the culture medium used for this purpose (Acién et al., 2016). Besides nutrients like phosphorus, nitrogen, carbon and various minor components essential for growth of microalgae, wastewaters may also contain toxic compounds like heavy metals and emerging pollutants (del Mar Morales-Amaral et al., 2015). Besides utilization of wastewater for nutrients, microalgae can also utilize polluted gaseous emissions as potential CO_2 sources required for their growth. CO_2 required for growth of microalgae may be obtained from gaseous emissions released from combustion of fossil fuels (Razzak et al., 2013). Microalgae selected for simultaneous wastewater treatment, CO_2 sequestration and production of valuable resources should essentially have rapid rate of growth and productivity, high content of lipids, high tolerance to widely occurring water pollutants, NH^{4+} and fluctuations in environmental conditions as well as good rates of CO_2 binding and O_2 release (Wang et al., 2016). Growth rates of algae has been considered as a significant limiting factor guiding efficient removal of nutrients and pollutants from wastewater. Indigenous species of microalgae found in wastewaters may be investigated for this purpose by making full use of their tolerance and resistance to exposed pollutants. However, this technology has only been implemented on small scales and is yet to be developed on a commercial scale (Acién et al., 2016). A comparative analysis of conventional and microalgae-based treatment of major pollutants present in wastewaters have been shown in Table 2.

In comparison to conventional WWT processes, microalgae based WWT processes provide advantages like reduced energy consumption, lowered emission of green house gases (as many of these gases are utilized by microalgal biomass for their own physiological activities), reduced formation of hazardous

Table 2. Comparative analysis of treatment of major pollutants via conventional and microalgae-based wastewater treatment (WWT) processes (Alcántara et al., 2015)

Nature of pollutant	Pollutant induced risk	Conventional WWT processes	Microalgae-based WWT processes
Biodegradable organic pollutants	· Uncontrolled biodegradation causes O_2 depletion in wastewaters; · Some pollutants may be hazardous as well.	· Aerobic or anaerobic biodegradation using heterotrophs; · Aerobic biodegradation is often applied for treatment of low-strength effluents like domestic wastewater; · Also applied for treatment of inhibitory and/or recalcitrant pollutants present in industrial wastewaters.	· Aerobic biodegradation; · Microalgae facilitate additional metabolic abilities and support biodegradation via substrate exchange.
Heavy metals	· Induces acute and chronic toxicity; · Bioaccumulation of pollutants at different trophic levels.	· Chemical precipitation using lime; · Electrochemical precipitation, · Ion exchange, · Adsorption using activated carbon or other biomaterials.	· Microalgal biomass can be used to prepare substantial amounts of cost-effective biosorbent; · Algal photosynthesis can also facilitate precipitation of heavy metals at high pH.
Nutrients	· Stimulate uncontrolled growth of algal biomass in ecosystems where the nutrient bearing effluents are discharged.	· Microbial assimilation; · Processes of nitrification and denitrification; · Biological removal of phosphate as a result of poly-P accumulation in bacterial cells; · Precipitation of chemical phosphate.	· Microalgal photosynthesis facilitates assimilation of nutrients by promoting productivity of biomass; · Controlled conditions of environmental favor volatilization of N and precipitation of P; · Few species can carry out intracellular accumulation of P in the form of stable poly-P.
Pathogens	· Can act as active vectors of different animal and human diseases	· Disinfection of maturation ponds; · Chemical disinfection via UV treatment, chlorination, ozonation, etc.	· Microalgal photosynthesis can facilitate disinfection of pathogens by raising pH and concentration of dissolved oxygen; · The high exposure to sunlight facilitated by high surface to volume ratio favors disinfection via UV irradiation.
Suspended solids (SS)	· Reduces clarity of water; · Causes clogging of waterways; · Many solids can also be biodegradable organic pollutants in nature.	· Gravity settling before or along with biological treatment.	· Not designed for treatment of non-biodegradable solids; · Initial settling of SS prior to microalgal treatment may often be required for avoiding operational difficulties and ensuring improved penetration of light in the influent.

sludge, cost effectivity and a parallel yield of useful algal biomass (Batista et al., 2015). This biomass produced by retrieving nutrients from wastewater is used for preparing biofuels and other valuable products like bioplastics, biofertilizers, biopolymers, dyes, paints, lubricants, etc. Simultaneous treatment of wastewater, utilization of green house gases and production of valuable algae-based products provide benefits in both environmental and economic aspects (Razzak et al., 2013). Hence, integrated processes for wastewater treatment may be considered as an extremely sustainable way for yielding bioenergy with simultaneous production of other bio-products (Brennan and Owende, 2010). Wijffels and Barbosa (2010) also suggested the significance of utilizing wastewater for cultivation of microalgae as production of biodiesel requires about 1.5 L of water for yielding a kg of biofuel in closed systems and even more for open systems where addition of fresh water is necessary to compensate water lost by evaporation.

For being preferred over other contemporary processes, microalgae based WWT processes must achieve the following objectives (Acién et al., 2016):

1. Increase in process tolerance to a wide range of environmental conditions;
2. Reduction in hydraulic retention time (from 7–11) days to that offered by conventional processes (0.3 day);
3. Reduction in power required for treatment of wastewater below 0.5 $kWhm^{-3}$;
4. Improvement in quality of treated water to meet discharge/ reuse regulations and standards.

Both biological and engineering concepts of the process are required to be reconsidered in order to achieve the aforementioned objectives. Moreover, the process must be carried out under optimum conditions and validated under real circumstances over a substantial period of time for obtaining maximum efficiency of the concerned process. Instances of different integrated wastewater treatment strategies have been shown in Table 3.

RECOVERY OF VALUABLE RESOURCES FROM CULTIVATED MICROALGAE

In order to meet the rising demand of consumers for natural and healthy products, researchers and industries have focused on the development of novel products having functional ingredients isolated from microalgae (Chen et al., 2014; Utomo et al., 2013; Wang et al., 2015, Yen et al., 2013). Microalgae reportedly yield several bioactive compounds like polysaccharides, PUFAs, peptides, enzymes, vitamins, essential minerals and natural pigments that have been found to exert positive health effects (Chang et al., 2017). Pigments and other valuable products prepared from microalgae are already being produced on a commercial scale. Development of new products require adequate research and upscaling of existing processes. While few products, such as biofuels and pigments, need to compete with their synthetic alternatives, some products like hydrocolloids and aquaculture feed are only prepared from microalgae. Some valuable resources obtained from microalgae have been discussed below.

Biofuels

The need for achieving global energy security and an unpolluted environment has urged the development of clean and renewable sources of energy. Of all other types of alternative energies reported till date, biomass energy capable of mitigating CO_2 emissions has received significant global attention. Microalgae has been considered as a feedstock for biofuel production as they are capable of synthesizing and accumulating huge amounts of lipids and neutral oils (precursors of biodiesel) as well as carbohydrates (precursors of bioethanol, biobutanol and biohydrogen) within their cells (Chen et al., 2014; Yeh et al., 2012). These oil and carbohydrate rich microalgae have been used for production of bioethanol or biodiesel by means of microbial fermentation or transesterification. Utilization of wastewater as nutrient source, CO_2 capture from flue gas as a carbon source and efficient land use have been found to be the primary advantages provided by production of microalgae-based biofuel (Singh and Gu, 2010). For competing with fossil fuels and meeting the existing demand for energy, production of microalgae-based

Table 3. Integrated wastewater treatment strategies reported in recent literature

Type of PBR	Type of Effluent	Microalgae used	Carbon source	Resource recovered	Reference
Pilot tubular PBR	Centrate from a wastewater treatment plant	A combination of *Tetraselmis chuii* and *Nanochloropsis gaditana*	Synthetic exhaust gas, with a composition of 24% of CO_2, 76% of N_2 and 498 ± 63 mg m^{-3} of toluene	Biomass	Anbalagan et al., 2017
Airlift bubble column PBR	Municipal wastewater	*Chlorella vulgaris* and *Scenedesmus obliquus*	Synthetic air mixture containing 5% (v/v) CO_2	Biodiesel, biogas, feed supplements for animals, etc.	Chaudhary et al., 2018
Bubble-column PBR	Brewery wastewater	*Scenedesmus obliquus*	10% (v/v) brewery CO2 supplement	Biohydrogen, bio-oil, bio-char and bio-gas	Ferreira et al., 2017
Membrane PBR	Treated sewage	*Chlorella vulgaris*	Simulated off-gas as released from a sewage treatment plant	Biomass	Honda et al., 2017
Annular PBR and Flat panel PBR	Palm oil mill effluent	*Chlorella vulgaris*	Normal air	Biodiesel	Idris et al., 2017
Hybrid Anaerobic Baffled Reactor coupled PBR	Domestic wastewater	A co-culture of *Chlorella vulgaris, Chlorella sorokiniana,* and *Scenedesmus simris 002*	Ambient air	Biofuel	Khalekuzzaman et al., 2019
Porous substrate PBR	Human urine	*Desmodesmus abundans*	Synthetic air mixture containing 5% (v/v) CO_2	Nutrients (N & P)	Piltz & Melkonian, 2018
Indigenous PBR	Effluents from a potato processing plant	A consortium of *Chlorella sorokiniana* and *Methylococcus capsulatus*	Synthetic biogas	Essential amino acids	Rasouli et al., 2018
Airlift circulation PBR	Anaerobic digested starch processing wastewater	*Chlorella pyrenoidosa*	Synthetic air mixture containing ~5–9% (v/v) CO_2	Polyunsaturated fatty acids (PUFAs)	Tan et al., 2014
PBR coupled with ultra-filtration	Anaerobic digestion effluent	*Chlorella vulgaris* (Suspended and immobilized)	Compressed air with 2.5% CO_2	Biomass	Xie et al., 2018

biofuels needs to be cost effective. This requires selection of a microalgal strain that has rich content of the target product and high productivity of biomass as it helps reduces cost incurred for harvesting and concentration of biomass. Genetically modified strains of microalgae having higher yield of lipids, enhanced tolerance to salt (for outdoor cultivation), etc. have also been investigated for meeting the needs of the process of production. Different microalgal strains and production technologies reported for production of biofuels have been enlisted in Table 4.

According to existing literature, 52% of all studies have investigated cultivation of microalgae using closed systems while the rest has mostly used natural conditions and open ponds (Singh and Gu, 2010).

Table 4. Industrial cultivation of microalgae for production of biofuels and bio-based chemicals (Chang et al., 2017)

Details of enterprise	Technology of production	Microalgal specie used	Target products
Algaelink, The Netherlands	· Horizontal and tubular PBRs · Computerized control over processes · Offers technical support for algae-based biofuel companies	· *Nannochloropsis* · *Isochrysis* · *Tetraselmis* · *Pavlova*	· Biomass, · Jet fuel
Algasol, Bangladesh	· Proprietary floating PBR · PBR made using thin, flexible polymer · Can be installed on land or in water.	-	Biomass
Algenol Biofuels, United States	· Indigenous flexible plastic film PBR · PBR facilitates both product creation and collection	· Genetically modified *Synechocystis* · *Anabaena* having heterologous genes for producing ethanol	Bioethanol
Aquaflow Bionomic, New Zealand	· Open ponds, · Large-scale harvesting of wild algae from open ponds	-	Biodiesel
Aquatic Energy, United States	Raceway ponds	· *Scenedesmus obliquus*	· Lipid precursors of biodiesel, · Omega-3 oils
Aurora Algae, United States	Open ponds having an indigenous system for pond information management	*Nannochloropsis*	· PUFAs · Omega 3 oils · Fuel · Feed · Protein
A2BE Carbon Capture, Colorado, United States	· Propriety PBR having rollers for mixing	-	Biofuel
Algae Tec. Ltd., Australia	· Closed, modular, PBR for industrial-scale algae cultivation	-	Biomass for biofuels
BFS Biofuel Systems, Spain	· Vertical tubular PBR equipped for capture of CO_2 emitted by industries	-	Biopetroleum
Cellana, United States	· Open ponds · Tubular systems with turbulent flow scheme	· *Lyngbya lagerheimii* · *Spirulina platensis* · *Haematococcus pluvialis* · *Chlorella sorokiniensis* · *Nanofrustulum* · *Tetraselmis*	· Biofuels, · Nutraceuticals, · Pharmaceuticals · Animal and aquaculture feeds · Omega-3 oils
Green star, United Kingdom	· Patented hybrid PBR, · Covered pond system	-	· Biodiesel, · Lubricants and engine additives · Algae growth enhancers
Heliae, United States	Spiral PBR built with novel oval glass tubes	*Haematococcus pluvialis*	· Astaxanthin, · Phycoterra
Live Fuels, United States	Open ponds for estuarine environments	-	Biodiesel
Muradel, Australia	Open ponds having paddlewheels for mixing	Salt-tolerant, lipid rich microalgal strains	Biofuels
Pond Biofuels, Canada	Unique PBR using exhaust gases as source of CO_2	-	· Biofuels · Biodiesel
Sapphire Energy, United States	Open pond systems	· *Chlamydomonas reinhardtii* · *Dunaliella salina* · *Haematococcus pluvialis* · *Scenedesmus dimorphus* · *Dunaliella viridis* · *Dunaliella tertiolecta*	· Biofuels, · Omega oils · High value proteins
Seambiotic, Israel	· Open pond systems · Raceway ponds · Systems equipped for using industrial gas as sources of CO_2	· *Skeletonema* · *Nannochloropsis* · *Chlorococcum* · *Dunaliella* · *Nannochloris* · *Tetraselmis*	Biodiesel
Solazyme, United States	· Heterotrophic cultivation of algae · Conventional PBRs using agriculture-based sugars as feed	· *Prototheca* · *Botryococcus braunii*	· Jet fuel · Renewable diesel · Biodiesel · Valuable algal oils and proteins
Solix Biosystems, United States	· Indigenous PBR · Offers light-extended surface area · Water supported PBR	*Nannochloropsis*	· Biodiesel · Green diesel · Jet fuel · Bioplastics · Omega oils · Personal care products
Subitec, Germany	Indigenous flat-panel air-lift PBRs having propriety static mixers	-	· Food · Feed · Fuel · Pharmaceuticals · Nutraceuticals

Most of the PBRs reported in previous studies have been customized for specific needs facilitating high yields of microalgal biomass required for further processing of valuable resources. These customized PBRs incur low cost of production, decrease usage of water and facilitate the reusage of waste water and gases released by different industries. Residues of microalgal biomass obtained after extraction of oils may be utilized further in biorefineries for carrying out anaerobic fermentation. Hence, an integrated system of waste management and biofuel production is more lucrative than a single-product approach.

Pigments

Microalgae are rich in different natural pigments like chlorophyll, phycocyanin, carotenoids and lutein. These pigments are extensively used as poultry and aquaculture feed as well as natural colorants in food, pharmaceuticals and cosmetic products. Many of these pigments exhibit anti-carcinogenic, neuroprotective, anti-angiogenic, anti-inflammatory and anti-obesity potentials (Pangestuti and Kim, 2011). The first algal pigment extracted on commercial scale was β-Carotene (a precursor of vitamin A). β-Carotene was obtained from successful outdoor cultivation of *Dunaliella salinae*, a microalgal strain having high salt tolerance (Borowitzka, 2013). β-Carotene is widely used as a food colorant and an antioxidant. It reportedly combats free radicals formed as a result of arthritis, aging and gastrointestinal cancer. Astaxanthin is another powerful antioxidant obtained from *Haematococcus pluvialis*. It reportedly reduces hyperpigmentation of the skin and is widely applied as a skin-lightening agent. Lutein, an essential human serum carotenoid, is present in microalgae as cis-lutein, trans-lutein, epoxy-lutein and protein linked lutein. However, this serum carotenoid is not synthesized by humans and only obtained by consuming lutein-rich food. Consumption of lutein reportedly helps prevent cancer, cardiovascular ailments, degenerative human diseases, etc. (Chang et al., 2017). *Chlorella protothecoides*, *Scenedesmus almeriensis* and *Murellopsis* sp. are examples of lutein producing microalgae (Fernańdez-Sevilla et al., 2010). Furthermore, fucoxanthins extracted from brown algae reportedly combat UV radiation induced oxidative stress and are therefore widely used in preparation of cosmetics (Wang et al., 2015).

Feed

Presently available provisions of animal protein are insufficient to fulfill the projected "protein gap" that may occur due to rapid exhaustion of resources like pastures and clean water (Devendra and Leng, 2011). Future projections suggest that rising trends of protein consumption and declining natural harvests will lead to the sourcing of large quantities of seafood from aquaculture. Cattle production will also increase for meeting such protein demand. In such scenarios, microalgal biomass may be utilized as a good food source for aquaculture and cattle. Owing to its rapid rate of growth, highest efficiency of photosynthesis carried out per unit volume, non-requirement of arable land, usage of brackish water/ wastewater streams for cultivation, microalgae are considered as promising feed for aquaculture and cattle alike. Algal proteins are reportedly of superior quality in comparison to vegetable proteins (Chang et al., 2017). Under natural growth conditions, some species of microalgae are reported to provide 28-71% protein content per dry weight (Chang et al., 2017). Hence, owing to their rich protein and PUFA content, microalgal biomass is considered as a nutritionally stable feed for aquaculture and cattle rearing.

Polyunsaturated Fatty Acids (PUFAs)

PUFAs are essential fatty acids that need to be supplemented in diets of animals and humans as they lack the enzymes required for synthesizing the same by themselves. Docosahexaenoic acid (DHA) is a type of omega-3 fatty acid that is extensively used as a supplement in maternal health and infant formulas. *Crypthecodinium cohini* is used for large scale production of DHA for usage in infant formulas. Eicosapentaenoic acid (EPA), linolenic acid and arachidonic acid are other types of PUFAs available from microalgal cultures.

Cosmetic Products

Microalgal proteins, lipids and pigments demonstrating antioxidant activity are widely used for developing cosmetic products. These cosmetics reportedly protect the skin from detrimental effects arising from exposure to UV irradiation and combat oxidative damage. Polysaccharides derived from microalgae reportedly demonstrate antiaging properties (Wang et al., 2015). Cosmetics developed using microalgae derived resources include sunscreens, toners, moisturizers and often used as skin brightening, tightening and antiaging treatments. *Chlorella* sp., *Nannochloropsis* sp., and *Spirulina* sp. are examples of microalgae that are widely used for preparation of cosmetics due to their rich protein content and antioxidant properties. Mycosporine-like amino acids (MAAs) that serve as natural sunscreens against UV radiation, have also been isolated from microalgae (Chang et al., 2017; Llewellyn and Airs, 2010). MAAs also reportedly protect cells against salt and thermal stress as well as serve as intracellular nitrogen reserves in some organisms.

CONCLUSION

This chapter has provided a detailed discussion on different aspects of integrated wastewater treatment strategies using microalgae based PBR processes. It has described how target products are decisive factors guiding the selection of environmental conditions and type of PBRs. Availability of light, nutrients and CO_2 required for large scale cultivation of microalgae depend widely on the configuration of the PBR selected for the concerned process. The primary parameters guiding efficient integrated wastewater treatment include selection of the ideal microalgal strain, efficient large-scale microalgae cultivation, utilization of wastewater and gaseous emissions for microalgae cultivation and extraction of valuable resources from obtained microalgal biomass. Large scale cultivation of microalgae using waste water and gases reduces environmental pollution, supports biofuel industries and commercial production of other commodities like cosmetics, health supplements, animal feed, etc. Though design of PBRs are specific to the process they are intended to perform, large scale microalgal cultivation is mostly carried out in open or raceway pond systems owing to their low installation and maintenance costs. However, hybrid PBRs have been found to be more efficient in comparison to single configuration ones. Hence, more investigations are required for developing versatile and cost effective PBRs that work efficiently under outdoor conditions for facilitating large scale cultivation of microalgae. The growing algae based chemical and biofuel industries testify the success of such investigations and are projected to achieve further growth with parallel advances in PBR technologies

REFERENCES

Acién, F. G., Gómez-Serrano, C., Morales-Amaral, M. D. M., Fernández-Sevilla, J. M., & Molina-Grima, E. (2016). Wastewater treatment using microalgae: How realistic a contribution might it be to significant urban wastewater treatment? *Applied Microbiology and Biotechnology*, *100*(21), 9013–9022. doi:10.100700253-016-7835-7 PMID:27645298

Acién, F. G., Molina, E., Reis, A., Torzillo, G., Zittelli, G. C., Sepúlveda, C., & Masojídek, J. (2017). Photobioreactors for the production of microalgae. In *Microalgae-based biofuels and bioproducts* (pp. 1–44). Woodhead Publishing.

Alcántara, C., Posadas, E., Guieysse, B., & Muñoz, R. (2015). Microalgae-based wastewater treatment. In *Handbook of Marine Microalgae* (pp. 439–455). Academic Press. doi:10.1016/B978-0-12-800776-1.00029-7

Anbalagan, A., Toledo-Cervantes, A., Posadas, E., Rojo, E. M., Lebrero, R., González-Sánchez, A., Nehrenheim, E., & Muñoz, R. (2017). Continuous photosynthetic abatement of CO_2 and volatile organic compounds from exhaust gas coupled to wastewater treatment: Evaluation of tubular algal-bacterial photobioreactor. *Journal of CO_2 Utilization, 21*, 353-359.

Batista, A. P., Ambrosano, L., Graça, S., Sousa, C., Marques, P. A., Ribeiro, B., Botrel, E. P., Neto, P. C., & Gouveia, L. (2015). Combining urban wastewater treatment with biohydrogen production–an integrated microalgae-based approach. *Bioresource Technology*, *184*, 230–235. doi:10.1016/j.biortech.2014.10.064 PMID:25453433

Borowitzka, M. A. (2013). High-value products from microalgae—Their development and commercialisation. *Journal of Applied Phycology*, *25*(3), 743–756. doi:10.100710811-013-9983-9

Brennan, L., & Owende, P. (2010). Biofuels from microalgae—A review of technologies for production, processing, and extractions of biofuels and co-products. *Renewable & Sustainable Energy Reviews*, *14*(2), 557–577. doi:10.1016/j.rser.2009.10.009

Chang, J. S., Show, P. L., Ling, T. C., Chen, C. Y., Ho, S. H., Tan, C. H., Nagarajan, D., & Phong, W. N. (2017). Photobioreactors. In *Current Developments in Biotechnology and Bioengineering* (pp. 313–352). Elsevier. doi:10.1016/B978-0-444-63663-8.00011-2

Chaudhary, R., Dikshit, A. K., & Tong, Y. W. (2018). Carbon-dioxide biofixation and phycoremediation of municipal wastewater using Chlorella vulgaris and Scenedesmus obliquus. *Environmental Science and Pollution Research International*, *25*(21), 20399–20406. doi:10.100711356-017-9575-3 PMID:28656576

Cheah, W. Y., Show, P. L., Chang, J. S., Ling, T. C., & Juan, J. C. (2015). Biosequestration of atmospheric CO_2 and flue gas-containing CO_2 by microalgae. *Bioresource Technology*, *184*, 190–201. doi:10.1016/j.biortech.2014.11.026 PMID:25497054

Chen, C. Y., Chang, C. Y., Lo, Y. C., Ho, S. H., & Chang, J. S. (2014). Enhancing biohydrogen production from *Chlorella vulgaris* FSP-E under mixotrophic cultivation conditions. *Energy Procedia*, *61*, 870–873. doi:10.1016/j.egypro.2014.11.984

Chen, C. Y., Chen, Y. C., Huang, H. C., Huang, C. C., Lee, W. L., & Chang, J. S. (2013a). Engineering strategies for enhancing the production of eicosapentaenoic acid (EPA) from an isolated microalga *Nannochloropsis oceanica* CY2. *Bioresource Technology, 147*, 160–167. doi:10.1016/j.biortech.2013.08.051 PMID:23994697

Chen, C. Y., Kao, P. C., Tsai, C. J., Lee, D. J., & Chang, J. S. (2013b). Engineering strategies for simultaneous enhancement of C-phycocyanin production and CO_2 fixation with *Spirulina platensis. Bioresource Technology, 145*, 307–312. doi:10.1016/j.biortech.2013.01.054 PMID:23664178

Christenson, L., & Sims, R. (2011). Production and harvesting of microalgae for wastewater treatment, biofuels, and bioproducts. *Biotechnology Advances, 29*(6), 686–702. doi:10.1016/j.biotechadv.2011.05.015 PMID:21664266

Craggs, R. J., Heubeck, S., Lundquist, T. J., & Benemann, J. R. (2011). Algal biofuels from wastewater treatment high rate algal ponds. *Water Science and Technology, 63*(4), 660–665. doi:10.2166/wst.2011.100 PMID:21330711

del Mar Morales-Amaral, M., Gómez-Serrano, C., Acién, F. G., Fernández-Sevilla, J. M., & Molina-Grima, E. (2015). Outdoor production of Scenedesmus sp. in thin-layer and raceway reactors using centrate from anaerobic digestion as the sole nutrient source. *Algal Research, 12*, 99–108. doi:10.1016/j.algal.2015.08.020

Demirbas, A., & Demirbas, M. F. (2011). Importance of algae oil as a source of biodiesel. *Energy Conversion and Management, 52*(1), 163–170. doi:10.1016/j.enconman.2010.06.055

Devendra, C., & Leng, R. A. (2011). Feed resources for animals in Asia: Issues, strategies for use, intensification and integration for increased productivity. *Asian-Australasian Journal of Animal Sciences, 24*(3), 303–321. doi:10.5713/ajas.2011.r.05

Fernández-Sevilla, J. M., Fernández, F. A., & Grima, E. M. (2010). Biotechnological production of lutein and its applications. *Applied Microbiology and Biotechnology, 86*(1), 27–40. doi:10.100700253-009-2420-y PMID:20091305

Ferreira, A., Ribeiro, B., Marques, P. A., Ferreira, A. F., Dias, A. P., Pinheiro, H. M., Reis, A., & Gouveia, L. (2017). Scenedesmus obliquus mediated brewery wastewater remediation and CO_2 biofixation for green energy purposes. *Journal of Cleaner Production, 165*, 1316–1327. doi:10.1016/j.jclepro.2017.07.232

Fulke, A. B., Mudliar, S. N., Yadav, R., Shekh, A., Srinivasan, N., Ramanan, R., Krishnamurthi, K., Devi, S. S., & Chakrabarti, T. (2010). Bio-mitigation of CO_2, calcite formation and simultaneous biodiesel precursors production using *Chlorella* sp. *Bioresource Technology, 101*(21), 8473–8476. doi:10.1016/j.biortech.2010.06.012 PMID:20580227

Gupta, P. L., Lee, S. M., & Choi, H. J. (2015). A mini review: Photobioreactors for large scale algal cultivation. *World Journal of Microbiology & Biotechnology, 31*(9), 1409–1417. doi:10.100711274-015-1892-4 PMID:26085485

Ho, S. H., Chan, M. C., Liu, C. C., Chen, C. Y., Lee, W. L., Lee, D. J., & Chang, J. S. (2014). Enhancing lutein productivity of an indigenous microalga *Scenedesmus obliquus* FSP-3 using light-related strategies. *Bioresource Technology, 152*, 275–282. doi:10.1016/j.biortech.2013.11.031 PMID:24296122

Ho, S. H., Chen, C. Y., & Chang, J. S. (2012). Effect of light intensity and nitrogen starvation on CO_2 fixation and lipid/carbohydrate production of an indigenous microalga *Scenedesmus obliquus* CNW-N. *Bioresource Technology, 113*, 244–252. doi:10.1016/j.biortech.2011.11.133 PMID:22209130

Ho, S. H., Chen, C. Y., Lee, D. J., & Chang, J. S. (2011). Perspectives on microalgal CO2-emission mitigation systems—A review. *Biotechnology Advances, 29*(2), 189–198. doi:10.1016/j.biotechadv.2010.11.001 PMID:21094248

Ho, S. H., Li, P. J., Liu, C. C., & Chang, J. S. (2013). Bioprocess development on microalgae-based CO_2 fixation and bioethanol production using Scenedesmus obliquus CNW-N. *Bioresource Technology, 145*, 142–149. doi:10.1016/j.biortech.2013.02.119 PMID:23566474

Honda, R., Teraoka, Y., Noguchi, M., & Yang, S. (2017). Optimization of hydraulic retention time and biomass concentration in microalgae biomass production from treated sewage with a membrane photobioreactor. *Journal of Water and Environment Technology, 15*(1), 1–11. doi:10.2965/jwet.15-085

Idris, N. A., Loh, S. K., Lau, H. L. N., Mustafa, E. M., Vello, V., Tan, C. Y., & Phang, S. M. (2017). Cultivation of microalgae in medium containing palm oil mill effluent and its conversion into biofuel. *Journal of Oil Palm Research, 29*(2), 291–299. doi:10.21894/jopr.2017.2902.13

Jacob-Lopes, E., Scoparo, C. H. G., Queiroz, M. I., & Franco, T. T. (2010). Biotransformations of carbon dioxide in photobioreactors. *Energy Conversion and Management, 51*(5), 894–900. doi:10.1016/j.enconman.2009.11.027

Juneja, A., Ceballos, R. M., & Murthy, G. S. (2013). Effects of environmental factors and nutrient availability on the biochemical composition of algae for biofuels production: A review. *Energies, 6*(9), 4607–4638. doi:10.3390/en6094607

Kao, C. Y., Chen, T. Y., Chang, Y. B., Chiu, T. W., Lin, H. Y., Chen, C. D., Chang, J. S., & Lin, C. S. (2014). Utilization of carbon dioxide in industrial flue gases for the cultivation of microalga Chlorella sp. *Bioresource Technology, 166*, 485–493. doi:10.1016/j.biortech.2014.05.094 PMID:24950094

Kargupta, W., Ganesh, A., & Mukherji, S. (2015). Estimation of carbon dioxide sequestration potential of microalgae grown in a batch photobioreactor. *Bioresource Technology, 180*, 370–375. doi:10.1016/j.biortech.2015.01.017 PMID:25616748

Khalekuzzaman, M., Alamgir, M., Islam, M. B., & Hasan, M. (2019). A simplistic approach of algal biofuels production from wastewater using a Hybrid Anaerobic Baffled Reactor and Photobioreactor (HABR-PBR) System. *PLoS One, 14*(12), e0225458. doi:10.1371/journal.pone.0225458 PMID:31805078

Khalil, Z. I., Asker, M. M., El-Sayed, S., & Kobbia, I. A. (2010). Effect of pH on growth and biochemical responses of *Dunaliella bardawil* and *Chlorella ellipsoidea. World Journal of Microbiology & Biotechnology, 26*(7), 1225–1231. doi:10.100711274-009-0292-z PMID:24026927

Kumar, K., Banerjee, D., & Das, D. (2014). Carbon dioxide sequestration from industrial flue gas by *Chlorella sorokiniana. Bioresource Technology, 152*, 225–233. doi:10.1016/j.biortech.2013.10.098 PMID:24292202

Kumar, K., Dasgupta, C. N., Nayak, B., Lindblad, P., & Das, D. (2011). Development of suitable photobioreactors for CO_2 sequestration addressing global warming using green algae and cyanobacteria. *Bioresource Technology*, *102*(8), 4945–4953. doi:10.1016/j.biortech.2011.01.054 PMID:21334885

Llewellyn, C. A., & Airs, R. L. (2010). Distribution and abundance of MAAs in 33 species of microalgae across 13 classes. *Marine Drugs*, *8*(4), 1273–1291. doi:10.3390/md8041273 PMID:20479978

Mata, T. M., Martins, A. A., & Caetano, N. S. (2010). Microalgae for biodiesel production and other applications: A review. *Renewable & Sustainable Energy Reviews*, *14*(1), 217–232. doi:10.1016/j.rser.2009.07.020

Pangestuti, R., & Kim, S. K. (2011). Biological activities and health benefit effects of natural pigments derived from marine algae. *Journal of Functional Foods*, *3*(4), 255–266. doi:10.1016/j.jff.2011.07.001

Piltz, B., & Melkonian, M. (2018). Immobilized microalgae for nutrient recovery from source-separated human urine. *Journal of Applied Phycology*, *30*(1), 421–429. doi:10.100710811-017-1266-4

Posten, C. (2009). Design principles of photo-bioreactors for cultivation of microalgae. *Engineering in Life Sciences*, *9*(3), 165–177. doi:10.1002/elsc.200900003

Rasouli, Z., Valverde-Pérez, B., D'Este, M., De Francisci, D., & Angelidaki, I. (2018). Nutrient recovery from industrial wastewater as single cell protein by a co-culture of green microalgae and methanotrophs. *Biochemical Engineering Journal*, *134*, 129–135. doi:10.1016/j.bej.2018.03.010

Raven, J. A. (2010). Inorganic carbon acquisition by eukaryotic algae: Four current questions. *Photosynthesis Research*, *106*(1-2), 123–134. doi:10.100711120-010-9563-7 PMID:20524069

Razzak, S. A., Hossain, M. M., Lucky, R. A., Bassi, A. S., & de Lasa, H. (2013). Integrated CO_2 capture, wastewater treatment and biofuel production by microalgae culturing—A review. *Renewable & Sustainable Energy Reviews*, *27*, 622–653. doi:10.1016/j.rser.2013.05.063

Ren, L. J., Ji, X. J., Huang, H., Qu, L., Feng, Y., Tong, Q. Q., & Ouyang, P. K. (2010). Development of a stepwise aeration control strategy for efficient docosahexaenoic acid production by *Schizochytrium* sp. *Applied Microbiology and Biotechnology*, *87*(5), 1649–1656. doi:10.100700253-010-2639-7 PMID:20445973

Rodolfi, L., Chini Zittelli, G., Bassi, N., Padovani, G., Biondi, N., Bonini, G., & Tredici, M. R. (2009). Microalgae for oil: Strain selection, induction of lipid synthesis and outdoor mass cultivation in a low-cost photobioreactor. *Biotechnology and Bioengineering*, *102*(1), 100–112. doi:10.1002/bit.22033 PMID:18683258

Singh, J., & Gu, S. (2010). Commercialization potential of microalgae for biofuels production. *Renewable & Sustainable Energy Reviews*, *14*(9), 2596–2610. doi:10.1016/j.rser.2010.06.014

Singh, R. N., & Sharma, S. (2012). Development of suitable photobioreactor for algae production–A review. *Renewable & Sustainable Energy Reviews*, *16*(4), 2347–2353. doi:10.1016/j.rser.2012.01.026

Slade, R., & Bauen, A. (2013). Micro-algae cultivation for biofuels: Cost, energy balance, environmental impacts and future prospects. *Biomass and Bioenergy*, *53*, 29–38. doi:10.1016/j.biombioe.2012.12.019

Spolaore, P., Joannis-Cassan, C., Duran, E., & Isambert, A. (2006). Commercial applications of microalgae. *Journal of Bioscience and Bioengineering, 101*(2), 87–96. doi:10.1263/jbb.101.87 PMID:16569602

Tan, X., Chu, H., Zhang, Y., Yang, L., Zhao, F., & Zhou, X. (2014). *Chlorella pyrenoidosa* cultivation using anaerobic digested starch processing wastewater in an airlift circulation photobioreactor. *Bioresource Technology, 170*, 538–548. doi:10.1016/j.biortech.2014.07.086 PMID:25164347

Tang, D., Han, W., Li, P., Miao, X., & Zhong, J. (2011). CO_2 biofixation and fatty acid composition of *Scenedesmus obliquus* and *Chlorella pyrenoidosa* in response to different CO_2 levels. *Bioresource Technology, 102*(3), 3071–3076. doi:10.1016/j.biortech.2010.10.047 PMID:21041075

Ting, H., Haifeng, L., Shanshan, M., Zhang, Y., Zhidan, L., & Na, D. (2017). Progress in microalgae cultivation photobioreactors and applications in wastewater treatment: A review. *International Journal of Agricultural and Biological Engineering, 10*(1), 1–29.

Toledo-Cervantes, A., Morales, M., Novelo, E., & Revah, S. (2013). Carbon dioxide fixation and lipid storage by *Scenedesmus obtusiusculus. Bioresource Technology, 130*, 652–658. doi:10.1016/j.biortech.2012.12.081 PMID:23334023

Tredici, M. R. (2004). Mass production of microalgae: photobioreactors. Handbook of Microalgal culture: Biotechnology and Applied Phycology, 1, 178-214.

Utomo, R. P., Chang, Y. R., Lee, D. J., & Chang, J. S. (2013). Lutein recovery from *Chlorella* sp. ESP-6 with coagulants. *Bioresource Technology, 139*, 176–180. doi:10.1016/j.biortech.2013.04.025 PMID:23660381

Van Den Hende, S., Vervaeren, H., & Boon, N. (2012). Flue gas compounds and microalgae: (bio-) chemical interactions leading to biotechnological opportunities. *Biotechnology Advances, 30*(6), 1405–1424. doi:10.1016/j.biotechadv.2012.02.015 PMID:22425735

Vasumathi, K. K., Premalatha, M., & Subramanian, P. (2012). Parameters influencing the design of photobioreactor for the growth of microalgae. *Renewable & Sustainable Energy Reviews, 16*(7), 5443–5450. doi:10.1016/j.rser.2012.06.013

Wang, H. M. D., Chen, C. C., Huynh, P., & Chang, J. S. (2015). Exploring the potential of using algae in cosmetics. *Bioresource Technology, 184*, 355–362. doi:10.1016/j.biortech.2014.12.001 PMID:25537136

Wang, S. K., Stiles, A. R., Guo, C., & Liu, C. Z. (2014). Microalgae cultivation in photobioreactors: An overview of light characteristics. *Engineering in Life Sciences, 14*(6), 550–559. doi:10.1002/elsc.201300170

Wang, Y., Ho, S. H., Cheng, C. L., Guo, W. Q., Nagarajan, D., Ren, N. Q., Lee, D. J., & Chang, J. S. (2016). Perspectives on the feasibility of using microalgae for industrial wastewater treatment. *Bioresource Technology, 222*, 485–497. doi:10.1016/j.biortech.2016.09.106 PMID:27765375

Wijffels, R. H., & Barbosa, M. J. (2010). An outlook on microalgal biofuels. *Science, 329*(5993), 796–799. doi:10.1126cience.1189003 PMID:20705853

Yadav, G., Karemore, A., Dash, S. K., & Sen, R. (2015). Performance evaluation of a green process for microalgal CO2 sequestration in closed photobioreactor using flue gas generated in-situ. *Bioresource Technology, 191*, 399–406. doi:10.1016/j.biortech.2015.04.040 PMID:25921786

Yeh, K. L., Chen, C. Y., & Chang, J. S. (2012). pH-stat photoheterotrophic cultivation of indigenous Chlorella vulgaris ESP-31 for biomass and lipid production using acetic acid as the carbon source. *Biochemical Engineering Journal, 64*, 1–7. doi:10.1016/j.bej.2012.02.006

Yen, H. W., Ho, S. H., Chen, C. Y., & Chang, J. S. (2015). CO_2, NO_x and SO_x removal from flue gas via microalgae cultivation: A critical review. *Biotechnology Journal, 10*(6), 829–839. doi:10.1002/biot.201400707 PMID:25931246

Yen, H. W., Hu, I. C., Chen, C. Y., Ho, S. H., Lee, D. J., & Chang, J. S. (2013). Microalgae-based biorefinery–from biofuels to natural products. *Bioresource Technology, 135*, 166–174. doi:10.1016/j.biortech.2012.10.099 PMID:23206809

KEY TERMS AND DEFINITIONS

Antioxidants: Compounds preventing damage of cellular components by inhibiting oxidation and preventing formation of free radicals.

Biodiesel: Biodiesel is a form of fuel containing long-chain fatty acid esters that is usually synthesized from reaction of lipids like animal fats or vegetable oils with an alcohol.

Bioremediation: Bioremediation refers to the application of indigenous or purposely introduced microorganisms for degradation of toxic compounds present at a polluted site.

Flue Gas: Exhaust gas produced as a result of combustion by power plants containing reaction products of fuel and combustion air as well as residual substances like particulate matter, SO_x, NO_x, and CO.

Pharmaceuticals: Compounds used for diagnosis, treatment or prevention of diseases and restoration, correction or modification of organic functions.

Photobioreactors: Photobioreactors are open, closed or partially closed containers prepared from transparent and water-resistant materials and are capable of facilitating ideal environment for growth of photosynthetic microorganisms.

Upscaling: Raising or upgrading a process for improved benefits.

Wastewater: Wastewater is contaminated water produced as a result of anthropogenic activities in agricultural, commercial, domestic, and industrial sectors.

Chapter 9
Research and Development in Automobile Windshield Waste (AWW)

Pawan Kumar Bharti
https://orcid.org/0000-0003-2456-8322
Shriram Institute for Industrial Research, India

R. K. Singh
Shriram Institute for Industrial Research, India

ABSTRACT

The world over, efforts are being made for the safe disposal through development of cost-effective technologies to convert fast growing solid waste into useful products. In India, one such solid waste causing concern is the automobile windshield waste. The main sources of waste are automobile windshield manufacturers, dismantlers, recyclers, processors, etc. Automobile windshield waste emanating from various sources is a mixture of glass, PVB, additives, etc. The glass layer is fused with the PVB plastic film. Automobile windshield plastic film cannot be separated and is also not being recycled. At present, the reclaiming of PVB materials from automobile windshield waste is not being carried out in India. It is presently being thrown in dumpsites and land fill sites. It has negative impact both on human health and environment.

INTRODUCTION

PVB can be a useful material for making protective primer/peelable coating for wooden and metal surface. The primer and strippable coating made using PVB recovered can be used in paint and coating industry. The use of protective primer and peelable paint made from PVB waste shall have the added advantages:

- PVB based primer for wooden surface prevent wood from damage by moisture and microbial attack. These primers provide excellent protection when applied prior to paints and top coat of

DOI: 10.4018/978-1-7998-4921-6.ch009

wooden surface. It also ultimately reduces the paint consumption. It will be a good substitute to conventional primers.
- Protective peelable coating based on PVB can be used for various metal substrates like mild steel, steel, aluminum, zinc, etc. Peelable coating provides protection to metal against corrosion, abrasion and scratch. It can be used for protection of machinery parts, equipment and instruments. After its use, peelable coating can be easily removed by hand or mechanical peeling.
- An effective management of automobile windshield waste
- A cheaper replacement of commercially used wooden primer and rubber based peelable coatings
- Lower production cost while saving the cost of conventional resin (NC/alkyd/PU/epoxy, etc.)
- To minimize the environmental impact of disposal of automobile windshield waste.

Figure 1. A snapshot of windshield waste

In recent years, many researchers have claimed that the PVB material from various windshield waste and other sources can be reused for making value added products. The PVB can be useful protective coating materials as it has unique properties such as film forming, adhesive, flexibility etc. It has good bonding to various substrates such as metal, glass, wood, ceramic, etc.

A process has been developed for making protective primer and adhesive for wood using recovered PVB. A process has also been optimized for recovery of PVB from automobile windshield waste. The recovered PVB were characterized for solubility, melting point, thermal stability, film forming and adhesiveness. A process has also been developed to make peelable coating using recovered PVB for aluminum and mild steel surfaces. It has good film forming properties and can be easily removed after use.

No effective technologies exist at present within the country for separation of PVB embedded in glass laminates of automobile windshield waste and to make protective primer and peelable coating for wooden and metal surface.

The proposed technology once scaled up will enable to use recovered PVB from automobile windshield waste as a cheap raw material in making protective primer and peelable coating for wooden and metal surfaces. Moreover, it will also replace conventionally used polymer resin at least to some extent in paint industry. Further, this will lead to an effective waste management system for automobile windshield waste generation in huge quantity and making useful protective coatings for wood and metal.

Keeping this in background, a proposal is being submitted so that the PVB and glass waste emanating from automobile windshield manufacturers, dismantlers as well as automobile recycling industry may be reused to make glass articles, protective primer for wooden surface and peelable coating for metal substrate. This will also provide an opportunity to the manufacturers and users of glass, wood, metal and paint industry to have an alternative material PVB recovered from AWW.

SRI and Laxmi Polymers has established an industrial process/knowhow for making protective primer and peelable coating useful for wooden and metal surface using recovered PVB from automobile windshield waste. The infrastructure support for the plant shall be provided by the industrial partner Laxmi Polymer. The industrial scale process would provide and channel the recycle and reuse the materials in an effective manner and solve the solid waste management disposal of windshield waste generated by automotive industry resulting into safe and clean environment.

The main aim of this work is to scale up of the process for separating PVB from glass automobile windshield waste for its use making protective primer and peelable coating useful for wooden and metal substrate.

Current Status of the Technology

The automobile industry is the most attractive and fastest growing industry in the world. PVB polymer is widely used in manufacture of automobile windshield. PVB has outstanding properties such as film forming, bonding, adhesive strength, flexibility etc for a wide range of application. The use of safety glass using PVB has also become an important part in building sectors, architectural sectors etc.

Automobile windshield waste is very difficult to recycle because it contains PVB plastic, glass, impurities etc. They have having different melting point and density. Glass layer is fused with PVB layer. Presently, glass from automobile windshield waste is removed by mechanical way leaving behind PVB by-product.

Presently in India, PVB emanating from various sources is not recyclable and cannot be reused in making laminated glass and automobile windshield. It can be seen that PVB plastic, especially those used in automobile windshield have a very good film forming, flexibility, bonding and adhesion to various substrates. They have outstanding film forming and coating properties and can be used for making protective coating for vide variety of substrates such as metal, glass, wood, ceramic, metal, etc.

PVB can be used as binder and converted into protective primer, sealer for wooden surface using solvent, plasticizers, filler, pigment and other performance additives (Market, 2015). Moreover, the PVB of automobile windshield waste can also be converted to peelable coating useful for metal substrate. The peelable paint using PVB protects the metal against corrosion, scratch and moisture. It can also be used in other industries to make wide variety of coating products.

In India this technology shall gain a lot of importance and shall be utilized as an alternative source of primer and metal strippable paint to replace conventionally used polymer resin (PU/epoxy/alkyd/rubber, etc.) used for making protective wood primer and strippable paint for metal substrate.

A process has been developed at SRI for recovery of PVB from windshield waste. The recovered PVB were characterized for solubility, melting point, thermal stability, film forming and adhesiveness. The recovered PVB have been used to make primer and wood for wooden surface along with performance additives. The protective primer and adhesive made with PVB have excellent scratch resistance, water resistance, gloss and good bonding to wood. A comparative evaluation of developed PVB primer vis-à-vis market wood primer has been undertaken. Similarly, peelable coating have been made with PVB for metal substrate. The peelable coating can be used in aluminum, steel, zinc etc.

Similarly, peelable coating have been made with PVB for metal substrates. The peelable coating can be used in protection of aluminum, steel, zinc etc.

Further, there is a need to scale up this technology for manufacture of protective primer and strippable paint using PVB polymeric portion of automobile windshield waste presently generated in India. Department of Science & Technology, New Delhi always encourage such work for utilization of automobile windshield waste and PVB waste.

The above proposed work will lead to great achievement in making protective primer and strippable/ or peelable coating from PVB polymer recovered from automobile windshield waste, which remains either on automobile manufacturers, workshop, accident sites, dismantle or dumpsites. This will also enable a great value addition to PVB plastic which are either thought as problematic waste or not fit for recycling. This outlet of utilization of the non-recyclable PVB plastic will thus result in a complete collection of automobile windshield waste from automobile windshield manufacturers, dismantlers, recyclers, workshop, dumpsites and landfill sites. Further, it will provide good earning to rag pickers, waste collectors, dismantlers and recyclers. Moreover, this will thus result in a complete collection and recycling of automobile windshield waste and PVB generated from various source.

To the best of our knowledge, no technology has been developed for making protective primer and peelable coating using PVB emanating from automobile windshield waste in our country except Shriram Institute for Industrial Research, Delhi.

In light of the above, it is proposed to scale up indigenous technology developed at SRI for manufacture of protective primer and peelable coating using recovered PVB from automobile windshield waste. The protective primer made using PVB can be used to protect wooden surface from water, scratch and minimize the risk of microbes. The main application of PVB based primer for coating in wooden industry: wooden furniture; wooden boxes; wood handicrafts etc. The process would be cost-effective. Once the high value targeted primer coating are made using PVB, it would become a valuable resource for wooden coating industries. Further, the primer made from PVB will also have export potential.

Under this work, SRI would provide technical support for designing and fabrication of industrial scale machineries for making protective primer and strippable paint using PVB. The infrastructure support is to be provided by the industrial partner. Industrial scale process for making protective primer and strippable coating shall be done.

International Status

A great deal of research and development for utilization of automobile windshield waste in making wide variety of products has been done in the world. There have been many successful application of automobile windshield waste and PVB to produce products having commercial value. The international status of development work are as follows:

E.I. DuPont De Nemours and Company have developed a process for recycling of PVB. Polyvinylbutyral was recovered from layer adhered to glass. It can be used in molding application when combined with polyolefin. An anhydride-modified polymer was used as compatibilizer in the blend. It has a melt temperature less than the decomposition temperature of the polyvinylbutyral. It can be used for making molded material. (WO 1993002141 A1).

United States Patent 2358355 describe a process to make interlayer sheeting for safety glass. It is simple and economical process for converting acetone-soluble polyvinyl butyral to an acetone-insoluble form suitable for use in interlayer sheeting for safety glass. The composition are polyvinyl ester 0-10, polyvinyl alcohol 15-22 and polyvinyl butyral 85-77. It can be used as adhesive for use in interlayer sheeting for safety glass. (US Patent 2358355)

United States Patent 3,392,130 discloses an anticorrosive coating compositions. It comprises polyvinyl butyral and hydrolyzed alkyl silicate. It also contains zinc as pigment and are incorporated into dissolve PVB solution. It is air cured and has good adhesion, toughness and flexibility. However, problem with PVB based zinc-rich paints is cracking. It take place when a thick coating is applied. Due to overlapping or poor application process, it also result in a cracking (US Patent 3392130 A).

Monsanto Co have developed a sprayable PVB based primer for wooden articles. A solution of polyvinyl butyral can be used as a primer and sealer for wood surfaces before coating with polyurethane and polyester finishes. The primer was prepared from the ingredients polyvinyl butyral, toluene and ethyl alcohol. It is brushable. The polyvinyl butyral used for making wooden primer had a molecular weight of 41,000 and a viscosity of 13-18 centipoises at 20°C (6% solution in methanol) (US Patent 3320203).

E.I. DuPont De Nemours and Company have developed adhesive composition using PVB. It comprises ABS 1-30%, 20-95% PVB and a coupling agent 1.0%. It has enhanced surface adhesion. However, it has low surface gloss (EP 1639041 A1)

United States Patent 3,320,203 describe a sprayable polyvinyl butyral wood primer. The PVB solution was prepared in solvent. It can be used to make primer and seal wood surfaces before coating with polyurethane and polyester finishes. These primer when applied to wood surface has high gloss surface of excellent appearance and good aging characteristics. They are resistance to heat, chemicals, water and abrasion. The primer can be applied by brush (US Patent 3,320,203).

Method for the recycling of polyvinyl butyral is described. It is based on the removal of polyvinyl butyral (PVB) from laminated glass which is shattered. The polyvinyl butyral (PVB) is separated from glass. The sheet is cut into pieces and subjected to a cleaning process to eliminate dirt and foreign matter. The purification of polyvinyl butyral (PVB) was carried out by way of a chemical treatment. (EP 2308919 A1)

A coating composition for adhesion to metal has the ability to inhibit the corrosion of metals. It is a mixture of ethyl cellulose, alchol and polyvinyl butyral. This primer has advantages over market primers. It has better stability and is also resistance to humidity. However, the primer is expensive. Moreover, its film is prone to blister formations under conditions of high humidity (US Patent 2798015 A)

A composition for making sprayable anti-corrosive primer consist of isopropanol, toluol and Polyvinyl butyral. The solvent dissolve up to about 12% of the polyvinyl butyral. It also contain 9.20 water, 5.40 chromic acid, 1.97 aluminum powder and 1.63 water. The aluminum employed was less than 100 mesh in order to ensure good dispersion throughout the resin solution. The amount of the aluminum vary from 1.3-1.9 parts by weight based on a total of 100 parts of primer (US Patent 3325432 A).

A process for making elastomeric blends contain PVB relates to elastomeric blends capable of forming cellular foams. A blend of 99-70% hexamethoxymethylmelamine, 1-30% polyvinylbutyral and 2-25% nitrogenous blowing agent. The blend was heated above room temperature to form a cellular structure. The blowing agents generally used are N, N-'dimethyl-N, N'-dinitrosoterephthalamide, N, N-dinitrosopentamethylenetetramine and sulfonyl hydrazides. The nitrogenous blowing agent is added in amounts 2-25% based on the weight of hexamethoxymethylmelamine and polyvinylbutyral. The acid catalyst used are from para-toluene sulfonic acid, formic, acetic acid, etc. The preferred range of catalyst are 0.25-2.5% based on the weight of polyvinylbutyral and hexamethoxymethylmelamine (US Patent 3282865 A).

A molding compositions containing polyvinylbutyral consist of 40-70 parts polyvinyl butyral, 60-30 poly(methylmethacrylate); 0-15 filler, 0-30 elastomer and 0-30 polyurethane. The filler may include clays, calcium carbonate, barium sulfate, titanium dioxide, hydrated alumina etc. Elastomeric polymers used in molding compositions comprise methyl methacrylate, alkyl acrylate, EPDM rubbers, etc. (US Patent 5130370).

The laminate are useful in the areas where transparency is required. The laminates comprise PVB bonded to polyester, polyamide, polyurethane polymethylmethacrylate, polyvinyl fluoride, polyvinylidene chloride, cellulose acetate or cellulose esters. The laminates have distinct advantageous. It is suitable for a wide range of uses. They are characterized by excellent adhesion between two layers. PVB has been used as an interlayer in optical laminates. The majority of windshields are made with a PVB interlayer. The PVB interlayer is laminated between two glass sheets. They are characterized by excellent adhesion between two layer (US Patent 4251591).

DuPont have made adhesive using PVB and phenol formaldehyde. A process is described for making adhesive using PVB and phenol formaldehyde. It can be used to bond metals to rubber. The ratio of the polyvinyl butyral to the phenol-formaldehyde is an important parameters. The polyvinyl butyral is 30-45% of the total weight of the adhesive. A catalyst such as NaOH, KOH, etc. in 0.5-1.5% was used (US Patent 2430053).

Monsanto Company have developed a process for preparing a sprayable anticorrosive primer for metal and wood. The anti-corrosive primer composition comprises 7.3-10.5 parts polyvinyl butyral, 76.5-85 solvent, 0.09-0.15 chromium trioxide; 1.4-1.8 aluminum and 6.15-9.05 water. The chromium trioxide was used to improve adhesion, water resistance and corrosion protection. The viscosity of the primer was10-500 centipoise. These anti-corrosive primers have excellent protection. The primer protect the metal and does not need to be removed prior to top coating of the metal. However, they must be removed prior to welding (US Patent 3325432 A).

A method is described for preparing a corrosion inhibiting coating composition A coating composition comprises 210 parts methyl ethyl ketone, 35 methyl isobutyl ketone, 101 xylene, 1.73 methanol, 0.072 sodium nitrite, 60 polyvinyl chloride and 24 dioctylphthalate plasticizer (US Patent 3035005).

A composition for making backcoating utilizes waste of polyvinyl butyral for making the backcoating. It can be used for coverings such as carpeting, pile carpets and tiles. It also utilizes plasticizer and filler along with PVB (US Patent 7521107).

A study has been carried out for recycling of automobile windshield waste. PVB was recovered from windshield recycling process. It has low amount of glass particles. Three stage process was carried out for recycling of PVB sheets. In the first-stage, laminated safety glass was mechanically cracked. In the second-stage, the adhesion reduction was carried out for separation of glass from PVB sheet using physical-chemical separation. In the third-stage, mechanical peeling of the remaining glass from the PVB sheet was carried out. Delamination process produces PVB sheet with residual glass content (up to 300 ppm). There was minimal change in PVB sheet properties. The process is environment friendly and the waste is recyclable. It could reduce the problem of PVB associated with windshield waste disposal (Michael, 2014).

A study has been undertaken in the preparation of composites with of PVB and PU foam waste. The composite were prepared using PVB waste and PU foam waste. The PU content in the composite were 20, 35 and 50%. They were mixed in an extruder and injection molded. The composites were characterized using DSC, MFI, TGA etc. The mechanical properties were satisfactory and composites can be used in the industries. This can also solve disposal problems with PVB and PU waste. The SEM micrographs showed that PVB/PU (50/50) composite exhibited the homogeneous and compact structure with fewer voids in the interface.

M/s. Escola Politécnica da Universidade de São Paulo have carried out work on utilization of the windshield after its use. The recycling of the laminated glass goes through a series of process. It is possible to obtain PVB from windshield waste. PVB sheet is used as a filler. It can solve disposal problem.

M/s. Institute of Technology of Building Materials and Components, Brno University of Technology, Czech Republic have carried out work on utilization of automobile windshield waste for concrete protection. The PVB layer can protect concrete and mortar against carbon dioxide carbonation. The barrier efficiency of PVB was verified by the measurement of diffusion characteristics. PVB sheet was obtained from recycled windshields waste. PVB sheets were exposed to CO_2 atmosphere (10%CO_2). The PVB exhibit excellent barrier capability against CO_2 permeability. This application of PVB can be used in civil engineering. This new application of PVB waste can bring significant improvement in solving environmental problem.

M/S. Shark Dispersion have made temporary peelable coatings using PVB. It was made using recycled PVB. It can be formulated as a ready to use temporary or peelable coating on numerous surfaces such as metal, coil coated surfaces, glass, glass fibers, polyesters or other plastics. The temporary coating can be removed e.g. when goods are ready for painting or assembly. It can be retained on a clean surface. It is a water based dispersion. The PVB form a uniform blended structures with a particle size < 50 µm. In this process, no precipitation occurs when stored under normal conditions 28^{0}C. However, damage may occur if exposed to temperatures below 1°C. It is easily applied by brush, dip, roller, spray, spreader, foam, or aerosol. However, when dispersion is exposed to air it begins to dry, skinning may occur.

An economical and environment friendly have been developed for recycling of waste automotive laminated glass. A process have been developed to separate PVB from glass. The separated PVB was characterized through various techniques such as SEM, EDS, IR, NMR etc. A nonionic surfactants D201 were used for the mechanochemical separation purpose. The optimize conditions are 301% D201, stirring speed 400 rpm, temperature 35°C and operation time 1 h. The ratio of the waste automotive laminated glass to surfactant is 1:25. The technology developed is sustainable, environmentally friendly, economical feasible and capable of mass recycling (Basudev, 2015).

A study has been undertaken on modification of PVB recovered from the automotive industry. PVB was chemically modified by melt mixing with a vinyl trimethoxysilane (VTMS). It react with the hydroxyls group present in the PVB and from crosslinked bonds between the chains. This chemical modification resulted in the improvement of the solvent resistance without impairing the mechanical properties of the polymer. The mixing of PVB was carried out with VTMS in an internal mixer equipped with roller rotors. The mixture was subsequently compression molded. Depending on the mixing time, the dynamic crosslinking reactions occurred during this time. Further, static crosslinking was observed during compression molding, which resulted in the crosslinking and solvent resistance (Marilia, 2016).

M/s. Technological Center Lurederra, Polígono Industrial Perguita Los Arcos, Navarra (España) have design and develop a demonstrative pilot plant for the recycling of polyvinyl butyral (PVB). The technology has been scaled up to pilot scale. The pilot plant for PVB recycling, which is located at Los Arcos (Navarra, Spain). It consist of three steps, each of them designed to perform one specific stage in the recycling process. The pilot plant technology for laminated glass recycling is environment friendly and economic efficiency.

M/s. Tandus have developed a process for utilization of polyvinyl butyral (PVB) film obtained from recycling of used automobile windshields. This can be used as backing carpet. It could be an attractive alternative to conventional PVC, EVA, PU and polyolefin material which are presently used in carpet based backing. It can be used for floor coverings.

M/s. Tomas Bata University, Czech Republic have developed a process for recycling of automobile windshield. The laminated safety glass from automotive was treated with NaOH solution. They were mixed thoroughly in a kneader. The roll mill was also used. The composites were prepared at 95 °C for 10 min at 10 rpm. The samples were then molded to a thickness of 125 × 125× 1 mm by heating in press for 5 min to 130°C. It was then cooled in a cold press. PVB were recovered were from a wide range of sources. The recycling of PVB are very important because it is widely used for laminated safety glass. The PVB could be used to make laminated safety glass. The recycled PVB materials have good mechanical and optical properties. The most suitable concentration for recycling PVB is 0.5% NaOH. It is apparent that PVB could be used to laminate without loss of optical clarity and mechanical properties.

Society per L'Elettronica Applicata (SEA), Villanova (Bologne) Italy have developed a dry process for separation of PVB from glass. This process introduce a separation of the PVB from glass. It consist of a dry filtering in melted state based on the different physical properties. The materials was separated using combined ultrasound and gravitational separation method (Flotation separation using fluid of a density > 1.0 g/cm3).

A process is described for making silicon carbide from waste of automobile glass. It describe a novel approach to synthesize silicon carbide (SiC) product by utilizing automotive glass and plastic waste. A significant amount of end-of-life plastics and glass are currently landfilled, incinerated or dumped even though conventional plastic and glass recycling practices are well established. The reduction of silica in glass by blend of graphite and plastic to produce SiC was made under inert condition at a temperature 1,550 °C. They were characterized by XRD, SEM, EDS to confirm the formation of SiC, This process has significant advantages. It is a major step toward transforming nonmetallic automotive waste into valuable resources for refractory materials. It reduces the industry's reliance on conventional raw materials including quartz and coke. These can be used as silicon and carbon bearing resources for making Si-C refractory materials.

Borno University of Technology, Czech republic have made blends of PVB. A blend consisting of PVB and PVC are studied. PVB waste was derived from windshields waste. PVC was plasticized with 38% of diisononyl phthalate (DINP) and PVB was plasticized with 28% of triethylene glycol and bis (2-ethylhexanoate) (3GO). PVB content of blend composition used was ranging from 0 to 100. Blend of PVB/PVC composition were prepared in a extruder with kneader. Due to a high rotation speed, the blending efficiency was very high. The amount of processed blend was 2 kg. The mechanical properties of recycled PVB and of the blend consisting of recycled PVB and PVC (1:1) are reduced about 10 to 15%, in comparison with the original PVB and its blends. PVB blends can be also used in applications like wires, carpets, window profiles, waterproofing and sheets, sewer pipes, adhesives and synthetic leather.

Thus, a great deal of research and development work on automobile windshield waste into have been done in the world. There has been a promising option of converting PVB plastic waste to produce coating products, which are in high demand such as protective primer and strippable paint etc instead of disposing it in landfill sites. With the emergence of new technology for recovery of PVB from automobile windshield waste and utilization of it in coating compounds will become a reality. The PVB plastic waste used to produce primer and strippable paint shall have advantages over the conventional market wooden primer and metal peelable paint. Strippable or peelable paint made using PVB can be used to protect metal from corrosion or damage during handling, storage and also in transportation from one place to another. The protective primer and strippable made using PVB polymer from automobile windshield waste will have following advantages:

- Utilization of PVB waste emanating from automobile windshield manufacturing; automobile repair shop; dismantlers, etc.
- Low cost of protective primer and strippable paint in comparison to market products
- Reduced Green House Emission
- Saves conventional polymer resin such as (NC, alkyd, epoxy, PU, etc.) in making primer and paint
- Substitute of conventional wooden primer and metal strippable coating.

Indian Scenario and Technological Gap

PVB is widely used in manufacture of windshield glass due to film forming, flexibility and bonding property (Michael, 2013). With the rapid growth of automobile industry in our country, a large number of windshield manufacturing plants are to be set up in future. As a consequence large quantities of automobile windshield waste will be produced. India will have massive quantities of PVB waste to be disposed off.

So far, most of the work undertaken by the recycling industry involve the mechanical recovery of glass from automobile windshield waste. A very little effort have been taken up to utilize PVB waste generated at automobile windshield manufacturer's end. PVB can be used as raw materials for paint and coating industry. The inbuilt PVB portion of automobile windshield waste will be used for making protective coating for wooden surface and strippable coating for metal surface.

PVB resin is not manufactured in India and is imported. PVB is mainly used in manufacture of laminated safety glass or automobile windshield. The price of PVB is much higher than price of common plastics such as LDPE, HDPE, PE, PP, PVC, PET etc. Moreover, the common plastics are recyclable and the PVB is not recycled in India by primary recycling.

In addition to the utilization of glass by way of primary recycling, the PVB waste has also potential to offer its utilization in making a value added products such as protective primer for wooden surface and peelable coating for metal substrate in our country.

However, PVB has very good film forming coating and bonding properties. PVB recovered from automobile windshield waste it can be utilized to make coating materials. This will solve the problem of PVB waste. PVB could be used as an alternative to conventional polymer resin used in making primer and peelable coating/ paint for wooden and metal substrate. This could allow substitution of conventional polymer resin (NC/alkyd/etc.) and market wooden primer.

Laxmi Polymers have optimized a process for recovery of glass from automobile windshield waste. During the recycling, glass is recovered leaving behind PVB-by-product.

PVB recovered from automobile windshield waste were characterized by SRI for melting point, solubility, film forming property, thermal stability etc. Based on the property of PVB, a research and development work has been undertaken to make protective primer for wooden surface and PVB based adhesive. The developed primer made using PVB were found suitable for wooden surface. PVB based primer were characterized in detail for adhesion to wood, flexibility, viscosity, scratch resistance, gloss etc. A process has also been made to make peelable coating using recovered PVB for aluminum and mild steel surfaces. However, the technology needs to be scaled up to plant scale for utilization of PVB recovered from windshield waste in making primer and peelable coating.

SRI has expertise in coating, paint and waste recycling. PVB based primer can be used for coating of wood. It protects the wood against moisture, microbe, abrasion etc. Example of PVB based wood coating can be in areas of wooden furniture, wooden boxes, wooden handicrafts etc. Further, PVB based peelable coating or paint can used on iron, steel, aluminum, zinc and other metal. It provide protection of metal against corrosion, abrasion and scratch. Further, the peeling paint can be removed from metal substrate whenever required or after use.

In view of above, recovered PVB could be used to make protective primer and peelable coating. The protective primer made using PVB could be used for protection of wooden surface.

The proposed work would be very helpful in utilizing PVB emanating from automobile windshield waste as a useful resource for manufacture of large number of coating and paint products. The wooden and metal industries dealing with primer, coating and paint would be the direct beneficiary of this work. Therefore, the present study is very much pertinent in order to develop and scale up a cost effective technology for manufacture of protective primer and peelable coating using PVB constituent of automobile windshield waste.

Development Status at the Participating Agencies

- SRI has developed a process for making protective primer and adhesive for wooden materials using PVB recovered from automobile windshield waste. PVB based primer can be used for coating and protection of wooden surface. It has good film forming and bonding property. It seals of any porosity of wooden surfaces and prevent microbial growth. Adhesive was also made using PVB material for bonding wooden materials. A comparative evaluation of developed protective primer made using PVB vis-à-vis market wooden primer have been carried out.
- A process for recovery of PVB from automobile windshield waste has been optimised. Evaluation of recovered PVB has been carried out with respect to melting point, thermal stability, solubility etc.
- M/s Laxmi Polymers have expertise and capabilities in dismantling of automobile windshield waste. Glass recovered from windshield waste are sent for recycling. However, during the dismantling of automobile windshield waste, PVB waste is also generated and is not finding any use.
- A laboratory scale study has been carried by SRI for utilization of recovered PVB in making protective peelable coating suitable for steel, aluminum and mild steel. It has good film forming and bonding properties. It can be removed after use.

- Various process parameters were optimized for making protective primer and peelable coating using PVB w.r.to solid content, drying time, film thickness, flexibility, consistency, scratch resistance, etc.
- PVB based primer for wooden surface shows better results in comparison to metal surface.

In view of the above, it is proposed to develop a scaled up/industrial scale process for making protective primer and peelable coating useful for wooden and metal substrate using PVB constituent of automobile windshield waste in India. Further, the technology can be used for entrepreneurship development in India for safe recycling of automobile windshield waste and PVB waste.

Significance of the Work

The current windshield recycling process in India is based on hand and mechanical peeling. This type of process provides glass scrap. The recycling processes produce a glass scrap which is re-used in the glass industry. However, the separated PVB waste is not recyclable. This PVB is a by-product and it cannot be reprocessed again into PVB sheet suitable for glass lamination. Further, PVB scrap at present is incinerated or land-filled.

PVB based primer for wood improves the top coat ability to resist surface moisture absorption and microbial growth. It shall allow the paint to adhere more tightly to the wooden surface. The application of PVB based coating shall be in coating of wooden boxes, wooden furniture, and wooden handicraft items.

PVB based peelable paint for metal substrate is flexible, scratch resistance and corrosion resistance. It provides a temporary and long term protection to the metal. It can be peeled away when required without damage to the metal substrate. It can be used in ferrous and non-ferrous metal like aluminum, iron, steel, zinc, etc.

The adoption of this industrial scale process will give a tremendous boost to coating, paint and other chemical industries in our country. At the same time the process once established, will showcase this new technology and will provide encouragement to automobile and plastic recyclers operating in different parts of India for adopting this technology.

So, we can say that utilization of PVB polymer recovered from windshield waste for making protective primer and peelable coating in India is a pioneering effort. It will also provide a starting point for adopting a new and environment friendly technology for wood and metal coating and benefit the recyclers of the country.

The wooden and metal coating industry dealing with wooden primer, coating and peelable paint would be the direct beneficiary adopting the scientific and better method of utilizable of PVB waste. In other words, the environment, society and ecology will be the indirect beneficiary.

In light of the above, it is proposed to utilize the recovered PVB from automobile windshield waste in making protective primer and peelable coating for wood and metal substrate. Further, process technology developed at SRI shall be used for setting up of a industrial scale for making protective primer and strippable/or peelable coating along with industrial partner Laxmi Polymer. A simple and environment friendly method shall be adopted to utilize PVB valuable materials emanating from automobile windshield waste in India to make wooden and metal paint as a cheaper replacement of commercially used wooden primer and metal peelable coating.

Attractive Properties of Wooden Primer Made Using PVB

- Moisture resistance
- Inter-coat adhesion
- Sub-base for final paint
- Block of stains
- Seals pores in the wood substrate
- Prevent growth of microbes and molds
- Provide protection to the wood substrate until it can be top-coated.

Application Areas of Proposed Technology

- Wood Coating Industry
- Wooden Furniture Manufacturers
- Wooden Box manufacturers
- Wooden Handicrafts

Attractive Properties of Peelable Paint for Metal Surface

- Easy and fast to remove
- Abrasion protection
- Protection against scratch, dirt and dust
- Excellent Surface adhesion
- To protect finished metal surface during transportation from one place to other
- Anticorrosive finish for metal surface

Application Areas of PVB Based Peelable Coating

- Automobile body
- Machine parts
- Equipment Packaging
- Aircraft industry
- Instruments Packaging, etc.

Peelable coating is ideal for packaging of auto parts, machinery, equipment and instrument where protection from corrosion, abrasion and scratch is important. The peelable coating can be removed easily after use.

In this proposed work, PVB shall be used in making protective primer for wood coating, and peelable paint for metal substrate due to its film forming, flexibility and good adherence to the wide substrate. The primer prepared with the PVB has good adherence, flexibility and stability when applied on wood. It could substitute the conventional wooden primers used nowadays that are made using (alkyd resin/ NC products), etc.

The end users shall comprise of small scale, medium scale and large scale coating/paint industry & new industrial ventures.

Scientific Basis and Methodology

Polyvinyl butyral (PVB) is an important polymer widely used in laminated safety glass. It exhibits a very film forming, flexibility and high adhesion to glass, wood, metal and other substrates. The main use of PVB is in safety glass laminates, particularly in automotive, building, architectural glass and solar panels.

PVB resin is not manufactured in India. PVB is imported and is extensively used in glass lamination. Automobile windshield is difficult to recycle. During recycling, glass is generally recovered and PVB waste is generated. They are presently disposed off into dumpsites and landfill. There is no effective technology in recovery and reuse of PVB from automotive windshield waste. Moreover, PVB cannot be reused in making laminated safety glass.

Instead PVB emanating from automotive sources disposed off in landfill, it is proposed to use PVB in coating industry as an alternative to laminated safety glass manufacture.

PVB is an imported and expensive material. The PVB possess coating properties very similar to those of the conventional paint. Due to this knowledge, PVB may be used in various coating and paint applications such as primer and strippable film in paint industry.

In India, presently common plastic such as LDPE, HDPE, PVC, PET, etc. waste is being recycled to make recycled pellets as well as in making low value added recycled products. The non-recyclable plastic waste is a cause of concern. Example multilayer packaging, metallized multilayer, windshield waste, etc.

Presently in India, windshield waste is generated from automobile windshield manufactures, dismantlers, automobile recyclers, workshop areas, etc. The glass is recovered from windshield waste using hand or mechanically technique. During the process, PVB by-product are generated. The PVB polymer emanating from automobile windshield waste contain glass and other constituents. It is not being recycled in India. It can also not be reused in making laminated safety glass. Therefore, alternatives for PVB waste utilization are needed to be developed particularly on adhesive, paint, primer, sealer etc. as it has good coating properties. PVB has good film forming, flexibility and adhesive properties. PVB has good bonding to variety of substrates such as metal, glass, ceramic, wood, etc.

In view of above, it is proposed to use PVB in manufacture of protective coating PVB in making primer and strippable paint. It could be valuable material. The primer made with PVB for wooden surface can be applied before application of final wood and paint. It has good water resistance and moisture resistance.

Not much work has been done in India for making of protective primer and peelable paint from PVB plastic. The technical viability at industry level needs to be studied in India so that the PVB waste can be used for manufacture of protective primer and peelable coating for wood or metal coating application. Use of PVB plastic as an alternative material for manufacture of primer and peelable coating can significantly fulfill the need of wooden primer and peelable paint consumption in India while at the same time decrease pollution load of PVB waste and automobile windshield waste.

PVB is a specialty polymer and is used for variety of application. It is manufactured from the condensation reaction of polyvinyl alcohol (PVA) and n-butyraldehyde. It contains minor amounts of polyvinyl acetate and polyvinyl alcohol. The unique properties of PVB are good adhesion strength, film-forming properties, water resistance, elasticity, bonding behavior, optical transparency, binders for a wide range of materials, good compatibility with organic solvents (Marilia, 2015).

The major areas of application of PVB are in sheets and films for laminated safety glass. It is used in the production of laminated safety glass as a windshield and safety glass (Michael, 2015). PVB polymer connects two sheets of glass resulting in a laminate with excellent optical and mechanical properties. The PVB has high adhesion and is used for automotive glass lamination.

Figure 2. **Methodology for making peelable coating**

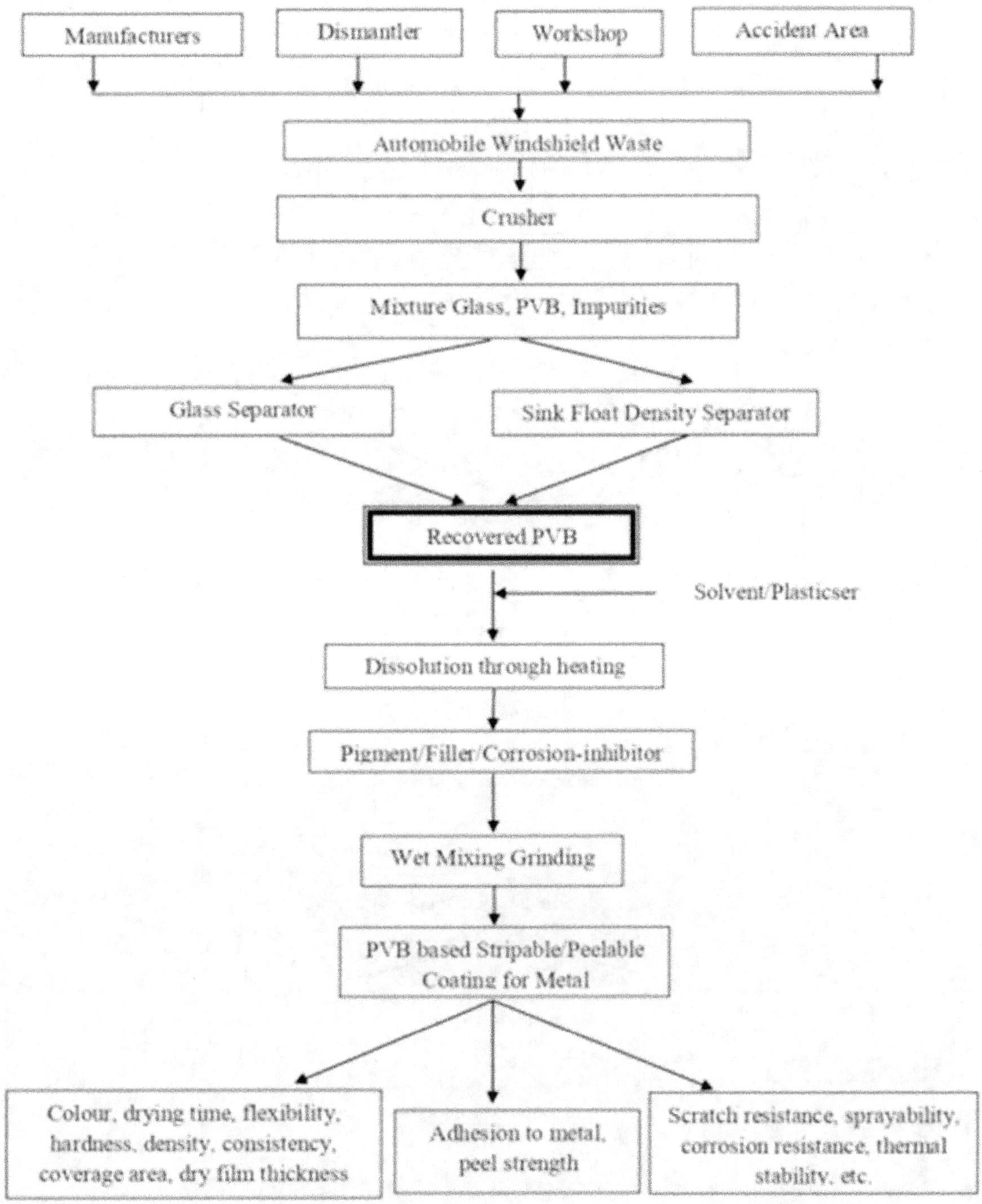

PVB waste is a part of automobile laminated safety scrap glass. There are huge quantities of waste PVB available in India. It is not possible to use PVB waste as raw material for manufacture of automobile windshield. Adherences of small glass particles and dust prohibit in the manufacture of automobile

Figure 3. Methodology for making PVB based primer for wooden materials

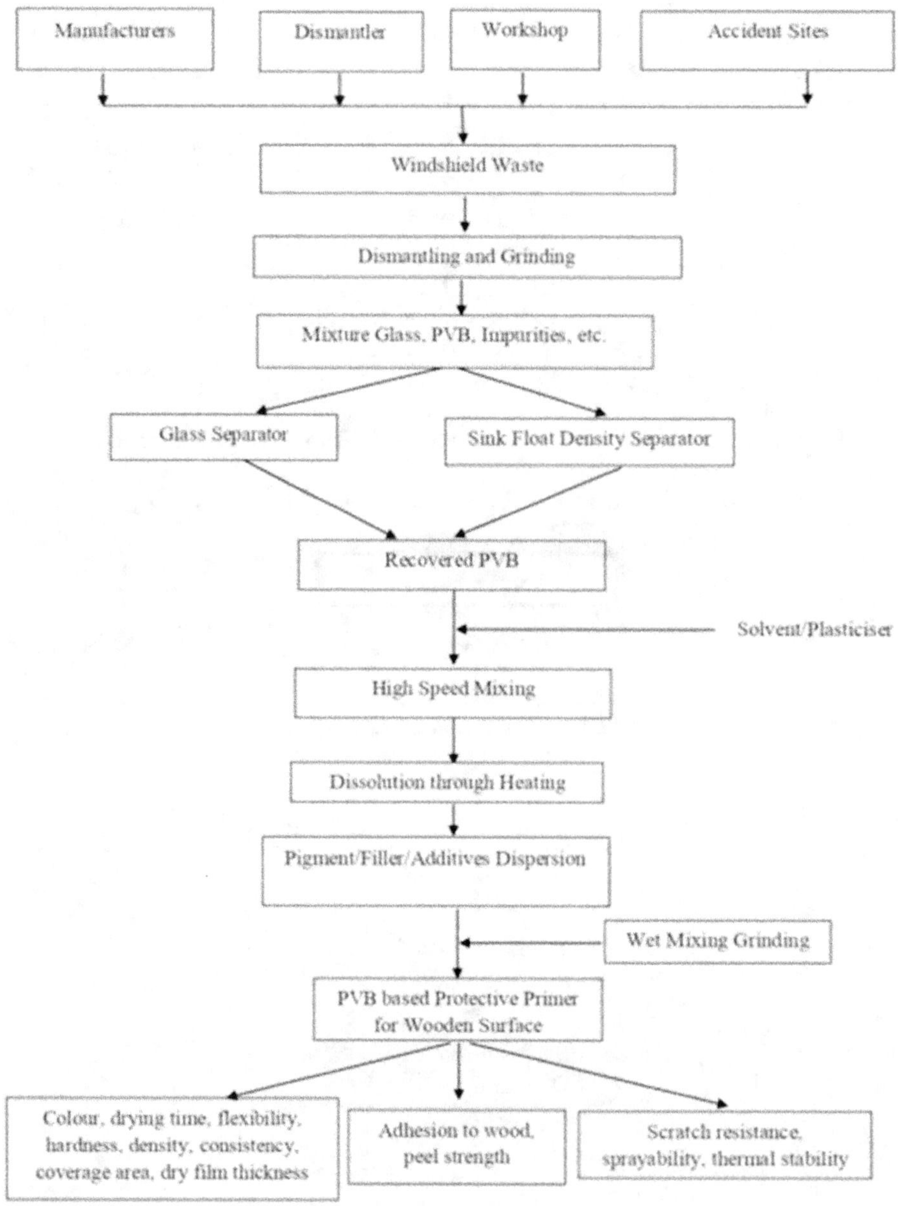

windshield using PVB. Moreover, PVB are non-recyclable plastic and cannot be recycled as common plastic. Moreover, PVB is not compatible with common plastic such as PVC, LDPE, HDPE, PP, PET, etc. It cannot also be used to make polymer blend because of non-compatible.

However, PVB has unique coating properties like film forming, toughness, and adherence. It can be used in coating application for wooden and metal industry.

PVB can be used for making protective primer, sealers, peelable coating and paint. These are useful because of their film forming, adhesiveness, water resistance, flexibility and other desirable characteristics. They can be used on various substrates such as glass, metal, ceramics, wood etc.

A protective coating material for metal and wood substrate may also be prepared from PVB, solvent, filler, pigment, additives etc. The drying time, flexibility, adhesion and other key properties can be modified by varying molecular weight of PVB, solvent, filler, additives etc.

PVB based primer can be used in indoor and outdoor application of wooden items. They protect the wood from damage by moisture, water and microbial attack. These primers provide excellent protection when applied prior to painting and top coating of wooden surfaces. It ultimately reduces the paint consumption. Example of wood coating using PVB is interior and exterior surfaces of furniture, wooden handicrafts, wooden floorings etc.

PVB based strippable/or peelable paint can be applied to metal for protection against corrosion, abrasion and scratch. These peelable paint provide excellent protection when applied prior to painting and top coating of metal. The peelable coating can be applied to machinery, instruments and equipment. It can be easily removed without leaving any residue in metal. The peelable coating using PVB can be used to protect metal ferrous and non-ferrous metals. Example of metal substrate where peelable coating can be used are iron, steel, zinc, aluminum, zinc and other metal.

PVB waste are generated from automobile windshield manufacturers, dismantlers, recyclers, etc. in India. It has serious human health and environmental impact. Presently, there is a need of technology to recycle these problematic PVB solid wastes and develop environment friendly technology.

In view of above, PVB will boost the demand for paints & coatings, in turn increasing the consumption and recycling of PVB waste. Moreover, the protective primer and peelable paint made using PVB will be cheaper in comparison to conventional market wooden primer and metal peelable paint.

In the proposed work, the PVB shall be used in the formulation of a protective primer for wooden substrate, due to its high flexibility, film forming and good adherence to the wood. The primer prepared with the PVB along with additives, when applied on wood shall have good adherence. Further, it could substitute the conventional wood primers used nowadays.

Targeted Level of Development

A process has been optimized for recovery of PVB from automobile windshield waste. The PVB was characterized for meting point, solubility, film forming, adhesiveness and coating properties. Keeping in this, SRI has developed a process for making PVB based primer and adhesive for wooden surface. Further, peelable coating based on PVB have also been made for aluminum and steel surface. The primer and peelable coating based on have been characterized in detail for viscosity, dry film thickness, scratch resistance, flexibility, adhesion to wood and metal substrate etc.

- A viable industrial scale process for use of PVB recovered from automobile windshield waste in making protective primer and peelable coating for wooden and metal substrate.
- Reduce the burden of automatic windshield waste and PVB disposal on dumpsite and landfill sites in India.
- Replacement of conventionally used wooden primer and metal peelable coating like resin alkyd/ epoxy/PU/rubber, etc.
- Lower production cost of protective primer and strippable coating made using PVB

Industrial scale process to make protective primer and peelable paint using recovered PVB from automobile windshield manufacturers waste, dismantlers, recyclers, etc. The scaled up technology for making protective primer and peelable paint from PVB waste will result in a new process suitable for paint manufacturing and coating industries.

Apart from the paint and coating manufacturers, the other beneficiary would be wood and metal coating industries. The PVB plastic which is presently being considered as a waste can be utilized in the manufacture of protective primer and peelable paint. Once the process of making primer and strippable paint using PVB waste is developed and scaled up under this work, the PVB plastic waste will get better value as it can be utilized for making protective primer and peelable coating instead of disposing it into landfill sites.

Target Specifications and Performance Standards

The proposed technology once scaled up will enable the process for making protective primer and peelable coating from PVB emanating during recycling of automobile windshield waste. Further, the use of PVB in manufacture of protective primer and peelable paint will preserve the environment by keeping the PVB polymer waste out of waste dump and landfill sites.

- Industrial scale process to make protective primer and peelable paint using PVB of automobile windshield waste.
- Scaled up process to make protective primer and peelable paint from PVB waste emanating from automobile windshield waste.
- Comparative evaluation of physico-chemical properties of PVB based primer at plant level vis-à-vis conventional market wooden primer such as density, viscosity, drying time, adhesion to wood, peel strength, covering power, hardness, flexibility, thermal stability etc.
- Comparative evaluation of physico-chemical properties of PVB based peelable coating for metal at industry level vis-à-vis conventional market peel able paint for metal substrate such as density, viscosity, coverage are, dry film thickness, corrosion resistance, thermal stability, etc.

The proposed technology once scaled up will enable the process for making protective primer and peelable paint from PVB of polymer automobile windshield waste. Further, the use of PVB automobile windshield waste in manufacture of primer and strippable paint will preserve the environment by keeping the PVB out of dump and landfill sites.

Innovative Elements/Components of the Work

The automotive industry in India is the seventh largest in the world with an annual production of 24.00 million vehicles in year 2014-15. It is growing at a rate of 8.0%. The automobile industry accounts for 7.0% of the country's gross domestic product (GDP). The automobile market is also growing in middle class. Moreover, the automobile industry are also exploring in the rural markets.

In order to keep growing demand of automobile, several industries has set up for manufacture of windshield manufacturers in India. They are made using glass, PVB, etc. During the process, PVB waste are also generated in large quantity and are not recycled.

In India huge quantity of automobile windshield waste per annum is generated. It is expected to grow further as rapid automobile growth would continue to remain a major source of PVB plastic in India in the years to come. Thus, the PVB of automobile waste is going to be huge volume to deal with. Efforts need to be made to develop methodologies either for safe disposal or utilize it for developing high value added products using PVB polymeric portion of automobile windshield waste.

Automobile windshield waste is a mixture of layer of PVB, glass and other additives. Glass layer is fused with PVB plastic sheet. Presently, PVB layer of automobile windshield waste is not being recycled in India. Based on the physical and chemical properties of PVB, it is possible to make protective primer and strippable/or peelable coating for wooden and metal surface as it has film forming, flexibility, adhesion and bonding properties. The necessity of this research was felt by the need to find new ways for the utilization of PVB plastic as the source of additional raw materials in the manufacture of paint products for different coating application.

Shriram Institute for Industrial Research have developed a process to make protective primer and adhesive for wooden surface using PVB. The PVB based primer can be used for protection of wood surface against moisture, abrasion and scratch. PVB based primer has good bonding to wooden materials and it seals of pores of wood and prevent growth of microbes and moulds. PVB based peelable coating have also been made. Strippable paint made using PVB can be used to protect finished metal surface against scratches, corrosion, dust, etc. It can be used in temporary protection of machine parts, equipment, instruments etc. PVB based primer, adhesive and peelable coating have also been evaluated at our Institute. However, the process has not been scaled up to pilot scale or industry scale.

This work is being started with a specific purpose to develop a scaled up process to make protective primer and peelable coating using PVB emanating from automobile windshield waste by way of coating technique. As an outcome of the study, we would be able to make protective primer and peelable coating/or paint using PVB waste in our country. This would solve the problem of non-recyclable plastic PVB waste.

To the best of knowledge of Investigators of this work except SRI, no work on the development of protective primer and peelable coating using PVB has been done in our country. Further, their exist requirements for scaling up technology for making protective primer and peelable coating for wood and metal substrate in India. This is mainly due to the demand from the existing and upcoming wood and metal coating industry, paint manufacturing, chemical industry and other applications in the country.

Researchers (Kim, et al., 2020; Guettler et al., 2020) are working on mechanical properties and decomposition performance and new spaying technologies of similar coatings respectively. A lot of work has been carried out in this direction in the developed countries, but hardly any work on commercial scale has been done in our country. The PVB based protective primer for wooden industry and peelable coating for metal substrates can be scaled up in India with active support from DST.

CONCLUSION

Presently, the PVB based protective primer for wooden surface and peelable coating for metal substrate are not being manufactured in India. With the emergence of coating and paint industry in our country in future, the proposed work will be of high commercial significance.

Under this work, a technology developed at laboratory shall be scaled up to make protective primer for wooden surface and peelable coating for metal substrate using PVB recovered from windshield waste so that the plastic recyclers, rubber processing industry and paint industry can derive immense benefit out of this use. The scientific and engineering data generated during the project shall be used for recycling of unutilized PVB polymer of automobile windshield waste.

REFERENCES

Agarwal, N. (2015). Recent Growth Trends of Automobile Industry in India. *Indian Journal of Applied Research, 5*(7), 376–378.

Alsaed, O., & Jalham, I. S. (2012). Polyvinyl Butyral. (PVB) and Ethyl Vinyl Acetate (EVA) as a Binding Material for Laminated Glass. *Jordan Journal of Mechanical and Industrial Engineering, 6*(2), 127–133.

Anupama, Rao, Saxena, & Rayabag. (2015). Challenges in manufacturing and end-of-life recycling or disposal of solar PV panels. *Journal of Electrical and Electronics Engineering (Oradea), 10*(4), 81–87.

Bharti, Sharma, Singh, Singh, Arora, Gogoi, Rai, Meenu, Tyagi, Tyagi, & Kaicker. (2019). *Waste Utilization and product development.* Demonstration of products Technology in exhibition showcase section of 'Ecothon-2019: Solutions and sustainability - conference & exhibition' organized by Vprospurs Singapore & Green Dream Foundation India, IGNCA, New Delhi, India.

Bharti, P. K. (2019). *Product Development and Waste Recycling*. Discovery Publishing House.

Dasgupta, A. M., David, D. J., & Misra, A. (1992). Synthesis and Characterization of Ionomeric Poly(Vinyl Butyral). *Journal of Applied Polymer Science, 44*(7), 1213–1221. doi:10.1002/app.1992.070440711

EP1639041 A1. (2004). *Polyvinylbutyral compositions and blends having enhanced surface properties and articles made therefrom.* European Patent Office Application.

EP2308919 A1. (2008). *Polyvinyl butyral recycling method.* European Patent Office Application.

EP2308919A1. (2011). *Polyvinyl Butyral Recycling Method.* WO2009/118426 (01.10.2009 Gazette 2009/40).

Farzana, R., & Sahajwalla, V. J. (2015). Novel Recycling to Transform Automotive Waste Glass and Plastics into SiC-Bearing Resource by Silica Reduction. *Journal of Sustainable Metallurgy, 1*(1), 65–74. doi:10.100740831-014-0004-2

Farzana, R. R., Rajarao, R., & Sahajwalla, V. (2016). Characteristics of waste automotive glasses as silica resource in ferrosilicon synthesis. *Waste Management & Research, 34*(2), 113–121. doi:10.1177/0734242X15617010 PMID:26628051

Jody, B. J., Daniels, E. J., Duranceau, C. M., Pomykala, J. A., & Spangenberger, J. S. (2010). End-of-Life Vehicle Recycling: State of the Art of Resource Recovery from Shredder Residue. Energy Systems Division, Argonne National Laboratory.

Kernbaum, S. (2014). Industrial scale purification of PVB of windshield waste. *Eco-Innovation.* https://ec.europa.eu/environment/eco-innovation/projects

Koji, N., Mizutani, K., Zhang, R., Sugimoto, K., & Ogata, N. (2012). Poly(Vinyl Butyral)-Zirconia Hybrid Films Formed by Sol-Gel Process. *World Journal of Engineering*, *9*(3), 233–238. doi:10.1260/1708-5284.9.3.233

Krishnaveni, M., & Vidya, R. (2015). Growth of Indian automobile industry. *International Journal of Current Research and Academic Review*, *3*(2), 110–118.

Kyung-Min, K., Park, H.-W., Shim, G.-S., Jang, S.-W., Kim, H.-J., Chae, G.-S., Shin, S., & Lee, J.-H. (2020). Mechanical properties and decomposition performance of peelable coating containing UiO-66 catalyst and waterborne silane-terminated polyurethane dispersions. *Journal of Materials Science*, *55*(6), 2604–2617. doi:10.100710853-019-04184-2

Lai, S. M., & Lin, Y. C. (2006). Mechanical Properties of Nylon 6,6/Polyvinyl Butyral Blends. *Journal Polymer-Plastics Technology and Engineering*, *45*(3), 421–428. doi:10.1080/03602550600553978

Market. (2015). *Polyvinyl Butyral (PVB) Market Poised to Surge from USD 3.0 Billion in 2015 to USD 3.9 Billion, Globally By 2021*. Retrieved from www.marketresearchstore.com

Michael, T., Mokrejas, P., Merinska, D., & Peter, S. (2013). Windshield Recycling Focused on Effective Separation of PVB Sheet. *Journal of Applied Polymer Science*, *131*, 39879.

Nico, G., Knee, P., Ye, Q., & Tiedje, O. (2020). Initial droplet conditions in numerical spray painting by electrostatic rotary bell sprayers. A framework for optimization of injection model coefficients. *Journal of Coatings Technology and Research*, *17*(5), 1091–1104. doi:10.100711998-020-00352-1

Pistek, D., Merinska, D., & Tupy, M. (2011). The Influence of Plasticizers on Mechanical and Optical Properties of Recycled Polyvinyl Butyral. *Proceedings of 13thWSEAS International Conference on Mathematical Methods and Techniques in Engineering and Environmental Science*, 305-307.

Pramendra, K., Nida, K., & Kumar, D. (2016). Polyvinyl Butyral (PVB), Versatile Template for Designing Nanocomposite/Composite Materials: A Review. *Green Chemistry and Technology Letters*, *2*(4), 185–194. doi:10.18510/gctl.2016.244

Ray. (2012). Economic Performance of Indian Automobile Industry: An Econometric Appraisal. *Business Intelligence Journal*, 151-153.

Sarangi, Bano, & Pant. (2014). Future Trend in Indian Automobile Industry: A Statistical Approach. *Apeejay-Journal of Management Sciences and Technology*, *2*(1), 28–32.

Singh, Singh, Arora, Bharti, Rathi, Tyagi, & Kaicker. (2019). Utilization of PVB windshield waste for making protective primer and peelable paint. Poster presentation in DST – Plastic Waste Management, Swachhta Hi Seva (SHS), Delhi, India.

Sivakumar, G. D., Barnabas, S. G., & Anatharam, S. (2013). Indian Automobile Material Recycling. Management. *International Journal of Innovative Research in Science, Engineering and Technology*, *3*(3), 2347–2671.

Sônego, Costab, & Ambrósio. (2016). Dynamic and Static Crosslinking of Polyvinyl Butyral with a Silane Agent. *Plastic Research Online*, 1-3.

Sônego, M., Costab, L. C., & Ambrósio, J. D. (2016). Polyvinyl Butyral Chemically Modified With a Silane Agent in the Molten State. *Polymer Engineering and Science*, *56*(9), 971–979. doi:10.1002/pen.24326

Sônego, M., Cristina, L. C., & Ambrósio, J. D. (2015). Flexible Thermoplastic Composite of Polyvinyl Butyral (PVB) and Waste of Rigid Polyurethane Foam. *Polímeros*, *25*(2), 175–180. doi:10.1590/0104-1428.1944

Swain, B., Park, J. R., Shin, D. Y., Park, K.-S., Hong, M. H., & Lee, C. G. (2015). Recycling of Waste Automotive Laminated Glass and Valorization of Polyvinyl Butyral through Mechanochemical Separation. *Environmental Research*, *142*, 615–623. doi:10.1016/j.envres.2015.08.017 PMID:26318256

Teotia, Meenu, & Soni. (2014). Polymer Interlayers for Glass Lamination-A Review. *International Journal of Science and Research*, *3*(8), 1264-1270.

Tupý, M., Měřínská, D., & Kašpárková, V. (2015). *PVB Sheet Recycling and Degradation*. Retrieved from www.intechopen.com

Tupy, M., Merinska, D., Tesarikova, A., & Peternek, V. (2014). Mechanical Properties of Recycled Plasticized PVB /PVC Blend. World Academy of Science, Engineering and Technology. *International Journal of Chemical and Molecular Engineering*, *8*(9), 981-986. Retrieved from www.waset.org

Tupý, M., & Petránek, V. (2014). Use of Recycled PVB as a Protection against Carbonation. *International Journal of Chemical, Molecular, Nuclear, Materials and Metaullargical Egg.*, *8*(6), 514–541.

US20080268206A1. (2006). *A process for making heat resistant adhesive.* United States Patent.

US2358355. (1944). *Treatment of polyvinyl butyral resins.* United States Patent.

US2430053. (1942). *Polyvinyl butyral-resorcinol-formaldehyde adhesive.* United States Patent.

US2798015 A. (1954). *Coating composition for Metals.* United States Patent.

US3035005. (1956). *Method for preparing a corrosion inhibiting resinous coating composition.* United States Patent.

US3282865 A. (1963). *Composition of hexamethoxymethyl-melamine with viscous plasticizers.* United States Patent.

US3320203A. (1963). *A sprayable Polyvinyl butyral wood primer for polyurethane or polyester finish.* United States Patent.

US3325432 A. (1965). *Sprayable weldable primer containing polyvinyl butyral, chromium trioxide and aluminum powder.* United States Patent.

US3392130 A. (1963). *Coating compositions comprising polyvinyl butyral-hydrolyzed alkyl silicate reaction product and zinc dust.* United States Patent.

US4251591. (1979). *A process for making laminates.* United States Patent.

US5130370. (1991). *Molding compositions of polyvinylbutyral blend.* United States Patent.

US5972176A. (1997). *A process for making adhesive using corona treatment.* United States Patent.

US7521107. (2003). *Recycled polyvinyl butyral compositions and uses.* United States Patent.

Vargas, I. M., & Wiebeck, H. (2007). *Formulation of Primers to Wood Floorings Covering Using PVB Films from the Recycling of Windshields.* Paper Presented at Ninth Conferences Brasilo De Polymers. Retrieved from www.iepn.org

Vulić, M., & Tomović, A. (2015). Recycling of Auto Glass from Aspect of Technological Impacts on Some Parts of Recycling. *Proceedings of 9th International Quality Conference held at University of Karagujevac.*

WO1993002141 A1. (1992). *Recycling of Polyvinyl butyral.* World Intellectual Property Organization Patent.

Yener & Jirsak. (2011). Improving Performance of Polyvinyl Butyral Electrospinning. *Nanocon 2011*, 21–23.

Chapter 10
The Plastic Waste Menace and Approaches to Its Management Through Biodegradation

Joan Nyika
https://orcid.org/0000-0001-8300-6990
Technical University of Kenya, Kenya

ABSTRACT

In contemporary society, the production and use of plastics is growing due to their assorted applications. However, these tendencies result to production of plastic waste whose effects to the environment are destructive. This chapter explores the production tendencies of plastics and plastic-waste chronologically, discusses its negative environmental effects, and proposes the adoption of biodegradation as a management approach. A number of effects including death, retarded growth and development, emergence of cancers, and altered physiology are associated with ingestion of plastics in animals and humans. Water and air pollution is associated with unscientific disposal of plastic waste resulting to leaching and release of greenhouse gases. The use of microbes such as bacteria and fungi that have inherent enzymes such as lipases, esterases, hydrolases, and amylases is discussed as a sustainable solution to the plastic menace. The successful use of biodegradation must be coupled with scientific research to optimise the factors influencing the process.

INTRODUCTION

The contemporary world characterised by industrial revolution has seen a surge in production and use of plastics (Howard, 2018). According to Sheth et al. (2019), plastics are an engineering marvel evident from their many applications. The trend is prompted by the suitable characteristics of plastics including corrosion resistance, affordability, durability, strength and lightweight according to Babayemi et al. (2019). By 2017, Geyer et al. (2017) noted that 8.3 billion metric tonnes of virgin plastics had be generated and out of this amount, 6.3 billion metric tonnes became waste. Another study estimated the yearly production of plastics to be approximately 140 million metric tonnes worldwide with a per capita plastic

DOI: 10.4018/978-1-7998-4921-6.ch010

consumption of 43 kilogrammes (Babayemi et al., 2019). Of concern is the waste resulting from plastic production and use (Howard, 2018; Alpizar et al., 2020). Caruso (2015) pointed out that plastic waste is an emerging global threat with grave environmental and social implications and must be prioritised in future research to enhance its management and controlled release to environmental media. Of the resultant waste, only 12% was recycled while the rest was released to the environment or disposed at landfill facilities (Geyer et al., 2017). Drzyzga and Prieto (2018) shared similar sentiments claiming that by 2016, the world was producing more than 335 million metric tonnes of plastics and the figure was projected to double in the next two decades though most of its resultant waste was disposed. Babayemi et al. (2019) also noted that 8,300 million tonnes of virgin plastics were produced between 1950 and 2015 though a majority at 79% were landfilled. A country like India has seen consistent increase in consumption of plastics with annual consumption rates being at 0.4, 4, 8 and 16.5 million tonnes per year in 1990, 2001, 2009 and 2019, respectively and so is its associated waste production (Venkatesh & Kukreti, 2018). The European Union, North America and Asia are also high plastic producers at 57 metric tonnes, 100 and 20 kg/capita annually in respective order (Babayemi et al., 2019).

The production of plastic waste and its problematic management is a human and environmental issue worldwide that is directly proportional to economic development (Howard, 2018). The situation is more prevalent in developing countries of Africa and Asia whose capacity to recycle is limited (Babayemi et al., 2019). According to Webb et al. (2012), plastic waste resilience to degradation and its increased use leading to pollution in terrestrial and aquatic environs is a global ecological threat. Two means ensure that the plastic pollution stream is continuous: 1) through carelessly maintained waste streams and 2) intentional illegal and unscientific disposal of industrial and domestic refuse (Webb et al., 2012). With the increased production of plastics and the precognition of its negative effects, devising novel techniques of lowering plastic waste impact is essential. This can be done through the adoption of biodegradation technologies as Wierckx et al. (2018) highlighted. This chapter explores the plastic menace, available approaches of management and measures to alleviate its environmental effects through biodegradation. This technique uses natural or artificially introduced microbes (mainly bacteria and fungi) to break down plastics and has unrealised potential to improve existing efforts to manage these pollutants as many authors have noted (Ahmed et al., 2018; Fesseha & Abebe, 2019; Sheth et al., 2019).

GLOBAL STATISTICS ON PLASTICS PRODUCTION AND THE FATE OF PLASTIC WASTE

Plastic Production

Initial production of plastics occurred in the early twentieth century though its commercialisation occurred in the 1950s. Since then, production and use has grown exponentially (Lebreton & Andrady, 2019). By the end of 1950, only about 2 million tonnes of plastics were produced annually. However, the production rates have increased 200 fold to 381 million by 2015 according to Ritchie and Roser (2018). Figure 1 illustrates this growth over the years. The fluctuations in production in (1972-1974, 2008-2010) could be associated to financial crises that corresponded to reduced production and consumption.

In 2015 alone, more than 330 million metric tonnes of plastics were produced worldwide according to PlasticsEurope (2016). Out of this figure, polypropylene (PP), low-density polyethylene (LDPE), poly-

Figure 1. Production of plastics between 1950 and 2015 (Ritchie & Roser, 2018)

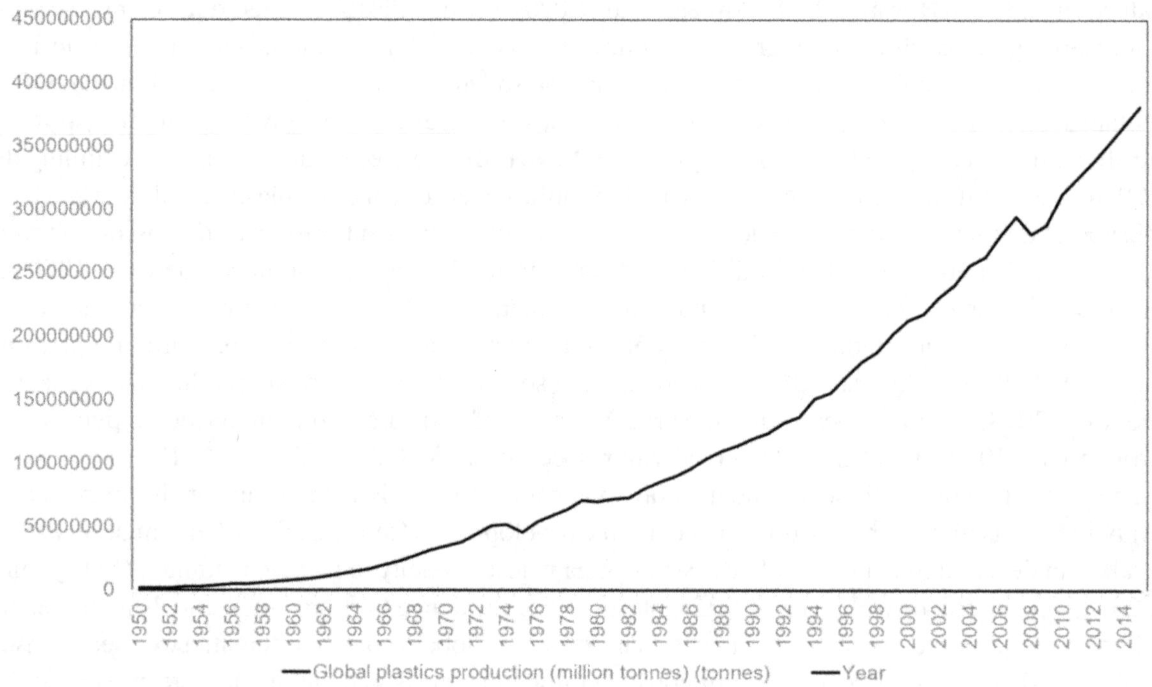

vinylchloride (PVC) and high-density polyethylene (HDPE) were a majority. The estimated percentage amounts of these and other plastic categories were as shown in Table 1.

Production and use of plastic worldwide differs and can be illustrated using Figure 2. The biggest producers and consumers of plastics are China, Western Europe and North America. These countries have a high population are involved in many industrial activities that correlate to high consumption and production of plastics.

Plastics are used in a variety of sectors including the transport, building and construction, electrical and electronics processing, industry and machinery processing and packaging among others (Morten et al., 2018). Packaging was the largest consumer of plastics while use in other sectors, the uses are distributed as shown in Table 2.

The Fate of Plastic Wastes

Of the total solid waste produced in North America, Western Europe, Japan, Asia, Africa, India, China and the Middle East annually, 5, 12, 9, 12, 5, 2, 7 and 10% of it is plastic, respectively (Morten et al., 2018). The industrial and packaging sectors are the largest producers of this waste as shown in Figure 3 that represents the 2015 statistics. Before the 1980s, 100% of plastic waste was discarded though by the 1990s, practices such as recycling and incineration were taken up at an average rate of 0.7% annually (Geyer et al., 2014). In 2015, massive progress in adopting recycling and incineration to the tune of 20 and 25%, respectively was evident though a large percentage of plastic waste at 55% was still being disposed off (Ritchie & Roser, 2018). Developed countries such as United States, Ireland, Netherlands, Germany and Kuwait have the highest daily per capita waste compared to developing countries such as

Table 1. Distribution of the different types of plastic produced in 2015 (PlasticsEurope, 2016; Boucher & Friot, 2017; Grand View Research, 2015)

Plastic Category	Percentage Amount
Bioplastics	0.5
Tyre-based elastomers	2
Polyamide (PA) fibres	1
Polyethylene terephthalate (PET) fibres	5
Polypropylene (PP) fibres	8
Coatings used to mark roads	0.2
Marine coatings	0.1
Coatings	1
Sealants	0.5
Adhesives	2
Thermosets	9
Elastomers excluding tyres	2
PA	1
Polycarbonate (PC)	1
Styrene-acrylonitrile, acrylonitrile styrene acrylate, acrylonitrile butadiene styrene	2
Other thermoplastics	3
Polyurethane (PUR)	4
Expanded polystyrene (EPS), polystyrene (PS)	5
PET	5
PP	16
LDPE and linear LDPE	12
HDPE	10
PVC	11
Total	100

Bangladesh, Mozambique and Tanzania (Jambeck et al., 2015). Populous countries such as China, US, Germany and Brazil produce high quantities of plastics (Jambeck et al., 2015).

The management of such plastic waste to avoid unscientific landfilling and littering varies based on regions. Developed countries have effective infrastructure to manage waste in closed landfills while a large percentage of such waste in developing and transition economies is disposed inadequately (Jambeck et al., 2015). A regional distribution of mismanaged plastic waste around the world is shown in Figure 4. This waste is likely to enter in environmental media (soil, surface and groundwater) and end up modifying food chains as a pollution hazard.

Table 2. Distribution of various uses of plastics in various sectors (Boucher & Friot, 2017; Geyer et al., 2017)

Application	Percentage Share
Packaging	30
Building and construction	17
Transportation	14
Institutional and consumer product	10
Electrical and electronics	4
Microplastics production	15
Industry and machinery uses	0.5
Textile	9
Road marking, marine coating and personal care products	1

Environmental Impacts of Plastic Waste

The myriad sources of plastics including wastewater, landfill leachates, particulates from worn out tires, agricultural and industrial wastes and their multifaceted nature are a threat to the environment. Such wastes spread to terrestrial, marine and freshwater ecosystems to access water bodies, soils, plants and animals through food chain modification (Alpizar et al., 2020). The ubiquitous nature of plastics is exacerbated

Figure 2. Variations in the use and production of plastics around the world (Morten et al., 2018)

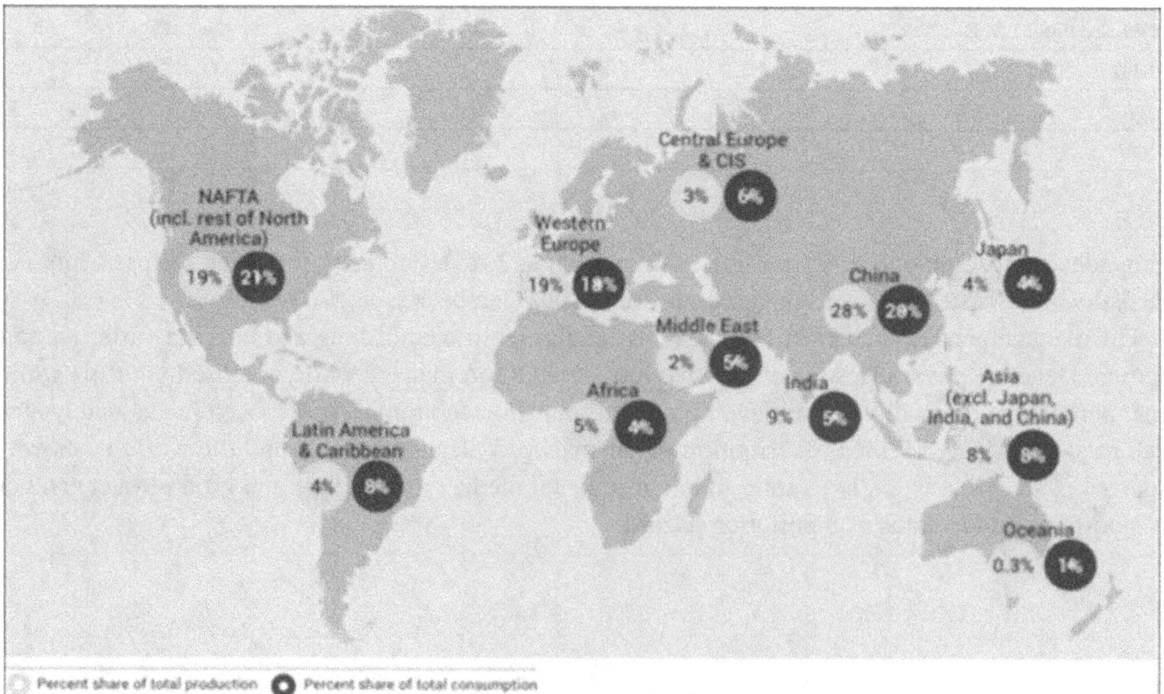

Figure 3. Production of plastic wastes at various sectors based on the 2015 waste production statistics (Ritchie & Roser, 2018)

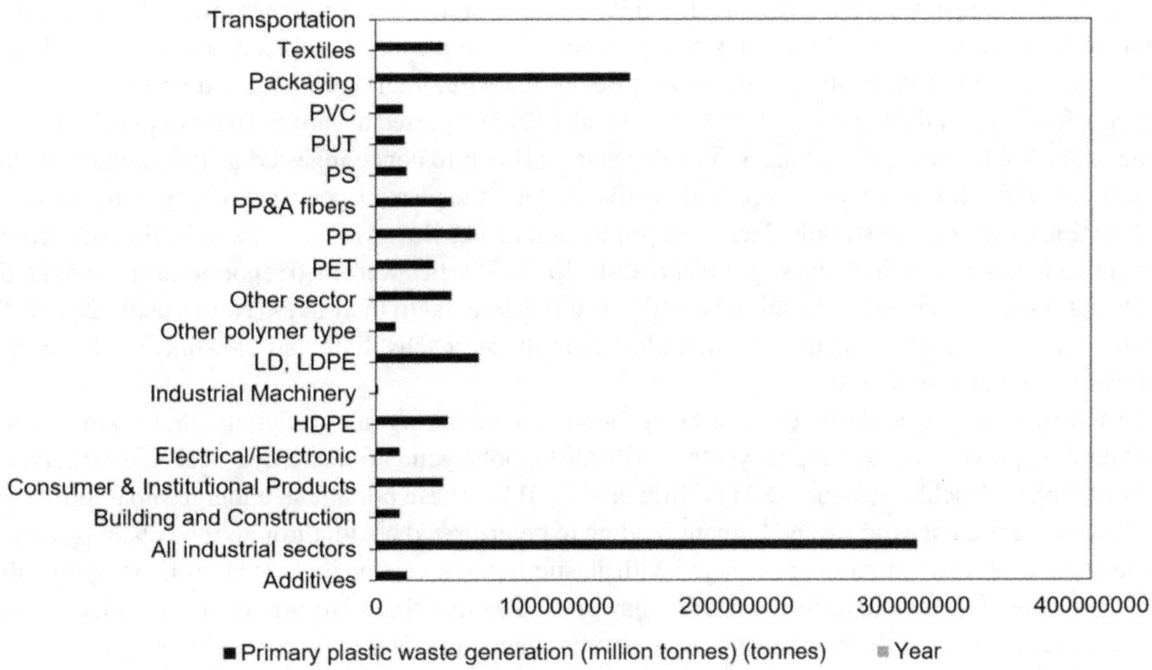

Figure 4. Distribution of mismanaged plastic waste based on regions (Jambeck et al., 2015)

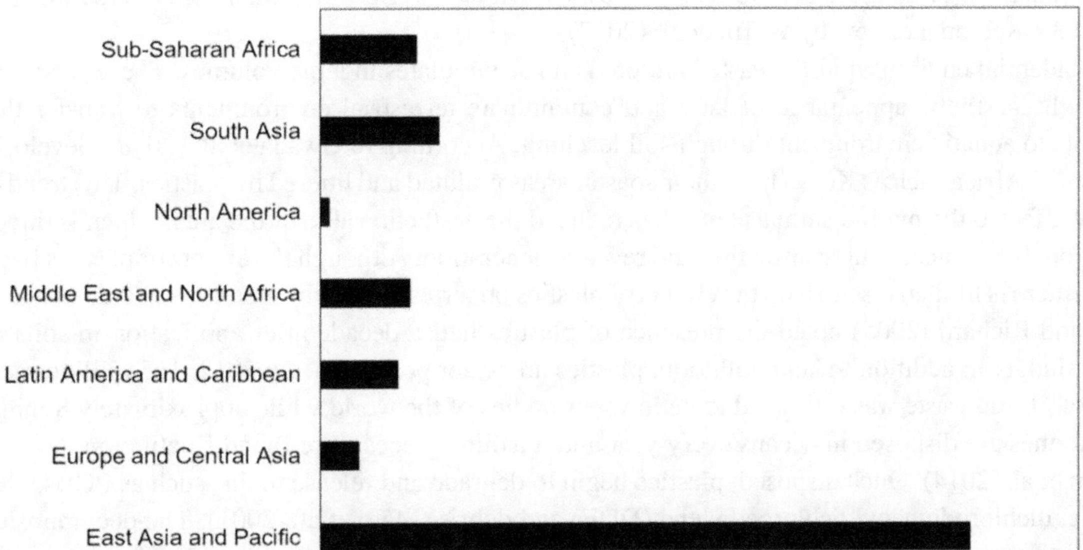

by historical mismanaged along with over-consumption and over-production (Koushal et al., 2014). Plastics have damaging effects on the ecology as reported in a number of studies. For wildlife, plastics are ingested and result to entanglement (Derraik, 2002). Entanglement hinders proper food digestion, compromises animal ability to feel hungry, reproduce, escape predators and locomotion. Entanglement restricts movement of juvenile animals, can injure them, restrict their breathing and prevent them from sourcing food and fodder (Gregory, 2009). Webb et al. (2013) reported negative effects of plastics on filter feeders, sharks, fur seals, cetaceans, sea turtles and marine birds once ingested as the animals mistake the wastes for food. Once in the animals' digestive system, the plastics persist resulting to reproduction dysfunctions, low feeding stimuli, decreased production of steroid hormones, low secretion of digestive enzymes and gastrointestinal blockage (Webb et al., 2013). Plastics were also reported to cause oxidative and hepatic stress among marine animals, which was detrimental to their lives (Nelms et al., 2019). The authors reported the accumulation of microplastics in the stomachs, liver and intestines of the animals and was attributable to their death.

In the oceans, plastic particles have been found to contain organic pollutants and toxins such as bisphenol A, polybrominated diphenyl ethers (PBDEs), polycyclic aromatic hydrocarbons (PAHs) and dichlorodiphenyltrichloroethane (DDT) (Hirai et al., 2011). These pollutants undergo bioamplification and can be ingested by wildlife and humans leading to deteriorated health through food chain processes. Some of the health abnormalities associated with plastic-based ingestion include DNA hypomethylation, diabetes, breast carcinoma, arthritis, neurobehavioral abnormalities, disruptions of the endocrine system, hormonal and growth abnormalities among other developmental impairments (Chung et al., 2011; Trudel et al., 2011). Plastic particles have been associated with the migration of destructive and invasive animals to new environs (Gregory, 2009). Molluscs, dinoflagellates, polychaeted, algae, bryozoans and barnacles adhere to plastic debris leading to their subsequent dispersal and colonisation as Barnes and Milner (2005) noted. Terrestrial animals are reported to transport the plastic debris to new areas as was reported in the case of ants that rode plastic debris to San Sebastial Island from the Brazilian mainland (Webb et al., 2013). Examples of the effects of plastics on a wide variety of animal species is summarised in Table 3 based on a review by Worm et al. (2017).

In residential environs, plastic waste is littered and accumulates in large volumes. These wastes can damage the aesthetic appearance of land and contaminate terrestrial environments to transfer these pollutants to aquatic environment through soil leaching. According to Gwada et al. (2018), developing countries of Africa such as Kenya have their coastal areas polluted and littered by plastics. This trend has not only affected the marine animals but it has reduced the aesthetic value of the area, which is directly proportional to reduced tourist attraction and revenue generation. Although the effects of plastics begins on land, there is limited research on the effects of plastics on terrestrial environment compared to oceans. Zubris and Richard (2005) noted the presence of plastics half a decade after application in soils and sewage sludge. In addition to land pollution, plastics are water pollutants. By 2012, 165 million metric tonnes of plastic waste was estimated to be in water bodies of the world while approximately 8 million metric tonnes are disposed in oceans every year and 5 trillion pieces were found floating on the ocean (Eriksen et al., 2014). Once disposed, plastics begin to degrade and release toxins such as PCBs, phenanthrene, dichlorodiphenyldichloroethylene (DDE) and debris (Mato et al., 2001). The decomposition and incineration of plastic wastes leads to the release of methane and carbon dioxide. These gases trap radiant heat and prevent its escape from the earth leading to global warming (Chandegara et al., 2015). Such air polluting tendencies result to deaths of millions of people and animals worldwide as Hamlet et al. (2018) noted suggesting that developing countries are more vulnerable compared to developed nations.

Table 3. Effects of plastics on various animals (Worm et al., 2017)

Species	Type of Plastic	Effects
Mammals Australian sea lion Sperm whale Fur seal	Plastic fishing gear Plastic debris/bags Plastic particles	Caused entanglement and eventual death. Starvation and stomach rupture. Bioaccumulation of the plastics from prey fish.
Seabirds Magellanic penguin Greater shearwater	Straws, lines and fragments Plastic bottle caps	Perforation of the stomach. Gastrointestinal obstruction leading to starvation.
Fish Bigeye tuna Larval perch Orchid dottyback Japanese medaka	Lines and fragments Microplastics debris Plastic bags Particulate plastic	Ingestion. Altered reproductive behaviour, retarded growth and inhibited hatching. Mortality resulting from leached plastic toxins. Liver dysfunction as a result of plastic accumulation.
Invertebrates Oyster Norway lobster Mussel Urchin larva	Microplastic particles Particles and strands Microplastic particles Polyethylene (PE) particles	Low energy uptake and reproduction dysfunction. Accumulation of plastics in the gut. Accumulation of plastics in blood. Abnormal development because of ingestion.
Sea turtles Leather turtle Green sea turtle	Plastic bags/debris Nylon strings, balloons and plastic bags/debris	Reduced egg laying due to injury and blockage of cloaca. Starvation due to gastrointestinal distress, exposure to predators.

Plastic polymers have different additives that are a health risk by being endocrine disrupters and carcinogens. A number of additives found in plastics, the plastic types they are found in and the health effects of their ingestion are summarised in Table 4.

Table 4. Types of plastic additives, the plastic types they are found and the effects of ingesting them (Halden, 2010)

Additive	Plastic Type where used	Health Effects of Ingestion
Nonylphenol	PVC	Hormonal disruptions.
Styrene monomer	Polystyrene	DNA mutations and cancer causing.
PCBs	All plastics	Thyroid hormone interference.
PAHs	All plastics	Low fertility and retarded development.
Dioxins	All plastics	Interferences of male sexual hormones.
Persistent organic pollutants (POPs)	All plastics	Reproductive and neurological damage.
Bisphenol A	PC and PVC	Ovarian disorders.
Phthalates	Polystyrene and PVC	Low male fertility.

Classification of Plastics

Plastics can be classified as 1) biodegradable or 2) non-biodegradable. The latter describes bio-based and fossil-based polymers that are petrochemical and hydrocarbon derivatives. They are highly stable, resist biospheric degradation and have high molecular weights due to repetition of monomeric units in their molecular structure (Ghosh et al., 2013). Most routinely used plastics such as PE, PUR, PET, PS, PP and PVC are non-biodegradable. Due to their stability, slow degradation and longevity in addition to mismanagement characterised by littering, these plastics are an environmental concern. A case example is the use of polyolefins such as PE used in making mugs, shopping and carry bags that have accumulated in environmental media (Shah et al., 2008). Additionally, these plastics resist microbial attack. Biodegradable plastics on the other hand describe polymers that can be broken down using enzymes or otherwise. These include polyhydroxylalkanoates (PHA), polylactic acid, polyethylene succinate (PES), polycaprolactone (PCL), starch/polyester and starch/PVA blends. This study discusses these biodegradable plastics in details in the following subsections.

Biodegradable Plastics

These plastics are derivatives of renewable resources and can be metabolised completely through biological means (Kale et al., 2007). They are made of cellulose and starch as well as their polymeric derivatives. These macromolecules are used due to their ubiquitous and inexpensive nature in addition to being degradable (Chattopadhyay et al., 2011). PHA and PLA are the commonest bio-based degradable plastics. PHA is a polyester that results from fermentation of lipids and sugars. It is used in pharmaceutical, medical and packaging sectors as a fast food service material, in disposal of medical tools and in some paints. In low energy and carbon levels, microorganisms such as *Cupriavidus, Nocardiopsis, Burkholderia* and *Bacillus* bacteria species as well as *Micromycetes* and *Mycobacterium* fungi species can degrade PHA to derive these energy sources (Boyandin et al., 2013). PLA on the other hand, is manufactured by NatureWorks of the USA and is a derivative of sugarcane, tapioca roots and/ or cornstarch. The biodegradable nature of these plastics by microbes such as *B. licheniformis* and *Amycolatopsis species* is an advantage to its usability (Anderson & Shive, 2012).

Fossil-based biodegradable polymers have a variety of uses in packaging of food items, make-up kits, various chemicals and pharmaceutical products. PES and polycaprolactone are examples of these plastics. PES is a thermoplastic polyester manufactured via copolymerisation of succinic anhydride and ethylene oxide or via poly-condensation of succinic acid and ethylene glycol (Hoang et al., 2007). The polymer manufactures paper coating agents, shopping bags and films for agriculture. A number of microorganisms such as *Pseudomonas species* AKS2, *Bacilllus* and *Paenibacillus species* are known to biodegrade PES (Tokiwa et al., 2009; Tribedi & Sil, 2014). Polycaprolactone is a crystalline polyester mixed with copolymers and used in manufacture of packaging materials. Aerobic and anaerobic microbes in the presence of esterases and lipases can degrade PCL (Karakus, 2016). A number of microbes including *Fusarium species, A. niger, A. flavus* and *R. delemar* degrade PCL in the presence of enzymes to result to a range of products including valeric, caproic, succinic and butyric acids (Shimao, 2001).

Biodegradable polymer blending is a cheaper technique to make eco-friendly plastics compared to copolymerisation (Tokiwa et al., 2009). Starch/polyester and starch/PVA blends are examples of these plastics. The former are renewable, easily accessible and affordable blends whose strength can be enhanced by adding anhydride polyesters. Lipase enzymes found in bacterial strains such as *R. delamar* and *R. arrhizus* degrade these polymers (Tokiwa & Calabia, 2004). Blending PVA with starch has been shown to enhance its solubility properties hence the extensive use of this mixture in packaging agricultural products (Tang et al., 2008). Through enzymatic hydrolysis, depolymerases secreted by *Alcaligenes faecalis* T1 strain were reported to degrade the polymer (Tokiwa et al., 2009)

The Management of Plastic Waste via Biodegradation

Conventional approaches of handling plastic wastes such as landfill disposal and recycling have their limitations in terms of sustainability. According to Siddiqui and Pandey (2013), the methods do not lead to full recovery of products, exacerbate the release of air pollutants as is the case of incineration and unscientific landfilling, lead to leaching of organic contaminants and plastic particles in natural resources and are not a long-term alternative to managing plastic wastes. There is a need for eco-friendly approaches of degrading plastics despite their recalcitrant nature, which has led to the wide use of biodegradation approaches (Siddiqui & Pandey, 2013; Webb et al., 2013; Caruso, 2015).

In bioremediation, microorganisms use a synthetic polymer as an exclusive energy and carbon source and thereby degrade them. Additionally, microbes such as fungi and bacteria have extracellular enzymes that catalyse the metabolism of various fossil-based and bio-plastics (Temoor et al., 2017). Catalytic activity and enzyme nature varies based on the species and strain of microorganism. As such, different microorganisms can degrade a variety of polymers. Case examples are the *Brevibacillus* and *Bacillus species* that breakdown various polymers (Sivian, 2011). Plastic degrading fungi that target lignin have laccases group of enzymes that degrade aromatic and non-aromatic compounds though redox reactions (Temoor et al., 2017).

A number of microorganisms and their enzymes that are capable of degrading various plastics are summarised in Table 5. These microbes attach themselves on polymers resulting to surface colonisation. This is followed by hydrolysis of plastics in two steps: 1) enzyme binding on the plastic substrate and 2) hydrolytic breakdown, which occurs to convert mono-, di- and oligo-mers to carbon dioxide and water through mineralisation of plastic polymers (Tokiwa et al., 2009). In aerobic processes of degradation, oxygen serves as the electron acceptor while in anaerobic processes, manganese, iron, nitrate and sulphate are the electron acceptors (Temoor et al., 2017).

A number of factors influence biodegradation of plastics: 1) exposure and 2) polymer characteristics. Exposure factors can be abiotic (pH, temperature and moisture) and biotic (specificity of enzymes and their hydrophobicity). Abiotic or weather factors influence the deterioration of exposed materials. When atmospheric temperature, non-ionising radiation and moisture conditions are combined with atmospheric and pollution gases as well as wind effects, biodegradation process is deteriorated. Polymer characteristics include aspects of plastic shapes and sizes, molecular weight, the presence/absence of bio-surfactants and additives. Moisture rich environs and microbes enhance plastic degradation processes such as hydrolytic breakdown in addition to supporting microbial growth and multiplication. pH influences hydrolysis and controls catalytic biodegradation while temperature increase enhances enzymatic metabolism of plastics. For instance, the hydrolysis of PCL by *R. delamar* was optimal at high temperature while the hydrolysis of PLA was optimal at a pH of 5 (Temoor et al., 2017). The specificity of enzymes at their active sites influences their biodegradation capacity. For example, enzymes of *A. flavus* and *A. niger* fungi species degrade diacid monomers of 6-12 carbon atoms rapidly compared to other monomers (Kale et al., 2007). Hydrophobicity is an impediment to plastic biodegradation because it interrupts microbial formation of biofilms and hinders hydrolysis (Sivian, 2011). Molecular weight is indirectly proportional to biodegradability of plastic polymers as highlighted by Tokiwa et al. (2009) using the microbe *R. delamar* used in PCL degradation using lipase enzyme. Although Stevens (2003) stated that biodegradation of plastic polymers of large and small surface area are similar, there are predetermined sizes and shapes used to optimise this process (ASTM, 2004). Bio-surfactants that have high affinity to

Table 5. Microorganisms with catalytic ability to degrade plastics

Microorganism	Enzyme	Plastic Degraded	Reference
Bacteria			
Pseudomonas stuzeri	Amylases	PHA	Muhamad et al., 2015
Bacillus brevis, Amycolatopsis species	Proteases, lipases	PLA	Kasirajan & Ngouahio, 2012
Ideonella sakaiensis	Amylases	PET	Yoshida et al., 2016
Streptomyces species		Polyester	Shah et al., 2008
Pseudomonas O-3	Serine hydrolase	Polyvinyl alcohol	Shimao, 2001
Rhodococcus rubber, Brevibacillus borstelensis	Hydratase, amidase	Polyethylene	Sivian, 2011
Pseudomonas and Flavobacterium species		Nylon	Tokiwa et al., 2009
Pseudomonas species		PBS	Tribedi & Sil, 2014
Clostridium botulinum, C. acetobutylicum		Polylcaprolactone	Abou-Zeid et al., 2001
Alcaligenes feacalis		PVA blends and starch	
Rhizopus delemar	Lipase	PCL	Tokiwa et al., 2009
C. botulinum	Lipase	PCL	
P. fluorescens B-22, P. putida, Ochrobactrum TD	Lipase	PHA	Danko et al., 2004
B. borstelensis		PET	Calabia & Tokiwa, 2006
P. stutzeri	Serine hydrolase	PHA	Muhamad et al., 2015
R. arrizus	Lipase	PCL, PBS, PHB	Tokiwa et al., 2009
Fungi			
Penicillium Roquefort		PLA	Kasirajan & Ngouahio, 2012
Phanerochaete chyrsosporium		Polyester	Shah et al., 2008
Fusarium solani		Polylcarprolactone	Abou-Zeid et al., 2001
P. chyrsosporium, Penicillium funiculosum, Aspergillus niger		Polyethylene and starch	Shah et al., 2008
A. flavus	Glycosidase	PCL	Tokiwa et al., 2009
P. funiculosum		PHB	
A. niger	Protease, catalase	PCL	
Fusarium	Cutinase	PCL	Muhamad et al., 2015
Amycolaptosis species	Manganese peroxidase	PE, PLA	
Fusarium solani, Curvularia senegalensis, Pestalotiopsis microspore	Serine hydrolase	PUR	Russell et al., 2011

water enhance biodegradability while filler and dyes among other additives affect the thermal stability of plastic polymers interfering with their degradation (Temoor et al., 2017).

Biodegradation is described using several steps as summarized by Francois-Heude et al. (2014). The preliminary step in biodegradation is bio-deterioration that involves the action of microorganisms. The

Figure 5. The mechanism of plastic biodegradation

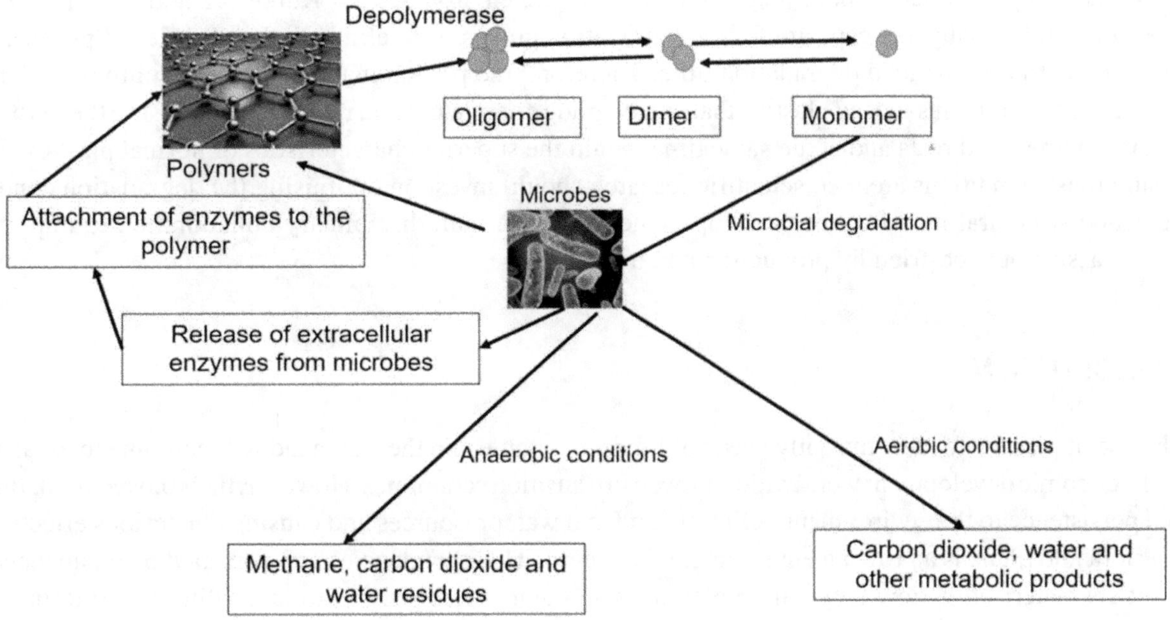

procedure entails superficial biodegradation that involves the modification of chemical, physical and mechanical features of a given plastic material (Lucas et al., 2008). The second step entails the cleavage of polymers of plastics to monomers, dimers and oligomers in the process of depolymerisation with the help of catalysts produced by microbes. The last two procedures involve assimilation and mineralization in respective order. Assimilation is the combining of molecules essential in formation of biomass, storage vesicles, production of energy, primary and secondary metabolites while mineralization is the excretion of complex metabolites and salts to extracellular surroundings. This mechanism of microbial degradation of plastics is summarised in Figure 5.

Current and Future Prospects in Plastic Biodegradation

To deal with the plastic menace particularly its littering and resistance to degradation, fossil and bio-based plastics are proving a sustainable solution as Iwata (2015) noted. Yoshida et al. (2016) shared similar sentiments in a study focusing of *I. sakaiensis* 201-F6 strain that degrades PET by using its carbon and energy in an enzymatic reaction based on hydrolysis. *P. putida* NBUS12 strain (Tan et al., 2015), *Achromobacter xylosoxidans* (Kowalczyk et al., 2016) and *Anoxybacillus rupiensis* Ir3 strain (JQ912241) (Mahdi et al., 2016) have been used in biodegradation of styrene, HDPE and aromatic polymers, respectively. As such, biodegradation of plastics is an innovative and eco-friendly approach to deal with its environmental effects sustainably. The use of these biodegradable plastics should be the focus of future research to replace them with non-biodegradable plastics in the packaging of foods, agricultural, pharmaceutical and surgical products. In events where separation of disposed waste is difficult and diffusion of plastic-based pollutants is imminent, the use of biodegradable plastics is essential. These are efforts to sustainable environmental management and could be enhanced if coupled with recycling, recovery and reuse practices. However, it is worth noting that even degradable plastics take a long time to completely biodegrade and continue to pose negative effects on the environment as Kubowicz and Booth (2017) reported. Additionally, the capacity for biodegradation processes to eliminate large scales of plastics is limited due to the extended degradation time. Therefore, the predominant challenge in eliminating the plastic menace remains to find plastics that can degrade completely and rapidly while posing the least or no environmental threats and at the same time, retain the superior characteristics of normal plastics. To ensure these conditions are met, scientific research should invest on optimising the degradation conditions and structural modifications of biopolymers through multidisciplinary collaboration to improve their massive but eco-friendly production and use.

CONCLUSION

The use of plastics to meet the daily needs of life is inevitable and their demand will continue to escalate with economic development worldwide and even in transition economies. However, their bioaccumulation and persistence to the environment polluting land and water resources and causing deleterious effects in the flora and fauna is a growing environmental concern. These tendencies are worsened by mismanagement characterised by low recycling preferences and tendencies skewed to landfilling and littering of such wastes leading to the leaching of their toxins and chemicals in environmental media. The use of biodegradation is however, a promising technique to reverse these predominant pollution trends from plastic wastes. These prospects will be enhanced through research and development to optimise on the mechanisms, microbes used, enzymatic conditions of reactivity and through better production capacity of biodegradable polymers.

REFERENCES

Abou-Zeid, D., Müller, R., & Deckwer, W. (2001). Degradation of natural and synthetic polyesters under anaerobic conditions. *Journal of Biotechnology*, *86*(2), 113–126. doi:10.1016/S0168-1656(00)00406-5 PMID:11245900

Ahmed, T., Shahid, M., Azeem, F., Rasul, I., Shah, A., Noman, M., Hameed, A., Manzoor, N., Manzoor, I., & Muhammad, S. (2018). Biodegradation of plastics: Current scenario and future prospects. *Environmental Science and Pollution Research International*, *25*(1), 1–12. doi:10.100711356-018-1234-9 PMID:29332271

Alpizar, F., Carlsson, F., Lanza, G., Carney, B., Daniels, R., Jaime, M., Ho, T., Nie, Z., Salazar, C., Tibesigwa, B., & Wahdera, S. (2020). A framework for selecting and designing policies to reduce marine plastic pollution in developing countries. *Environmental Science & Policy*, *109*, 25–35. doi:10.1016/j.envsci.2020.04.007

Anderson, J., & Shive, M. (2012). Biodegradation and biocompatibility of PLA and PLGA microspheres. *Advanced Drug Delivery Reviews*, *64*, 72–82. doi:10.1016/j.addr.2012.09.004

ASTM. (2004). 6400-04 Standard specification for compostable plastics. *Journal of ASTM International*.

Babayemi, J., Nnorom, I., Osibanjo, O., & Weber, R. (2019). Ensuring sustainability in plastics use in Africa: Consumption, waste generations and projections. *Environmental Sciences Europe*, *31*(1), 60. doi:10.118612302-019-0254-5

Barnes, D., & Milner, P. (2005). Drifting plastic and its consequences for sessile organism dispersal in the Atlantic Ocean. *Marine Biology*, *146*(4), 815–825. doi:10.100700227-004-1474-8

Boucher, J., & Friot, D. (2017). *Primary Microplastics in the Oceans: a Global Evaluation of Sources*. International Union for Conservation of Nature and Natural Resources. doi:10.2305/IUCN.CH.2017.01.en

Boyandin, A., Prudnikova, S., Karpov, V., Ivonin, V., Đo, N., Nguyen, T., Lê, T. M. H., Filichev, N. L., Levin, A. L., Filipenko, M. L., Volova, T. G., & Gitelson, I. I. (2013). Microbial degradation of polyhydroxyalkanoates in tropical soils. *International Biodeterioration & Biodegradation*, *83*, 77–84. doi:10.1016/j.ibiod.2013.04.014

Calabia, B., & Tokiwa, Y. (2006). A novel PHB depolymerase from a thermophilic *Streptomyces sp. Biotechnology Letters*, *28*(6), 383–388. doi:10.100710529-005-6063-5 PMID:16614903

Caruso, G. (2015). Plastic degrading microorganisms as a tool for bioremediations of plastic contamination in aquatic environments. *Journal of Pollution Effects and Control*, *3*(3), e112. doi:10.4172/2375-4397.1000e112

Chandegara, V., Cholera, S., Nandasana, J., Kumpavat, M., & Pate, K. (2015). Plastic Packaging Waste Impact on Climate Change and its Mitigation. In R. Subbaiah & G. Prajapati (Eds.), *Water management and climate smart agriculture. Adaptation of Climatic Resilient Water Management and Agriculture*. Gyan Publishing House.

Chattopadhyay, S., Singh, S., Pramanik, N., Niyogi, U., Khandal, R., Uppaluri, R., & Ghoshal, A. (2011). Biodegradability studies on natural fibers reinforced polypropylene composites. *Journal of Applied Polymer Science*, *121*(4), 2226–2232. doi:10.1002/app.33828

Chung, S. Y., Yettella, R. R., Kim, J. S., Kwon, K., Kim, M. C., & Min, D. B. (2011). Effects of grilling and roasting on the levels of polycyclic aromatic hydrocarbons in beef and pork. *Food Chemistry*, *129*(4), 1420–1426. doi:10.1016/j.foodchem.2011.05.092

Danko, A., Luo, M., Bagwell, C., Brigmon, R., & Freedman, D. (2004). Involvement of linear plasmids in aerobic biodegradation of vinyl chloride. *Applied and Environmental Microbiology*, *70*(10), 6092–6097. doi:10.1128/AEM.70.10.6092-6097.2004 PMID:15466555

Derraik, J. (2002). The pollution of the marine environment by plastic debris: A review. *Marine Pollution Bulletin*, *44*(9), 842–852. doi:10.1016/S0025-326X(02)00220-5 PMID:12405208

Drzyzga, O., & Prieto, A. (2018). Plastic waste management, a matter for the community. *Microbial Biotechnology*, *0*(0), 1–3. PMID:30411497

Eriksen, M., Lebreton, L., Carson, H., Thiel, M., Moore, C., Borerro, J., Galgani, F., Ryan, P. G., & Reisser, J. (2014). Plastic pollution in the world's oceans: More than 5 trillion plastic pieces weighing over 250,000 tons afloat at sea. *PLoS One*, *10*(12), e111913. doi:10.1371/journal.pone.0111913 PMID:25494041

Fesseha, H. & Abebe, F. (2019). Degradation of plastic materials using microorganisms: A review. *Public Health Open Journal, 4*(2), 57-63.

Francois-Heude, A., Richard, E., Desnoux, E., & Colin, X. (2014). Influence of temperature, UV-light wavelength and intensity of polypropylene photo-thermal oxidation. *Polymer Degradation and Stabilization, 100*, 10–20. doi:10.1016/j.polymdegradstab.2013.12.038

Geyer, R., Jambeck, J., & Law, K. (2017). Production, use and fate of all plastics ever made. *Science Advances, 3*(7), 1–5. doi:10.1126ciadv.1700782 PMID:28776036

Ghosh, S., Pal, S., & Ray, S. (2013). Study of microbes having potentiality for biodegradation of plastics. *Environmental Science and Pollution Research International, 20*(7), 4339–4355. doi:10.100711356-013-1706-x PMID:23613206

Grand View Research. (2015). *Synthetic Fibers Market Analysis, By Type (Acrylics, Polyester, Nylon, Polyolefin), By Application (Clothing, Home Furnishing, Automotive, Filtration), By Region, And Segment Forecasts, 2014 – 2025*. https://www.grandviewresearch.com/industry-analysis/synthetic-fibers-market

Gregory, M. R. (2009). Environmental implications of plastic debris in marine settings—Entanglement, ingestion, smothering, hangers-on, hitch-hiking and alien invasions. *Philosophical Transactions of the Royal Society of London. Series B, Biological Sciences, 364*(1526), 2013–2025. doi:10.1098/rstb.2008.0265 PMID:19528053

Gwada, B., Ogendi, G., & Trott, S. (2018). Composition of plastic waste discarded by household and its management approaches. *Global Journal of Environmental Science and Management, 5*(1), 83–94.

Halden, R. (2010). Plastics and health risks. *Annual Review of Public Health, 31*(1), 179–194. doi:10.1146/annurev.publhealth.012809.103714 PMID:20070188

Hamlet, C., Matte, T. & Mehta, S. (2018). *Combating plastic air pollution on earth's day. Vital strategies environmental health division*. Academic Press.

Hirai, H., Takada, H., Ogata, Y., Yamashita, R., Mizukawa, K., Saha, M., Kwan, C., Moore, C., Gray, H., Laursen, D., Zettler, E. R., Farrington, J. W., Reddy, C. M., Peacock, E. E., & Ward, M. W. (2011). Organic micropollutants in marine plastic debris from the open ocean and remote and urban beaches. *Marine Pollution Bulletin, 62*(8), 1683–1692. doi:10.1016/j.marpolbul.2011.06.004 PMID:21719036

Hoang, K., Tseng, M., & Shu, W. (2007). Degradation of polyethylene succinate (PES) by a new thermophilic Microbispora strain. *Biodegradation, 18*(3), 333–342. doi:10.100710532-006-9067-5 PMID:17109189

Howard, M. (2018). Plastic pollution of the world's seas and oceans as a contemporary challenge in ocean governance. *Nature Communications, 9*(1), 667. doi:10.103841467-018-03104-3 PMID:29445166

Iwata, T. (2015). Biodegradable and bio-based polymers: Future prospects of eco-friendly plastics. *Angewandte Chemie International Edition, 54*(11), 3210–3215. doi:10.1002/anie.201410770 PMID:25583677

Jambeck, J., Geyer, R., Wilcox, C., Siegler, T., Perryman, M., Andrady, A., Narayan, R., & Law, K. L. (2015). Plastic waste inputs from land into the ocean. *Science, 347*(6223), 768–771. doi:10.1126cience.1260352 PMID:25678662

Kale, G., Kijchavengkul, T., Auras, R., Rubino, M., Selke, S., & Singh, S. (2007). Compostability of bioplastic packaging materials: An overview. *Macromolecular Bioscience*, *7*(3), 255–277. doi:10.1002/mabi.200600168 PMID:17370278

Karakus, K. (2016). Polycaprolactone (PCL) based polymer composites filled wheat straw flour. *Kastamonu University Journal of Forestry Faculty*, *16*(1), 264–268. doi:10.17475/kujff.03251

Kasirajan, S., & Ngouajio, M. (2012). Polyethylene and biodegradable mulches for agricultural applications: A review. *Agronomy for Sustainable Development*, *32*(2), 501–529. doi:10.100713593-011-0068-3

Koushal, V., Sharma, R., Sharma, M., Sharma, R., & Sharma, V. (2014). Plastics: Issues Challenges and Remediation. *International Journal of Waste Resources*, *4*, 134.

Kowalczyk, A., Chyc, M., Ryszka, P., & Latowski, D. (2016). *Achromobacter xylosoxidans* as a new microorganism strain colonizing high-density polyethylene as a key step to its biodegradation. *Environmental Science and Pollution Research International*, *23*(11), 11349–11356. doi:10.100711356-016-6563-y PMID:27072033

Kubowicz, S., & Booth, A. (2017). Biodegradability of plastics: Challenges and misconceptions. *Environmental Science & Technology*, *51*(21), 12058–12060. doi:10.1021/acs.est.7b04051 PMID:29022342

Lebreton, L., & Andrady, A. (2019). Future scenarios of global plastic waste generation and disposal. *Palgrave Communications*, *5*(6), 1–11. doi:10.105741599-018-0212-7

Lucas, N., Bienaime, C., Belloy, C., Queneudec, M., Silvestre, F., & Nava-Saucedo, J.-E. (2008). Polymer biodegradation: Mechanism and estimation techniques. *Chemosphere*, *73*(4), 429–442. doi:10.1016/j.chemosphere.2008.06.064 PMID:18723204

Mahdi, M., Ameen, R., & Ibrahim, H. (2016). Study on degradation of nylon 6 by thermophilic bacteria *Anoxybacillus rupiensis* Ir3 (JQ912241). *International Journal of Advanced Research in Biological Sciences*, *3*(9), 200–209. doi:10.22192/ijarbs.2016.03.09.027

Mato, Y., Isobe, T., Takada, H., Kanehiro, H., Ohtake, C., & Kaminuma, T. (2001). Plastic resin pellets as a transport medium for toxic chemical in the marine environment. *Environmental Science & Technology*, *35*(2), 318–324. doi:10.1021/es0010498 PMID:11347604

Morten, W., Ryberg, A., & Michael, H. (2018). *Mapping of global plastics value chain and plastics losses to the environment*. United Nations Environmental Program.

Muhamad, W., Othman, R., Shaharuddin, R., & Irani, M. (2015). Microorganism as plastic biodegradation agent towards sustainable environment. *Advances in Environmental Biology*, *9*, 8–14.

Nelms, S., Brownlow, B., Davidson, N., Deaville, R., Galloway, T., & … . (2019). Microplastics in marine mammals stranded around the British coast: Ubiquitous but transitory? *Scientific Reports*, *9*(1), 1075. doi:10.103841598-018-37428-3 PMID:30705316

RitchieH.RoserM. (2018). *Plastic pollution*. https://ourworldindata.org/plastic-pollution

Russell, J., Huang, J., Anand, P., Kucera, K., Sandoval, A., Dantzler, K., Hickman, D. S., Jee, J., Kimovec, F. M., Koppstein, D., Marks, D. H., Mittermiller, P. A., Núñez, S. J., Santiago, M., Townes, M. A., Vishnevetsky, M., Williams, N. E., Vargas, M. P. N., Boulanger, L.-A., ... Strobel, S. A. (2011). Biodegradation of polyester polyurethane by endophytic fungi. *Applied and Environmental Microbiology*, *77*(17), 6076–6084. doi:10.1128/AEM.00521-11 PMID:21764951

Shah, A., Hasan, F., Hameed, A., & Ahmed, S. (2008). Biological degradation of plastics: A comprehensive review. *Biotechnology Advances*, *26*(3), 246–265. doi:10.1016/j.biotechadv.2007.12.005 PMID:18337047

Sheth, M., Kwartler, S., Schmaltz, E., Hoskinson, S., Martz, E., Dunphy-Daly, M., Schultz, T. F., Read, A. J., Eward, W. C., & Somarelli, J. A. (2019). Bioengineering a future free of marine plastic waste. *Frontiers in Marine Science*, *6*, 1–10. doi:10.3389/fmars.2019.00624

Shimao, M. (2001). Biodegradation of plastics. *Current Opinion in Biotechnology*, *12*(3), 242–247. doi:10.1016/S0958-1669(00)00206-8 PMID:11404101

Siddiqui, J., & Pandey, G. (2013). A review of plastic waste management strategies. *International Research Journal of Environmental Sciences*, *2*(12), 84–88.

Sivian, A. (2011). New perspectives in plastic biodegradation. *Current Opinion in Biotechnology*, *22*(3), 422–426. doi:10.1016/j.copbio.2011.01.013 PMID:21356588

Stevens, E. (2003). What makes green plastics green? *BioCycle*, *44*, 24–27.

Tan, G., Chen, C., Ge, L., Li, L., Tan, S., & Wang, J. (2015). Bioconversion of styrene to poly (hydroxyalkanoate) (PHA) by the new bacterial strain *Pseudomonas putida* NBUS12. *Microbes and Environments*, *30*(1), 76–85. doi:10.1264/jsme2.ME14138 PMID:25740622

Tang, S., Zou, P., Xiong, H., & Tang, H. (2008). Effect of nano-SiO2 on the performance of starch/polyvinyl alcohol blend films. *Carbohydrate Polymers*, *72*(3), 521–526. doi:10.1016/j.carbpol.2007.09.019

Temoor, A., Shahid, M., Azeem, F., Rasul, I., Shah, A., Noman, M., & ... . (2017). Biodegradation of plastics: Current scenario and future prospects for environmental safety. *Environmental Science and Pollution Research International*, *25*(8), 7287–7298. PMID:29332271

Tokiwa, Y., & Calabia, B. (2004). Review degradation of microbial polyesters. *Biotechnology Letters*, *26*(15), 1181–1189. doi:10.1023/B:BILE.0000036599.15302.e5 PMID:15289671

Tokiwa, Y., Calabia, B., Ugwu, C., & Aiba, S. (2009). Biodegradability of plastics. *International Journal of Molecular Sciences*, *10*(9), 3722–3742. doi:10.3390/ijms10093722 PMID:19865515

Tribedi, P., & Sil, A. (2014). Cell surface hydrophobicity: A key component in the degradation of polyethylene succinate by *Pseudomonas sp.* AKS2. *Journal of Applied Microbiology*, *116*(2), 295–303. doi:10.1111/jam.12375 PMID:24165295

Trudel, D., Scheringer, M., von Goetz, N., & Hungerbühler, K. (2011). Total consumer exposure to polybrominated diphenyl ethers in North America and Europe. *Environmental Science & Technology*, *45*(6), 2391–2397. doi:10.1021/es1035046 PMID:21348481

Venkatesh, S., & Kukreti, I. (2018). *An Indian consumes 11 kg plastic every year and an average American 109 kg. Down to Earth.* https://www.downtoeart h.org.in/news/waste /an-indian-consumes-11-kg-plastic-every-year-and-an-average-american-109-kg-60745

Webb, H., Arnott, J., Crawford, R., & Ivanova, E. (2012). Plastic degradation and its environmental implications with special reference to poly(ethylene terephthalate). *Polymers*, *5*(1), 1–18. doi:10.3390/polym5010001

Wierckx, N., Narancic, T., Eberlein, C., Wei, R., Drzyzga, O., Magnin, A., & … . (2018). *Plastic biodegradation: challenges and opportunities*. Springer International Publishing AG.

Worm, B., Lotze, H. K., Jubinville, I., Wilcox, C., & Jambeck, J. (2017). Plastic as a Persistent Marine Pollutant. *Annual Review of Environment and Resources*, *42*(1), 1–26. doi:10.1146/annurev-environ-102016-060700

Yoshida, S., Hiraga, K., Takehana, T., Taniguchi, I., Yamaji, H., Maeda, Y., Toyohara, K., Miyamoto, K., Kimura, Y., & Oda, K. (2016). A bacterium that degrades and assimilates poly (ethylene terephthalate). *Science*, *351*(6278), 1196–1199. doi:10.1126cience.aad6359 PMID:26965627

Zubris, K., & Richards, B. (2005). Synthetic fibres as an indicator of land application of sludge. *Environmental Pollution*, *138*(2), 201–211. doi:10.1016/j.envpol.2005.04.013 PMID:15967553

KEY TERMS AND DEFINITIONS

Bioamplification: Biological magnification of toxin levels in tissues of organisms that have developed resistance to such pollutants.

Biospheric Degradation: The breakdown of contents found in the earth and atmosphere where living things live through natural nutrient cycling processes.

Carcinogen: A substance that causes mutation of cells in living things leading to development of cancerous tissues.

Copolymerisation: The manufacture of polymers using a variety of monomers whose physicochemical characteristics are different.

Endocrine Disrupters: Chemicals that alter the production and release of hormones in living things.

Hypomethylation: A modification of the genetic material, DNA through the loss of its methyl group on the cytosine and guanosine bases especially in mammalian organisms.

Nature Works: An international company that specialises in the manufacture of biodegradable plastic polymers from plants.

Polyelefins: A common industrial polymer used to produce packaging films, pipes, bottles and automobile parts.

Chapter 11
Review on Plastic Waste Disposal and Role of Microorganisms in Bioremediation of Plastics

Vyshnavi V. Rao
MES College of Arts, Commerce, and Science, India

Sonashree R.
MES College of Arts, Commerce, and Science, India

Rashmi R. Halbavi
MES College of Arts, Commerce, and Science, India

ABSTRACT

Plastics are the most commonly used polymers for conventional applications. Plastic wastes accumulating the environment are presenting an ever-increasing ecological threat. Low density polyethylene is major cause of long-term environmental pollution. An eco-friendly approach to resolve this ever-growing persistent menace is bioremediation. The diverse metabolic capability of microbes can be effectively exploited for breakdown of plastic wastes. Another approach is to synthesise biodegradable or organic plastics which is soluble in the natural environment. There are several advantages associated with these biodegradable plastics or plastic degrading microbes. They can be composted with organic wastes and returned to enrich the soil. Their use will not only reduce stress and loss of habitat of wild animals caused by dumping of conventional plastics but will also lessen the labor expenses for the removal of plastic wastes in the environment because they degrade naturally. This review describes bioremediation/ biodegradation process, production, types of and advantages of bioplastics.

DOI: 10.4018/978-1-7998-4921-6.ch011

BACKGROUND

One of the major environmental threats is the least rate of degradation or non-biodegradability of synthetic organic polymers like plastics under natural condition. The various forms of plastics such as nylon, polycarbonate, polyethylene-terephthalate, polyethylene, polypropylene, polystyrene, polytetra-flouro ethylene, polyurethane and polyvinyl chloride are being continuously used in our day-to-day life (Smith, 1964). The word plastic is derived from the Greek word "plastikos", which literally translates to 'able to be moulded in to varied shapes'. Plastics are polymers that consist of monomers linked together by chemical bonds (Vignesh *et al.,* 2016). Plastics are defined as the polymers (solid materials) which on heating become mobile and can be casted into specific shapes. They are also called as non-metallic mouldable compounds and the materials that are made from them can be moulded into any desired shape and size (Seymour, 1989). These have greater advantage as they are strong, lightweight, and durable. However, they do have grave disadvantages as they are resistant to biodegradation, leading to pollution, and pose to be harmful to the natural environment (Amal *et al.,* 2015).

Low density polyethylene (LDPE) is viewed to be one of the major causes of environmental pollution. Polyethylene is a polymer made up of several repeating units of monomers called ethylene. The use of this synthetic polymer is growing at a rate of 12% per year and about 140 million tons of synthetic polymers are produced worldwide each year (Sharma et al., 2015). It causes serious threats to the environment during its production and after its disposal. Microorganisms that can degrade plastics are over 90 genera from bacteria, fungi, among them *Bacillus megaterium, Pseudomonas spp, Azotobacterial spp, Ralstonia eutropha, Halomonas spp* etc (Chee *et al.,* 2008).

Utilization of plastic materials or articles has found wide application in virtually all aspects of human life in both domestic and commercial environments. There is hardly anything that one will do without encountering plastics or its products (Raziyafathima *et al.,* 2016). The high consumer demand drove global production to approximately 140 million tons of synthetic polymers, which has increased by 1.74-fold over the past 15 years to about 243 million tons (Sharma *et al.,* 2015). The rise in production and use of synthetic polymers has on the other hand increase in the amount of global plastic wastes with numerous adverse effects on the environment. This is a concern put forward by a large sector of public including environmental advocates, cultivators, welfare organisations and researchers (Boris Worm *et al.,* 2017).

TYPES OF PLASTICS

Plastics can be broadly classified into two types. These are the two types of plastics widely in use: thermoplastics and thermosetting polymers. Thermoplastics are plastics that do not undergo chemical change in their composition upon heating and can be moulded to any shape repeatedly. While the thermosetting polymers melt upon heating and can be moulded into various shapes. After they are solidified, they retain the same shape and remain solid forever. In the thermosetting process, a chemical reaction occurs which is irreversible. Thermosets are assumed to have infinite molecular weight. These are made of long polymeric chains with repeating molecular units, known as or derivatives of monomers; each polymer chain will have several thousand repeating units. Vulcanization of rubber is one such thermosetting process. The poly-isoprene is a tacky, slightly runny material, before heating with sulphur, but after vulcanization the product is rigid, non-tacky and retains its shape.

There are several other classifications of plastics based on the properties and chemical nature of the monomer units or the parameters that are relevant for manufacturing or production design. Plastics can also be classified depending on various physical properties, such as density, high tensile strength, and resistance to various chemical products.

PLASTIC DEGRADING MICROORGANISMS

Numerous microorganisms exist and have been confirmed by many researchers globally which have the ability to use polyethylene (one kind of plastic) as a carbon and energy source. This physiochemical ability has been studied only recently. In general, the attachment of microorganisms to the surface of the plastics, followed by the colonization of the exposed surface, is the first stage of growth leading to degradation (Tokiwa *et al.,* 2009). Degradation of plastics by microorganisms is primarily accomplished by hydrolytic activities of enzymes which specifically cleave the bonds in the polymer releasing monomers and oligomers (Lucas *et al.,* 2008). Bacteria and fungi are involved in the degradation of plastics (Gu *et al.,* 2000). For example, Streptomyces spp. and other wood degrading fungi (e.g., *Piptoporus botulinus, Irpex lacteus* etc.) can attach to the surface of the polyethylene and produce extracellular enzymes. These extracellular hydrolytic enzymes degrade polyethylene into respective ethene monomers (Pometto *et al.,* 1992) (Kim *et al.,* 2005). Other studies showed that bacteria, such as *Pseudomonas, Streptococcus, Staphylococcus, Micrococcus, Moraxella, Brevibaccillus, and Rhodococcus* (Gilan *et al.*, 2004; Hadad *et al.,* 2005), and fungi, including *Aspergillus niger, A. glaucus, Saccharomonospora spp.* and *Actinomycetes spp.* were associated with plastic and plastic like polymer degradation (Swift, 1997; Raaman *et al.,* 2012).

Studies on biodegradation of plastics such as High Density Polyethene (HDPE), Low Density Polyethene (LDPE), Polypropylene (PP) and Polystyrene (PS) in soils revealed that some fungi and bacteria species are capable of degrading plastics into simple compounds carbon and energy molecules for cellular metabolism. For instance, Yamada-Onodera (2001) reported that the fungus, *Penicillium simplicissimum*, degraded HDPE up to molecular weight of 2800 g/mol (S Yamada-Onodera *et al.,* 2001). Thermophilic bacteria *Brevibaccillus borstelensis and Rhodococcus ruber* were also reported to assimilate LDPE as a carbon source (Gilan *et al.*, 2004; Hadad *et al.,* 2005). Soil inhabiting fungi (e.g., *A. niger, A. japonicas, A. terreus, A. flavus and Mucor sp.*) were also found to have the potential to decompose synthetic plastics although the results were inconclusive (Raaman N *et al.,* 2012). Also, *Bacillus stearothermophilus and B. pumilus* were reported to degrade Poly Butylene Succinic Acid (PBSA) and Poly lactic acid in soils. This potential is used to compost and activate sludge (Tomita *et al.,* 2000).

PLASTIC DISPOSAL METHODS

There are four methods of plastic disposal currently being practices all over the world (Webb *et al.,* 2013):

1. **Incineration:** Technically incineration refers to the burning in order to dispose the plastic waste. Incineration has an advantage that it does not require space to be carried out and there is even the capability for energy recovery in the form of heat (Sinha *et al.,* 2010). However, burning plastic on a large scale releases Polycyclic Aromatic Hydrocarbons, Poly Chlorinated Biphenyls, heavy

metals, toxic carbon- and oxygen-based free radicals, significant quantities of greenhouse gases, especially carbon dioxide which lead to major environmental crisis.

2. **Landfill:** Landfill refers to burying the plastic waste in excavated pits. This has been a traditional practice which over several years has shown an impact on the soil quality. Limited oxygen supply and the persistence of anaerobic conditions upon landfill of plastics leads to the release of benzene, toluene, xylenes, ethyl benzenes and trimethyl benzenes, released both as gases and dissolved state (Urase *et al.,* 2008).
3. **Recycling:** Recycling of plastics can be classified into mechanical processing and chemical processing. Mechanical processing also called mechanolytic recycling involves subjecting the plastic waste to pressure and heat to melt and remould them desired shape. Chemical processing is performed by carrying out chemolysis with a number of compounds, resulting in depolymerisation of the plastic (Awaja *et al.,* 2005). Depolymerisation can be carried out by hydrolysis (using water), methanolysis (methanol), glycolysis (EG) or aminolysis (e.g., methylamine,) (Awaja *et al.,* 2005; Sinha *et al.,* 2010). Each results in different monomer units that can be recovered, but all of the monomers can be used as polymerisation materials to produce new plastic.
4. **Biodegradation:** It is the process of degrading the plastic waste by employing suitable microbes. The metabolic diversity of microorganisms makes them a useful biodegradation (Iranzo *et al.,* 2001). Biodegradation is an inviting alternative to all the above practices for waste disposal. A few of the several advantages include that it is an economical process, potentially much more efficient and does not produce secondary pollutants. On the contrary, it may even be possible to obtain useful end products with economic benefit from bacterial metabolism of pollutants, for example, ethanol for use in biofuels (Iranzo *et al.,* 2001).

DEGRADATION OF PLASTICS

Any physical or chemical change in polymer as a result of several environmental factors, such as light, temperature, moisture, chemical conditions or biological activity can be termed degradation. Processes that induce changes in polymer properties due to physical, chemical or biological reactions resulting in subsequent chemical transformations (formation of structural homogeneities) are categorized as polymer degradation. Degradation with respect to plastics is referred to as the changes in mechanical, optical or electrical characteristics or properties of any plastic material resulting in cracking, erosion, discoloration, phase separation and delamination.

Plastic degradation can be classified into the following based on the causative agents

1. **Biodegradation**: This is a natural process by which microorganism (bio-decomposers) breakdown plastics into small units in order to use the carbon and energy sources for subsequent growth and energy utilization (Sharma *et al.,* 2015; Sen *et al.,* 2015). This process can occur through hydrolytic biodegradation and oxidative biodegradation (Bonhomme, *et al.,* 2003).
2. **Photo Degradation**: This is brought about by the action of sunlight. Photo degradation can result in an alteration in the physical and chemical structure of the plastic, usually as a result of sunlight in outdoor exposure (Yousif *et al.,* 2013).
3. **Thermo-Oxidative Degradation**: This is caused by slow oxidative breakdown at moderate temperatures.

4. **Thermo-Degradation**: This is caused by the degrading action of plastics at high temperatures.
5. **Hydrolysis**: This involves breakdown of polymers or oligomers of plastic on reaction with water molecules. This releases monomers to the environment (Hakkarainen *et al.*, 2004; Andrady, 2011).

Biodegradation of Plastics

Plastics are used as potential substrates by heterotrophic microorganisms (Glass and Swift G., 1989). Microorganisms such as bacteria and fungi are prominently involved in the degradation of both natural and synthetic plastics (Gu *et al.,* 2000).

The biodegradation of plastics proceeds actively and differ indefinitely depending on parameters like different soil conditions, properties and physical characteristics of the plastic disposed and the microbial composition of the soil (because the microorganisms that are responsible for the process of degradation differ from each other and they have their own optimal growth conditions in the soil). Biodegradation is specifically governed by different factors that include characteristics of polymer, type of organism, and nature of pre-treatment.

The characteristics of polymer such as mobility, crystallinity, molecular weight, functional groups and substituents present in its structure, and plasticizers or additives when added to the polymer all play a significant role in the time and extent of degradation (Gu *et al.,* 2000) **(Fig.1).**

Several organizations are involved in setting standards for biodegradable and compostable plastics: the U.S. Composting Council (USCC), the American Certification System of Biodegradable Products Institute (BPI), the Environment & Plastics Industry Council, the American Society for Testing and Materials (ASTM), the European Committee for Standardization (CEN), Japan's Greenplum program, and the British Plastics Federation. The restrictions and regulations from these organizations help create biodegradable and compostable products that meet the increasing worldwide demand for more environmentally friendly plastics. Biodegradable polymers decompose into carbon dioxide, methane, water, inorganic compounds or biomass by the actions of microorganisms (Greene, 2007). Biodegradable polymers used for producing biodegradable agricultural films include inter alia, copolyesters, poly(vinyl) alcohol, poly(vinyl)chloride, acylated starch-plastic, modified starch and vegetable oil (Kyrikou, 2007).

Biodegradation of Synthetic Plastics

The degradation of most synthetic plastics in nature is a very slow process that involves environmental factors, which follows the action of wild microorganisms (Albertsson 1980; Cruz-Pinto *et al.,* 1994; Albertsson *et al.,* 1994). The oxidation or hydrolytic cleavage by enzyme generates functional groups that improve the hydrophilicity of polymer. This is the primary mechanism for the biodegradation of high molecular weight polymer. Subsequently, the main chain of polymer is degraded resulting in polymer of low molecular weight with feeble mechanical properties. This makes them more accessible for further microbial assimilation (Albertsson *et al.,*1990; Albertsson *et al.,* 1987; Huang *et al.,* 1990). Poly (vinyl alcohol), poly (lactic acid), polycaprolactone, and polyamides are some examples of synthetic polymers with oligomeric structures that can succumb to biodegradation. The rate of degradation is affected by several physical properties such as crystallinity and orientation of the monomers and morphological properties such as surface area of the polymers (Huang *et al.,*1992) **(Fig.2)**

Figure 1.

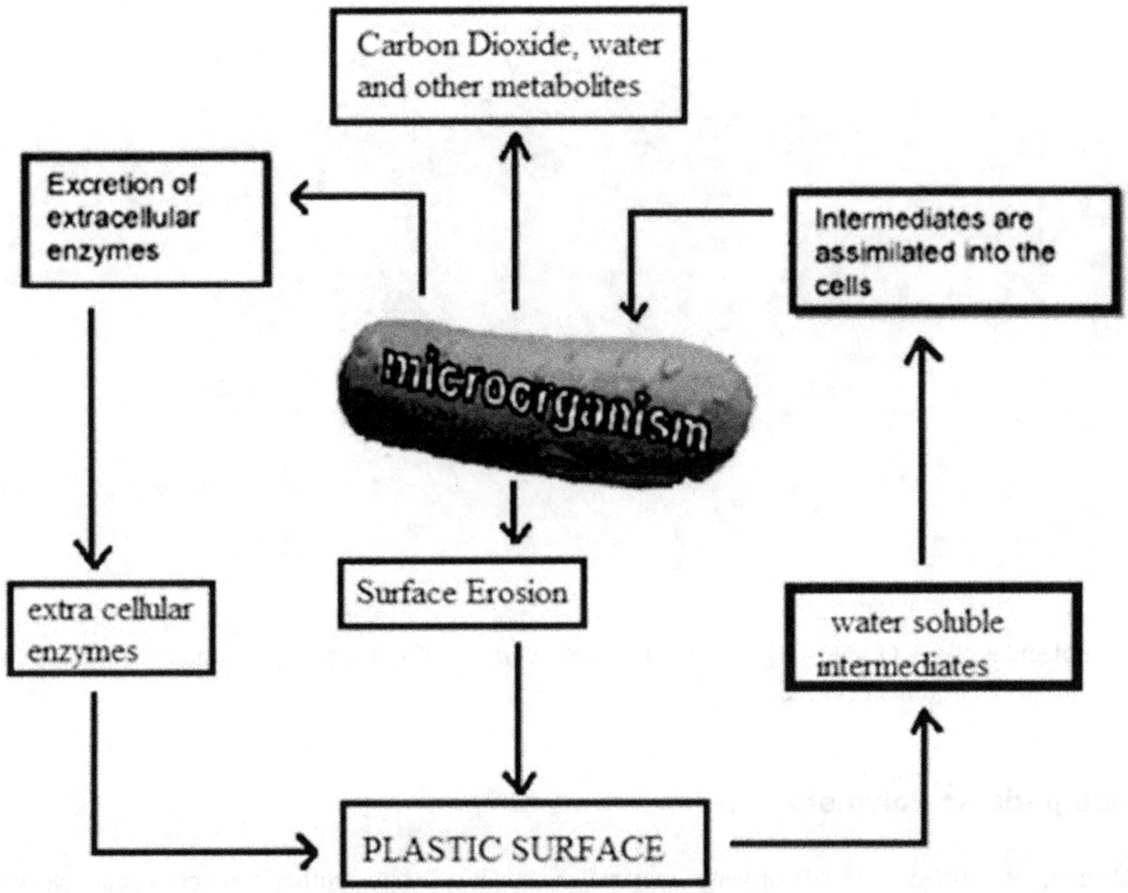

STEPS OF PLASTIC BIODEGRADATION BY MICROORGANISMS

There are four steps involved in the biodegradation of plastic:

1. **Bio-Deterioration:** This refers to the superficial degradation of plastics by physical and chemical deterioration that is a result of the activities of microbial communities and other bio-decomposers.
2. **Bio-Fragmentation**: This refers to the catalytic activity of extracellular hydrolytic enzymes secreted by microorganisms that breaks down polymeric plastics into smaller oligomers, dimers or monomers. Some of the enzymes secreted include Alkane hydroxylase, Lipases, Esterase, Protease and Lactase.
3. **Assimilation:** This refers to the integration, transport or assimilation of molecules or monomers from the surroundings into the cytoplasm of microbes during metabolism.
4. **Mineralization:** This is the last phase where microorganisms utilize the assimilated monomers in the energy pathways to break them down completely to simple oxidised metabolites or micro-

Figure 2.

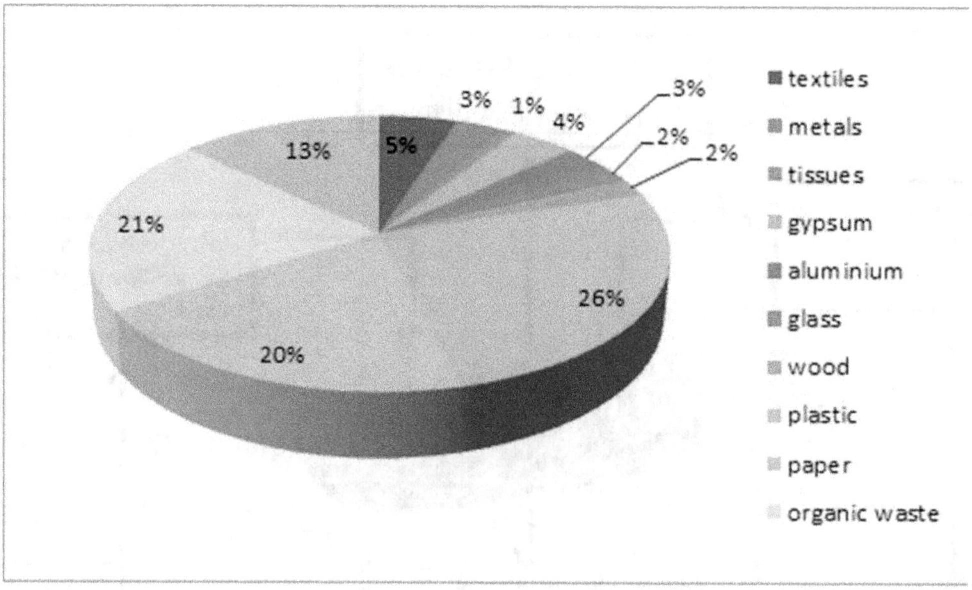

molecules like CO_2, N_2, CH_4 and H_2O (Lucas *et al.,* 2008). These are ultimately released to the natural environment **(Fig. 3).**

Biodegradable Polymers

In Poland, 3 to 10% of the mass of waste materials is constituted by synthetic materials and every year it rises by 3%. This results in the increase in the volume of waste dumps. The most efficient solution to the problem is the replacement of conventional synthetic fibres with biodegradable polymers.

Biodegradable materials may be divided into three groups: the mixtures of synthetic polymers and substances which are easy digestible by microorganisms (inter alia, modified natural polymers, natural polymers such as starch and cellulose), synthetic materials with groups that are susceptible to hydrolytic microbial attack (for example, polycaprolactone) and the bio polyesters from bacterial sources. The most popular and important biodegradable polymers are aliphatic polyesters [e.g., polylactide, poly(ε-caprolactone), polyethylene oxide, poly(3-hydroxybutyrate), polyglycolic acid] and thermoplastic proteins (Liao *et al.,* 2009).

The major advantages of biodegradable plastics are that they can be composted with organic wastes and returned to enrich the soil, their use will not only reduce ecological stress and loss of habitat to wild animals caused by dumping of conventional plastics, but will also lessen the labour cost for the removal of plastic wastes in the environment because they are degraded naturally, their decomposition will help increase the longevity and stability of landfills by reducing the volume of garbage, they could be recycled to useful monomers and oligomers by microbial and enzymatic treatment (Tokiwa *et al.,* 2007).

Figure 3.

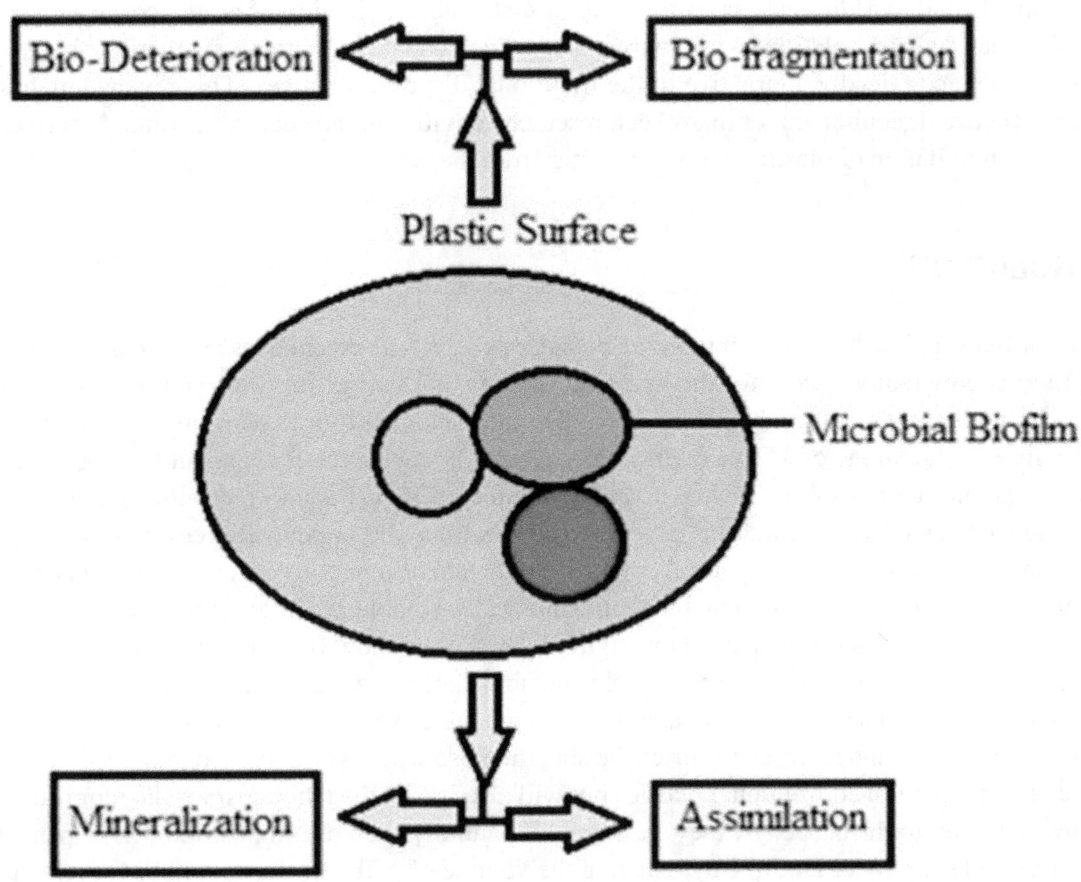

FUTURE PROSPECTIVE

Recent trend in research has observed, the development of the methods for manufacture of such polymers. Using biodegradable polymers in many fields of industry, instead of synthetic materials, may significantly help protect the natural environment. On the other hand, it is important to isolate several microorganisms like bacteria, fungi, zooplanktons, phytoplankton from the natural environment which possesses the enzymatic machinery that suitable degrade the disposed plastics. Such microorganisms can be found in several sources like soil, dumps, ocean bed and even the gut of few arthropods which feed on plastic. Such a modern approach developed for remediation combined and applied with such microorganisms will help solve plastic waste problem through bioremediation. This will ensure an eco-friendly and sustainable development of mankind where in all the other life forms can coexist.

Use of plastic in general has gained a subtle predominance over the past three decades. With increase in the use of plastic enrooted the need to effectively dispose the waste generated by it. Plastic waste disposal is gaining importance with higher awareness about the ill-effects of plastics among several sectors of the society. Industries largely depend upon plastics for various purposes and hence turn out to be major pollutants. Considering the hazards of its disposal, there is an immediate need to replace the conventional plastics with the bio-degradable ones. Our planet needs to be protected by this menace as effectively and expeditiously so as to ensure a sustainable wellbeing of all the species on it. Biodegradable materials or isolation of microorganisms which can degrade these synthetic polymers can be viewed as an eco-friendly alternative to the traditional disposal practices. This review emphasises on the relevance and requirement of many such research activities in the near future aimed at effectual and powerful annihilation of plastics and their wastes from the world.

CONCLUSION

In recent years, public has shown interest and concern over environmental issues. The problem of plastic pollution is now really a mess for the mankind. There is no part of the world untouched from its impacts. In the present era of globalization some thought must be given to plan safe disposal of products before making it commercial. This review has covered the major concerns about the plastics and their types, uses and degradability. Technical and economic problems connected with materials recycling have induced scientists to search for new materials that may be organically recycled. Synthetic polymers' negative influence on the natural environment creates many problems with waste deposition and utilization. Nowadays, there are few biodegradable water-soluble polymers available commercially, in spite of the obvious and pressing need for environmental protection. Biodegradable polymers have been researched, but polymers based on renewable and abundantly available sources (especially like starch) are most desirable. The major advantages of biodegradable polymers are that they can be composted with organic wastes and returned to enrich the soil, their use will not only reduce injuries to wild animals caused by dumping of conventional plastics but will also lessen the labor cost for the removal of plastic wastes in the environment because they are degraded naturally, their decomposition will help increase the longevity and stability of landfills by reducing the volume of garbage, they could be recycled to useful monomers and oligomers by microbial and enzymatic treatment. To solve plastic waste problem there is an immediate need to fund research for the production of biodegradable plastics.

REFERENCES

Albertsson, A. C. (1980). The shape of the biodegradation curve for low- and high-density polyethylene's in prolonged series of experiments. *European Polymer Journal*, *16*(7), 623–630. doi:10.1016/0014-3057(80)90100-7

Albertsson, A. C., Andersson, S. O., & Karlsson, S. (1987). The mechanism of biodegradation of polyethylene. *Polymer Degradation & Stability*, *18*(1), 73–87. doi:10.1016/0141-3910(87)90084-X

Albertsson, A. C., Barenstedt, C., & Karlsson, S. (1994). Abiotic degradation products from enhanced environmentally degradable polyethylene. *Acta Polymerica*, *45*(2), 97–103. doi:10.1002/actp.1994.010450207

Albertsson, A. C., & Karlsson, S. (1990). The influence of biotic and abiotic environments on the degradation of polyethylene. *Progress in Polymer Science*, *15*(2), 177–192. doi:10.1016/0079-6700(90)90027-X

Alshehrei, F. (2017). Biodegradation of synthetic and natural plastic by microorganisms. *Journal of Applied & Environmental Microbiology*, *5*(1), 8–19.

Andrady, A. L. (2011). Microplastics in the marine environment. *Marine Pollution Bulletin*, *62*(8), 1596–1605. doi:10.1016/j.marpolbul.2011.05.030 PMID:21742351

Awaja, F., & Pavel, D. (2005). Recycling of PET. *European Polymer Journal*, *41*(7), 1453–1477. doi:10.1016/j.eurpolymj.2005.02.005

Bonhomme, S., Cuer, A., Delort, A. M., Lemaire, J., Sancelme, M., & Scott, G. (2003). Environmental biodegradation of polyethylene. *Polymer Degradation & Stability*, *81*(3), 441–452. doi:10.1016/S0141-3910(03)00129-0

Chee, J. W., Amirul, A. A., Muhammad, T. T., Majid, M. I. A., & Mansor, S. M. (2008). The influence of copolymer ratio and drug loading level on the biocompatibility of P (3HB-co-4HB) synthesized by Cupriavidus sp.(USMAA2-4). *Biochemical Engineering Journal*, *38*(3), 314–318. doi:10.1016/j.bej.2007.07.018

Cruz-Pinto, J. J. C., Carvalho, M. E. S., & Ferreira, J. F. A. (1994). The kinetics and mechanism of polyethylene photo-oxidation. *Die Angewandte Makromolekulare Chemie: Applied Macromolecular Chemistry and Physics*, *216*(1), 113–133. doi:10.1002/apmc.1994.052160108

Glass, J. E., & Swift, G. (1989). Agricultural and synthetic polymers, biodegradation and utilization. In ACS symposium series (Vol. 433, pp. 9-64). Academic Press.

Greene, J. (2007). Biodegradation of compostable plastics in green yard-waste compost environment. *Journal of Polymers and the Environment*, *15*(4), 269–273. doi:10.100710924-007-0068-1

Gu, J. D., Ford, T. E., Mitton, D. B., & Mitchell, R. 2000. Microbial degradation and deterioration of polymeric materials. In The Uhlig Corrosion Handbook (2nd ed.). New York: Wiley.

Hadad, D., Geresh, S., & Sivan, A. (2005). Biodegradation of polyethylene by the thermophilic bacterium Brevibacillus borstelensis. *Journal of Applied Microbiology*, *98*(5), 1093–1100. doi:10.1111/j.1365-2672.2005.02553.x PMID:15836478

Huang, J. C., Shetty, A. S., & Wang, M. S. (1990). Biodegradable plastics: A review. *Advances in Polymer Technology*, *10*(1), 23–30. doi:10.1002/adv.1990.060100103

Huang, S. J., Roby, M. S., Macri, C. A., & Cameron, J. A. (1992). The effects of structure and morphology on the degradation of polymers with multiple groups. *Special Publication - Royal Society of Chemistry*, *109*(1), 149–149.

Iranzo, M., Sainz-Pardo, I., Boluda, R., Sanchez, J., & Mormeneo, S. (2001). The use of microorganisms in environmental remediation. *Annals of Microbiology*, *51*(2), 135–144.

Hussein, Al-Mayaly, & Khudeir. (2015). Isolation, Screening and Identification of Low-Density Polyethylene (LDPE) degrading bacteria from contaminated soil with plastic wastes. *Journal*, *1*(4), 1–14.

Kim, Y., Yeo, S., Kum, J., Song, H. G., & Choi, H. T. (2005). Cloning of a manganese peroxidase cDNA gene repressed by manganese in Trametes versicolor. *Journal of Microbiology (Seoul, Korea)*, *43*(6), 569–571. PMID:16410775

Kyrikou, I., & Briassoulis, D. (2007). Biodegradation of agricultural plastic films: A critical review. *Journal of Polymers and the Environment*, *15*(2), 125–150. doi:10.100710924-007-0053-8

Liao, H. T., & Wu, C. S. (2009). Preparation and characterization of ternary blends composed of polylactide, poly (ε-caprolactone) and starch. *Materials Science and Engineering A*, *515*(1-2), 207–214. doi:10.1016/j.msea.2009.03.003

Lucas, N., Bienaime, C., Belloy, C., Queneudec, M., Silvestre, F., & Nava-Saucedo, J. E. (2008). Polymer biodegradation: Mechanisms and estimation techniques–A review. *Chemosphere*, *73*(4), 429–442. doi:10.1016/j.chemosphere.2008.06.064 PMID:18723204

Luckachan, G. E., & Pillai, C. K. S. (2011). Biodegradable polymers-a review on recent trends and emerging perspectives. *Journal of Polymers and the Environment*, *19*(3), 637–676. doi:10.100710924-011-0317-1

Mierzwa-Hersztek, M., Gondek, K., & Kopeć, M. (2019). Degradation of Polyethylene and Biocomponent-Derived Polymer Materials: An Overview. *Journal of Polymers and the Environment*, *27*(3), 600–611. doi:10.100710924-019-01368-4

Orr, I. G., Hadar, Y., & Sivan, A. (2004). Colonization, biofilm formation and biodegradation of polyethylene by a strain of Rhodococcus ruber. *Applied Microbiology and Biotechnology*, *65*(1), 97–104. PMID:15221232

Pometto, A. L. III, Lee, B. T., & Johnson, K. E. (1992). Production of an extracellular polyethylene-degrading enzyme (s) by Streptomyces species. *Applied and Environmental Microbiology*, *58*(2), 731–733. doi:10.1128/AEM.58.2.731-733.1992 PMID:1610196

Raaman, N., Rajitha, N., Jayshree, A., & Jegadeesh, R. (2012). Biodegradation of plastic by Aspergillus spp. isolated from polythene polluted sites around Chennai. *Journal of Academia and Industrial Research*, *1*(6), 313–316.

Raziyafathima, M., Praseetha, P. K., & Rimal, I. R. S. (2016). Microbial degradation of plastic waste: A review. *Chemical and Biological Sciences*, *4*, 231–242.

Sen, S. K., & Raut, S. (2015). Microbial degradation of low density polyethylene (LDPE): A review. *Journal of Environmental Chemical Engineering*, *3*(1), 462–473. doi:10.1016/j.jece.2015.01.003

Seymour, R. B. (1989). Polymer science before and after 1899: Notable developments during the lifetime of Maurits Dekker. *Journal of Macromolecular Science. Chemistry*, *26*(8), 1023–1032. doi:10.1080/00222338908052032

Sharma, J., Gurung, T., Upadhyay, A., Nandy, K., Agnihotri, P., & Mitra, A. K. (2015). Isolation and characterization of plastic degrading bacteria from soil collected from the dumping grounds of an industrial area. *International Journal of Advance & Innovative Research*, *3*, 2278–7844.

Sharma, M., Sharma, P., Sharma, A., & Chandra, S. (2015). Microbial Degradation of Plastic—A Brief Review. *CIBTech Journal of Microbiology*, *4*(1), 85–89.

Sinha, V., Patel, M. R., & Patel, J. V. (2010). PET waste management by chemical recycling: A review. *Journal of Polymers and the Environment*, *18*(1), 8–25. doi:10.100710924-008-0106-7

Smith, W. M. (Ed.). (1964). *Manufacture of plastic* (Vol. 1). Reinhold Pub. Corp.

Swift, G. (1997). *Non-medical biodegradable polymers: environmentally degradable polymers. Handbook of biodegradable polymers*. Hardwood Academic.

Tokiwa, Y., & Calabia, B. P. (2007). Biodegradability and biodegradation of polyesters. *Journal of Polymers and the Environment*, *15*(4), 259–267. doi:10.100710924-007-0066-3

Tokiwa, Y., Calabia, B. P., Ugwu, C. U., & Aiba, S. (2009). Biodegradability of plastics. *International Journal of Molecular Sciences*, *10*(9), 3722–3742. doi:10.3390/ijms10093722 PMID:19865515

Tomita, K., Kuroki, Y., Hayashi, N., & Komukai, Y. (2000). Isolation of a thermophile degrading poly (butylene succinate-co-butylene adipate). *Journal of Bioscience and Bioengineering*, *90*(3), 350–352. doi:10.1016/S1389-1723(00)80096-1 PMID:16232871

Urase, T., Okumura, H., Panyosaranya, S., & Inamura, A. (2008). Emission of volatile organic compounds from solid waste disposal sites and importance of heat management. *Waste Management & Research*, *26*(6), 534–538. doi:10.1177/0734242X07084321 PMID:19039069

Vignesh, R., Deepika, R. C., Manigandan, P., & Janani, R. (2016). Screening of plastic degrading microbes from various dumped soil samples. *Int Res J Eng Tech*, *3*(4), 2493–2498.

Webb, H. K., Arnott, J., Crawford, R. J., & Ivanova, E. P. (2013). Plastic degradation and its environmental implications with special reference to poly (ethylene terephthalate). *Polymers*, *5*(1), 1–18. doi:10.3390/polym5010001

Worm, B., Lotze, H. K., Jubinville, I., Wilcox, C., & Jambeck, J. (2017). Plastic as a persistent marine pollutant. *Annual Review of Environment and Resources*, *42*(1), 1–26. doi:10.1146/annurev-environ-102016-060700

Yamada-Onodera, K., Mukumoto, H., Katsuyaya, Y., Saiganji, A., & Tani, Y. (2001). Degradation of polyethylene by a fungus, Penicillium simplicissimum YK. *Polymer Degradation & Stability*, *72*(2), 323–327. doi:10.1016/S0141-3910(01)00027-1

Yousif, E., & Haddad, R. (2013). Photodegradation and photostabilization of polymers, especially polystyrene. *SpringerPlus*, *2*(1), 398. doi:10.1186/2193-1801-2-398 PMID:25674392

Chapter 12
Sustainable Solid Waste Management via Biological Treatment:
A Critical Study

Arunima Nayak
Graphic Era University, Dehradun, India

Brij Bhushan
https://orcid.org/0000-0002-9360-3083
Graphic Era University, Dehradun, India

ABSTRACT

The solid wastes generated as a result of human activity have become very complex. Unscientific management of such wastes has resulted in environmental pollution, thereby putting a serious risk to the ecosystem and human health. A sustainable waste management is required for reduced pollution levels as well as for safeguarding the environment for future generation. Biological methods have gained in importance mainly because of its environmentally friendly option for recovering materials and energy. This is true as per the guidelines of "Hierarchy of Sustainable Waste Management." Composting and anaerobic digestion are two treatment options that is commonly followed in both developing as well as developed countries. The aim of this chapter is thus to address the technical background, advantages, as well as shortfalls related to the waste treatment via biological methods. An up-to-date literature thus presented may be beneficial to the scientists and researchers for working towards further improvement of the present system of waste management.

DOI: 10.4018/978-1-7998-4921-6.ch012

INTRODUCTION

Growing population, fast economic growth, rapid urbanization and changing lifestyle patterns has resulted in a catastrophic increase in the production of solid wastes. Undesirable solid by-products resulting from human activities are known as solid wastes, which have no economic value and hence are disposed off (Ojoawo and Agbede, 2011). Depending on the human activities, solid wastes are categorized into municipal solid wastes (MSW), industrial wastes, clinical wastes and hazardous wastes (Sridhar, 2000). MSW are generated from household activities, gardening activities, commercial activities, agricultural activities, sewage facilities etc. Solid wastes like sludge generated from domestic and industrial wastewater treatment plants are also included under MSW. Organic wastes generated from industries and hospital generated wastes are also classified under MSW but in some countries such has been segregated under industrial or clinical wastes. Inorganic wastes like ash, metal and wastes with hazardous nature, exhibiting properties like flammability, explosiveness, causticity, and toxicity are included in hazardous waste. Thus, MSW is the highest contributor to solid wastes.

The volume of solid wastes generated has been observed to vary depending on various factors like population density, income, life style etc (Fantola, 1997). As per a survey conducted during 2018 by T.Wang on global population and solid waste generation (Fig 1), China and India lead the share of global population (approximately 18% each) and are also the highest contributors of solid waste generation (15.55% and 11.95% respectively). A similar research conducted by Verisk Maplecroft (Idris et al., 2004) on waste generation and recycling indices in 2019 revealed that over 2.1 billion tonnes of solid waste are generated globally each year. Out of which only 16% is recycled each year and the remaining 46% is disposed off unscientifically and unsustainably. As per data projected by the World Bank(datatopics.worldbank.org), waste generation has been expected to increase to 3.40 billion tonnes by 2050 which is more than double the population growth expected across the same time period. As per Fig 2 (source: The World Bank), it is estimated that the daily rate of waste generation in developed countries would increase by 19% while the same for low income countries would increase by 40%.In India, as per Central Pollution Control Board (CPCB, 2000), a survey conducted in 2000 by Ministry of Urban Development showed that total solid waste generated was 100,000MT and the per capita waste generation rate was 0.3-0.76kg/day. The amount of waste generated per capita was estimated to increase at the rate of 1.33% annually (Sharholy et al., 2008). As compared to data in 1947, Indian cities generated 8 times more solid waste and the major reason cited was increased urbanization and changing life styles.

BACKGROUND

Composition and Characteristics of Solid Wastes

The data related to both the volume and composition of the solid wastes is crucial and is required for formulating wastemanagement plans and also for designing resource recovery and material cycles. Thus waste minimization strategies can only be carried out after having reliable waste composition data. Irrespective of any place or region, the solid wastes are classified into compostables, recyclables and the inerts (www.seas.columbia.edu; Buenrostro et al. 2001; www.worldbank.org). While the compostables represent the organic fraction consisting of food, agricultural wastes and by-products, recyclables comprise of paper, plastic, glass, metal parts etc. The inerts consist of stones, pebbles, ash, sand, glass etc

Figure 1. Share of global population in contribution to solid waste generation

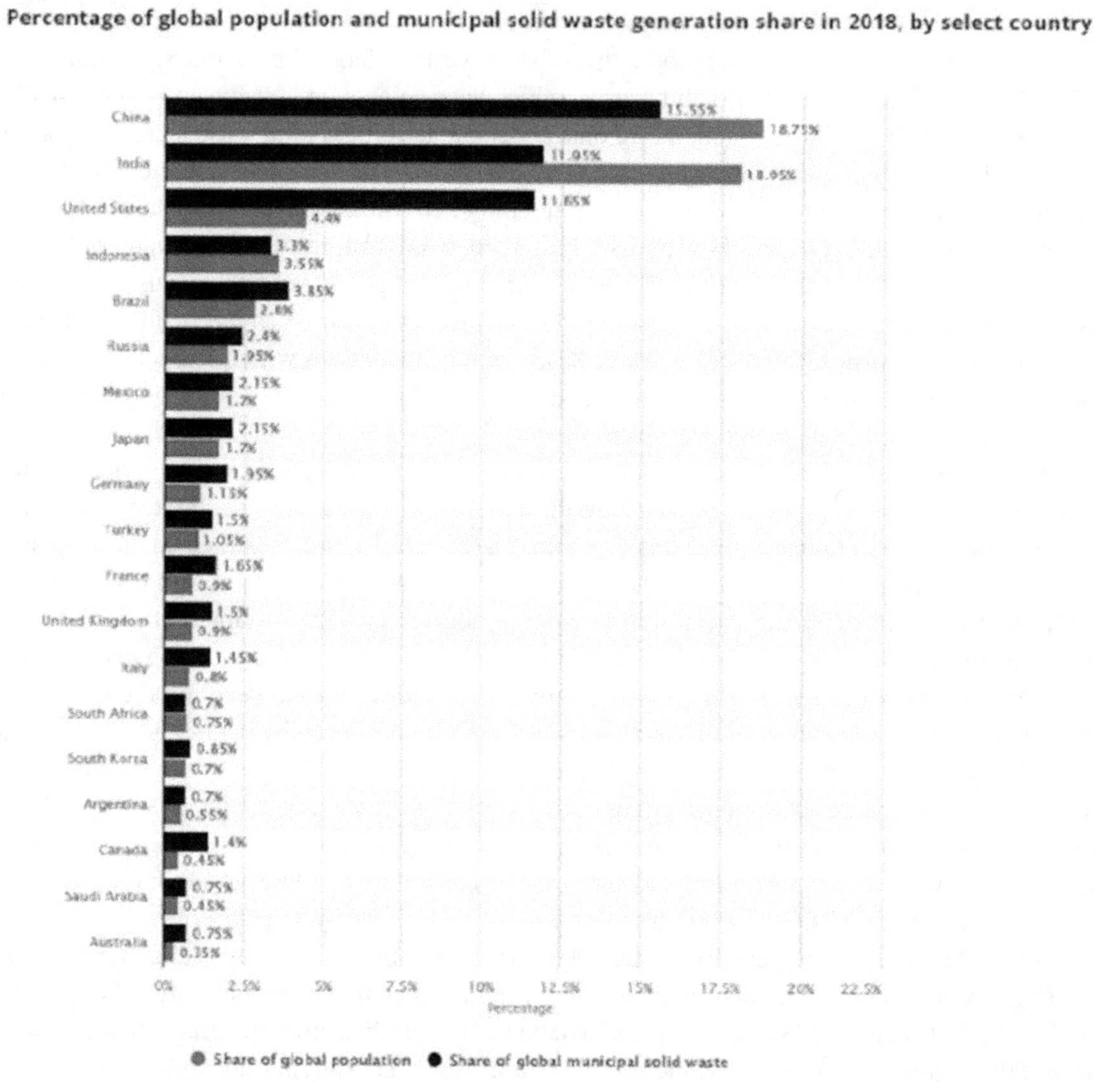

and hence can neither be recycled nor composted. The organic fraction consist of approximately 50-70% of the solid waste generated and may vary from region to region as well as can depend on the lifestyle of a particular region.

The composition of solid wastes has undergone a dramatic change with growing urbanization, changing lifestyle and nature of agricultural, biomedical or industrial activities. As per the data from World Bank (www.siteresources.worldbank.org), compostables make up to 60% of the solid wastes in low income countries, while the same in high income countries is 30%. The recyclables showed an increase to 30%. Majority of solid waste in India is organic (40-60%)(Joshi and Ahmed, 2016; IPCC, 1996). Recyclables and inerts contribute to 10-30% and 30-50% respectively. Studies reveal that the C/N ratio of the solid waste was 26%.

Figure 2. Projected waste generation in millions of tonnes/year (source: The World Bank)

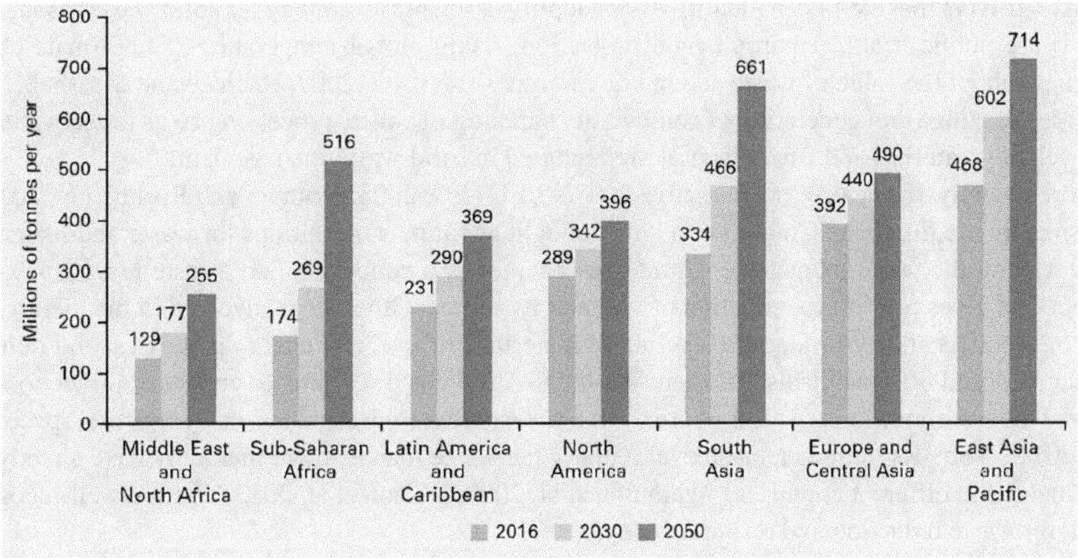

Improper Handling of Solid Wastes and Impact on Environment

The burgeoning volume of solid waste generated has become one of the biggest challenges faced by urban cities. Improper handling and unscientific disposal methods have added risk to environment, economy and society. Unscientific ways of dumping of MSW onto land bodies has not only destroyed the landscape but also has led to the breeding ground of different microbes and diseases. Leaching of organic and inorganic pollutants from the solid waste dumping has led to soil pollution. Plastics, metals parts, glass, rubber tyres, obsolete electronic parts, clinical wastes etc are non-biodegradable and are the biggest contributors of pollution and health hazards. Various toxic gases are emitted from solid waste dumpings of which methane gas has more potential to trap solar radiation; thereby causing increased global temperatures (IPCC, 2007; Henry et al. 2006). Uncontrolled burning of solid wastes has caused an increase in carbon mono-oxide, hydrocarbons, nitrogen oxides, sulphur oxides as well as particulate matter. Such emissions have a major impact on climate change. Unscientific handling and disposal of wastes have thus resulted in spread of human diseases causing premature deaths. Stringent environmental laws for safeguarding the environment, health and economy as well as a growing awareness among the population regarding the adverse effects of increased pollution levels of the environment due to indiscriminate and unscientific handling of solid wastes has resulted in adopting measures for a sustainable waste management.

Sustainable Solid Waste Management Strategies and its Need

The main drivers that have been identified for adopting a scientific strategy for solid waste management (SWM) are public health, environmental protection, rising global temperatures and climate change, awareness about the value of waste and public awareness (Wilson, 2007; Rachael and Marshall, 2013). Activities starting from generation of solid waste, handling of waste, processing so as to recover energy or recyclable materials and final disposal are included in solid waste management.

More recently, the new Waste Directive 2008/98/EU (EU, 2008; Brunner and Rechberger, 2015), as published by the European Commission have brought about new definitions for waste and its management. Among the waste management strategies adopted is a concept called "waste hierarchy". It is a concept that gives priority to reduction of waste at its source. Other steps involved in the hierarchy are reuse of products and by-products, recycling of materials, recovery of energy processes, and minimum amount disposed off to landfills. The new Waste Directive has direct link to enhanced protection ofhuman and the environment, as well as conservation of resources such as materials, energy and space (Pires et al. 2011). The 3Rs (reduce reuse recycle) under the waste management hierarchy have already been implemented in different countries (Agamuthu et al. 2009; Wilson et al. 2013) and is a well-recognised slogan for waste reduction and resource recovery.

Biological Treatment of Solid Wastes

Choice of the treatment method is dependent on the composition of the solid waste. In view of the high proportion of the compostables (50-70% of organics), biological methods are a viable option for the processing of municipal solid wastes as well as the sewage sludge (Kiely et al. 1997; Velmurugan and Arathy, 2010). Biological methods are applicable only after source separation of compostables from solid waste. Composting and Anaerobic Digestion are two widely preferred biological methods for treatment of solid wastes. The advantages are that both are natural processes involving the use of microrganisms and need less energy consumption and are environmentally friendly as compared to other technologies like as incineration, pyrolysis, etc. The increasing popularity of the biological treatment methods involve not only reduction in the volume, destruction of pathogens in the waste material, stabilisation of the waste but also ensures resource recovery in the form of biogas and bio-fertilizer. It thus can also ensure the reduction in the volume of wastes going to landfills as well as can prevent the environmental degradation due to the breakdown of organic wastes dumped in landfills. Recovery of biogas as a renewable fuel is highly desirable mainly on account of rising fossil fuel prices and depleting fossil fuel reserves in the face of ever increasing demand for energy consumption. Anaerobic Digestion and Composting are two widely popular technologies used for biological treatment of solid wastes. The aim of this chapter is not only to address the technical background, advantages related to the waste treatment via biological methods but also a critical discussion is presented so as to address the short falls and challenges of each technology. An up-to-date literature thus presented may be beneficial to the scientists and researchers for working towards further improvement of the present system of waste management.

Anaerobic Digestion

Methane and carbon dioxide gases released as a result of uncontrolled *anaerobic digestion* of the organic solid waste in landfills lead to global warming. As per literature studies, it is estimated that methane gas has a Global Warming Potential (GWP) which is over 20 times higher as compared to that for CO_2 (Mata-Alvarez et al. 2000; Themelis and Ulloa, 2007). But the same "Anaerobic Digestion" technology when performed in an enclosed chamber under controlled conditions (of temperature, pH and waste composition) and on the same substrate (organic solid waste) has the potential to provide useful products like biogas and a stabilized digestate. The major product of anaerobic digestion is biogas whose composition is 50-70% methane (CH_4), 20-50% carbon dioxide (CO_2) with trace amounts of hydrogen (H_2) and other gases. The major use of biogasis for production of electricity and heat. The demand for biogas is increasing on account of its environmentally friendly nature in comparison to that for fossil fuels. Excessive dependence on fossil fuel has not only led to fuel crisis but also has resulted in exorbitant price rise. The demand for biogas for use as a renewable energy source is hence economically viable. The digestate having a moisture content of 80% is dewatered to produce solid compost. The quality and composition of the compost depends on the composition of the solid waste, the pre-treatment methods employed and on the digestion process parameters (Ostrem, 2004). Thus a post aerobic treatment is preferred to improve the quality of the compost for use as a soil amendment.

Process Technique

The technology of the Anaerobic Digestion involves the biological degradation of wastes in the absence of oxygen. The biological degradation occurs via a series of bio-based chemical reactions like hydrolysis, acidogenesis, acetogenesis and methanogenesis. Each of these reactions takes place by the action of individual class of bacteria which work in close consortium (Behera et al. 2010). The process of digestion begins with the breakdown of the organic waste substrate into organic polymers like carbohydrates, proteins and lipids. The breakdown efficiency depends on the physical and chemical nature of the waste substrate. In many instances, a pre-treatment method is required to facilitate the process of breakdown. During the hydrolysis stage, the extracellular enzymes released from the hydrolytic bacteria help in breaking down the glycosidic bond in carbohydrates, ester bonds in lipids, and peptide bonds in proteins to sugars, fatty acids and amino acids respectively. The process is followed by the action of acidogenic bacteria which producescarbon dioxide, hydrogen, ammonia, alcohols and organic acids. Unlike the acidogenic bacteria, the acetogenic bacteria are slow and take long time to convertthe organic acids into acetic acid, hydrogen, and carbon dioxide. Finally, the process of methanogenesis takes place by the action of two separate classes of bacteria which work on different substrates. Methane is produced from acetic acid by the slow growing acetoclastic methanogens while the same is produced by the hydrogenotrophic methanogens which utilize hydrogen and carbon dioxide. A strict control of pH and correct nutrient availability is highly essential so as to ensure and maintain the high growth rate of the acidogenic bacteria as well as the sluggish metabolic growth rate of acetogenic and acetoclastic bacteria.

Based on the biodegradation kinetics of carbohydrates, proteins and lipids, it has been reported that the hydrolysis stage is the slowest and rate limiting step among all the stages of anaerobic digestion (Gunaseelan, 1997). Thus high efficiency of biogas production and the presence of higher proportion of methane in biogas depend on various factors ranging from waste substrate type and quality but also on a close synergistic activity of various classes of bacteria (Carlsson et al. 2012).

Pre-Treatment Methods

It is well known that hydrolysis which is the first stage of anaerobic digestion is the rate limiting step. Thus for complex substrates, anaerobic digestion suffers from long solid residence time and low conversion efficiency. Research studies have shown that a prior treatment of the organic waste substrate accelerates and also ensures high efficiency in biogas generation via anaerobic digestion (Cesaro and Ariunbaatar et al. 2014; Brummeler and Koster, 1990; Belgiorno, 2014). It is well known that solid wastes have complex physical and chemical characteristics. Pre-treatment can result in an increase in surface contact area of the substrate for enhanced microbial and/or enzymatic activity. The different pre-treatments undertaken so far are classified as under chemical, physicaland biological methods. Peroxide (Dewil et al. 2007; Shahriari et al. 2012), ozone (Chu et al. 2008; Cesaro, and Belgiorno, 2013) and alkali (Delgenès et al. 2003;Neves et al. 2006; Torres and Lloréns 2008)have been used as chemicals to not only bring about enhanced biodegradation but also to remove toxic compounds present in the waste substrate. Ozonation of solid organic waste resulted in 55% increase in solubilisation and 37% enhanced biogas yield (Cesaro and Belgiorno, 2013). The mechanistic action is that such chemicals have helped in rupture of cell membranes of polysaccharides, proteins and lipids; thereby bringing about faster degradation and conversion to simple and small molecular weight compounds (Delgenès et al. 2003).

Thermal and mechanical methods are the widely used physical pre-treatment of waste substrates having high solid content. The hard to degrade solid wastes can undergo better solubilisation and can exhibit better de-waterability characteristics as a result of thermal treatment (Haug et al., 1978; Bougrier et al., 2007). The treatment has shown improved handling of the digestate after the anaerobic digestion process. Studies have shown that thermal pre-treatment of sludge brought about enhanced hydrolysis and reduced retention time during its anaerobic digestion (Ferrer et al., 2008).

Mechanical pre-treatment is undertaken to bring about a reduction in particle size and crystallinity of various lignocellulosic based waste substrates and wastewater treatment sludge (Mshandete et al., 2006). Depending on the final particle size, the methods used were grinding, milling, chipping, maceration, liquefaction, high pressure homogenizer etc. Reduction of particle size results in increased contact area for the bacterial community for bringing about better hydrolysis, biodegradation and enhanced biogas yield. A study conducted by Izumi et al. (Izumi et al., 2010) revealed that a grinding pre-treatment of food waste substrate brought about 40% higher solubilisation and 28% enhanced methane production from the anaerobic digestion process. The study further highlighted the importance of size reduction of the waste substrate on bio-methanization. The major advantage of mechanical pre-treatment includes its environmental friendliness. But size reduction needs to be optimized as the mechanical treatment is a highly energy intensive process.

More recently the use of microwave andultrasoundas emerging pre-treatment technologies has resulted in better anaerobic digestibility using lesser energy consumption as compared to the use of thermal pre-treatment methods. Use of microwave to pre-treat results in selective and improved disruption of the hard to degrade waste substrate; thereby bringing about improved hydrolysis and solubilization (Jackowiak et al., 2011). Ultrasonication brings about cavitations at lower frequencies and chemical reactions due to the release of free radicals at higher frequencies (Tiehmet al., 1997; Tiehm et al., 2001). Such physical and chemical changes can improve bacterial activity and hence anaerobic digestion performance. In a study conducted by Tiehm et al. (Tiehm et al., 1997), ultrasound pre-treatment of sewage sludge resulted in reduced retention time by 12 days during anaerobic digestion as well as a reduction in the volatile solid removal of 44%.

Biological pre-treatment methods include the use of microrganisms under both aerobic as well as anaerobic conditions that can enable the enhancement of breakdown of the substrate. Various studies have (Fdéz-Güelfo et al., 2011a; 2011b) demonstrated that irrespective of substrate type, addition of a mature compost resulted in enhanced solubilisation and better biogas yields. The performance of biological pre-treatment was found to be better as compared to thermal and chemical methods in bringing about higher anaerobic digestion rate (Fdéz-Güelfo et al., 2012). High degradation efficiency of lignocellulosic based waste substrates under anaerobic conditions was observed while using white rot fungi (Keller et al., 2003). More recently, a microbial consortium was used to pre-treat the lignocellulosic based municipal solid waste (Yuan et al., 2014). Anaerobic digestion tests conducted on the pre-treated and untreated solid waste showed higher solubilisation of the organic content as well as higher biogas yields from the pre-treated wastes. Enhanced hydrolysis of the solid substrate was also brought about by addition of external hydrolytic enzymes (Delgenès et al. 2003). But because of the complex characteristics of organic municipal and food solid waste, the purpose of bringing about specific hydrolysis by using enzymes may appear to be challenging. Masse et al. (Masse et al., 2003) demonstrated an enhanced fat degradation when using an enzyme pre-treatment during the anaerobic digestion of poultry slaughterhouse effluent.

Although biological pre-treatment is environmentally and economically friendly, but effective hydrolysis rate is much lower as compared to the other methods. The other drawbacks are low methane yields and application to complex substrates is challenging.

Environmental Factors

The efficiency of the digestion process depends on the metabolism efficiency of individual class of bacteria. The growth of the bacterial microrganisms are highly sensitive to various environmental factors like temperature, pH, availability of nutrient levels i.e the waste substrate composition, the inoculum type, organic loading rate etc (Jain et al., 2015). A continuous monitoring of the pH, alkalinity, total solids as well as the presence of metabolites like volatile fatty acids, free ammonia etc is a must for ensuring stability of the reactor in which the digestion process is conducted. It is a proven fact that maximum biogas yields are obtained under optimum operating conditions which result in enhanced microbial metabolic activity. Extensive research work (Castillo et al., 2006; Kiran et al., 2016) has identified two temperature ranges for optimum operation of anaerobic digestion: 35-40°C (mesophilic) and50-65°C (thermophilic)temperature range and a neutral pH range of 6.3-7.8. Zhu et al., (Zhu et al., 2009) reported that as compared to the digestion process taking place under mesophilic conditions, thermophilic process has more advantages like a faster degradation rate of organic waste, higher methane production and higher pathogen. But the methanogens thriving under thermophilic conditions are very sensitive to temperature changes and take longer time to adapt to new temperature (Kiran et al., 2016). It has been reported (Briski et al., 20007) that temperatures above 65°C cause denaturation of enzymes and hence are detrimental for anaerobic biodegradation. The metabolic activity of the different bacterial community during the different stages of anaerobic digestion is highly pH dependent (Khalid et al., 2011). Various authors have reported that optimum pH levels for methanogenesis in an anaerobic digester is 6.5–8.2, while the pH levels for acidogenesis is 5.5-6.5 (Kim et al., 2003; Lee et al., 2009; Mao et al., 2015). Zhang et al (Zhang et al., 2009) reported a positive correlation between hydrolysis rate constant and pH; with highest hydrolytic enzyme activity observed at pH 5.5. Accumulations of metabolites like free ammonia and volatile fatty acids cause pH imbalances which then can inhibit the activity of the methanogens. Continuous supply of nutrients especially carbon and nitrogen from the waste substrate is required for healthy growth and activity of the microrganisms which inturn ensures reactor (digester) stability and also results in enhanced biogas yield. An optimum ratio of carbon and nitrogen (25-30:1) has been identified as very crucial to ensuring digester stability. Certain substrates having low nitrogen content can cause slowdown of the breakdown process leading to lower biogas yields (Fountoulakis et al., 2008). Whereas, higher nitrogen content in the waste substrate can lead to accumulation of ammonia which inturn can destabilize the process. pH levels within the reactor and simultaneously the specific activity of methanogenic bacteria has been found to decrease with increasing concentrations of ammonia (Chen et al., 2008). Thus continuous and appropriate nutrient availability is an important factor which is dependent on the waste substrate quality. Besides carbon and nitrogen, availability of macronutrients and trace elements is essential for healthy activity of the bacterial community. For an enhanced anaerobic digestion performance, macronutrients and trace elements are supplemented. Lipid content in waste substrate is another factor that can affect the anaerobic digestion process. Although the biochemical methane potential of lipids is much higher as compared to that for carbohydrates, but the long chain fatty acids (LCFA) formed as a result of the hydrolysis of lipids is detrimental to the digestion process (Chen et al., 2009; Kiran et al., 2016). High concentration of LCFA can lead to system failure. Thus the composition of waste substrate has a major effect on the performance of the anaerobic digestion process. Co-digestion or use of more than one substrate during anaerobic digestion has helped in circumventing the shortfall of a single substrate digestion, has helped in preventing system failure as well as has helped in improving the biogas yield.

Reactor/Digester Configuration

Various research works has cited the correlation between well designed reactors with high performance of anaerobic digestion leading to higher biogas yields (William and David, 1999; Lina et al., 2018; Salama et al., 2019). A well designed anaerobic digester must allow an optimum environment for the growth of the bacterial community. To ensure maximum biomass retention within the reactor, the solid retention time must be greater than the hydraulic retention time. This will also allow sufficient contact between the waste substrate and the biomass (Letting, 1995; Ward et al., 2008).

Various reactor designs have been proposed so as to ensure high biomass retention and subsequently higher reactor efficiency for biogas generation. Among the more conventional ones is the anaerobic sequencing batch reactor (ASBR), continuous stirred tank reactor (CSTR) and anaerobic plug flow reactor (APFR)(Singh and Srivastava, 2011; Mao et al., 2015). While ASBR is used solely for the degradation of high strength wastewater, CSTR has been also been applied for treatment of high strength liquid manure as well as for organic industrial wastes. Both ASBR and CSTR operate in a single stage digestion process where all the treatment steps and process take place in a single chamber. While ASBR has the potential to handle wide range of influent volumes, a CSTR requires a fixed influent flow rate (Boe and Angelidaki, 2009). The major drawback is that both reactor types do not ensure biomass retention and hence maximum microrganisms are washed out of the reactor along with the effluent. Also, continuous stirring operation in CSTR has resulted in accumulation of volatile fatty acids leading to inhibition and lesser biogas generation.

The APFR has been used for treatment of a wide variety of substrates like pig manure, distillery waste water, cattle residue, organic solid waste of municipalities etc (Sharma et al., 2000). It has offered many advantages over CSTR like ensuring high biomass retention, low volatile fatty acid and a stable reactor performance. On the other hand the high rate anaerobic reactors like anaerobic contact reactor

(ACR), up-flow anaerobic sludge bed reactor (UASB), Up-flow anaerobic solid-state reactor (UASS) and anaerobic baffled reactor (ABR) ensure enhancedbiomass retention (Singh and Prerna, 2009; Şentürket al., 2012). The ACR has two main components: the reactor and the settling tank in which the settled sludge is allowed to recycle back to the reactor; thereby allowing further contact between the biomass and the waste substrate. But, the major drawback is the poor sludge settlement due to gas formation by the bacterial community in the settlement tank. The biomass in the UASB is retained within the reactor as a blanket and is kept in suspension by controlling the upflow velocity of waste substrate (Chong et al., 2012). Another feature of UASB is the gas and solid phase seperation system located at the top of the reactor. This helps in the collection of biogas and also allows internal recycling of sludge. The solid retention time is much higher as compared to hydraulic retention time thereby allowing a dense growth of microrganisms. The ABR consists of a series of compartments each equipped with baffles that force the waste substrate to make up flow or down flow movement which in turn ensures greater contact of biomass with the waste substrate as well as high active biomass content (Barber and Stuckey, 1999). Besides ensuring a higher solid retention time, the other significant advantage of the ABR is the ability to separate acidogenesis stage from the methanogenesis. In contrast to UASB, ABR does not require separate sludge seperation equipment.

More recently developed reactors are the anaerobic membrane reactors (AMB) which exhibit dual advantages of enhanced biogas production as well as a high effluent quality that is devoid of solids and pathogens (Switzenbaum, 1983; Omil et al., 2003; Bodkhe, 2008; Andalib, 2014). Membranes help in retaining higher biomass concentrations (leading to higher solid retention time) and also ensure higher removal of volatile fatty acids and soluble organics. In a recent review (Kim et al., 2010), it is stated that AMB has demonstrated great efficiency in dealing with inhibitory substrates. Besides dealing with inhibitory substrates, AMB has demonstrated high operational stability, high treatment efficiency as well as stable biogas production under extreme conditions like high salinity, high solid content wastewater or poor biomass properties. The greatest disadvantage is due to membrane fouling which can be minimized by high liquid velocities across the membrane and gas agitation systems (Gao et al., 2014). Advancement in membrane design and measures to overcome fouling can make AMB a viable technology for treatment of industrial as well as municipal wastes.

Anaerobic Co-Digestion

As the name suggests, co-digestion is the combination of two or more substrates and research has demonstrated that digestion of two or more substrates simultaneously helps to acquire a stable anaerobic digestion performance (Agdag et al., 2007). Single substrate is known to have demonstrated some negative aspects like imbalance in nutrients or increased inhibitory contents. Mata-Alvarez et al., 2000 (Mata-Alvarez et al., 2000) have suggested that co-digestion have led to improved biogas yields which may have been due to positive synergism established between the microrganisms inside the reactor and also may be due to the replenishment of missing nutrients by the co-substrates (Mata-Alvarez et al., 2014). Similar advantages of improved nutrient availability and enhanced biogas production was reported by Braun, 2002 (Braun, 2002), but the disadvantages cited are additional pre-treatments and mixing requirements, increased organic load in effluent etc (Neves et al., 2006). Some of the substrates that were co-digested and have demonstrated improved anaerobic digestion performance are sludge and food waste, sewage sludge and municipal solid waste (Dai et al., 2013), wastewater and food waste (Kim et al., 2011), cattle manure and food waste (Zhang et al., 2011), municipal waste and livestock waste (Zhang et al., 2013), food waste and yard waste (Kim and Oh, 2011), food waste and green waste (Brown and Li, 2013). The compositions of the co-substrates as well as their proportion in the digester medium are crucial parameters that will dictate the success of the co-digestion process (Kumar et al., 2010).

Composting

The process of controlled, aerobic microbial degradation of organic solid waste leading to a stable end-product is composting (Wang et al., 2004). Compost has great utility as a soil conditioner or a biofertilizer. Composting has resulted in killing of pathogens, prevention of odors/insect breeding, bulk reduction as well as resource recovery via compost. Various aerobic microrganisms like bacteria, actinomycetes, fungi, protozoa etc break down the complex organic matter in the waste substrate to simpler stable organic by-products (Gajalakshmi and Abbasi, 2008). The rate of the breakdown depends on the quality of the organic substrate as well as the process parameters.

Process Overview

The microbial action on the organic waste substrate during composting is dependent on temperature (Bharadwaj, 1995, Epstein, 1997). Initially the microbial community act at temperatures of 10-45°C to break down the sugars and amino acids present in the waste via oxidation. Formation of volatile fatty acids results in the rise of temperature with a consequent fall in pH. As per Hallman et al., thermophilic microrganisms start to operate at increased temperatures of 45-50°C (Hellmann et al., 1997). In the thermophilic stage, maximum of the organic matter is degraded via consumption of more oxygen. During the cooling phase, the microbial activity starts to decrease resulting in a decrease in temperature. When the temperature fall to that of the surroundings, the compost maturation phase begins in which the mesophilic microrganisms start to degrade the complex organic matter like the lignin. The carbon in substrate is converted to carbon dioxide which gets integrated into humus like substances to form mature compost. Nitrogenous compounds like urea, protein etc get degraded to ammonium ion which in the presence of oxygen get further degraded to nitrate. The process of mineralization as well as humification takes place simultaneously during the composting process. Various studies have shown that the microbial community produce specific hydrolytic extracellular enzymes that in turn help in degradation of the complex constituents like cellulose, hemicellulose, lignin, carbohydrates etc into smaller water soluble compounds (Garcia et al., 1992; Ayuso et al., 1996; Hankin et al., 1976). The enzymes identified as responsible for hydrolysis of cellulose, hemicellulose and proteins are cellulase, xylanase, and protease respectively (Tiquia and Tam, 2000).

Factors Affecting the Composting Process

The quality and stability of the compost is dependent on not only the quality of the initial substrate (solid waste) but also on other abiotic and biotic factors (Gajalakshmi and Abbasi, 2008). Quality of the solid organic waste is a vital factor affecting the composting process as it provides all nutrients for the healthy growth and activity of the different microbial community (Zucconi et al., 1981). Plant based solid organic wastes are composed of cellulose, hemicellulose and lignin (Barker, 1997). Biobased wastes consist of carbohydrates, lignin and nitrogenous compounds. It was reported that during the mesophilic phase, carbohydrates and sugars underwent degradation to form carbon dioxide and water, while nitrogenous compounds degraded to form ammonia. Cellulose, hemicellulose and lignin degradation took place during the thermophilic phase. The rate of degradation is much slower. Degradation is known to take place by the action of specific extracellular enzymes released by the microbial community. Besides mineralization, humification also takes place simultaneously. Besides, the constituents present in the initial substrate, the particle size and porosity also plays a dominant role in providing aeration for enhanced microbial metabolism. The substrate is also the source of nutrients in the form of carbon and nitrogen. An optimum ratio of carbon to nitrogen is required for matured compost. Studies (Onwosi et al., 2017) have demonstrated that excessive carbon to nitrogen ratio can lead to a deficiency of nutrients to the microbial community while undesirable odours can result from lower carbon to nitrogen ratio. A moisture content of 50-70% is required during the composting process as it provides a medium for transport of nutrients from the substrate to the microbial community. While low moisture content can lead to inhibition of metabolic activity, excess of it can create anaerobic conditions resulting in putrefaction (Bertoldi and Vallini, 1983). Both temperature and oxygen supply are linked to aeration which ensures the healthy growth of microbial community (Barrington et al., 2003). The various stages of the composting process are linked directly to temperature. Oxygen is vital for aerobic microrganisms to grow and multiply. pH is another important parameter that affects the composting process. Various authors have cited an optimum pH level of 6-7 for bacterial growth while fungi prefer a pH environment of 5-8. Lower pH values observed during the initial phase of composting has been cited due to the formation of carbon dioxide and volatile fatty acids due to the degradation of sugars, starch etc. pH levels increase during the later phases due to depletion of volatile fatty acid and accumulation of ammonia Kapetanois et al., 1993; Zorpas et al., 2003). Studies (Nagasaki et al., 1993; Cerda et al., 2018) have proved that the rate of breakdown of organic matter is faster in the pH-controlled experiment as compared to the rates observed in the experiment without pH control.

CRITICAL DISCUSSION

Anaerobic digestion and composting are biological processes that operate using the fundamentals of microbiology, chemistry and physics. Research work conducted till date provides evidence of the process viability for the sustainable treatment of solid waste to produce useful by-products in the form of digestate, bio-fuels in the form of biogas (from anaerobic digestion) and compost (from composting). The economic viability of both processes depends on the commercial usage of the end products as well as the net energy requirement (Fig 3). A representation of the advantages and disadvantages of the two biological methods of degradation of solid waste viz. anaerobic digestion and composting is depicted in Figure 3. The utilisation efficiency for biogas as an alternative fuel has put the anaerobic digestion technique in a more economically favorable position as compared to that for composting. Also, through anaerobic digestion, a significant proportion of input energy is recovered via biogas. In this respect, composting is a net energy consumer. Kusch et al. (Kusch et al., 2011) conducted a comparative study in which it was revealed that about 60% of the energy from organic waste is lost as heat during composting while approximately 85% of energy is recovered via biogas in anaerobic digestion.

The digestate and the compost are a good source of nutrients. But the probability of the direct use of digestate as fertilizer is limited because of the high volatile acid and pathogen content. Compost, on the other hand has demonstrated better results during its application on land because of the absence of phytotoxicity and higher germination indices. From an environmental perspective, composting is known

Figure 3. Advantages and Disadvantages of Anaerobic Digestion and Composting technologies

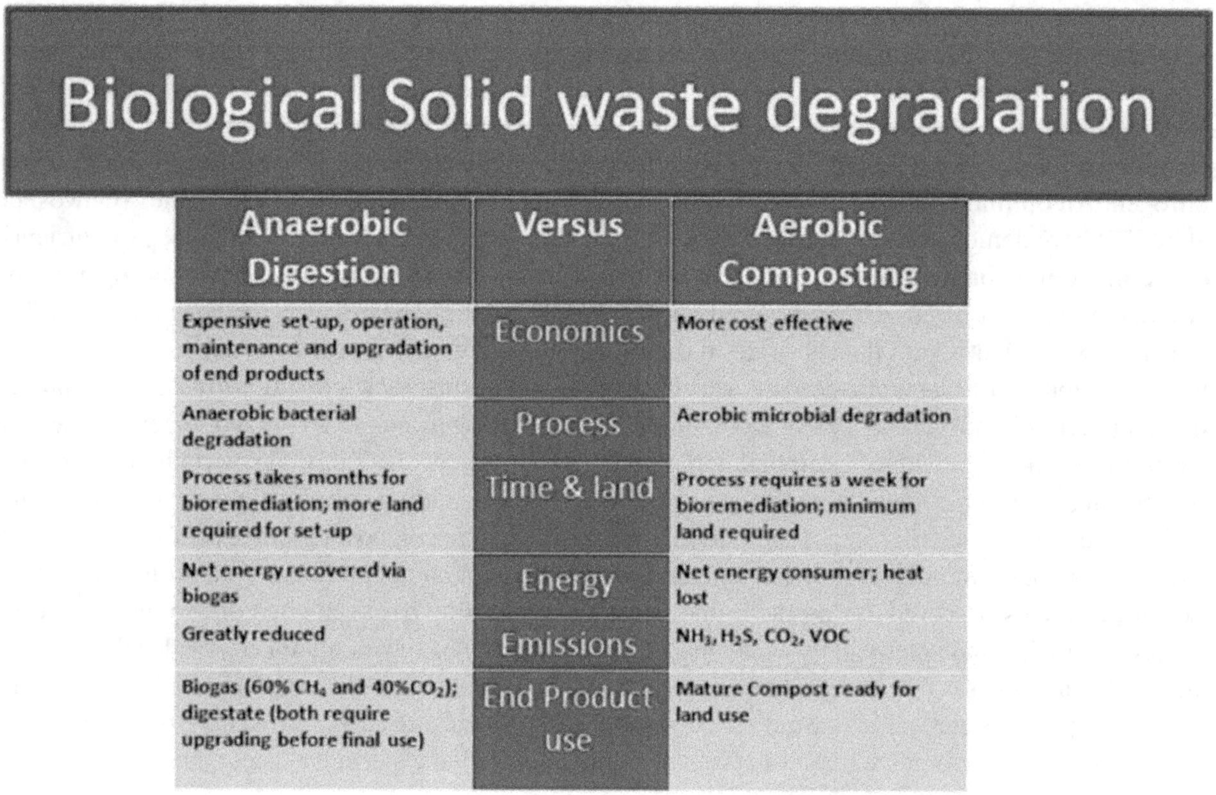

Biological Solid waste degradation

Anaerobic Digestion	Versus	Aerobic Composting
Expensive set-up, operation, maintenance and upgradation of end products	Economics	More cost effective
Anaerobic bacterial degradation	Process	Aerobic microbial degradation
Process takes months for bioremediation; more land required for set-up	Time & land	Process requires a week for bioremediation; minimum land required
Net energy recovered via biogas	Energy	Net energy consumer; heat lost
Greatly reduced	Emissions	NH_3, H_2S, CO_2, VOC
Biogas (60% CH_4 and 40%CO_2); digestate (both require upgrading before final use)	End Product use	Mature Compost ready for land use

to generate large volumes of emissions of ammonia, hydrogen sulphide, volatile organic compounds as well as green house gas emissions. Large scale application of composting is thus limited but research is ongoing on various methods to control such emissions (Ahn et al., 2011; Steiner et al., 2010; Zhang et al., 2016). Emissions from digestate have also to a large extent been controlled via appropriate valorisation techniques. Pyrolysis conducted on the digestate under reducing atmospheres has helped to recover value added by-products like syngas, bio-oil and biochar (Monlau et al., 2015). A thorough characterisation of the solid wastes is beneficial for proper selection of the parameters affecting the biodegradability performance.

A wide range of solid waste of biological origin has been tested as substrate with different levels of efficacy from anaerobic digestion and composting processes. Wastes from agricultural and forest residues, yard trimmings etc. have demonstrated a high C/N ratio, with significant amounts of cellulose, hemicellulose and lignin. The biochemical methane potential was recorded at 107-482L/kg VS (volatile solids) (Raposo et al., 2012). But the biodegradability of such lignocellulosic substrates has been observed to be lower in anaerobic digestion as compared to that in composting (Lin et al., 2014). While the anaerobic digestion of cellulose and hemicellulose was reported to be 20-50%, the same during composting was 40-80%. Lignin has been reported to be degraded by aerobic microorganisms (Tuomela et al., 2000). Based on the higher proportion of lignin, pre-treatment is an essential prerequisite for improving the biodegradability and hence enhanced methane yields from lignocellulosic based solid waste. Studies related to the need for pre-treatment during composting has been scarce. Food wastes, found in abundance in municipal solid wastes, are characterised by low C/N (USEPA, 2012) and possess soluble organic constituents in the form of carbohydrates, fats and proteins. The biochemical methane potential has been observed to be in the range of 245-968L/kg VS (Raposo et al., 2012). Nevertheless, very few studies have been conducted on the anaerobic digestion of food wastes. As per Xu et al., (Xu et al., 2017), the challenges faced during their anaerobic digestion are the inherent high solubility of its organic fraction; thereby requiring strict process control and optimization. Composting on the other hand causes substantial sulphur and ammonia emissions due to the degradation of proteins and fats (Masebinu et al., 2019). Third category of solid wastes are the sewage sludge which are the biosolids generated as a result of municipal waste water treatment plant. They are characterised by low C/N (Zhang et al., 2014), and may contain high amounts of fats and proteins as well as pathogens. As per a study conducted by Chynoweth et al., (Chynoweth et al., 1993), the biochemical methane potential of sewage sludge was 590 L/kg VS (volatile solids). High odorous emissions from such wastes have restricted its valorisation via composting. Studies have shown that the digestate from anaerobic digestion of such wastes operating under mesophilic temperature conditions have an increased incidence of pathogens which lead to its restricted usage for land applications (Fu et al., 2014). Mixing of such different categories of organic solid substrates, a term referred to as co-digestion has resulted in balance of nutrients and C/N ratio as well as has helped in reducing the action of inhibitors towards the microbial growth. Studies have also shown that anaerobic digestion operating with co-digestion of different solid substrates have resulted in efficient usage of digester volume thereby causing reduced economic burden as compared to the same using single substrate digestion (Koupaie et al., 2014). The correct proportion of mixing the substrates has also played a vital role in maintaining balanced conditions within the digester. Various studies have shown the use of mathematical models and experimental designs so as to optimise the mixing ratio and which in turn have demonstrated better biodegradability. The mixing proportion of three types of kitchen waste was optimized and the same substrate revealed better composting as per studies conducted by Abdullah and Chin (Abdullah and Chin, 2010). Experimental design and modelling studies revealed that an optimum mixing ratio provided better synergistic interactions between the different organic substrates under anaerobic conditions (Astals et al., 2014).

Individual process parameters need optimisation so as to enable a conducive environment for the healthy metabolism of the microrganisms which thus can enhance the efficiency of both anaerobic digestions as well as composting process. Correct selection of inoculum, temperature, moisture and C/N ratio is priority factors that highly influence the process efficiency.pH, alkalinity as well as volatile acids are indicators of process stability and hence strict monitoring is essential (Nayak and Bhushan, 2019).

CONCLUSION

Solid wastes generated due to human activities are characterised by large volumes as well as diversified chemical composition. Unscientific means of disposal have led to global warming, climate changes and health issues arising from various water and air borne diseases. Based on the high organic fraction of such wastes, the chapter has put a major focus on the scientific importance of adopting biological methods for its sustainable management. Both anaerobic digestion and composting involve not only volume reduction and stabilization of the waste, but also ensure resource recovery in the form of biogas and bio-fertilizer. The other advantages are that both technologies have prevented the environmental degradation which occurred due to the breakdown of organic solid wastes that were dumped in landfills.

Detailed literature review on the technical details, process parameters, advantages of each technique reveals that anaerobic digestion has favorable characteristics both from economic as well as environmental point of view as compared to composting. Research is focussed now more on adopting different pre-treatment methods or on co-digestion of different categories of solid waste or on different digester designs so as to bring about enhanced anaerobic digestion efficiency. But success rates can be assumed only after large scale application of the anaerobic digestion technique. The shortfall in composting is the voluminous toxic emissions and presence of inorganic materials which may affect the quality of the compost. Major research is focussed on curtailing such emissions and inorganic materials for improving the environmental viability and sustainability of the composting technique.

REFERENCES

Abdullah, N., & Chin, N. L. (2010). Simplex-centroid mixture formulation for optimised composting of kitchen waste. *Bioresource Technology*, *101*(21), 8205–8210. doi:10.1016/j.biortech.2010.05.068 PMID:20624604

Agamuthu, P., Khidzir, K. M., & Hamid, F. S. (2009). Drivers of sustainable waste management in Asia. *Waste Management & Research*, *27*(7), 625–633. doi:10.1177/0734242X09103191 PMID:19470545

Agdag, O. N., & Sponza, D. T. (2007). Co-digestion of mixed industrial sludge with municipal solid wastes in anaerobic simulated landfilling bioreactors. *Journal of Hazardous Materials*, *140*(1-2), 75–85. doi:10.1016/j.jhazmat.2006.06.059 PMID:16884847

Ahn, H. K., Mulbry, W., White, J. W., & Kondrad, S. L. (2011). Pile mixing increases greenhouse gas emissions during composting of dairy manure. *Bioresource Technology*, *102*(3), 2904–2909. doi:10.1016/j.biortech.2010.10.142 PMID:21111610

Andalib, M., Elbeshbishy, E., Mustafa, N., Hafez, H., Nakhla, G., & Zhu, J. (2014). Performance of an anaerobic fluidized bed bioreactor (AnFBR) for digestion of primary municipal wastewater treatment biosolids and bioethanol thin stillage. *Renewable Energy*, *71*, 276–285. doi:10.1016/j.renene.2014.05.039

Annepu, R. R. K. (n.d.). *Sustainable solid waste management in India, Waste-to-Energy Research and Technology Council (WTERT)*. Columbia University. Retrieved from http://www.seas.columbia.edu/earth/wtert/sofos/Sustainable%20Solid%20Waste%20Management%20 in%20India_Final.pdf

Ariunbaatar, J., Panico, A., Esposito, G., Pirozzi, F., & Lens, P. N. L. (2014). Pre-treatment methods to enhance anaerobic digestion of organic solid waste. *Applied Energy*, *123*, 143–156. doi:10.1016/j.apenergy.2014.02.035

Astals, S., Batstone, D. J., Mata-Alvarez, J., & Jensen, P. D. (2014). Identification of synergistic impacts during anaerobic co-digestion of organic wastes. *Bioresource Technology*, *169*, 421–427. doi:10.1016/j.biortech.2014.07.024 PMID:25079207

Ayuso, M., Hernandez, T., Garcia, C., & Pascual, J. A. (1996). Biochemical and chemical structural characterization of different organic materials used as manures. *Bioresource Technology*, *57*(2), 201–207. doi:10.1016/0960-8524(96)00070-3

Barber, W. P., & Stuckey, D. C. (1999). The use of the anaerobic baffled reactor (ABR) for wastewater treatment: A review. *Water Research, 33*(7), 1559–1578. doi:10.1016/S0043-1354(98)00371-6

Barker, A. V. (1997). Composition and uses of compost. In *Agricultural uses of by-products and wastes.* ACS Symposium Series 668. 10.1021/bk-1997-0668.ch010

Barrington, S., Choiniere, D., Trigui, M., & Knight, W. (2003). Compost convective airflow under passive aeration. *Bioresource Technology, 86*(3), 259–266. doi:10.1016/S0960-8524(02)00155-4 PMID:12688469

Behera, S. K., Park, J. M., Kim, K. H., & Park, H. (2010). Methane production from food waste leachate in laboratory-scale simulated landfill. *Waste Management (New York, N.Y.), 30*(8-9), 1502–1508. doi:10.1016/j.wasman.2010.02.028 PMID:20227867

Bertoldi, M. (1983). The biology of composting: A review. *Waste Management & Research, 1*(1), 157–176. doi:10.1177/0734242X8300100118

Bharadwaj, K. K. R. (1995). Improvements in microbial compost technology: A special reference to microbiology of composting. In S. Khanna & K. Mohan (Eds.), *Wealth from waste* (pp. 115–135). Tata Energy Research Institute.

Bodkhe, S. (2008). Development of an improved anaerobic filter for municipal wastewater treatment. *Bioresource Technology, 99*(1), 222–2226. doi:10.1016/j.biortech.2006.11.026 PMID:17207998

Boe, K., & Angelidaki, I. (2009). Serial CSTR digester configuration for improving biogas production from manure. *Water Research, 43*(1), 166–172. doi:10.1016/j.watres.2008.09.041 PMID:18976792

Bougrier, C., Delgenès, J. P., & Carrère, H. (2007). Impacts of thermal pre-treatments on the semi-continuous anaerobic digestion of waste activated sludge. *Biochemical Engineering Journal, 34*(1), 20–27. doi:10.1016/j.bej.2006.11.013

Braun, R. (2002). *Potential of Co-digestion.* https://www.novaenergie.ch/iea-bioenergytask37/Dokumente/final.PDF

Briski, F., Vukovic, M., Papa, K., Gomzi, Z., & Domanovac, T. (2007). Modelling of compositing of food waste in a column reactor. *Chemical Papers, 61*(1), 24–29. doi:10.247811696-006-0090-0

Brown, D., & Li, Y. (2013). Solid state anaerobic co-digestion of yard waste and food waste for biogas production. *Bioresource Technology, 127*, 275–280. doi:10.1016/j.biortech.2012.09.081 PMID:23131652

Brunner, P. H., & Rechberger, H. (2015). Waste to energy – key element for sustainable waste management. *Waste Management (New York, N.Y.), 37*, 3–12. doi:10.1016/j.wasman.2014.02.003 PMID:24630214

Buenrostro, O., Bocco, G., & Cram, S. (2001). Classification of sources of municipal solid wastes in developing countries. *Resources, Conservation and Recycling, 32*(1), 29–41. doi:10.1016/S0921-3449(00)00094-X

Carlsson, M., Lagerkvist, A., & Morgan-Sagastume, F. (2012). The effects of substrate pre-treatment on anaerobic digestion systems: A review. *Waste Management (New York, N.Y.), 32*(9), 1634–1650. doi:10.1016/j.wasman.2012.04.016 PMID:22633466

Castillo, E. F. M., Cristancho, D. E., & Arellano, V. A. (2006). Study of the operational conditions for anaerobic digestion of urban solid wastes. *Waste Management (New York, N.Y.), 26*(5), 546–556. doi:10.1016/j.wasman.2005.06.003 PMID:16321515

Cerda, A., Artola, A., Font, X., Barrena, R., Gea, T., & Sánchez, A. (2018). Composting of food wastes: Status and challenges. *Bioresource Technology, 248*, 57–67. doi:10.1016/j.biortech.2017.06.133 PMID:28693949

Cesaro, A., & Belgiorno, V. (2013). Sonolysis and ozonation as pre-treatment for anaerobic digestion of solid organic waste. *Ultrasonics Sonochemistry, 20*(3), 931–936. doi:10.1016/j.ultsonch.2012.10.017 PMID:23231941

Cesaro, A., & Belgiorno, V. (2014). Pre-treatment methods to improve anaerobic biodegradability of organic municipal solid waste fractions. *Chemical Engineering Journal, 240*, 24–37. doi:10.1016/j.cej.2013.11.055

Chen, Y., Cheng, J. J., & Creamer, K. S. (2008). Inhibition of anaerobic digestion process: A review. *Bioresource Technology, 99*(10), 4044–4064. doi:10.1016/j.biortech.2007.01.057 PMID:17399981

Chong, S., Sen, T. K., Kayaalp, A., & Ang, H. M. (2012). The performance enhancements of upflow anaerobic sludge blanket (UASB) reactors for domestic sludge treatment—A state-of-the-art review. *Water Research, 46*(11), 3434–3470. doi:10.1016/j.watres.2012.03.066 PMID:22560620

Chu, L., Yan, S.-T., Xing, X. H., Yu, A.-F., Sun, X.-L., & Jurcik, B. (2008). Enhanced sludge solubilisation by microbubble ozonation. *Chemosphere, 72*(2), 205–212. doi:10.1016/j.chemosphere.2008.01.054 PMID:18328535

Chynoweth, D., Turick, C., Owens, J., Jerger, D. E., & Peck, M. W. (1993). Biochemical methane potential of biomass and waste feedstocks. *Biomass and Bioenergy, 5*(1), 95–111. doi:10.1016/0961-9534(93)90010-2

CPCB. (2000). *Status of solid waste generation, collection, treatment and disposal in metero cities* (Series:CUPS/46/1999-2000). CPCB.

Dai, C., Duan, N., Dong, B., & Dai, L. (2013). High-solids anaerobic co-digestion of sewage sludge and food waste in comparison with mono-digestions: Stability and performance. *Waste Management (New York, N.Y.), 33*(2), 308–316. doi:10.1016/j.wasman.2012.10.018 PMID:23177568

Delgenès, J. P., Penaud, V., & Moletta, R. (2003). Pretreatments for the enhancement of anaerobic digestion of solid waste. In J. Mata-Alvarez (Ed.), Bio-methanization of the Organic Fraction of Municipal Solid Waste. IWA Publishing. doi:10.1002/chin.200313271

Dewil, R., Appels, L., Baeyens, J., & Degrève, J. (2007). Peroxidation enhances the biogas production in the anaerobic digestion of biosolids. *Journal of Hazardous Materials, 146*(3), 577–581. doi:10.1016/j.jhazmat.2007.04.059 PMID:17532118

Epstein, E. (1997). *The science of composting*. Technomic Publishing.

EU (European Union). (2008). Directive 2008/98/EC of the European Parliament and of the Council of 19 November 2008 on Waste and Repealing Certain Directives.

Fantola, A. (1997). *Introduction to Solid Waste Management Engineering*. Bibis Press.

Fdéz-Güelfo, A., Álvarez-Gallego, C., & Sales-Márquez, D., & Romero, G. L. I. (2011). Biological pre-treatment applied to industrial organic fraction of municipal solid waste (OFMSW): Effect on anaerobic digestion. *Chemical Engineering Journal, 172*, 321–325. doi:10.1016/j.cej.2011.06.010

Fdez-Güelfo, A., Álvarez-Gallego, C., Sales-Márquez, D., & Romero García, L. I. (2012). New parameters to determine the optimum pre-treatment for improving the biomethanization performances. *Chemical Engineering Journal, 198–199*, 81–86. doi:10.1016/j.cej.2012.05.077

Fdez-Güelfo, L. A., Álvarez-Gallego, C., Sales Márquez, D., & Romero García, L. I. (2011). The effect of different pretreatments on biomethanation kinetics of industrial Organic Fraction of Municipal Solid Waste (OFMSW). *Chemical Engineering Journal, 171*, 411–417. doi:10.1016/j.cej.2011.03.095

Ferrer, I., Ponsá, S., Vázquez, F., & Font, X. (2008). Increasing biogas production by thermal (70°C) sludge pre-treatment prior to thermophilic anaerobic digestion. *Biochemical Engineering Journal, 42*(2), 186–192. doi:10.1016/j.bej.2008.06.020

Fountoulakis, M. S., Drakopoulou, S., Terzakis, S., Georgaki, E., & Manios, T. (2008). Potential for methane production from typical Mediterranean agro-industrial by-products. *Biomass and Bioenergy, 32*(2), 155–161. doi:10.1016/j.biombioe.2007.09.002

Fu, B., Jiang, Q., Liu, H., & Liu, H. (2014). Occurrence and reactivation of viable but non-culturable E. coli in sewage sludge after mesophilic and thermophilic anaerobic digestion. *Biotechnology Letters, 36*(2), 273–279. doi:10.100710529-013-1361-9 PMID:24101245

Gajalakshmi, S., & Abbasi, S. A. (2008). Solid Waste Management by Composting: State of the Art. *Critical Reviews in Environmental Science and Technology, 38*(5), 311–400. doi:10.1080/10643380701413633

Gao, D.-W., Hu, Q., Yao, C., & Ren, N.-Q. (2014). Treatment of domestic wastewater by an integrated anaerobic fluidized-bed membrane bioreactor under moderate to low temperature conditions. *Bioresource Technology, 159*, 193–198. doi:10.1016/j.biortech.2014.02.086 PMID:24650533

Garcia, C., Hernandez, T., Costa, F., & Ayusho, M. (1992). Evaluation of maturity of municipal waste compost using simple chemical parameters. *Communications in Soil Science and Plant Analysis, 23*(13-14), 1501–1512. doi:10.1080/00103629209368683

Gunaseelan, V. N. (1997). Anaerobic digestion of biomass for methane production: A review. *Bio Bioener, 1997*(13), 83–114. doi:10.1016/S0961-9534(97)00020-2

Hankin, L. H., Poincelot, R. P., & Anagnostakis, S. L. (1976). Microorganisms from composting leaves: Ability to produce extracellular enzymes. *Microbial Ecology, 2*(4), 296–308. doi:10.1007/BF02011649 PMID:24241392

Haug, R. T., Stuckey, D. C., Gossett, J. M., & Mac Carty, P. L. (1978). Effect of thermal pre-treatment on digestibility and dewaterability of organic sludges. *Journal - Water Pollution Control Federation, 23*, 73–85.

Hellmann, B., Zelles, L., Palojarvi, A., & Bai, Q. (1997). Emission of climate-relevant trace gases and succession of microbial communities during open-windrow composting. *Applied and Environmental Microbiology*, *63*(3), 1011–1018. doi:10.1128/AEM.63.3.1011-1018.1997 PMID:16535535

Henry, R. K., Yongsheng, Z., & Jun, D. (2006). Municipal solid waste management challenges in developing countries – Kenyan case study. *Waste Management (New York, N.Y.)*, *26*(1), 92–100. doi:10.1016/j.wasman.2005.03.007 PMID:16006111

Hosseini Koupaie, E., Barrantes Leiva, M., Eskicioglu, C., & Dutil, C. (2014). Mesophilic batch anaerobic co-digestion of fruit-juice industrial waste and municipal waste sludge: Process and cost-benefit analysis. *Bioresource Technology*, *152*, 66–73. doi:10.1016/j.biortech.2013.10.072 PMID:24280084

Idris, A., Inanc, B., & Hassan, M. N. (2004). Overview of waste disposal and landfills/dumps in Asian countries. *Journal of Material Cycles and Waste Management*, *6*(2), 104–110. doi:10.100710163-004-0117-y

IPCC. (1996). *Intergovernmental panel on climate change*. http:/www.ipcc-nggip-.org.jp.

IPCC. (2007). *Intergovernmental panel on climate change*. https://www.ipcc.ch/pdf/assessment-report/ar4/wg1/ar4-wg1-spm.pdf

Izumi, K., Okishio, Y., Nagao, N., Niwa, C., Yamamoto, S., & Toda, T. (2010). Effects of particle size on anaerobic digestion of food waste. *Int. Biodeterior. Biodegr*, *64*(7), 601–608. doi:10.1016/j.ibiod.2010.06.013

Jackowiak, D., Sfrigon, J. C., Ribeiro, T., Pauss, A., & Guiot, S. (2011). Enhancing solubilisation and methane production kinetic of switchgrass by microwave pre-treatment. *Bioresource Technology*, *102*(3), 3535–3540. doi:10.1016/j.biortech.2010.11.069 PMID:21147525

Jain, S., Jain, S., Wolf, I. T., Lee, J., & Tong, Y. W. (2015). A comprehensive review on operating parameters and different pre-treatment methodologies for anaerobic digestion of municipal solid waste. *Renewable & Sustainable Energy Reviews*, *52*, 142–154. doi:10.1016/j.rser.2015.07.091

Joshi, R., & Ahmed, S. (2016). Status and challenges of municipal solid waste management in India: A review. *Cogent Environmental Science*, *2*(1), 1139434. doi:10.1080/23311843.2016.1139434

Kapetanois, E. G., Loizidu, M., & Vakanas, G. (1993). Compost production from Greek domestic refuse. *Bioresource Technology*, *43*(1), 13–16. doi:10.1016/0960-8524(93)90201-L

Keller, F. A., Hamilton, J., & Nguyen, Q. A. (2003). Microbial pre-treatment of biomass. *Appl. Biotechnol. Bioeng*, *105*, 27–41.

Khalid, A., Arshad, M., Anjum, M., Mahmood, T., & Dawson, L. (2011). The anaerobic digestion of solid organic waste. *Waste Management (New York, N.Y.)*, *31*(8), 1737–1744. doi:10.1016/j.wasman.2011.03.021 PMID:21530224

Kiely, G., Tayfur, G., Dolan, C., & Tanji, K. (1997). Physical and mathematical modelling of anaerobic digestion of organic wastes. *Water Research*, *31*(3), 534–540. doi:10.1016/S0043-1354(96)00175-3

Kim, D. H., & Oh, S. E. (2011). Continuous high-solids anaerobic co-digestion of organic solid wastes under mesophilic conditions. *Waste Management (New York, N.Y.)*, *31*(9-10), 1943–1948. doi:10.1016/j.wasman.2011.05.007 PMID:21684733

Kim, H. W., Nam, J. Y., & Shin, H. S. (2011). A comparison study on the high-rate co-digestion of sewage sludge and food waste using a temperature-phased anaerobic sequencing batch reactor system. *BioresTechn*, *102*(15), 7272–7279. doi:10.1016/j.biortech.2011.04.088 PMID:21600764

Kim, J., Kim, K., Ye, H., Lee, E., Shin, C., McCarty, P. L., & Bae, J. (2010). Anaerobic fluidized bed membrane bioreactor for wastewater treatment. *Environmental Science & Technology*, *45*(2), 576–581. doi:10.1021/es1027103 PMID:21158433

Kim, J., Park, C., Kim, T. H., Lee, M., Kim, S., Kim, S. W., & Lee, J. (2003). Effect ofvarious pretreatmentsforenhancedanaerobicdigestionwithwasteactivatedsludge. *Journal of Bioscience and Bioengineering*, *95*(3), 271–275. doi:10.1016/S1389-1723(03)80028-2 PMID:16233404

Kumar, M., Ou, Y. L., & Lin, J. G. (2010). Co-composting of green waste and food waste at low C/N ratio. *Waste Management (New York, N.Y.)*, *30*(4), 602–609. doi:10.1016/j.wasman.2009.11.023 PMID:20034778

Kusch, S., Schäfer, W., & Kranert, M. (2011). Dry digestion of organic residues. *Integr Waste Manag*, *I*, 115–134.

Lee, D. H., Behera, S. K., Kim, J., & Park, H. S. (2009). Methane production potential of leachate generated from Korean food waste recycling facilities: A lab scale study. *Waste Management (New York, N.Y.)*, *29*(2), 876–882. doi:10.1016/j.wasman.2008.06.033 PMID:18796348

Lettinga, G. (1995). Anaerobic digestion and wastewater treatment systems. *Antonie van Leeuwenhoek*, *67*(1), 3–28. doi:10.1007/BF00872193 PMID:7741528

Lin, L., Yang, L., Xu, F., Michel, F. C. Jr, & Li, Y. (2014). Comparison of solid-state anaerobic digestion and composting of yard trimmings with effluent from liquid anaerobic digestion. *Bioresource Technology*, *169*, 439–446. doi:10.1016/j.biortech.2014.07.007 PMID:25079209

Lina, L., Xu, F., Ge, X., & Li, Y. (2018). Improving the sustainability of organic waste management practices in the food-energy-water nexus: A comparative review of anaerobic digestion and composting. *Renewable & Sustainable Energy Reviews*, *89*, 151–167. doi:10.1016/j.rser.2018.03.025

Lopez Torres, M., & Espinosa Lloréns, M. (2008). Effect of alkaline pretreatment on anaerobic digestion of solid wastes. *Waste Management (New York, N.Y.)*, *28*(11), 2229–2234. doi:10.1016/j.wasman.2007.10.006 PMID:18068345

Mao, C., Feng, Y., Wang, X., & Ren, G. (2015). Review on research achievements of biogas from anaerobic digestion. *Renewable & Sustainable Energy Reviews*, *45*, 540–555. doi:10.1016/j.rser.2015.02.032

Masebinu, S. O., Akinlabi, E. T., Muzenda, E., & Aboyade, A. O. (2019). A review of biochar properties and their roles in mitigating challenges with anaerobic digestion. *Renewable & Sustainable Energy Reviews*, *103*, 291–307. doi:10.1016/j.rser.2018.12.048

Masse, L., Masse, D. I., & Kennedy, K. J. (2003). Effect of enzymatic pre-treatment on fat degradation during anaerobic digestion of slaughterhouse wastewater. *Process Biochemistry (Barking, London, England)*, *38*(9), 1365–1372. doi:10.1016/S0032-9592(03)00020-7

Mata-Alvarez, J., Dosta, J., Romero-Güiza, M. S., Fonoll, X., Peces, M., & Astals, S. (2014). Romero-G€uiza, M.S., Fonoll, X., Peces, M., Astals, S., (2014). A critical review on anaerobic co-digestion achievements between 2010 and 2013. *Renewable & Sustainable Energy Reviews*, *36*, 412–427. doi:10.1016/j.rser.2014.04.039

Mata-Alvarez, J., Mace, S., & Llabres, P. (2000). Anaerobic digestion of organic solid wastes. An overview of research achievements and perspectives. *Bioresource Technology*, *74*(1), 3–16. doi:10.1016/S0960-8524(00)00023-7

Monlau, F., Sambusiti, C., Antoniou, N., Barakat, A., & Zabaniotou, A. (2015). A new concept for enhancing energy recovery from agricultural residues by coupling anaerobic digestion and pyrolysis process. *Applied Energy*, *148*, 32–38. doi:10.1016/j.apenergy.2015.03.024

Mshandete, A., Bjornsson, L., Kivaisi, A. K., Rubindamayugi, M. S. T., & Mattiasson, B. (2006). Effect of particle size on biogas yield from sisal fiber waste. *Renewable Energy*, *31*(14), 2385–2392. doi:10.1016/j.renene.2005.10.015

Nakasaki, K., Yaguchi, H., Sasaki, M., & Kubota, H. (1993). Effects of pH control on composting of garbage. *Waste Management & Research*, *11*(2), 117–125. doi:10.1177/0734242X9301100204

Nayak, A., & Bhushan, B. (2019). An overview of the recent trends on the waste valorization techniques for food wastes. *Journal of Environmental Management*, *233*, 352–370. doi:10.1016/j.jenvman.2018.12.041 PMID:30590265

Neves, L., Oliveira, R., & Alves, M. M. (2006). Anaerobic co-digestion of coffee waste and sewage sludge. *Waste Management (New York, N.Y.)*, *26*(2), 176–181. doi:10.1016/j.wasman.2004.12.022 PMID:16310117

Neves, L., Ribeiro, R., Oliveira, R., & Alves, M. M. (2006). Enhancement of methane production from barley waste. *Biomass and Bioenergy*, *30*(6), 599–603. doi:10.1016/j.biombioe.2005.12.003

Ojoawo, S., Agbede, O., & Sangodoyin, A. (2011). On the Physical Composition of Solid Wastes in Selected Dumpsites of Ogbomosoland, South-Western Nigeria. *Journal of Water Resource and Protection*, *3*(09), 661–666. doi:10.4236/jwarp.2011.39076

Omil, F., Garrido, J. M., Arrojo, B., & Méndez, R. (2003). Anaerobic filter reactor performance for the treatment of complex dairy wastewater at industrial scale. *Water Research*, *37*(17), 4099–4108. doi:10.1016/S0043-1354(03)00346-4 PMID:12946891

Onwosi, C., Igbokwe, V., Odimba, J., Eke, I., Nwankwoala, M., Iroh, I., & Ezeogu, L. (2017). Composting technology in waste stabilization: On the methods, challenges and future prospects. *Journal of Environmental Management*, *190*, 140–157. doi:10.1016/j.jenvman.2016.12.051 PMID:28040590

Ostrem, K. (2004). *Greening Waste: Anaerobic Digestion For Treating The Organic Fraction Of Municipal Solid Wastes*. Earth Engineering Center Columbia University.

Pires, A., Martinho, G., & Chang, N. B. (2011). Solid waste management in European countries: A review of systems analysis techniques. *Journal of Environmental Management*, *92*(4), 1033–1050. doi:10.1016/j.jenvman.2010.11.024 PMID:21194829

Rachael, E., & Marshall, K. F. (2013). Systems approaches to integrated solid waste management in developing countries. *Waste Management (New York, N.Y.)*, *33*(4), 988–1003. doi:10.1016/j.wasman.2012.12.023 PMID:23360772

Raposo, F., De La Rubia, M. A., Fernández-Cegrí, V., & Borja, R. (2012). Anaerobic digestion of solid organic substrates in batch mode: An overview relating to methane yields and experimental procedures. *Renewable & Sustainable Energy Reviews*, *16*(1), 861–877. doi:10.1016/j.rser.2011.09.008

Salama, E.-S., Saha, S., Kurade, M. B., Dev, S., Chang, S. W., & Jeon, B.-H. (2019). Recent trends in anaerobic co-digestion: Fat, oil, and grease (FOG) for enhanced biomethanation. *Progress in Energy and Combustion Science*, *70*, 22–42. doi:10.1016/j.pecs.2018.08.002

Şentürk, E., İnce, M., & Engin, G. O. (2012). The effect of transient loading on the performance of a mesophilic anaerobic contact reactor at constant feed strength. *Journal of Biotechnology*, *164*(2), 232–237. doi:10.1016/j.jbiotec.2012.08.016 PMID:22960727

Shahriari, H., Warith, M., Hamoda, M., & Kennedy, K. J. (2012). Anaerobic digestion of organic fraction of municipal solid waste combining two pre-treatment modalities, high temperature microwave and hydrogen peroxide. *Waste Management (New York, N.Y.)*, *32*(1), 41–52. doi:10.1016/j.wasman.2011.08.012 PMID:21945550

Sharholy, M., Ahmad, K., Mahmood, G., & Trivedi, R. C. (2008). Municipal solid waste management in Indian cities – A review. *Waste Management (New York, N.Y.)*, *28*(2), 459–467. doi:10.1016/j.wasman.2007.02.008 PMID:17433664

Sharma, V., Testa, C., Lastella, G., Cornacchia, G., & Comparato, M. (2000). Inclined-plug-flow type reactor for anaerobic digestion of semi-solid waste. *Applied Energy*, *65*(1-4), 173–185. doi:10.1016/S0306-2619(99)00084-7

Singh, M., & Srivastava, R. (2011). Sequencing batch reactor technology for biological wastewater treatment: A review. *Asia-Pacific Journal of Chemical Engineering*, *6*(1), 3–13. doi:10.1002/apj.490

Singh, S., & Prerna, P. (2009). Review of recent advances in anaerobic packed-bed biogas reactors. *Renewable & Sustainable Energy Reviews*, *13*(6-7), 1569–1575. doi:10.1016/j.rser.2008.08.006

Sridhar, M. K. C. (2000). *Nature of Solid Wastes and Their Management*. Paper Presented at *a one Day Seminar Organized by NINNAFEH*, Ibadan, Nigeria.

Steiner, C., Das, K. C., Melea, N., & Lakly, D. (2010). Reducing nitrogen loss during poultry litter composting using biochar. *Journal of Environmental Quality*, *39*(4), 1236–1242. doi:10.2134/jeq2009.0337 PMID:20830911

Switzenbaum, M. A. (1983). Comparison of the anaerobic filter and the anaerobic expanded/fluidized bed processes. *Water Science and Technology*, *15*(8-9), 345–358. doi:10.2166/wst.1983.0177 ten Brummeler, E., & Koster, I. W. (1990). Enhancement of Dry Anaerobic Batch Digestion of the Organic Fraction of Municipal Solid Waste by an Aerobic Pre-treatment Step. *Biological Wastes, 31*(3), 199–210. doi:10.1016/0269-7483(90)90159-P

Themelis, N. J., & Ulloa, P. A. (2007). Methane generation in landfills. *Renewable Energy, 32*(7), 1243–1257. doi:10.1016/j.renene.2006.04.020

Tiehm, A., Nickel, K., & Neis, U. (1997). The use of ultrasound to accelerate the anaerobic digestion of sewage sludge. *Water Science and Technology, 36*(11), 121–128. doi:10.2166/wst.1997.0402

Tiehm, A., Nickel, K., Zellhorn, M., & Neis, U. (2001). Ultrasonic waste activated sludge disintegration for improving anaerobic stabilization. *Water Research, 35*(8), 2003–2009. doi:10.1016/S0043-1354(00)00468-1 PMID:11337847

Tiquia, S. M., & Tam, N. F. Y. (2000). Microbiological and chemical parameters for compost maturity evaluation of spent pig litter disposed from the pig-on litter system. In *Proceedings of the International Composting Symposium (ICS 1999)*. CBA Press.

Tuomela, M., Vikman, M., Hatakka, A., & Itävaara, M. (2000). Biodegradation of lignin in a compost environment: A review. *Bioresource Technology, 72*(2), 169–183. doi:10.1016/S0960-8524(99)00104-2

Uçkun Kiran, E., Stamatelatou, K., Antonopoulou, G., & Lyberatos, G. (2016). Production of biogas via anaerobic digestion. Handbook of Biofuels Production, 259–301.

USEPA. (2012). *Industrial food processing waste analyses*. USEPA.

Velmurugan, B., & Arathy, E. C. (2010). Anaerobic co-digestion of fruit and vegetables wastes and primary sewage sludge. *Environ. Sc. And Engg, 52*, 19–22. PMID:21114101

Wang, J. Y., Stabnikova, O., Tay, S. T., Ivanov, V., & Tay, J. H. (2004). Biotechnology of intensive aerobic conversion of sewage sludge and food waste into fertilizer. *Water Science and Technology, 49*(10), 147–154. doi:10.2166/wst.2004.0631 PMID:15259949

Ward, A. J., Hobbs, P. J., Holliman, P. J., & Jones, D. L. (2008). Optimization of the anaerobic digestion of agricultural resources. *Bioresource Technology, 99*(17), 7928–7940. doi:10.1016/j.biortech.2008.02.044 PMID:18406612

William, P. B., & David, C. S. (1999). The use of the anaerobic baffled reactor (ABR) for wastewater treatment: A review. *Water Research, 33*(7), 1559–1578. doi:10.1016/S0043-1354(98)00371-6

Wilson, D. C. (2007). Development drivers for waste management. *Waste Management & Research, 25*(3), 198–207. doi:10.1177/0734242X07079149 PMID:17612318

Wilson, D. C., Velis, C. A., & Rodic, L. (2013). Integrated sustainable waste management in developing countries. *Proceedings of the Institution of Civil Engineers - Waste and Resource Management, 166*(2), 52–68. 10.1680/warm.12.00005

Xu, F., Li, Y., Ge, X., Yang, L., & Li, Y. (2017). Anaerobic digestion of food waste – challenges and opportunities. *Bioresour Technol.*

Yuan, X., Wen, B., Ma, X., Zhu, W., Wang, X., Chen, S., & Cui, Z. (2014). Enhancing the anaerobic digestion of lignocellulose of municipal solid waste using a microbial pre-treatment method. *Bioresource Technology*, *154*, 1–9. doi:10.1016/j.biortech.2013.11.090 PMID:24365784

Zhang, B., Li, S., Michel, F., Li, G., Luo, Y., & Zhang, D. (2016). Effects of mix ratio, moisture content and aeration rate on sulfurodor emissions during pig manure composting. *Waste Management (New York, N.Y.)*, *56*, 498–505. doi:10.1016/j.wasman.2016.06.026 PMID:27363616

Zhang, C., Xiao, G., Peng, L., Su, H., & Tan, T. (2013). The anaerobic co-digestion of food waste and cattle manure. *Bioresource Technology*, *129*, 170–176. doi:10.1016/j.biortech.2012.10.138 PMID:23246757

Zhang, J., Lü, F., Shao, L., & He, P. (2014). The use of biochar-amended composting to improve the humification and degradation of sewage sludge. *Bioresource Technology*, *168*, 252–258. doi:10.1016/j.biortech.2014.02.080 PMID:24656550

Zhang, L., Lee, Y. W., & Jahng, D. (2011). Anaerobic co-digestion of food waste and piggery wastewater: Focusing on the role of trace elements. *BioresTech*, *102*(8), 5048–5059. doi:10.1016/j.biortech.2011.01.082 PMID:21349706

Zhang, P., Chen, Y., & Zhou, Q. (2009). Waste activated sludge hydrolysis and short-chain fatty acids accumulation under mesophilic and thermophilic conditions: Effect of pH. *Water Research*, *43*(15), 3735–3742. doi:10.1016/j.watres.2009.05.036 PMID:19555988

Zhu, B., Gikas, P., Zhang, R., Lord, J., Jenkins, B., & Li, X. (2009). Characteristics and biogas production potential of municipal solid wastes pretreated with a rotary drum reactor. *Bioresource Technology*, *100*(3), 1122–1129. doi:10.1016/j.biortech.2008.08.024 PMID:18849162

Zorpas, A. A., Arapoglou, D., & Panagiotis, K. (2003). Waste paper and clinoptilolite as a bulking material with dewatered anaerobically stabilized primary sewage sludge (DASPSS) for compost production. *Waste Management (New York, N.Y.)*, *23*(1), 27–35. doi:10.1016/S0956-053X(02)00042-9 PMID:12623099

Zucconi, F., Pera, M.A., & Forte, M., & Bertoldi, M. d. (1981). Evaluating toxicity of immature compost. *BioCycle*, *22*, 54–57.

Chapter 13
A Review on Isolation of Acetobacter and Production of Vinegar From Pineapple and Papaya Waste (Peels) Using Acetobacter aceti

B. Praveena
MES College of Arts, Commerce, and Science, India

R. Sonashree
MES College of Arts, Commerce, and Science, India

Rashmi R. Halbavi
MES College of Arts, Commerce, and Science, India

J. Bhavana
MES College of Arts, Commerce, and Science, India

Shamsiya Rizwana
MES College of Arts, Commerce, and Science, India

Vyshnavi V. Rao
MES College of Arts, Commerce, and Science, India

ABSTRACT

India is the second-largest producer of contemporary fruits and vegetables. This leads to generation of a large amount of waste, both solid and liquid. This waste is incautiously or carelessly being thrown without any treatment promoting environmental deterioration. The chapter deals with how these fruit wastes, especially pineapple and papaya peels, which are usually discarded during the processing or

DOI: 10.4018/978-1-7998-4921-6.ch013

consumption of the fruit, are useful in preparation of vinegar, which is termed as a useful product from waste. It also discussed the general characteristics of Acetic acid bacteria, importance of the bacteria, and how it can be isolated by natural sources, which is cost effective. So, the waste peels generated after consuming the fruits and vegetables need not be discarded as waste; instead, it can be used in so many ways to produce useful products.

BACKGROUND

Vinegar was known as a seasoning or food preserving agent. Vinegar is defined as "a liquid fit for human consumption, produced from suitable raw material of agricultural origin, containing starch and sugars by the process of double fermentation, alcoholic and acetous, and contains specific amount of acetic acid. Vinegar is an important preservative and condiment and it is being produced since centuries. It is produced through the action of acetic acid bacteria on dilute solution of ethyl alcohol derived from yeast fermentation. Vinegar plays an important role in salad dressings, pickling of fruits and vegetables, ketchup, hot sauce and other sauces. To overcome the demand industrial fermentation systems should be capable of producing a large amount of vinegar. Many techniques have been developed to improve industrial production of vinegar. Today, the most common technology for the vinegar industry is based on the submerged culture with diverse technical modifications which try to improve the general fermentation conditions (aeration, stirring, heating, etc.,) (Tan, 2005). Pineapple (*Ananas comosus*) by-products are not exceptions. Indeed, several efforts have been made in order to utilize pineapple wastes, which have already been used as the substrate for the production of bromelain and organic acids (Dacera *et al*., 2009). Pineapple peels as agricultural wastes represent around 35% of the whole fruit mass. If these wastes are discharged to the environment untreated, they could cause a serious problem. So, it is necessary to have recycled waste raw material into useful product of higher value-added products or even as a raw material for other industries or feed after biological treatment. Papaya (*Carica papaya*) is an oblong tropical fruit, often seen in orange red, yellow green and yellow orange hues with a rich orange pulp. Papaya has high content of vitamin A, B and C, proteolytic enzymes like papain and chymopapain which have antiviral, antifungal and antibacterial properties.

Most nations, whether economically advanced or at different stages of development are faced with the issue of disposal and treatment of wastes (Itelima *et al*., 2013). Agro-industrial wastes are generated in large amounts every year and their reuse in processes is of particular interest due to their availability, low cost, and characteristics that allow at obtaining different value-added compounds (De Freitas Borghi *et al*., 2009). Tropical and subtropical fruits processing have considerably higher ratios of by-products than the temperate fruits (Schieber *et al*., 2001) **[Figure 1]**.

Figure 1.

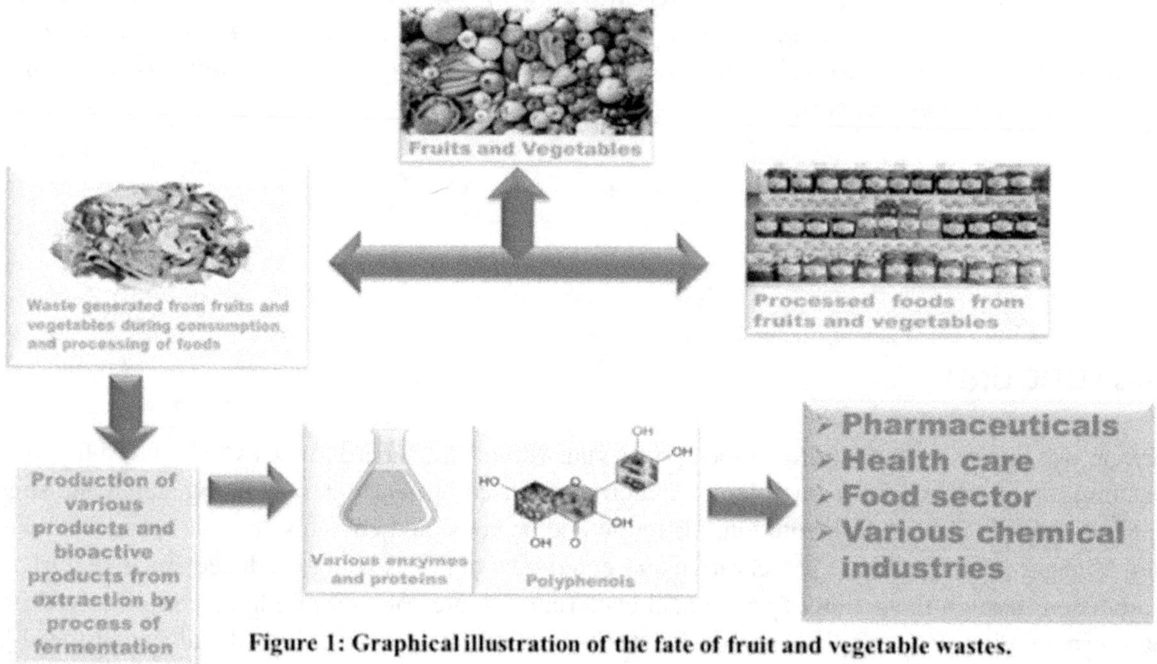

Figure 1: Graphical illustration of the fate of fruit and vegetable wastes.

GENERAL PROPERTIES OF VINEGAR

Vinegar is a dilute solution of acetic acid; it has same chemical formula as acetic acid. The table below showed general properties of vinegar **[Table 1].**

Table 1. General properties of Vinegar

Parameter	Value
Density and phase	1.01gcm-3, liquid; 1.266gcm-3, solid
Solubility in water	fully miscible
In toluene, hexane	fully miscible
In ethanol, acetone	fully miscible
Freezing point	-20C (28 ^{o} F)
Acidity (pKa)	4.76 at 25°C
pH	2.4
Viscosity	0.92 cp at 27°C
Dipole moment	Dipole moment
Refractive index	1.370

Types of Vinegars

Different type of vinegar is classified based on the raw material used for its production. Malt vinegar, Wine vinegar, Apple cider vinegar, Balsamic vinegar, Fruit vinegar and many other types of vinegar exist in today's global market (Hailu *et al.,*2012). These are some of the most common types of vinegars discussed below

- **Wine Vinegar:** This is a type of vinegar which is made from red or white wine and is the most commonly used vinegar in Mediterranean countries and Central Europe. As with wine, there is a considerable range in quality. Better quality wine vinegars are matured in wood for up to two years and exhibit a complex, mellow flavor. Wine vinegar will have a lower acidity than that of white vinegars or cider vinegars. Vinegar produced by individual varieties of wine like Champagne, Sherry, or pinot grigio are more expensive.
- **Beer Vinegar**: Vinegar produced from beer is called as beer vinegar mostly produced in the United Kingdom, Germany, Austria, and the Netherlands. Its flavor mostly depends on the particular type of beer from which it is made, it is often described as having a malty taste. In Beer vinegar and Wine vinegar production only one type of fermentation takes place because already ethanol is present in them and then conversion of ethanol in to acetic acid. But other types of vinegars, discussed later, are grouped under a production process where there are two types of fermentations; one for ethanol formation and another conversion of the ethanol in to acetic acid.
- **Fruit Vinegars**: Fruit vinegars are made from fruit wines, fruit peels, fruits pulp usually without any additional flavoring and the flavors of the original fruits remain in the final product. Common flavors of fruit vinegar includes apple, pear, black current, strawberry, raspberry, fig, pineapple and tomato. It is considered to have medicinal values for treating stomach, spleen and diabetic ailments.
- **Balsamic Vinegar:** Balsamic vinegar is originated in Italy and is produced from grape must. Grape must be nothing but a freshly prepared grape juice in which it also contains skin, seeds, stem. This vinegar is an aromatic, aged type of vinegar traditionally produced in the Modena and Reggio Emilia regions of Italy. It is very dark brown in color and its flavor is rich, sweet, and complex, with the finest grades being the product of years of aging in a successive number of casks made of various types of woods. True balsamic vinegar is mostly aged for 12 to 25 years. Balsamic vinegars that have been aged for up to 100 years are available, though they are usually very expensive. The commercial balsamic available in supermarkets is typically made up of concentrated grape juice which is mixed with strong vinegar and is laced with caramel and sugar. Regardless of how it is produced, balsamic vinegar must be made from a grape product. Balsamic vinegar has a high acidity level but the tart flavor is usually hidden by the sweetness of the other ingredients, making it very mellow.
- **Malt Vinegar:** Malt vinegar is made by the process of malting barley, causing the starch in the grain to turn to maltose. Then ale is brewed from the maltose and allowed to turn into vinegar, which is then aged. It is typically light brown in color.
- **Rice Vinegar:** Rice vinegar is usually produced by using rice as raw material and is most popular in the cuisines of East and Southeast Asia. It is mostly available in "white" (light yellow), red, and black varieties. The people in Japan mostly prefer light rice vinegar for the preparation of sushi rice and salad dressings. Red rice vinegar traditionally is colored with red yeast rice. Black rice

vinegar (made with black glutinous rice) is most popular in China, and it is also widely used in other East Asian countries. White rice vinegar has a mild acidity and a somewhat "flat", uncomplex flavor. Some varieties of rice vinegar are sweetened or otherwise seasoned with spices or other added flavorings.

Steps Involved in Vinegar Production: Budak *et al.*, 2014, studied functional properties of vinegar and flow chart represents the steps and types of fermentation involved in preparation of vinegar **[Figure 2]**.

Figure 2.

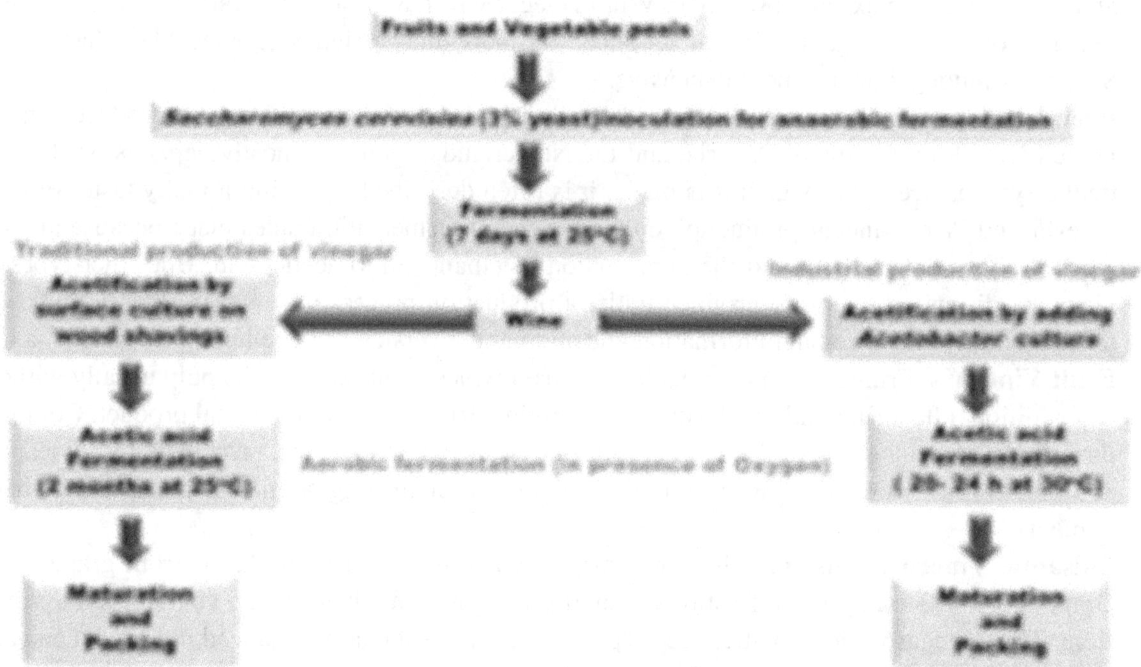

There are 2 types of fermentation in vinegar preparation process which are mentioned below:

1. **Anaerobic fermentation:** It is a method in which cells use to extract energy from carbohydrates when oxygen or other electron acceptors are not available in the surrounding environment.

Example: Ethanol fermentation converts two pyruvate molecules, the products of glycolysis, to two molecules of ethanol and two molecules of carbon dioxide.

2. **Aerobic fermentation:** is a metabolic process by which cells metabolize sugars via fermentation in the presence of oxygen and occurs through the repression of normal respiratory metabolism.

In the first step anaerobic fermentation will takes place in which glucose or sugars present in the raw materials are converted to ethanol in the presence of yeast. After the formation of ethanol second step aerobic fermentation will takes place in the presence of Acetobacter species. The process continues with conversion of ethanol which is formed in anaerobic fermentation to acetic acid. Formed acetic acid is to be diluted and is used as vinegar.

Glucose

Box 1.

$$\text{Glucose} \xrightarrow[\text{Yeast}]{\text{Anaerobic}} \underset{\text{Ethanol}}{2C_2H_5O} \xrightarrow[\text{Acetic acid bacteria}]{\text{Aerobic}} \underset{\text{Acetic Acid}}{2CH_3COOH} + 2H_2O$$

DIFFERENT METHODS OF VINEGAR PREPARATION

During the making of vinegar, there are three known methods, which are The Orleans Method, The Tricking, Quick Process, and The Submerged Fermentation Method. Although those are the main methods known, vinegar could also be created differently depending on the time period taking place and the location.

Orleans Method

This is one of the slower and older methods of creating a high-quality vinegar which is used by our ancestors. This method involves the fermentation of vinegar inside a case that has holes drilled into it to allow oxygen let inside. These holes are also covered with screen filters that prevent insects and bugs from entering the cask which will disturb the entire production process. Oxygen is necessary for the production of vinegar as it should undergo aerobic fermentation, due to the fact that the bacteria that turns the solution into vinegar requires oxygen. To create the vinegar, alcoholic liquid is poured into the cask and then about 20% of fresh vinegar is poured into the cask to begin the fermentation process. Once the acetobacter bacteria oxidizes the acetic acid the vinegar is now finished. There is a plug on the cask to collect the finished vinegar, and also a tube to add more substances without destroying the film of vinegar bacteria.

Manufacturing Steps 1) Wooden barrels are laid on their sides. Bungholes are drilled into the top side and plugged with stoppers. Holes are also drilled into the ends of the barrels. 2) The alcohol is poured into the barrel via long-necked funnels inserted into the bungholes. Mother of vinegar is added at this point. The barrel is filled to a level just below the holes on the ends. Netting or screens are placed over the holes to prevent insects from getting into the barrels. 3) The filled barrels are allowed to sit for several months. The room temperature is kept at approximately 85°F (29°C). Samples are taken periodically by inserting a spigot into the side holes and drawing liquid off. When the alcohol has converted to vinegar, it is drawn off through the spigot. About 15% of the liquid is left in the barrel to blend with the next batch.

Trickling, Quick Process

As Orleans method is very slow, newer methods are tried by many have tried to increase production rate in less time. In this method of producing vinegar it involves spraying the alcoholic substrate in the top layer of the fermentation chamber where it is filled with materials that carry a slime made of acetic bacteria so that the bacteria could react with the substrate and create vinegar. Due to the heat that is made during the fermentation of vinegar, air is forced through the chamber to keep it cool. The vinegar is recirculated for about two-three times until the desired concentration of vinegar is achieved. Once the vinegar is achieved it is then collected through the collection chamber.

Submerged Fermentation Method

Submerged fermentation method is advanced method which is newer, faster of creating vinegar. It is mainly used in industrial businesses where the required equipment is available. In this method A high speed motor breaks down air that is brought down from a stainless-steel tank into tiny bubbles and is forced into the solution of alcoholic liquid and the bacteria for even faster oxidization. The final steps are filtration and pasteurization of the vinegar that helps to stop any more bacteria growth and enzyme actions. This is a fastest process usually takes one to two days to produce vinegar, which is why this method is mainly used by big industries.

Vinegar Preparation Using Fruit Peels

It is studied that Vinegar is very important food preservative in food industry. Vinegar can be obtained from various raw materials such as various fruits and fruit peels (Bhavana *et al.*,2019). Various agricul-

Figure 3.

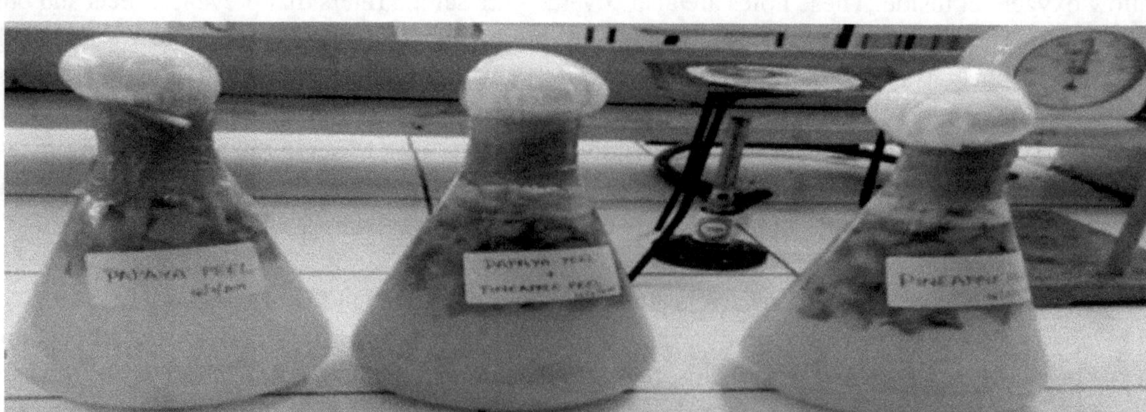

tural waste materials can also be used as raw materials. It was observed during various investigations that enzymatic treatments of pineapple wastes had a significant effect on the saccharification process **[Figure 3]**. Pineapple peel vinegar had comparatively high total phenol content and antioxidant activity. Apple vinegar was also obtained through classical methods by few investigators.

Minh (2014) investigated Fermentation of star fruit juice to produce vinegar using the yeast, *Saccharomyces cerevisiae* and bacteria, *Acetobacter aceti*. He used Star fruits collected in rural area of Tra Vinh province, Vietnam. He observed that at ethanol concentration of 10%, the acetic acid formation was maximum. At pH value of 3.5 fermentation was optimum. According to his studies, the most common technology used in the vinegar industry is the feed-back method to increase the speed of the acetic acid biological reaction.

Vinegar From Pineapple Peels: Pineapples are cultivated and produced in many tropical regions. In continent like Africa where rainfall is adequate and most are consumed locally, it is an important luxury fruit of tropical regions. Study of pineapple peeling processing into vinegar by biotechnology done by Sayram *et al.*,2009. The study was done to find the possibility of using peelings of pineapple fruit for vinegar production trough biotechnology to reduce the post-harvest losses of pineapple that is utilization of pineapple peels which are usually discarded during the processing or consumption of the fruit.

The flavors in pineapple are due to the presence of small quantity of esters and essential oils distributed throughout pulp. The presence of organic acids also contributes to the flavors. nature and quantity of sugar present in fruits depends largely upon the ripeness and the degree of acidity. The table below showed various composition of pineapple pulp and peel (Hemalatha *et al.*,2013) **[Table 2]**.

Table 2. Physical and Chemical constituents of pineapple pulp and peel

Sl.No.	Parameters	Pineapple pulp	Pineapple waste
1.	Moisture (%)	87.3	91.35
2.	Ash content (mg/100g)	1.8	0.04
3.	Titratable acidity (%)	2.03	1.86
4.	Ascorbic Acid (mg/100g)	21.5	26.5
5.	Reducing Sugars (%)	10.5	8.2
6.	Non-Reducing Sugars (%)	7.4	8.8
7.	Total Soluble Solids (%)	13.3	10.2
8.	Total Sugars (%)	8.66	9.75
9.	Crude Fibre (g/100g-fw)	0.41	0.60
10.	Protein (mg/100g)	7.2	10

Production of vinegar from pineapple peels using Acetobacter species isolated from soil sample and its antimicrobial activity (Sarkar *et al.*, 2012). Firstly, wine is prepared by *Saccharomyces cerevisiae*, which is confirmed by performing CO_2 production and iodoform test. The wine produced was inoculated with *Acetobacter* species. It is incubated for 11days by aerobic fermentation at 37°C. by this it was calculated that 4.60% of vinegar was produced. The antibacterial activity of vinegar was tested against *Escherichia coli, staphylococcus aureus, salmonella Para typhi* and *pseudomonas aeruginosa*. The result of antibacterial activity of vinegar against *Escherichia coli* -16mm, Staphylococcus *aureus* -20mm,

Salmonella Para typhi -19mm, and *pseudomonas aeruginosa* -19mm. Subsequently up on this research they concluded that materials like pineapple peel considered generally as waste can be bio converted in to important value- added materials thus aiding environmental safety **[Figure 4].**

Vinegar production from pineapple wastes was carried out by Roda *et al.*,2014. Their research was aimed at completely processing pineapple wastes in to vinegar which may be then used as dressing, food

Figure 4.

preservative, and disinfectant. They cut and chopped the fruit waste then divided it in to samples of peel and core. Then distilled water was added. For changing yield, they arranged physical treatments in order to disaggregate the fibrous structures followed by enzyme treatments to breakdown cellulose polymers and to hydrolyse sucrose also invertase was added. They obtained more than 100g of reducing sugars per kg fresh peels and about 330g of reducing sugars per kg of fresh core. Praveena and Estherlydia carried out studies on phytochemical screening and antioxidant capacities of vinegar made from peel and fruit of pineapple. They produced vinegar from the peel and fruit of pineapple, sugar and starter culture. They assessed the phytochemical properties and antioxidant activity of pineapple peel versus fruit vinegar. They obtained yellow coloured vinegar from the pineapple fruit mixture. They observed that antioxidant content of peel vinegar (2077mg acetate equivalence/100ml) was higher compared to that of fruit vinegar. They concluded that pineapple peel vinegar can be produced in large scale and marketed for its therapeutic effects. Producing vinegar from fermented pineapple peel Yusuf et al,.2012. the study revealed that pineapple peel produced the desired and optimum yield of vinegar. The parameters density, refractive index, viscosity, % acetic acid and acid value were evaluated and recorded as; 2.80, 1.80g/ml, 1.390, 0.94cp, 4.77 and 0.0477 respectively.

Vinegar from Papaya Peels: Bioconversion of papaya peel waste in to vinegar using *acetobacter aceti* carried out by Vikas.O.V. and Mridual Umesh (2014). The study revealed that dilute acid hydrolysis resulted in the conversion of complex sugars in papaya hydrolysate in to simpler fermentable sugars. During the anaerobic fermentation, amylase from yeast initially break the starchy residues further into monomeric residues. These residues are utilized for ethanol production. The total ethanol content was found to be 8.11%. Finally, the acetic acid fermentation is carried out via the conversion of ethanol to hydrate acetaldehyde and dehydrogenation of acetaldehyde to acetic acid by aldehyde dehydrogenase produced by *acetobacter aceti*. They concluded that production of vinegar from papaya peel suggested an adoptable methodology for turning a potential waste in to a commercially important organic acid. The Table below showed various composition of papaya peel **[Table 3]**.

Table 3. Proximate chemical composition of flours from peels of solo papaya 8

Parameter	Value (g/100 g)
Moisture (%)	9.82 ± 0.09
Total sugars	1.78 ± 0.63
Reducing sugars	1.03± 0.42
Total fiber	32.51 ± 0.03
Ash	5.98 ± 0.03
Proteins	11.67 ± 0.04
Fat	2.51 ± 0.13
Carbohydrates	47.33 ± 0.08

(Carica papaya) (Martial-Didier *et al.*,2017).

They concluded that a strain of yeast named *saccharomyces cerevisiae* and an acetic bacterium named acetobacter species and acetic bacteria have been isolated in pineapple juice. The production of pineapple vinegar to at least 4.5, brix 5.3%, pH 2.8 requires 23 to 25 days for alcohol and acetic fermentation. Acidity stressful and fungicide on yeast during the production of vinegar and is no longer necessary to develop in the process, an elimination of yeast before beginning the acetic fermentation. The sugar tolerance is an important technological factor for the growth of acetic bacteria during pineapple juice. Acetic fermentation unlike that of ethanol which has no limit to the growth of acetic bacteria. From all these results they concluded that post- harvest losses of pineapple fruits may be used to vinegar and have a commercial value.

Isolation of *Acetobacter aceti* from Sugar Cane

Acetic acid bacteria are a large group of obligate aerobic gram-negative bacteria with the ability to oxidize ethanol to acetic acid (Matsushita K *et al.*,1994) **[Figure 5 and 6].** They are widely distributed in natural habitats and classified into the family *Acetobacteraceae*. Members of this family are useful in industrial production of vinegar (Adachi O *et al.*,2003).

Figure 5.

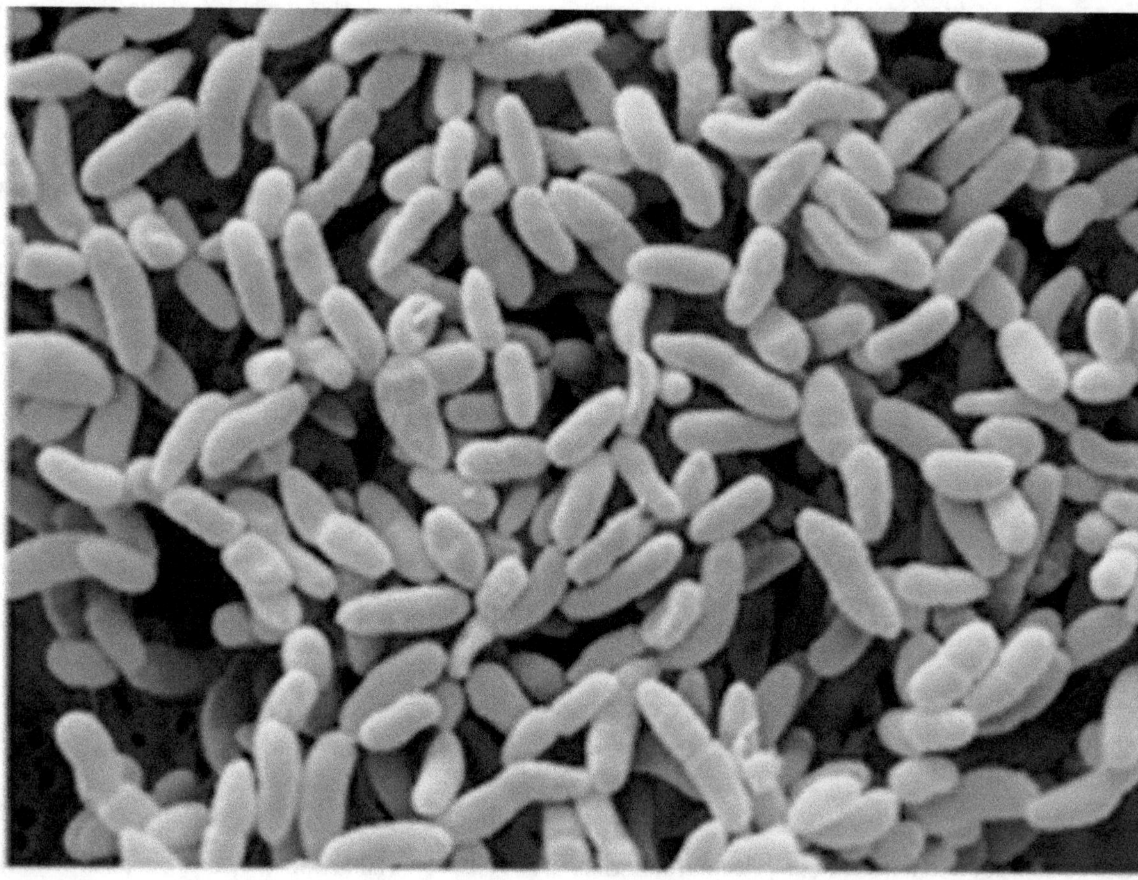

Acetic acid bacteria (AAB) can use substrates such as glucose, ethanol, lactate or glycerol as energy sources. However, most of these compounds are not completely oxidized into CO_2, and water and several metabolites, especially acetic acid, are accumulated in the growth medium. AAB are commonly found in nature because of their high resistance to acidity and the variety of substrates that they can use (De Ley J *et al.*,1984). These bacteria have been isolated from alcoholic beverages, vinegar, fruits, flowers, honey, sugarcane, fruit juice, soil and water (Maal KB *et al.*,2010; Sharafi, SM *et al.*,2010). Among the most important acetic acid bacteria, the strains of genus Acetobacter are mainly involved in vinegar (Sokollek, S.J *et al.*,1998; Kadere, T.T *et al.*,2008).

Figure 6.

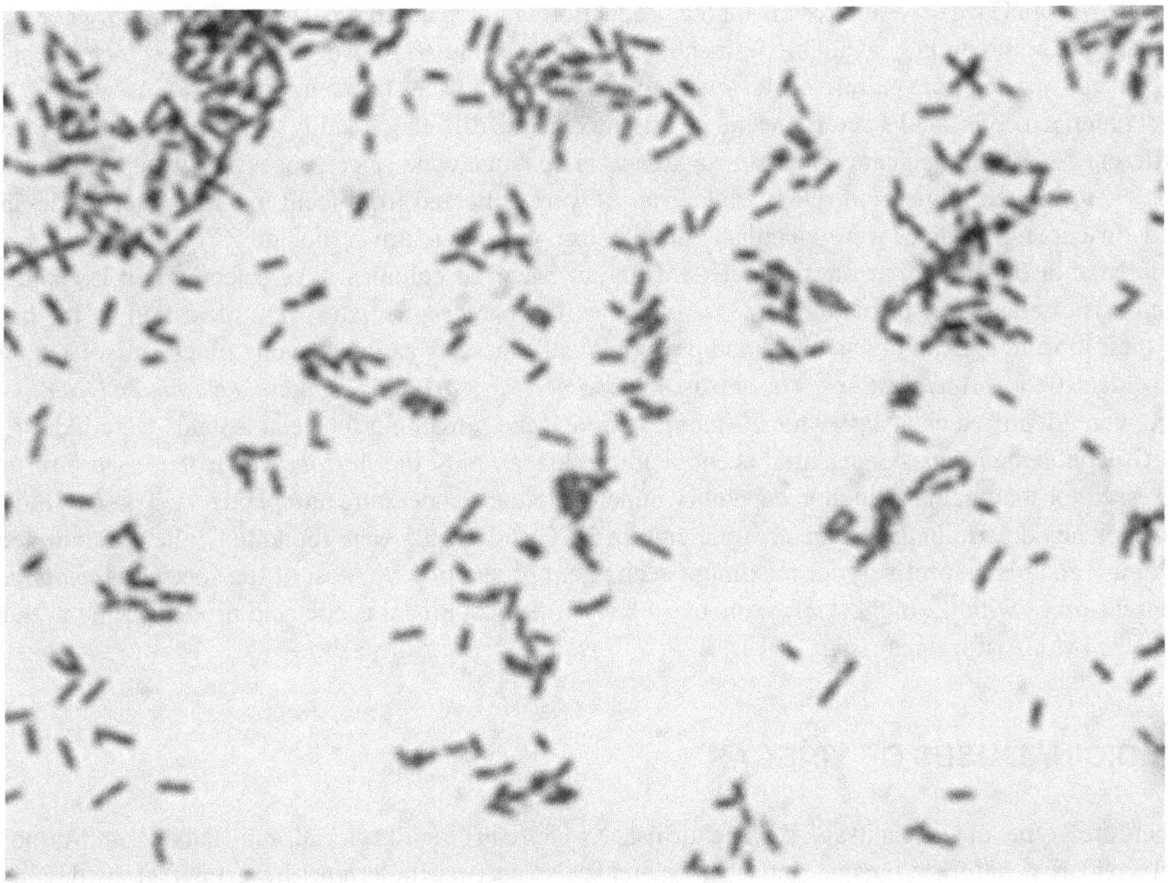

Tharinee *et.al*, in 2015, referred isolation of acetic acid bacteria from various kinds of fruits and fermented fruit juices. They collected thirty varieties of fruits: apple, black grape, cantaloupe, cherry, Chinese pear, dragon fruit, green grape, guava, lorgan, longkong, lychee, mango, mangosteen, mulberry musckmelon, papaya, peach, persimmon, pineapple, pisang mas, plum, mango, rakum plam, rambutan, red grape, rose apple, santol, strawberry, sugarcane and watermelon and 4 fermented juices of kaffir lime, Indian gooseberry, pineapple and star fruit.

Tahir Zahoor *et al.,* in 2006 made a study conducted by to isolate vinegar culture *(Acetobacter aceti)* from sugar cane juice, rotten apples, flowers, wine, canal water and vinegar as a primary source for *Acetobacter* by continuous sub-culturing on standard medium glucose, yeast extract and calcium carbonate (GYC) (Sonashree *et al.,*2020). GYC Agar described by Swings in 1992 detects the presence of acid-producing microorganisms and is regarded as "standard growth medium". The culture was identified on the basis of colony characteristics and morphology. It was finally confirmed by different biochemical/ enzymatic tests and further specified by nutritional and temperature requirements for the growth. The isolated strain was later used for the production of vinegar through fermentation. Among canal water, crushed apples, sugar cane juice, alcohol, vinegar and flowers, the alcohol and vinegar were found to be the most suitable sources for isolation of *Acetobacter spp*. The colonies of purified culture were found to be pale to off-white, circular, raised, convex, smooth and not 3 mm in diameter with morphology of Gram -ve, ellipsoidal, rods, squat bacilli, roundish, single, in pairs and in chains. The isolated and identified spp. Gave excellent results for the production of vinegar and this vinegar was more acceptable rather than commercially available fermented vinegar. At industrial level good quality vinegar can be produced by using pure culture of *Acetobacter aceti* for acetic acid fermentation.

Potential acetic acid bacteria were investigated from different readily available sources. Seven different samples (sugarcane bagasse, sugarcane juice, sugarcane juice processing water, soil, rotten apples, rotten red grapes and rotten white grapes) were collected from local market. After processing and enrichment, samples were inoculated on Glucose Yeast Calcium carbonate (GYC) agar plates and incubated at 30°C for four days. Nineteen different bacterial colonies were selected and isolated on the basis of clear zone formation on GYC medium. The bacterial isolates were identified on the basis of their morphological, biochemical and physiological characterization. Among nineteen isolates, one was identified as *Acetobacter aceti*, one as *Acetobacter Pasteurian us*, one as *Acetobacter Orleans* is, two were identified as *Acetobacter fibrinogenesis*, and the remaining fourteen isolates were identified as *Gluconobacter spp*. As potential acetic acid producers, only the *Acetobacter* isolates were further assessed for their acid production capability under different temperature and pH using 'Potency Index' as a potency determining parameter. Temperature 30°C and pH 5.5 were found to be the optimum temperature and pH respectively for maximum acetic acid production by most of the species. *Acetobacter Pasteurian us* with the highest P.I. value of 3.78 was the most potent acetic acid producer among these isolates (Md. Arifuzzaman *et al.,*2014).

MEDICINAL USES OF VINEGAR

There are 2 types of vinegar usage, therapeutic usage which can be antibacterial, anti-diabetic, anti-tumour, anti-oxidative, antihypertensive, anti-obesity, and is also used as cholesterol-lowering therapy due to its organic compounds including gallic acid, acetic acid, chlorogenic acid, ferulic acid caffeic acid, and p-coumaric acid. The second use of vinegar is commercial usage like used as the preservation of different food, pickling of vegetables and fruits preparation of salad dressing, mayonnaise and mustard. Insulin is present in type 2 diabetes but tissues are resistant to the insulin sensitivity has been improved in type 2 through vinegar uses and 34% in an individual with pre-diabetic. Postprandial blood glucose and insulin response can be influenced by dose and type of vinegars. Acetic acid is an active component in organic vinegar which can penetrate the cell membrane of bacteria and bring the death of cells of bacteria. It is also used in many conditions like oral hygiene, cleaning agent, killing head Lice, ear infection, treating nails fungus and warts. Vinegar also has the ability to treat the site infection caused by Pseudomonas in peritoneal dialysis patients and it also reduces antibiotic use, prevent the antibiotics resistance in PD patients by using organic vinegar containing 4% acetic acid (Budak *et al*.., 2014).

FUTURE PROSPECTIVES

Vinegar is very important food preservative in food industry. The vinegar can be obtained various raw materials such as various fruits and fruit peels. It was observed during various investigations that enzymatic treatments of pineapple and papaya wastes had a significant effect on the saccharification process. The effort of waste minimization would substantially reduce the large quantities of fruit and vegetable wastes accumulated globally. This formed vinegar from waste is again used as a preservative to store various fruits and vegetables. By this waste minimising process, a useful product can be produced. Acetic Acid Bacteria are most commonly known for their role in vinegar production. It is necessary to develop pure vinegar cultures for vinegar production so that the use of synthetic vinegar may be avoided as it is prohibited in most countries. This strain could be a potential strain for production of vinegar type with a new and desirable taste. So, there is a need to isolate *Acetobacter* from the natural sources to maintain chemical free life.

CONCLUSION

Vinegar is very important food preservative in food industry. The vinegar can be obtained from various raw materials such as various fruits and fruit peels. Various agricultural waste materials can also be used as raw materials. It was observed during various investigations that enzymatic treatments of pineapple wastes had a significant effect on the saccharification process. It was also found that that fermentation is a better method for obtaining higher antioxidant activity of Roselle products. Even the production of vinegar from papaya peels suggested an adaptable methodology for turning a potential waste in to a commercially important organic acid which significantly reduces production cost.

REFERENCES

Adachi, O., Moonmangmee, D., Toyama, H., Yamada, M., Shinagawa, E., & Matsushita, K. (2003). New developments in oxidative fermentation. *Applied Microbiology and Biotechnology*, *60*(6), 643–653. doi:10.100700253-002-1155-9 PMID:12664142

Arifuzzaman, Md., Zahid Hasan, Md., Badier Rahaman, S. M., & Kamruzzaman Pramanik, Md. (2014). Isolation and characterization of *Aectobacter and Gluconobacter spp* from sugarcane and rotten fruits. *Research and Reviews in Bio Sciences*, *8*(9), 359–365.

Bhavana, Sonashree, Rashmi, Rao, & Praveena. (2019). A review on the preparation of vinegar from fruit peels. *World Journal of Pharmacy and Pharmaceutical Sciences, 8*(9), 227-233.

Budak, N. H., Aykin, E., Seydim, A. C., Greene, A. K., & Guzel-Seydim, Z. B. (2014). Functional Properties of Vinegar. *Journal of Food Science*, *79*(5), 757–765. doi:10.1111/1750-3841.12434 PMID:24811350

Claudio, & Estibaliz, Cerezo, Torija, & Albert. (2010). Production of fruit vinegars: Technological process for persimmon and strawberry vinegars. *Int.J. of Wine Research*, *2*, 55–61.

Dacera, D. D. M., Babel, S., & Parkpian, P. (2009). Potential for land application of contaminated sewage sludge treated with fermented liquid from pineapple wastes. *Journal of Hazardous Materials*, *167*(1-3), 866–872. doi:10.1016/j.jhazmat.2009.01.064 PMID:19232826

De Freitas Borghi, D., Guirardello, R., & Cardozo Filho, L. (2009). Storage logistics of fruits and vegetables: Effect of temperature. *Chemical Engineering Transactions*, *17*, 951–956.

De Ley, J., Gillis, M., & Swings, J. (1984). *Baltimore: Williams and Wilkins Baltimore*. Bergey's Manual of Systematic Bacteriology.

Hailu, S., Admassu, S., & Jha, Y. K. (2012). *Vinegar Production Technology–An Overview*. Beverage & Food World.

Hemalatha, R., & Anbuselvi, S. (2013). Physicohemical constituents of pineapple pulp and waste. *Journal of Chemical and Pharmaceutical Research*, *5*(2), 240–242.

Hidalgo, C., Vegas, E., Mateo, W., Tesfaye, A. B., Cerezo, R. M., Callejón, M., Poblet, J. M., Guillamón, A., Mas, M. J., & Torija, M. J. (2010). Effect of barrel design and the inoculation of Acetobacter pasteurianus in wine vinegar production. *International Journal of Food Microbiology*, *141*(1-2), 56–62. doi:10.1016/j.ijfoodmicro.2010.04.018 PMID:20478638

Itelima, J., Onwuliri, F., Onwuliri, E., Onyimba, I., & Oforji, S. (2013). Bio-Ethanol production from banana, plantain and pineapple peels by simultaneous saccharification and fermentation process. *International Journal of Environmental Sciences and Development*, *4*, 213–216. doi:10.7763/IJESD.2013.V4.337

Jasmine Praveena, R., & And Estherlydia, D. (2014). Comparative Study of Photochemical Screening and Antioxidant Capacities of Vinegar Made from Peel and Fruit of Pineapple (Ananas Comosus L.). *Int. Pharm. Bio. Sci*, *5*(4), 394–403.

Kadere, T. T., Miamoto, T., Oniang'o, R. K., Kutima, P. M., & Njoroge, S. M. (2008). Isolation and identification of genera *Acetobacter* and *Gluconobacter* in coconut toddy (mnazi). *African Journal of Biotechnology*, *7*, 2963–2971.

Klawpiyapamornkun, T., Bovonsombut, S., & Bovonsombut, S. (2015). Isolation and Characterization of Acetic acid Bacteria from Fruits and Fermented fruit juices for Vinegar Production. *Food and Applied Bioscience Journal*, *3*(1), 30–38.

Kowser, J., Aziz, M. G., & Uddin, M. B. (2015). Isolation and characterization of Acetobacter aceti from rotten papaya. *Bangladesh Agril.Univ.*, *13*(2), 299–306. doi:10.3329/jbau.v13i2.28802

Lagha-Benamrouche, S., & Madani, K. (2013). Phenolic contents and antioxidant activity of orange varieties (*Citrus sinensis* L. and *Citrus aurantium* L.) cultivated in Algeria: Peels and leaves. *Industrial Crops and Products*, *55*, 723–730. doi:10.1016/j.indcrop.2013.07.048

Maal, K. B., Shafiei, R., & Kabiri, N. (2010). Production of Apricot vinegar using an Isolated acetobacter strain from Iranian Apricot. *World Academy of Science, Engineering and Technology*, *71*, 177–180.

Martial-Didier, A. K., Hubert, K. K., Parfait, K. E. J., & Kablan, T. (2017). Phytochemical properties and proximate composition of papaya (Carica papaya L. var solo 8) peels. *Turkish Journal of Agriculture-Food Science and Technology*, *5*(6), 676–680. doi:10.24925/turjaf.v5i6.676-680.1154

Matsushita, K., Toyama, H., & Adachi, O. (1994). Respiratory Chains and Bioenergetics of Acetic Acid Bacteria. In Advances in Microbial Physiology. Academic Press. doi:10.1016/S0065-2911(08)60181-2

Minh, N. P. (2014). Utilization of Ripen Star Fruit for Vinegar Fermentation. *International Journal of Multidisciplinary Research and Development*, *1*(4), 82–93.

Roda, A., De Faveri, D. M., Dordoni, R., & Lambri, M. (2014). Vinegar production from pineapple wastes- Preliminary Saccharification Trials. *Chemical Engineering Transactions*, *37*, 607–612.

Sarkar, N., Ghosh, S. K., Bannerjee, S., & Aikat, K. (2012). Bioethanol production from agricultural wastes: An overview. *Renewable Energy*, *37*(1), 19–27. doi:10.1016/j.renene.2011.06.045

Schieber, A., Stintzing, F. C., & Carle, R. (2001). By-products of plant food processing as a source of functional compounds - recent developments. *Trends in Food Science & Technology*, *12*(11), 401–413. doi:10.1016/S0924-2244(02)00012-2

Seyram, S. K., Yaovi, A., Simplice, K. D., & Comlan, D. S. (2009). Study of pineapple peelings processing into vinegar by biotechnologies. *Pakistan Journal of Biological Sciences*, *12*(11), 859–865. doi:10.3923/pjbs.2009.859.865 PMID:19803120

Sharafi, S. M., Rasooli, I., & Beheshti-Maal, K. (2010). Isolation, characterization and optimixation of indigenous acetic acid bacteria and evaluation of their preservation methods. *Iranian Journal of Microbiology*, *2*(1), 41–48. PMID:22347549

Sokollek, S. J., Hertel, C., & Hammes, W. P. (1998). Cultivation and preservation of vinegar bacteria. *Journal of Biotechnology*, *60*(3), 195–206. doi:10.1016/S0168-1656(98)00014-5

Sonashree, R., Bhavana, J., Rashmi, R. H., Shamsiya, R., & Praveena, B. (2020). Isolation and Characterization of *Acetobacter aceti* sugar cane- A review. *Indo American Journal of Pharmaceutical Sciences, 7*(3), 478-481.

Tan, S. C. (2005). *Vinegar Fermentation*. Department of Food Science, Louisiana State University, Agricultural and Mechanical College.

Vegas, E., Mateo, W., Tesfaye, A. B., Cerezo, R. M., Callejón, M., Poblet, J. M., Guillamón, A., & Mas, M. J. (2010). Population dynamics of acetic acid bacteria during traditional wine vinegar production. *International Journal of Food Microbiology*, *138*(1-2), 130–136. doi:10.1016/j.ijfoodmicro.2010.01.006 PMID:20117853

Vikas, O.V., & Umesh, M. (2014). Bioconversion of papaya peel waste into vinegar using Acetobacter aceti. *Int. J. Sci. Res, 3*(11), 409-411.

Zahoor, T., Siddique, F., & Farooq, U. (2006). Isolation and characterization of Vinegar Culture (*Acetobacter aceti)* from indigenous sources. *British Food Journal*, *108*(6), 429–439. doi:10.1108/00070700610668405

Chapter 14
Bioremediation and Biodegradation Technologies for Removal of Pollutants and Wastes from the Environment

Amit Kumar
Swami Vivekanand Subharti University, Meerut, India

ABSTRACT

Bioremediation is defined as the process whereby organic wastes are biologically degraded under controlled conditions to an innocuous state, or to levels below concentration limits established by regulatory authorities. For bioremediation to be effective, microorganisms must enzymatically attack the pollutants and convert them to harmless products. As bioremediation can be effective only where environmental conditions permit microbial growth and activity, its application often involves the manipulation of environmental parameters to allow microbial growth and degradation to proceed at a faster rate. Bioremediation techniques are typically more economic than traditional methods such as incineration, and some pollutants can be treated on site, thus reducing exposure risks for clean-up personnel, or potentially wider exposure as a result of transportation accidents. Bioremediation is based on natural attenuation. The public considers it more acceptable than other technologies.

BACKGROUND

Bioremediation is defined as "The use of biological mechanisms to destroy, transform, or immobilize environmental contaminants in order to protect potential sensitive receptors." In the past, thermal, chemical, and physical treatment technique have didn't eliminate the pollution drawback as a result of those technique solely shift the pollution to a replacement part like air pollution (Kleijntjens et al., 2000 and Klein 2000). Bioremediation technology, which leads to degradation of pollutants, may be a lucrative and environmentally beneficial alternative that could produce economic profit (Hlihor et al., 2017).

DOI: 10.4018/978-1-7998-4921-6.ch014

Bioremediation has been used at a number of sites worldwide, with varying degrees of success (Boopathy et al., 2000). Techniques are rising as larger data and knowledge are gained, and there's little doubt that bioremediation has nice potential for coping with certain kinds of website contamination. Unfortunately, the principles, techniques, advantage, and downsides of bioremediation don't seem to be wide identified or understood, particularly among those that can ought to deal directly with bioremediation proposal, like site owners and regulators. (Van Deuren et al.,2002). We have a tendency to mean to help by providing a simple, pragmatic read of the processes concerned in bioremediation, the proposal and cons of the technique, and therefore the problems to be thought of once dealing with a proposal and cons of the techniques, and therefore the problems to be thought of once dealing with a proposal for bioremediation. Some tests make an exhaustive examination of the literature of bioremediation of organic and inorganic pollutants, and another test takes a look at pertinent field application case histories.

Pollutants is also within the kind of any chemical (radionuclide, organo phosphorus compound or trace gases) or geochemical substances (dust, sediment, grit etc.), biotic component or product or physical agent (heat, sound etc.), that's free by man in to the atmosphere in such quantities that will have adverse effects.

For an environmentally compatible sustainable development, the core problems such as management of wastes and other pollutants require application of clear technologies. The use of microbes in the degradation of different types of pollutants and wastes leading to bioremediation, is one such way. Microbes may be deployed as voracious scavengers, removing all types of pollutants, similarly within the world utilisation of matter, since they unloose carbon, nitrogen, phosphorus and sulfur from a large range of advanced organic compounds(Iranzo et al., 2001).

On the basis of their behaviour in the environment, organic compounds are classified as the following:

1. **Biodegradable:** Those organic compounds which undergo biological transformations are termed as biodegradable.
2. **Persistent:** Those organic compounds which do not undergo biodegradation in certain environment, are termed as persistent.
3. **Recalcitrant:** Those which resist biodegradation in a wide range of environments are termed as recalcitrant.

Xenobiotics are synthetic compounds not formed by natural biosynthetic processes, and can be recalcitrant. Therefore, xenobiotic compounds are foreign substrates in our ecosystem and may have toxic effects when accumulated above threshold value(Chojnacka, 2010).

BIOREMEDIATION STRATEGIES

Bioremediation can be achieved by different techniques (Fig-1). Bioremediation has two strategies, *In-situ* bioremediation "can be used at the site of contamination" and *Ex-situ* Bioremediation means "on contamination removed from the original site" (Shannon et al., 2000). Details of *Ex-situ* and *In-situ* bioremediation are given below:

Figure 1. Man Assisted Bioremediation

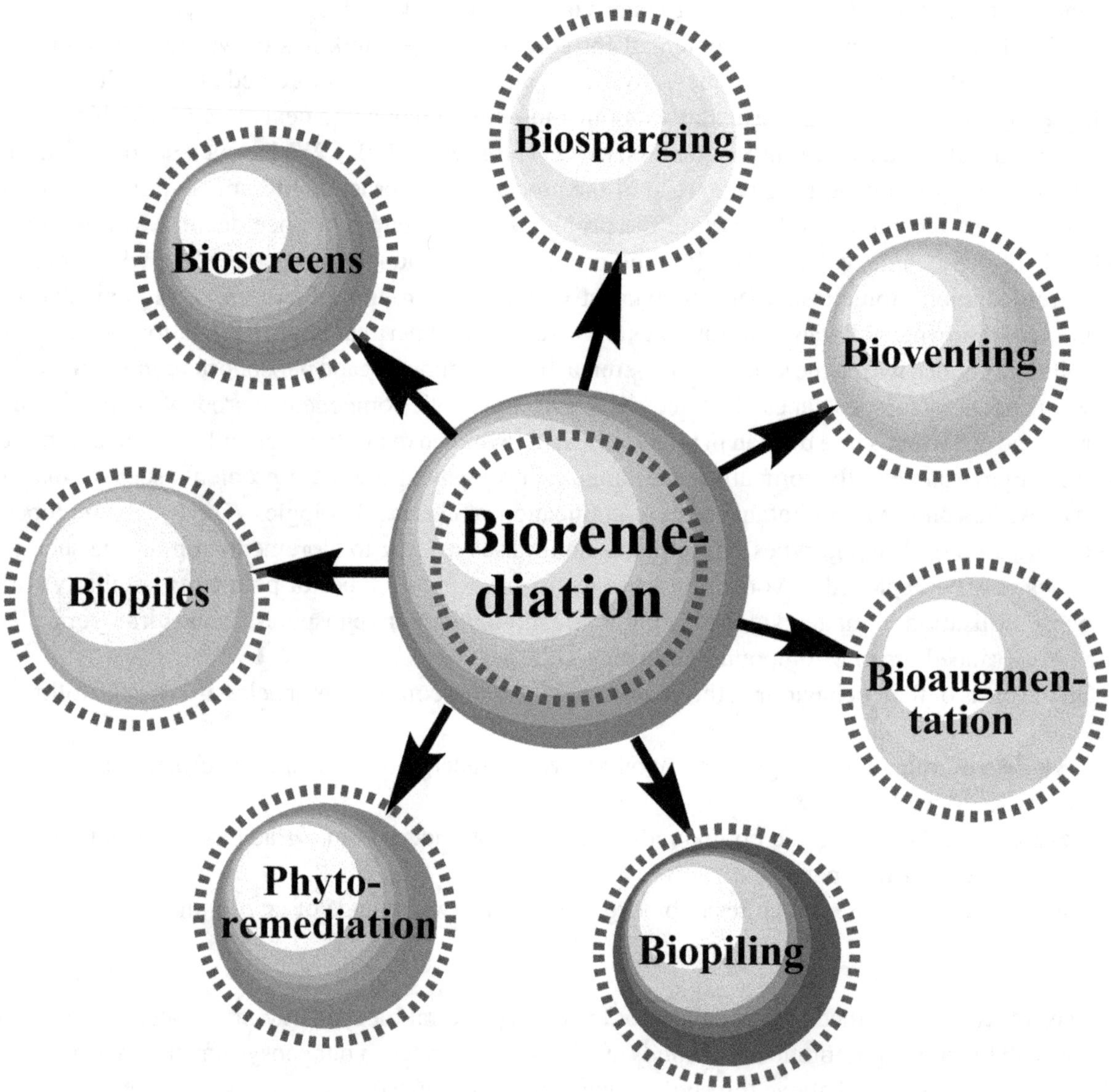

IN – SITU BIOREMEDIATION

In situ bioremediation is the application of biological treatment to the cleanup of hazardous chemicals present in the subsurface (Held, T. and Don, H. 2000). The optimisation and management of microorganism transformations of organic contaminants need the mixing of the many scientific and engineering disciplines.

BIOSCREENS

Bioscreens are a lovely treatment choice as a result of they need high longevity, no important maintenance, and no nutrient renewal. These screens are composed of organic wastes and sedimentary rock. The organic wastes not solely function a high permeableness structure, however conjointly as a supply of nutrients and bacterium. The bioscreens even have high material retentions and inflated bioactivity. The hydraulic conductivity of the material within the bioscreens is ten times over that of the encompassing formation. The thickness of the bioscreens depends on groundwater flow rates, the pollute concentrations, and therefore the degradation rates.

BIOSPARGING

Biosparging involves the injection of air harassed below the formation to extend groundwater O2 concentrations and enhance the speed of biological degradation of contaminants by in-vivo bacterium. Biosparging will increase the blending within the saturated zone and thereby will increase the contact between soil and groundwater. The benefit and low price of putting in small-diameter air injection points permits significant flexibility within the style and construction of the system.

BIOVENTING

Bioventing could be a promising new technology that stimulates the natural unchanged biodegradation of any aerobically degradable compounds in NAPL among the soil by providing O_2 to existing soil microorganisms. In distinction to soil-vapor extraction (SVE), bioventing uses low air-flow rates to supply simply enough O_2 to sustain microorganism activity. O_2 is most ordinarily provided through direct air injection into residual contamination in soil by means that of wells. adsorbable fuel residuals are biodegraded, and volatile compounds are biodegraded as vapors inch through biologically active soil.

BIOAUGMENTATION

Bioaugmentation is that the introduction of a bunch of natural microorganism strains or a genetically built variant to treat contaminated soil or water (Akhtar et al., 2013). it's ordinarily employed in municipal waste treatment to restart activated sludge bioreactors. Most cultures accessible contain a research based association of microorganism cultures, containing all necessary microorganisms At sites wherever soil and groundwater are contaminated with chlorinated ethenes, like tetrachloroethylene and trichloroethane, bioaugmentation is employed to make sure that the naturally occuring microorganisms will fully degrade these contaminants to chloride, that are non-toxic observation of this technique is tough (Guidi et al., 2018).

BIOPILING

Biopile treatment could be a complete technology within which excavated soils are mixed with soil amendments, placed on a treatment space, and bioremediated exploitation forced aeration. The contaminants are reduced to CO2 and water. The essential biopile system includes a treatment bed, water irrigation, nutrient management and a leach ate assortment system. Temperature, Moisture, micro and macro nutrient nutrients, O2, and pH scale are controlled to increase biodegradation. The irrigation/nutrient system is buried below the soil to pass air and nutrients either by vacuum or positive pressure. Soil piles may be up to twenty feet high and will be lined with plastic to manage runoff, evaporation and volatilization, and to enhance solar heating system. If volatile organic compounds (VOCs) within the soil change into the air stream, the air move out the soil is also treated to get rid of or destroy the VOCs before they're discharged into the atmosphere. Treatment time is usually three to six months.

EX-SITU BIOREMEDIATION

Ex situ Bioremedy includes techniques like Composting, land-farming, biopiling, and process by bioreactors along side thermal, chemical, and physical processes. *Ex situ* bioremedy could be a lot of thorough remedy technique, however because of the prices associated not solely with the remediation processes, however conjointly can the excavation and transportation of the soil, many persons are trying towards *in situ* remediation techniques.

Ex situ thermal Technique involves the transfer of pollutants from the soil to a gaseous part. The pollutants are free by vaporization and therefore the burned at high temperatures. Ex situ thermal remedy is completed in three phases: soil condition, temperature treatment, and exhaust gas purification. Soil condition could be a method within which broken into little grains and sieved in preparation for thermal treatment. In temperature treatment, temperature of the soil in order to convert volatile pollutants to a gas phase. Heating is completed by employing a sintering strand, fluid bed, or rotary oven plants. The heating range of the soil is typically vary between 350-550^{0}C. Combustion of the gases occurs on the uppermost of the soil, but the volatile gases are not terminated. The gases are then burned in an after-burner chamber at more than the 1000^{0}C dioxins are destroyed.

Ex situ thermal remediation processes are ideal for use when removing polycyclic aromatic hydrocarbons (PAH), petroleum hydrocarbons (TPH), benzene, toluene, ethylbenzene, xylenes (BTEX), phenolic compounds, cyanides, and chlorinated compounds like polychlorinated biphenyls (PCB), pentchlorphenol (PCP), chlorinated hydrocarbons, chlorinated pesticides, polychlorinated dibenzodioxins (PCDD), and polychlorinated dibenzofurans (PCDF) (Alegbeleye et al., 2017, Gogoi et al.,2002).

Ex situ biological processes include; composting, landfarming, Heap Technique and the use of bioreactors which are given below:

COMPOSTING

Composting could be a method by which organic wastes are degraded by microorganisms, generally at elevated temperatures. Typical compost temperatures are within the range of 55° to 65° C. The inflated temperatures result from heat created by microorganisms throughout the degradation of the organic material within the waste.

LAND FARMING

Land farming could be a method by which the soil is excavated and mechanically separated via sieving. The impure soil is then place in layers no quite 0.4 meters thick. An artificial, concrete, or clay membrane is then used the contaminated soil layer. O2 is supply and mixing occurs via harrowing, or milling. Nutrients and wetness also be used to assist the remedy method. The pH scale of the soil is additionally regulated (keeping it approximately 7.0) using crushed sedimentary rock or agricultural lime. Land farming is most felicitous in removing PAH and PCP. Figure-2 illustrates the land farming technique. (Schnoor et al., 2000).

Figure 2. Landfarming Technique

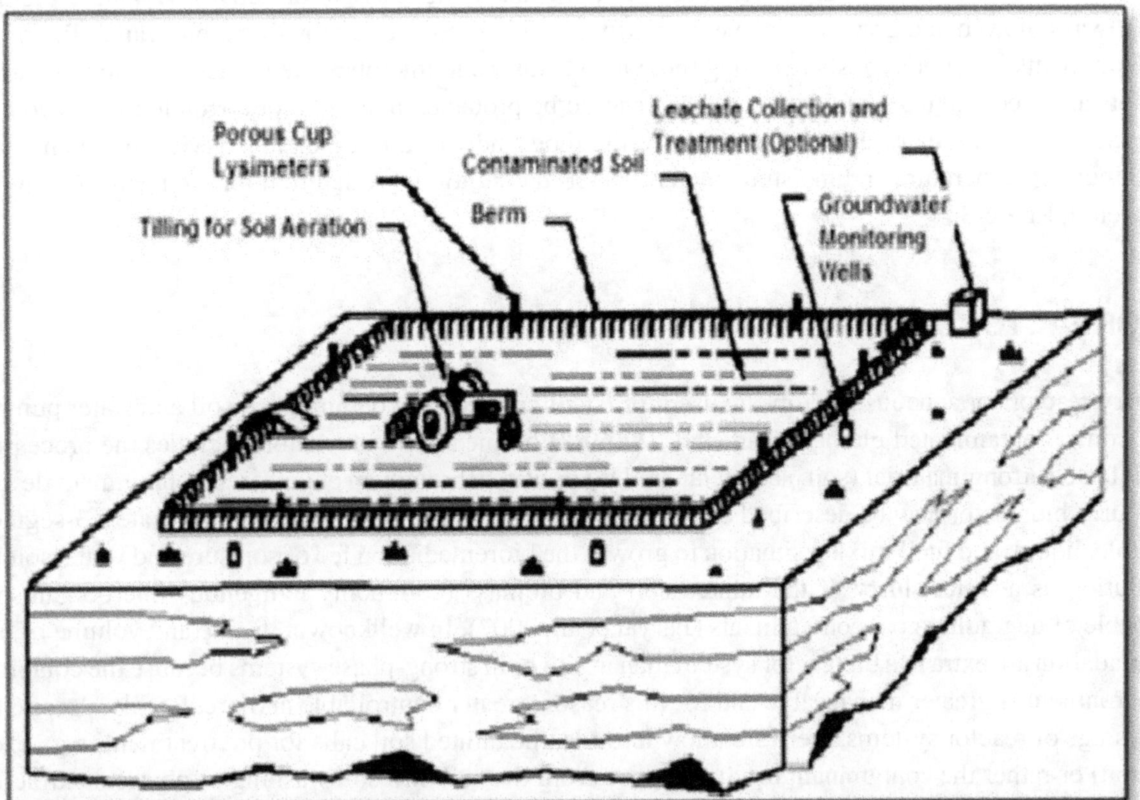

HEAP TECHNIQUE

Biopiling is an In *situ* process that is also known as the heap technique (Schulz-Berendt, 2000). The primary step in the biopiling method is to perform qualitative tests that may confirm the biological degradation capabilities of the soil sample. The next step involves the mechanical separation of the soil, which which is able to homogenize the sample and remove any disruptive material such as plastics, metals, and stones (Brooks et al., 1998 and pandey et al., 2011). The stones can then be crushed into smaller particle and then based on the degree of impurities can either be additional to a pile or sent out for utilize. The soil is then homogenized, meaning that the pollution concentration is averaged out across the entire soil sample (Congeevaram et al.,2007). Homogenization allows for biopiling to be more effective.

As soon as the soil is piled, micro and macro nutrients, microbes, O_2, and substrate are introduced to begin the biological degradation of the contaminants. The results of the initial laboratory exams illustrate to the operators which substrates consisting of bark, lime, or composts desires to be delivered to the soil. nutrients such as mineral fertilizers will also be delivered. Moreover, microorganisms along with fungi, bacteria, or enzymes might be added.

Static piles are normally within the form of pyramids or trapezoids. Their heights vary between zero.6 and 1.8m depending at the form of aeration used (either passive or active). Dynamic biopiles are always plowed and grew to become to maximize their exposure to growth the bioavailability of the contaminants via continuously exposing them to O_2, water and microbes. Regardless of which varieties of lots are used, the region under every heap should be protected in asphalt or concrete to save you the seepage of contaminants and the place above the thousands need to be included with the intention to manipulate temperature and moisture content material conditions. A diagram for the heap strategies is proven in figure-3.

BIOREACTORS

Slurry reactors or aqueous reactors are used for *Ex situ* remedy of contaminated soil and water pumped up from a contaminated plume (Khan et al., 2009). Bioremediation in reactors includes the processing of infected strong material (soil, sediment, sludge) or water through an engineered containment device. A slurry bioreactor may be described as a containment vessel and apparatus used to create a 3-segment (stable, liquid, and fuel) mixing situation to growth the bioremediation fee of soil sure and water-soluble pollution as a water slurry of the tainted soil and biomass (commonly indigenous microorganisms) capable of degrading goal contaminants (Faryal et al., 2007). In wellknown, the fee and volume of biodegradation are extra in a bioreactor system than in situ or in strong-phase systems because the contained environment is greater manageable and for this reason greater controllable and predictable. despite the blessings of reactor systems, there are a few hazards. the tainted soil calls for pre treatment (e.g., excavation) or rather the contaminant can be stripped from the soil thru soil washing or physical extraction (e.g., vacuum extraction) earlier than being positioned in a bioreactor (Jain et al.,2012). Bioremediation techniques and its utility shown in Table- 1.

Figure 3. Heap Technique Diagram
(Source: Schulz-Berendt, 2000)

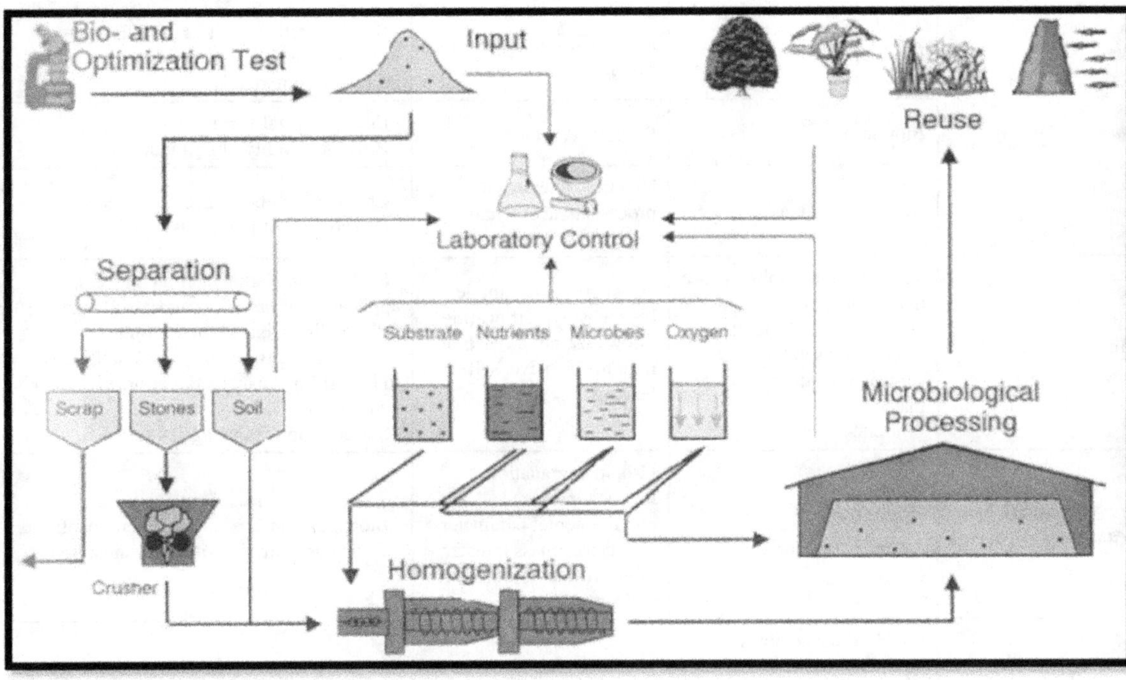

MECHANISM OF BIOREMEDIATION

Microbes are the play the key role in bioremediation as they generate the enzymes that catalyze the degradative reactions. Why the microbes carry out degradative reactions? It is because the microbes use organic substances as a source of carbon and energy. Thus while transforming the contaminant microbes gain energy and raw material for their multiplication and maintenance.

Based totally on the mechanism by using which microbes gain energetically, they're extensively labeled into three classes (Table-2). But, numerous xenobiotic contaminants won't be amenable to one of the above described categories and other mechanisms are hired by way of the microorganisms for degradation of such compounds.

REDUCTIVE DEHALOGENATION

It plays very important role in the detoxification of halogenated organic contaminants. Microorganisms catalyze a reaction in which halogen atom of contaminant is replaced by hydrogen atom. Thus the reaction adds 2 electrons to contaminant and reduces it. It yields no energy but seems to be detoxification mechanism as dehalogenated derivatives are less toxic and susceptible to further microbial decay.

Table 1. Bioremediation techniques, benefits and its applicaiton

Techniques	Examples	Benefits	Applications
In situ	Biosparging	Most efficient & Non Invasive	· Biodegradative abilities of indigenous microorganisms presence of metals and inorganic compounds
	Bioventing	Relative Passive	· Environmental Parameters · Biodegradability of pollutants
	Bioaugmentation	Naturally attenuated process, treat soil and water	· Chemical solubility geological factors · Distribution of pollutants
Ex-situ	· Land farming (solid –phase treatment system) · Composting (Anaerob ic, converts solid organic wastes into humus – like material) · Biopiles	· Cost efficient, Simple, Inexpensive, self heating · Low cost rapid reaction rate, Inexpensive, self heating · Can be done on site	· Surface application, aerobic process, application of organic materials to natural soils followed by irrigation and tilling · To make plants healthier good alternative to land filling or incinerating practical and convenient. · Surface application, agricultural
Bioreactors	· Slurry reactors · Aqueous reactor	· Rapid degradation kinetic Optimized environmental parameters · Enhances mass transfer effective use of inoculants and surfactant	· Bioaugmentat Toxicity of amendments · Toxic concentration of contaminants
Precipitation or flocculation	· Non – directed physic – chemical complex-ation reaction between dissolved contaminants and charged cellular components (dead Biomass)	· Cost effective	· Removal of heavy Metals
Microfiltration	· Microfiltration membranes are used at a constant pressure	· Remove dissolved solids rapidly	· Waste water treatment: recovery and reused of more than 90% of original waste water
Electrodialysis	· Use cation and anion exchange membrane pairs	· Withstand high temperature and can be reused	· Removal of dissolved solid efficiently

COMETABOLISM

In this example additionally the conversion response yields no benefit to the cell. This non-useful transformation is regularly termed as secondary utilization, cometabolism or gratuitous metabolism. Co-metabolism is one form of secondary substrate transformation wherein enzymes produced for num-

Table 2. Mechanisms of energy generation by the microbes

Mechanism	Electron donor	Electron acceptor	Product
Aerobic respiration	Organic compound	Oxygen	CO_2, H_2O
Anaerobic respiration	Organic compound	NO_3, SO_4, Fe^{+3}, Mn^{+4}, CO_2	N_2, H_2S, CH_4, Reduced metal
Fermentation	Organic compound	Organic compound	Organic acids, alcohols, H_2 & CO_2

ber one substrate oxidation are capable of degrading the secondary substrate happily, even though the secondary substrates do not have enough money enough strength to maintain the microbial populace. it's miles consequently defined as degradation of a compound most effective in presence of other natural compound that serves as a primary energy source. numerous contaminants such as Poly-nuclear Aromatic Hydrocarbons (PAHs) and pesticides are degraded through this mechanism. A rising application entails the injection of water containing dissolved primary substrate (e.g. methane, toluene) and oxygen into ground water to guide the co-metabolic breakdown of targeted natural contaminants (Chen et al., 2015). The addition of methane or methanol helps methanotrophic pastime, which has been validated powerful to degrade chlorinated solvents, along with vinyl chloride and Trichloroethane (TCE), via co-metabolism (Fig. 4).

Figure 4. Schematic diagram of the In-situ ground water remediation via cometabolic activity

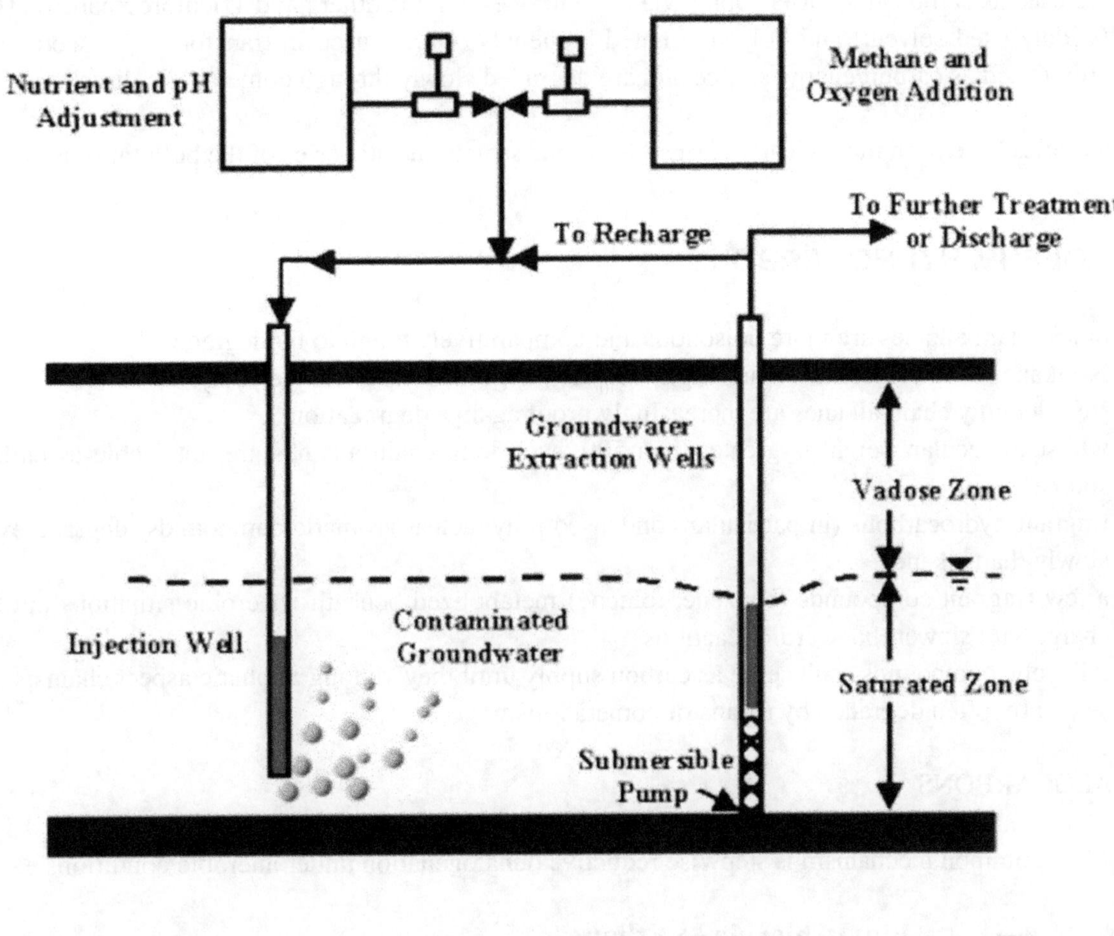

CONTAMINANT STRUCTURE V/S BIODEGRADABILITY RELATION

Biodegradability is important for bioremediation of organic pollution. Chemical shape of pollutant governs the ability of microorganisms to metabolize them, specially the rate and quantity of biodegradation. As we mentioned above, some compounds are conveniently biodegradable while others are not. let us study the general rules that determine this.

- Low to mid molecular weight hydrocarbons and alcohols constitute easily biodegradable chemicals.
- Branched and polynuclear compounds are greater tough to degrade that the instantly-chain and easy non-fragrant compounds.
- Halocarbons are immune to biodegradation and are often termed as xenobiotic compounds. Dioxns are very tough to degrade. Increasing degree of halogenation decreases the charge of biodegradation. as an example Dichloromethane as well as Monochloro biphenyls are degraded fast as these are used as carbon and power supply via the microbes. On the other hand Trichloroethane or TCE (widely used solvent) and Polychlorinated biphenyls (widely used in transformers/insecticides) aren't used as carbon/energy source and are degraded slowly through cometabolic direction.

Such rules have been more clearly defined for broad structural categories of the pollutants as below:

PETROLEUM HYDROCARBONS

- quick chain alkanes are more poisonous and comparatively tough to biodegrade
- N-alkanes of intermediate chain length (C_{10}-C_{24}) degraded most unexpectedly
- Very lengthy chain alkanes are increasingly proof against degradation
- whilst molecular weight is greater than 500, such long chain alkanes are not viable as carbon source
- fragrant hydrocarbons (in particular condensed polynuclear aromatic compounds) degrade extra slowly than alkanes
- a few fragrant compounds (benzene, toluene) metabolized beneath anaerobic situations but the charges are slower than cardio reactions
- Alicyclic compounds can't serve as carbon supply until they've long aliphatic aspect chains, consequently often degraded by means of cometabolism

HALOCARBONS

- Most common mechanism is stepwise reductive dehalogenation under anerobic conditions e.g.

$$\textbf{TCE} \xrightarrow[\textit{step wise dechlorination}]{} \textbf{Vinyl Chloride} \rightarrow \textbf{Ethane}$$

- Cometabolic degradation by using methanogenic consortium additionally said
- cardio degradation of TCE by way of methane making use of consortium is also possible
- Microcosm research record both cardio/anaerobic ameliorations

- In case of haloaromatics including chlorobenzenes, cardio biodegradability decreases with wide variety of halosubstituents
- substantially chlorinated haloaromatics degraded by way of stepwise anaerobic dechlorination. The consecutive dechlorination steps are greater hard, sluggish and often incomplete but the mono and dichloro substitutes can be aerobically degraded.

NITRO-AROMATICS

- Nitro-substitution makes the compounds less biodegradable
- Excessively nitro-substituted aromatics transformed under anaerobic conditions by stepwise conversion of nitro groups to amino groups

BIOREMEDIATION MICROORGANISMS/AGENTS

Distinctive microorganisms are capable of degrade unique contaminants relying upon the nature and concentration of contaminant and the metabolic needs of the microorganisms. Scientists round the sector are constantly operating to find out novel and more efficient biodegraders. Allow us to discuss the crucial bioremediation microorganisms underneath the following;

PURE CULTURES

For the reason that conventional techniques of microbiology are designed for natural lifestyle studies and also it is easy to exercise session the efficiency of microbes in such systems, maximum of the laboratory research recognition on biodegradative capability of pure cultures. desk three indicates the famous bioremediation microorganisms. Pseudomonas species are most extensively detected microbes inside the contaminated website online due to their sizeable biodegradation capacities (Naz, 2015). most of the fugal groups, phanerochaete has been proposed for the biodegradation of numerous pollution including DDT, TNT, excessive molecular weight polynuclear aromatics like benzo (a) pyrene and plastics which includes polyethylene (Benson et al., 1990).

ACCLIMATIZED MICROORGANISMS & MOLECULAR BREEDING

The metabolic variety of obviously going on microbiota won't be capable of degrading certain compounds or sure training of compounds. There it could be necessary to complement with the specialised microbes Table-3. One way of developing such specialised microbes is via repeatedly exposing them to better attention of contaminants. Frequently the microorganisms with specialized degradation capability also can be enriched from the tainted site.

Figure 5. stepwise anaerobic dechlorination

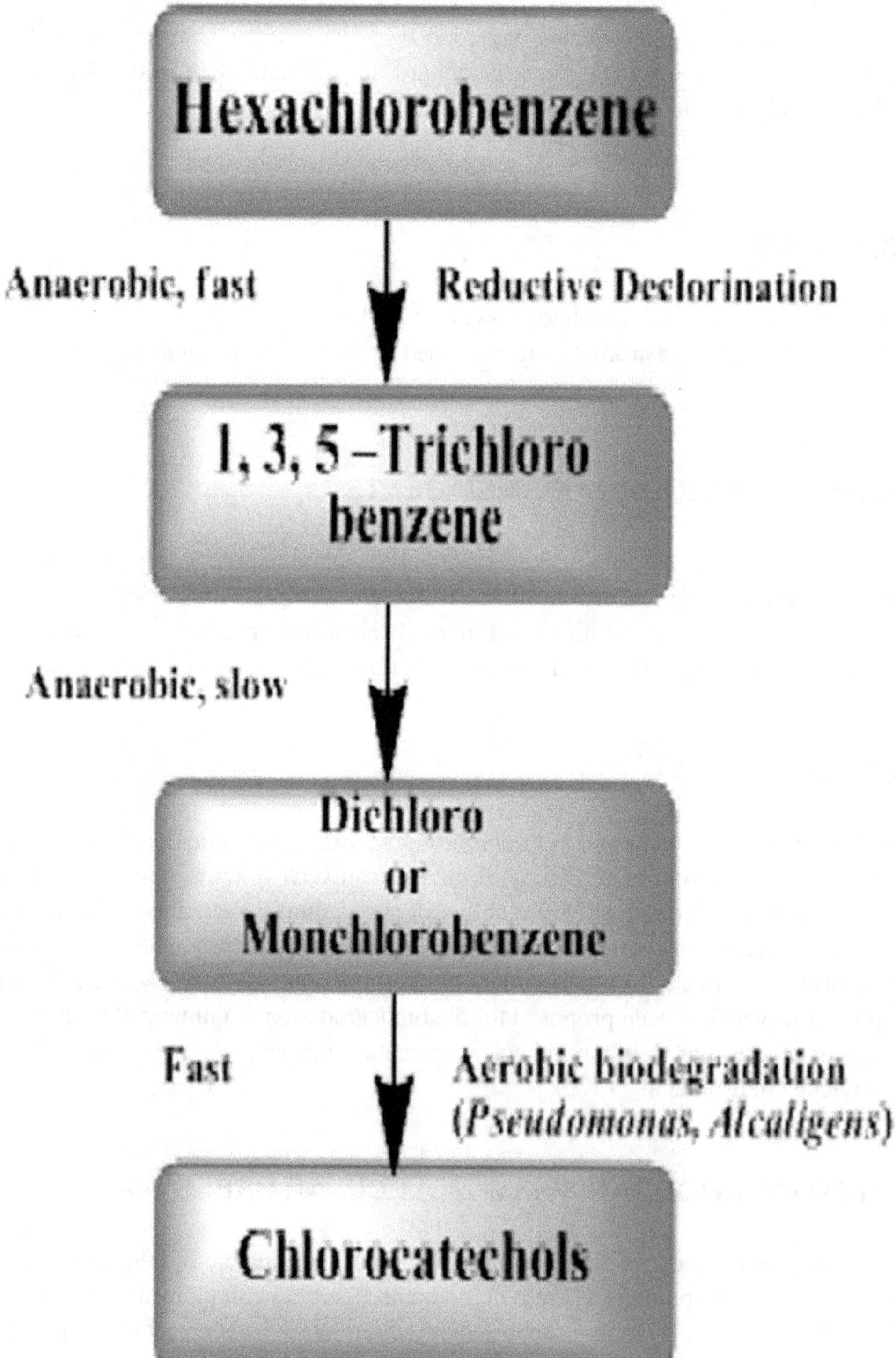

Table 3. Micro-organisms (Pure culture) helpful in bioremediation

S.No.	Pollutant	Microorganism (s)
1.	Atrazine	*Pseudomonas* sp. strain ADP, Acinetobacter species
2.	Chlorpyrifos	*Aspergillus niger, Trichoderma viride,* *Bacterium strain* B-14
3.	2,4,6-Trichlorophenol	*Alcaligenes eutrophus* TCP
4.	2,4-Dichlorophenoxyacetic acid	*Ralstonia eutropha* (pJP4), *Ralstonia eutropha* JMP134
5.	Carbon tetrachloride	*Pseudomonas stutzeri* KC
6.	Methyl *tert*-butyl ether	*Hydrogenophaga flava* ENV735, β-proteobacterium strain PM1
7.	BTEX	B. sp. Strain JS150 B. Cepacia G4 R. Pickettii PKO1 S. yanoikuyae B1
8.	Orange 3, 4-(4-nitrophenylazo) aniline	Pleurotus ostreatus

MICROBIOLOGY OF BIODEGRADATION

Microbial degradation technologies aim to render harmless those substances which cannot be utilizes profitably and includes both natural and chemically synthesized substrates. This decomposition of substance is brought about by microbial activities either by single organisms or more often by microbial consortia. All natural organic compounds should be degraded under suitable environmental conditions.

Microorganisms found in soil and water utilize the organic substance encountered as sources of energy and carbon by enzymatically breaking them downinto somple molecules which can be absorbed and utilized.

If alteration is brought about by a single reation, it is referred to as primary biodegradation, and if more extensive chemical changes are involved, then partial biodegradation. When these is complete degradation to end products of CO_2, water and other inorganic compounds, it is refereed to as mineralization.

Natural microbial communities are in the form of com.plex assemblages in which various microorganisms are highly interdependent thought interactions like commensalism and mutualism. Commensalism may take different forms; eg,many soil microorganism require aminos acids or vitamins for growth, which many other such organisms produce these compounds. This crossfeeding is a general feature of natural microbial communities. Similarly, the ability of anaerobes to survive in surface layer of soil depends on the efficient consumption of oxygen by aerobes. Ethanol, which produced by yeast from sugars in fruit is oxidized to acetic acid by *Acetobacter* species. Methane, the product of anaerobic breakdown of various organic compounds by methane producing methanogens, is oxidized aerobically by methane oxidizing bacteria. Similarly, mutualistic interactions are diverse; e.g. Where nitrogen fixers and cellulose decomposer coexist, each organism utilizes compounds produced by h other. *Lactobacillus arabinosus* and *Strepococcus faecalis* depend on each other to satisfy nutntional requirements; L. arabinosus makes the folic acid required *by S. faecalis* and *S. faecalis* makes the *phenlalanine* that *L. arabinosus needs.*

Organic compounds as a rule, are more effectively degraded in environments containing many microorganisms that in a pure culture of a single organism. This is because the range of degradative capabilities represented in a complex community of many bacteria and fungi is far greater than the capabilities of any single organism. The concerted action of several different organisms may lead to complete mineralization of the xenobiotic. Furthermore, a microbial community is likely to be more resistant than a solitary species to a toxic product of biodegradation, since one of its members may be able to detoxify it.

Several physical, chemical and biological factors may determine the fate of organic compounds introduced into the soil. Particular molecules may be removed by volatilization or leaching, or adsorption. The molecules may be degraded photo chemically, or undergo biodegradation though the action or bacteria and fungi. Sometime the products of nonbiological ad biological degradation may be same or otherwise nonbiological degradation may be very slow compared to biotransformation.

In nature, microbial communities are likely to be exposed to mixture of organic compounds and heavy metals (Abbas et al., 2014). Many strains are unable to grow in the presence of heavy metals (Zhiqi et al.,1999). The order of resistance to heavy metals in generally, fungi > acticomycetes > Gram – negative bacteria > Gram – positive bacteria. Therefore, the bacteria which can degrade certain persistent organic chemicals, such as polycyclic aromatic hydrocarbons or chlorinated organic compounds, may disappear from the environment, when heavy metals are also present. Consequently those organic chemicals may persist much longer in such an environment.

FORTUITOUS BIODEGRADATION AND CO – METABOLISM

These are two key factors influencing the degradation of *xenobiotics*. Since the stereo specificity of enzymes may not always be exact, biodegradation of *xenobiotic* compounds can be possible by naturally occurring enzymes (Koning et al., 2000). If the functional groups of these enzymes do not greatly alter the charge make up of active site, then it is possible for the enzyme to catalyze its specific reaction on the *xenobiotic* compound. This appropriation of an existing enzyme toward a novel substrate is called fortuitous or gratuitous biodegradation, i.e., when an enzyme is able to transform a compound other than its natural substrate. Since bacteria and fungi are so diverse in their metabolic capabilities, they produce enzymes able to act on a wide range of organic molecules. If these organisms use the energy released and can utilize the resulting carbon skeletons for growth, the degradation will proceed as in natural substrates.

If bacteria are unable to extract energy from the reactions, they can effect degradation of xenobiotic compound only through the use of additional carbon and energy sources provided in the environment. The ability of an organism to transform a nongrowth substrate as long as a growth substrate or other transformable compound, called co-substrate, is also present, is defined as Co-metabolism. A non growth substrate is one that cannot serve as the sole source of carbon and energy for a pure culture of a bacterium and therefore cannot support cell division (Santini et al., 2002).

A single reaction step catalyzed by an enzyme is called transformation, and the transformation product may be transformed by a second species until, finally, a product results which is a normal metabolite and can be completely utilized as an energy and carbon source. Therefore complete pathways of degradation are more likely to be operational through the combined effects of different types of organisms within a community, than a single species. Biodegradable analogues of the pollutant, may be even the pollutant itself, may induce the enzymes needed and may enrich microorganisms capable of metabolic degradation of the organic pollutant.

An attempt to estimate the role of co-metabolism in biological oxidation of synthetic compounds was made by Venkataramani and Ahlert (1985) using high strength complex industrial wastewater, and a model was developed with couples co – metabolism countributes substantially to over degradation. Also bacteria growing on mixed substrates have the ability to select for a specific substrate, which is preferentially utilized for the energy and carbon requirement of the cell.

Bioremediation is defined as the spontaneous or managed process in which biological catalysis acts on pollutants and thereby remedies or eliminates environmental contamination.

BIODEGRADATION OF WASTES

The most important and challenging practical application of microbes is in the treatment of different types of waste materials, since they occur in highly diversified and time variable composition. By weight and volume, the solid wastes (urban garbage, crops and food processing wastes and manures) are the most significant (Rani et al., 2010). These are either burnt for energy, normally decomposed, anaerobically digested, dumped in landfills or bio converted into diverse products Fig-6.

Some examples of groups of organisms capable of degrading different solids are given in **Table-4**. However, compared to solid and gaseous wastes, liquid wastes are the most troublesome because of their unusual contents of non retractable substances. The amounts of waste materials discharged around the

Figure 6. General Waste Treatment Processes

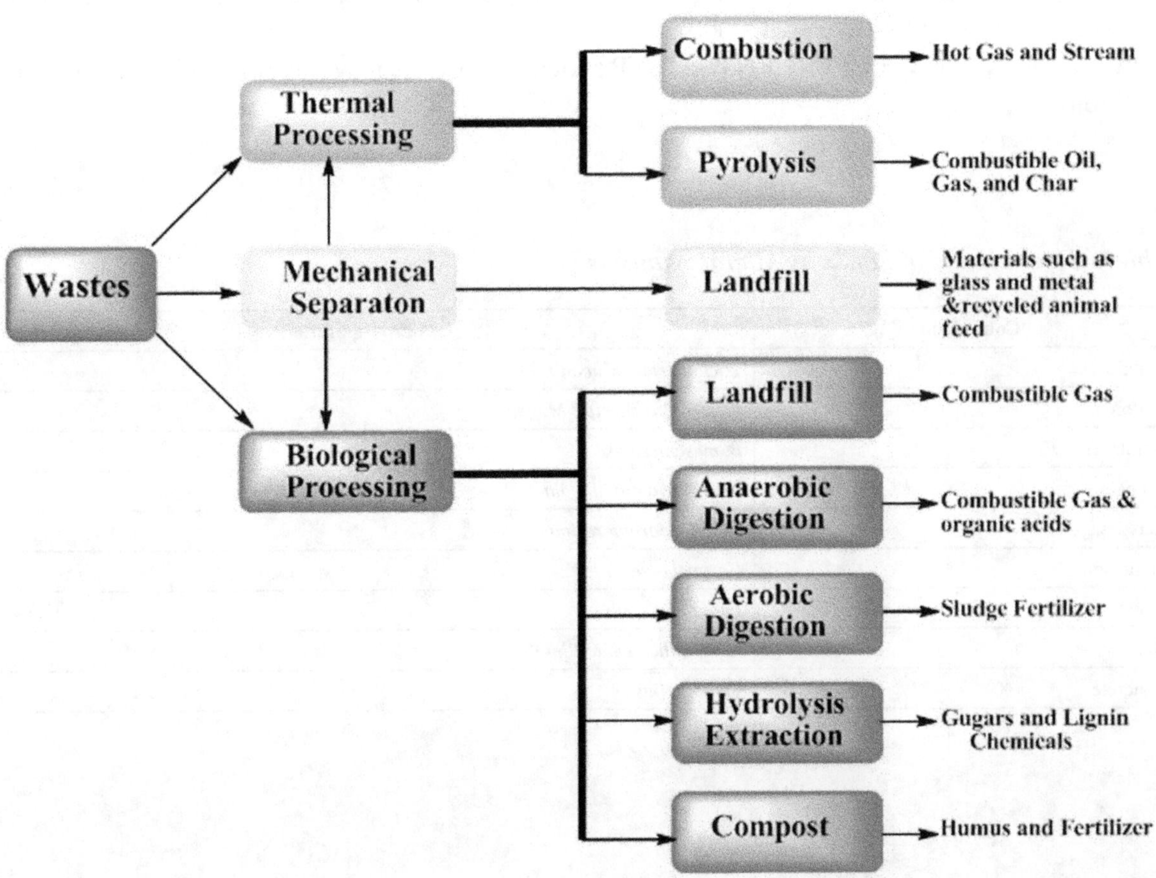

Figure 7. Three stages involved in waste water treatment system

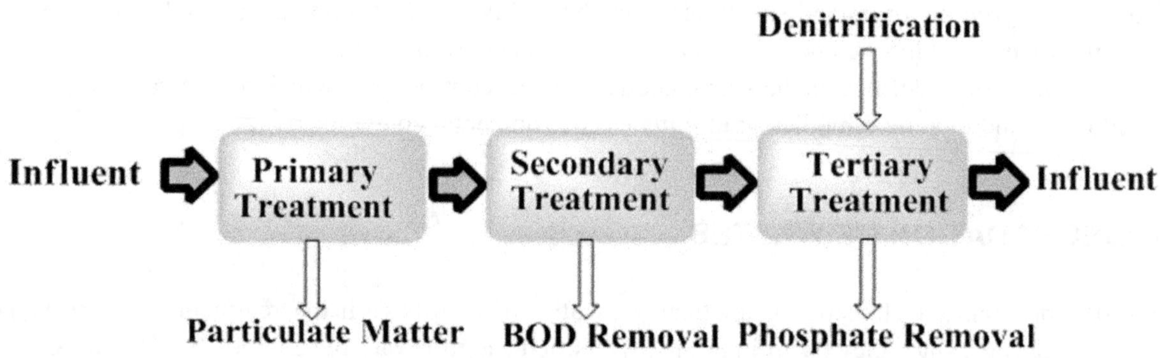

world are rising steadily as population increases and dependence of agriculture on fertilizer and pesticides increases. Also, the expansion of food industry and growth of other industrial processes contribute to volume of sewage and waste water. Thre treatment processes involve primary, secondary and tertiary treatment operations, of which the secondary and tertiary are multistage biological system Fig-7.

The secondary treatment exploits the degradative capabilities of various bacteria and fungi, whereas tertiary treatment is necessary to prevent eutrophication (i.e., reduction of dissolved oxygen in the water, resulting from bacterial and algal growth promoted by discharge of phosphate and nitrate), and to render the treated water potable(Jones et al.,2000).

PURIFICATION OF EFFLUENTS: The Purification process may involve aerobic and anaerobic condition.

Table 4. Degradation of Solids by Microorganisms

Compounds	Microorganisms
Cellulose	*Trichoderma, Mucor*
Cotton	*Aspergillus species, Mucor*
Animal fibers	*B. mesentericus*
Oil paints	*Altermaria dianthicola*
Kerosine	*Claudosporium resinae*
Picric acid	*Corynebacteria*
Wood	*Fungi*
Glass	*Aspergillus, Penicillim*
Concrete	*Thiobacillus*

PURIFICATION OF EFFLUENTS WITH AEROBIC PROCESSES

In the aerobic phase of the treatment process, a mixed microbial population degrades the complex mixture of organic compounds in liquid waste of carbonate, nitrate, ammonia, phosphate and sulphate. For this purpose either **A.** activated sludge processes, **B.** fixed film processes are used.

Activated Sludge: The principle of activated sludge is that in a reactor a community of microorganisms is constantly supplied with organic matter and oxygen. The microorganisms consume the organic matter and transform it by means of aerobic metabolism, partly into new microbial biomass and partly into carbon dioxide, water minerals (Dixit et al., 2015). The reactions involved are:

1. Sorption of soluble colloidal and suspended organic in and on the sludge flocs.
2. Biodegradation of the organics resulting in end products (i.e., carbondioxide, water, minerals) and synthesis of new microbial biomass.
3. Ingestion of bacteria and possibly of other suspended matter b protozoa and other predators
4. Oxidation of ammonium to nitrite and further to nitrate by nitrifying bacteria.
5. When there is insufficient supply of energy (or waste) – oxidation of cell reserves resulting in sludge mineralization and lysis.

There are several process types in which three fundamental aspects can be varied independently i.e.

1. The layout (convential completely mixed or gradient in substrate or aeration supply)
2. Loading rate (high rate, conventional and low rate or very low i.e. extended aeration)
3. The aeration system (surface aeration or submerged aeration.)

MICROBIAL COMMUNITY OF ACTIVATED SLUDGE

The incoming water contains a multitude or organic and inorganic substrates. A diverse assemblage of bacteria, fungi, protozoa and rotifers, living directly on the incoming substrates using metabolites thus produced, or preying on primary users, come into existence. The activated sludge community is normally dominated by organotrophic bacteria. The major part is aggregated in sludge flocs, but some are freely dispersed in the liquor. The bacteria, together with some saprobic protozoa, from the lowest trophic level, on which holozoic protozoa and rotifers feed.

Some, commonly encountered bacterial genera in activated sludge, and their niche specialities are;

1. *Pseudomonas* - Removal of carbohydrates, and denitrificaiton.
2. *Arthrobacter* – Removal of carbohydrates
3. *Baccillus* – Breakdown of Protein
4. *Cytophaga* – Breakdown of Polymericsubstrates
5. *Zoogloea* – Formation of slime matrices
6. *Acinetobacter* – Storage of Polyphosphate
7. *Nitrosomonas* and *Nitrobacter* – Nitrification
8. *Sphaerotolus* – Filamentous growth

Fungi and yeasts are usually present, but are not capable of dominating, except in unusual conditions, e.g. High levels of sugars or constant influx of recralicitrant organo chemicals. Some commonly found protozoa are; *Aspidisca costata* (crawling), *Vorticella convallaria* (sessile), *Trachelophyllum pusillum* (free swimming).

Fixed Film Processes: These attached growth or fixed film treatment systems are amongst the oldest of the currently used waste water treatment processes i.e., those in which microorganisms for waste treatment are attached to stationary media or rotating media, e.g. Trickling filters, rotating biological contractors, activated biofilters, biological towers, submerged filters and biological fluidized beds. The objective of all these processes is the removal of dissolved and suspended organic matter. Some processes may be capable of providing nitrification or denitrification and also removal of phosphorus and may be used in association with tertiary treatment. The organisms responsible for treatment are present in the form of a biofilm, attached to a solid support surface. Waste water either flows over the film or the support surface moves through waste water. Organics are transported to microorganisms in the biofilm and subsequently biodegraded.

The biological slime layer or biofilm contains bacteria, protozoa and fungi and may also contain algae, sludge worms, filter fly larvae, rotifers or other metazoan. As waste water passes over biofilm, soluble organic diffuse into the film and are sorbed on to the microorganisms and metabolized. Colloidal organics adsorb onto the biofilm surface and can be solubilized by extracellular enzymes. Molecular oxygen also diffuses into the film, providing a terminal electron acceptor for aerobic metabolism. Heterotrophic organisms in upper part of film, obtain energy and nutrients by oxidizing organic matter and therefore multiply and new film is formed.

PURIFICATION OF EFFLUENTS WITH ANAEROBIC PROCESSES

After aerobic digestion, sewage sludge is generally digested under anaerobic condition. Materials with high content of insoluble organic matter e.g. Cellulose, concentrated industrial wastes e.g. those from food processing industry, are commonly treated by anaerobic digestion. The capacity of anaerobic bacteria for degrading organic substances to methane and carbon dioxide has been utilized since a very long time. The other products produced during anaerobic degradation (sludge and effluent) can be reused and nutrients recovered. Because of increasing energy costs, methane fermentation or *Biomethanation* with high loads of organic matter has attracted great interest. It was assumed for a long time that only two groups of bacteria are involved in the anaerobic conversion of organic substances into methane and carbon dioxide. It is well known now that at least three different groups of bacteria are involved i.e., (i) Hydrolytic and Acid forming, (ii) Acetogenic (acetic acid and hydrogen forming), (iii) Methanogenic (Methane Forming).

In the degradational chain, the various organic compounds are degraded to acids, alcohols, hydrogen, and carbon dioxide in the first step by the first group of microorganisms Figure-8.

The second group of bacteria then converts these alcohols, acids etc., into acetic acid, hydrogen and carbon dioxide, which are in turn used as substrate by the last group, i.e. methane bacteria, and are converted into methane and carbon dioxide. Some examples of fermentative, hydrolytic bacteria are *Clostridia, Bacteroides, Bifidobacteria, Enterobacteria and Streptococci. Syntrophomonas* and *Desulphovibrio* belong to the second group i.e. the Acetogenic bacteria and the third group include *Methanabacterium, Methanobrevibacter, methanococcus, Methanomicrobium, Methanospirillu.* This

Figure 8. Scheme of Fermentation of organic substances to biogas

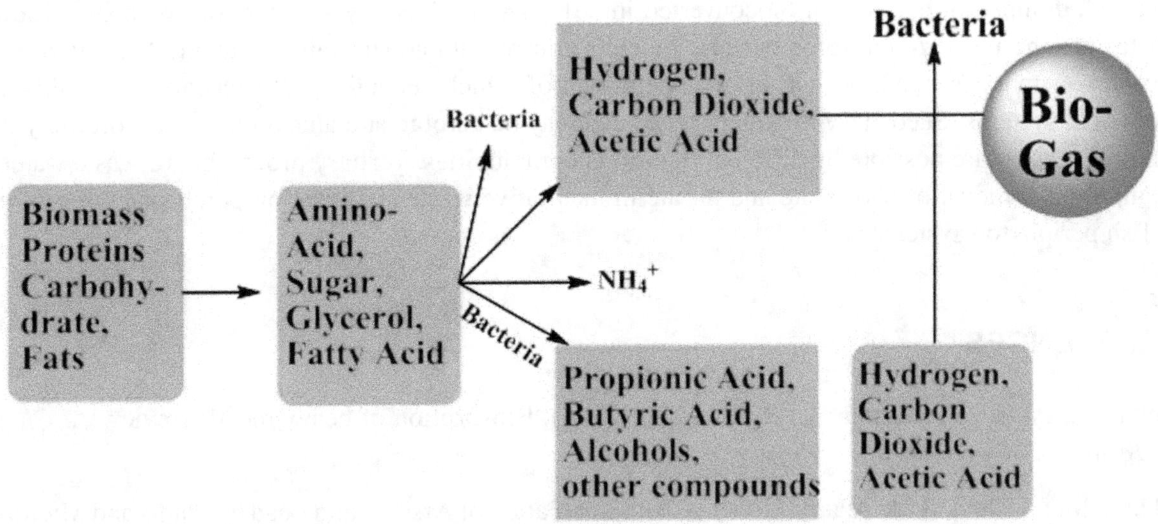

last group is the most oxygen sensitive and require a redox potential of about – 300 mV and some get killed very rapidly by oxygen. They have a very limited substrate spectrum – most species assimilate hydrogen as the sole source of energy. For anaerobic degradation of butyric acid and propionic acid, very low pH_2 values (< 2 X 10^{-3} and < 9 X 10^{-3}) are necessary. Therefore the hydrogen produced must be consumed immediately at the position where it arises, for which an intimate contact between cells of acetogenic and methanogenic bacteria is necessary.

Thus maintenance of pH and constant temperature are important requirements for the processes. An important practical aspect of anaerobic digestion is that most of the free energy present in the substrate is conserved in the methane that is produced. Using the methane for energy generation offsets the cost of the sewage treatment process. The other advantages are that higher organic loadings are possible, a smaller volume of sludge is produced, and the energy consuming aeration is unnecessary.

CONCLUSION

The outcome of consumption and production activities is the generation of large amounts of waste and release of pollutants. For an environmentally compatible sustainable development, application of clean technologies such as, use of microbes in degradation and bioremediation, is urgently reqired. For understanding these knowledge of their microbiological processes and the microbiological communities involved in different types of waste treatment processes aerobic and anaerobic, is very important. Recent development in the bioremediation strategies and the use of specialized microbial strains for degradation of recalcitrant compounds are important aspects to be looked into.

Microbial degradation technologies render harmless substances which cannot be utilized profitably and include both natural and chemically synthesized substrates. Primary or particle biodegradation may be possible depending on interactions such as communalism and mutualism. Understanding of fortuitous biodegradation and co- metabolism, is required for the degradation of xenobiotics. By weight and volume solid wastes are significant and are either burnt for energy, decomposed naturally or are anaerobically digested, dumped in landfills or bioconverted into diverse products by a variety of organisms. Liquid wastes are the most troublesome because of their non retraitable substance content. Their treatment involves primary, secondary and teriary operations of which secondary and tertiary are multistage biological systems. Secondary treatment processes may be aerobic and anaerobic and accordingly different processes are possible by different microbial communities. Tertiary processing of waste water is required for removal of phosphate and nitrate in facultative waste stabilization ponds or by recent use of fish periphyton system.

REFERENCES

Abbas, S., Ismail, I., Mostafa, T., & Abbas, S. (2014). Biosorption of heavy metals [review]. *J ChemSciTechnol*, *3*, 74–102.

Akhtar, M. S., Chali, B., & Azam, T. (2013). Bioremediation of Arsenic and Lead by Plants and Microbes from Contaminated Soil. *Res. Plant Sciences.*, *1*(3), 68–73.

Alegbeleye, O. O., Opeolu, B. O., & Jackson, V. A. (2017). Polycyclic aromatic hydrocarbons: A critical review of environmental occurrence and bioremediation. *Environmental Management*, *60*(4), 758–783. doi:10.100700267-017-0896-2 PMID:28573478

Benson, H. J. (1990). *Microbiological Application – A Laboratory Manual in General Microbiology. C.* Brown Publishers.

Boopathy, R. (2000). Factors Limiting Bioremediation Technologies. *Bioresource Technology*, *74*(1), 63–67. doi:10.1016/S0960-8524(99)00144-3

Brooks, R. R. (1998). *Plants that hyperaccumulate heavy metals*. Cambridge University Press.

Chen, M., Xu, P., Zeng, G., Yang, C., Huang, D., & Zhang, J. (2015). Bioremediation of soils contaminated with polycyclic aromatic hydrocarbons, petroleum, pesticides, chlorophenols and heavy metals by composting: Applications, microbes and future research needs. *Biotechnology Advances*, *33*(6), 745–755. doi:10.1016/j.biotechadv.2015.05.003 PMID:26008965

Chojnacka, K. (2010). Biosorption and bioaccumulation – the prospects for practical. *Appl. Environ.*, *36*, 299–307. PMID:20051290

Congeevaram, S., Dhanarani, S., Park, J., Dexilin, M., & Thamaraiselvi, K. (2007). Biosorption of chromium and nickel by heavy metal resistant fungal and bacterial isolates. *Journal of Hazardous Materials*, *146*, 270–277. PMID:17218056

Dixit, R., Wasiullah Malaviya, D., Pandiyan, K., Singh, U. B., Sahu, A., Shukla, R., Singh, B. P., Rai, J. P., Sharma, P. K., Lade, H., & Paul, D. (2015). Bioremediation of heavy metals from soil and aquatic environment: An overview of principles and criteria of fundamental processes. *Sustainability*, *7*, 2189–2212.

Faryal, R., Sultan, A., Tahir, F., Ahmed, S., & Hameed, A. (2007). Biosorption of lead by indigenous fungal strains. *Pakistan Journal of Botany*, *39*, 615–622.

Gogoi, B. K., Dutta, N. N., Goswami, P., & Krishna Mohen, T. R. (2002). A Case Study of Petroleum hydrocarbon Contaminated Soil at a Crude Oil Site. *Advances in Environmental Research, 7*(4), 767–782.

Guidi, W., Palm, E., Mancuso, S., & Azzarello, E. (2018). Trace element phytoextraction from contaminated soil. a case study under Mediterranean climate. *Environmental Science and Pollution Research International, 18*, 1197–1205.

Held, T., & Don, H. (2000). *In Situ* Remediation. *Biotechnology: An Indian Journal, 11b*, 350–370.

Hlihor, R. M., Gavrilescu, M., Tavares, T., Fvier, L., & Olivieri, G. (2017). Editorial bioremediation: An overview of current practices, advances, and new perspectives in environmental pollution treatment Hindawi. *BioMed Research International, 2*, 3–5.

Iranzo, M., Sainz-Padro, I., Boluda, R., Sanchez, J., & Mormeneo, S. (2001). The Use of Microorganisms in Environmental Engineering. *Annals of Microbiology, 51*, 135–143.

Jain, R., Ahmad, P., Gadgil, K., & Sarma, S. (2012). Heavy metal toxicity: Effect of plant growth, Biochemical parameters and metal accumulation by Brasica juncea. *L. Int.J. Plant Production., 3*(3), 65–75.

Jones, C. A., Langer, H. W., Anderson, K., McDermott, T. R., & Inskeep, W. P. (2000). Rates of microbially mediated arsenate reduction and solubilization. *Soil Science Society of America Journal, 64*, 600–608.

Khan, M. S., Zaidi, A. A., Wani, A. P. A., & Mohammad, A. (2009). Role of plant growth promoting rhizobacteria in the remediation of metal contaminated soils. *Environmental Chemistry Letters, 7*, 1–19.

Kleijntjens, R. H., & Luyben, K. C. A. M. (2000). Bioremediation. *Biotechnology: An Indian Journal, 11b*, 329–347.

Klein, J. (2000). Possibilities, Limits and Future Developments of Soil Bioremediation. *Biotechnology: An Indian Journal, 11b*, 466–476.

Koning, M., Hupe, K., & Stegmann, R. (2000). Thermal Processes, Scrubbing/Extraction, Bioremediation and Disposal. *Biotechnology: An Indian Journal, 11b*, 306–317.

Konopka, A., Zakharova, T., Bischoff, M., Oliver, L., Nakatsu, C., & Turco, R. F. (1999). Microbial biomass and activity in lead-contaminated soil. *Applied and Environmental Microbiology, 65*, 2256–2259. PMID:10224032

Naz, T., Khan, M. D., Ahmed, I., Rehman, S. U., Rha, E. S., & Malook, I. (2015). Biosorption of heavy metals by Pseudomonas species isolated from sugar industry. *Toxicology and Industrial Health, 10*, 525–534. PMID:25739395

Pandey, S., Saha, P., Biswas, S., & Maiti, T. K. (2011). Characterization of two heavy metal resistant strains isolated from slag disposal site at Burnpur. *Indian J. Environ Biol., 32*, 773–779. PMID:22471215

Rani, A., Ray, K., & Goel, R. (2010). *Strategies for crop improvement in contaminated soils using metaltolerant bioinoculants*. Springer.

Santini, J. M., Stolz, J. F., & Macy, J. M. (2002). Isolation of a new arsenate respiring bacterium –physiological and Phylogenetic studies. *Geomicrobiology Journal, 19*, 41–52.

Schnoor, J. L. (2000). Degradation by Plants-Phytoremediation. *Biotechnology: An Indian Journal, 11b*, 372–384.

Schulz-Berendt, V. (2000). Bioremediation with Heap Technique. *Biotechnology: An Indian Journal, 11b*, 320–328.

Shannon, M. J. R., & Unterman, R. (2000). Evaluation Bioremediation: Distinguishing Fact from Fiction. *Annual Review of Microbiology*, *47*, 715–738. PMID:8257113

Van Deuren, J., Lloyd, T., Chhetry, S., Raycharn, L., & Peck, J. (2002). Remediation Technologies Screening Matrix and Reference Guide. *Federal Remediation Technologies Roundtable, 5*, 1-4.

Zhiqi, H., Reiske, H. R., & Wilson, D. B. (1999). Chatacterization of cadmium uptake in Lacto plantarum and isolation of cadmium and manganese uptake system. *Applied and Environmental Microbiology*, *65*, 4741–4745. PMID:10543780

Chapter 15
Use of Wastage Biomaterial/ Biomass for Wastewater Treatment and Modeling:
Mathematical Modeling for Adsorption Processes

Biswajit Singha
Camellia Institute of Technology, India

ABSTRACT

Recently, the existence of toxic materials in our environment has caused severe concern to the living organism. The main causes for the contamination of such toxic materials in the environment are rapid industrialization and agricultural activities. The different toxic entities present in the environment mainly depend on the nature of the industry. However, some of the common generated toxic pollutants present in the water bodies are heavy metal ions, phenol, dyes, pesticides, insecticides, detergents, and a wide spectrum of aromatic compounds. Among those substance, heavy metals have been reported the major one for the last few decades. So, these heavy metals must be removed from the environment by proper techniques to save the society. The chapter deals with the different toxic material removal technologies and mathematical modeling.

INTRODUCTION

Arsenic, cadmium, copper, chromium, zinc, lead, nickel, mercury, etc. are such type of heavy metals which have lethal effects to the living beings. Due to its non-biodegradability nature and tend to accumulate in living organism, thus enter into the food chain. The concentration of such type of toxic heavy metals in the water bodies must be reduced and the industries are advised to treat their effluents before discharged to acceptable level into the environment.

DOI: 10.4018/978-1-7998-4921-6.ch015

Arsenic is one of the important toxic heavy metal that contaminated in water, especially groundwater. It has been recognized as a major water pollutant. The main source of arsenic in the environment from smelting, rock sediments, energy production from fossil fuels and mining industries etc. The toxicological effect of it in the living beings are gastrointestinal symptoms, bone marrow depression, disturbances of cardiovascular and nervous system functions, hepatomegaly, hemolysis, melanosis and liver tumor etc. So it is very important to eliminate from our environment.

U.S. Environmental Protection Agency classified that cadmium is a probable human carcinogen. It is entered into the environment from refining, mining, pigments, plastic, alloy manufacturing, smelting and electroplating industries etc. The discharged cadmium in effluent absorbed and accumulated by microorganisms. Eventually, it is transferred to living beings through food chain. It causes *Itai-Itai* diseases, renal disturbances, lung insufficiency, bone lesions, cancer, hypertension, weight loss etc.

Catalytic activity of a number of enzymes is controlled by copper. But it is toxic when deficient or in excess in the living beings. Electronics plating, printed circuit board manufacturing, copper polishing, wire drawing, paint manufacturing wood preservatives and printing operations industries are the main source of copper in wastewater. Excessive copper accumulation causes neurotoxicity, reproductive and developmental toxicity, acute toxicity, dizziness and diarrhea etc.

Chromium exits mainly trivalent and hexavalent oxidation states in the aqueous solution. Microorganism accumulates trivalent chromium as an essential trace metal nutrient but hexavalent state is a great concern because of its toxicity. The major source of hexavalent chromium are leather tanning, electroplating, metal processing, textile dyeing, paints and pigments, wood preservatives, steel fabrication and canning industries etc. It is reported as carcinogenic, teratogenic, mutagenic, epigastric pain nausea, vomiting, severe diarrhea, producing lung tumors etc.

Zinc is a micronutrient, essential for physiological and biochemical processes of living tissue. It enters into the water bodies from mining and manufacturing process industries. Too much concentration of zinc causes gastrointestinal distress, short term ''metal-fume fever'', diarrhea, nausea, etc.

There is large number of non-biodegradable toxic heavy metal existing in our environment but lead is one of the them which enter into the living being through the food chain. It damages brain, causes anorexia, anaemia, malaise, loss of appetite, diminishing IQ etc. The main source of this toxic heavy metal in the environment is battery manufacturing, electroplating, ammunition, pigments industries etc.

Trace amount of nickel is required for humans and animals but medical science reported that in excess concentration it can be very toxic, even carcinogenic. Nickel containing effluents enter in the environment from electroplating, paint formulation, mineral processing, non-ferrous metal, porcelain enameling, copper sulphate manufacture and steam-electric power plants industries etc. Lung cancer, chronic bronchitis, reduced lung function etc. are caused by the toxic effect of nickel.

One of the highly toxic heavy metal in our environment is mercury. It has high tendency to effects the neurological and renal disturbances, kidney damage, impairment of pulmonary function, corrosive to skin, eyes, muscles etc. Once mercury enters into the food chain, it accumulates into the humans and animals. Minamata disease was caused by mercury pollution in Japan (1953-1960). The most toxic species are, however, organic mercury, especially CH_3Hg^+ as it is soluble in fat, lipid of membranes and brain tissue. The main source of this toxic heavy metal are volcanic eruptions, weathering of mercuriferous areas, battery production, naturally-caused forest fires, mining and metallurgical processes, fossil fuel burning, biogenic emissions, paint and chloralkali industries etc.

BACKGROUND

Environmental contamination and exposure due to heavy metals is a serious growing problem throughout the world. Metal ions and their complexes exhibit a wide range of toxicity to the organisms that ranges from sub lethal to lethal depending on the time of exposure and the prevailing conditions in the ambient medium. The toxicity of metals can be greatly modified by physico-chemical and biological factors, which can interfere with the toxicity in antagonistic or synergistic manners. Hardness has been reported as an antagonistic factor in reducing toxicity of heavy metals. Other factors like temperature, pH, dissolve oxygen and salinity, can modify the physiology of organisms and the chemical from the metals. Factors affecting the toxicity may be broadly classified into two groups: metal-related and host-related. Metal related factors include abundance, chemical form (elemental, inorganic or organic), speciation of oxidation state, ionizability particle size of exposure, and irritant, corrosive, cytotoxic, mutagenic and carcinogenic properties of metallic compound. Host related factor include susceptibility route of exposure, (oral inhalatory, dermal, parenteral, implants, etc.), bioavailability, adsorption, bindining, excretion ability to penetrate blood-brain and plasmental barriers, interactions with dietary nutrients, other elements, other elements, alcohol and drugs, critical concentration nutritional and immune status, target organ/tissue, and critical effect/host response. Sub lethal toxicity may be in the form of changes in morphology, histology, growth, development, swimming performance, respiration, blood chemistry, enzyme activity, endocrinology, reproduction and behavior of the organs. Hence, there is a need for the removal of heavy metals from waste water.

Rapid globalization and industrialization have lead to an increased in the generation of various types of toxic heavy metals, which are the main cause of water pollution on the global scale. If the industrial discharges are emitted without treatment, they may have an adverse effect on living being and the environment, even at very low concentration. Due to the toxic effect of such heavy metals, it is necessary to eliminate those entities from the contaminated water bodies as per permissible level.

Tolerance Limits of Toxic Heavy Metals

The different National/International agency recommended the permissible limit of different toxic heavy metals concentration in the drinking water (EPA, 1990; IS 10500, 1992, WHO, 1993). It is reported in Table 1. Therefore, the concentration of this toxic pollutants discharged from different industries must be reduced to level that satisfy environmental regulations and maintaining the healthy condition of our environment.

Wastewater Treatment Procedure

Presently, it is challenging task to the environmental scientists and pollution control agencies for economical treatment of these hazardous metal containing wastewater. Different water treatment technologies are being used such as ion-exchange, chemical precipitation, coagulation & flocculation, membrane filtration, electrochemical treatment, cementation, photocatalysis, adsorption using activated carbon and natural /agricultural wastes etc. (Babel and Kurniawan, 2004; Bernardo et al., 2009; Eckenfelder and Ford, 2009; Naiya et al., 2009a; Patterson and Jancuk, 1977; Chowdhury et al., 2014). All methods have some disadvantages and limitations. Adsorptions using wastage biomaterial/biomass have some advantage such as low-cost and locally available, although they are sensible to pH and temperature.

Table 1. Tolerance limits for toxic heavy metal concentration in drinking water as per National/International agency

Metal ion	EPA [1990]	IS 10500, 1992 [Reaffirmed 1993]	WHO [1993]
Arsenic (mg/L)	0.01	0.05	0.06
Cadmium (mg/L)	0.005	0.01	0.003
Copper (mg/L)	1.3	0.05	2.0
Hexavalent chromium (mg/L)	0.1	0.05	0.05
Zinc (mg/L)	5.0	5.0	5.0
Lead (mg/L)	0.015	0.05	0.01
Nickel (mg/L)	0.1	-	0.02
Mercury (mg/L)	0.002	0.001	0.001

Ion Exchange

Environmental scientists developed different physicochemical method to treat the toxic heavy metal containing wastewater, ion exchange method is one of them. It is widely used due to, high removal efficiency, high treatment capacity and fast kinetics. Ion-exchange resin, natural solid or either synthetic resin has the specific ability to exchange its cations with the toxic heavy metals present in the wastewater. Synthetic resins are the more preferable material used in the ion exchange process for the wastewater treatment.

Chemical Precipitation

One of the effective water treatment methods is chemical precipitation. In this process chemicals react with the toxic metal ions present in the released industrial effluents to form insoluble precipitates, these precipitates can be separated from the water by sedimentation or filtration, and the treated water is then decanted and appropriately discharged or reused. This method is inexpensive to operate and relatively simple. Hydroxide and sulfide precipitation are the conventional method of chemical precipitation.

Coagulation and Flocculation

Removal of toxic heavy metal from industrial effluents also carried out using coagulation and flocculation followed by sedimentation and filtration. Coagulation is the destabilization of colloidal particles by adding a coagulant. Ferrous sulfate, ferric chloride, alam etc are widely used as coagulants for the conventional wastewater treatment process. It results in the effective removal of wastewater particulates and impurities by charge neutralization of particles and by enmeshment of the impurities on the formed amorphous metal hydroxide precipitates. Flocculation is the action of polymers to form bridges between the flocs and bind the particles into large agglomerates or clumps. Once suspended particles are flocculated into larger particles, they can usually be removed or separated by filtration, straining or floatation. Polyferric sulfate and polyacryl amide are widely used as flocculants for the treatment of wastewater.

Membrane Filtration

Another important water treatment technology is membrane filtration. In this process different types of membranes show great promise for heavy metal removal for their high efficiency, easy operation and space saving. The membrane processes used are reverse osmosis, ultrafiltration, electrodialysis and nanofiltration etc. This process is very much expansive compare to other one due to high cost of membrane and its operational cost.

Electrochemical Treatment

The toxic metal ions are plating out on a cathode surface by electrochemical method. In this process metals can be recovered in the elemental state. The relatively large capital investment and expensive electricity supply is required for this process. So it has not been widely applied. However, with the stringent environmental regulations regarding the wastewater discharge, electrochemical technologies have regained their importance worldwide during the past few decades.

Cementation

It is one type of electrochemical precipitation of metal ion by another more electropositive metal to zero valence state. A common example is cementation of copper. In this process copper ions from solution is recover by electrochemical reduction to the elemental state in presence of solid iron. The reaction is as follows

$$Cu^{2+} + Fe^{0} \rightarrow Cu^{O} + Fe^{2+}$$

This process is used industrially to recover a variety of heavy metals. The advantages of this process are simple to control, low energy utilization and recovery of very high purity metals.

Photocatalysis

Photocatalysis is a promising method for the industrial wastewater treatment. Many different semiconductors are used like ZnS, ZnO, TiO2, CdS, and CeO_2.

Adsorption Using Activated Carbon

Metal transfer from solution phase to solid phase occurred by adsorption process. Basically, it is mass transfer process by which a substance is transferred from the liquid phase to the surface of a solid, and becomes bound by physical and or chemical interactions. This process is widely used to remove toxic heavy metal ions from industrial effluents. Activated carbon is one of the important adsorbent used for this purpose. It is a crude form of graphite with a random or amorphous structure, which is highly porous, exhibiting a broad range of pore sizes, from visible cracks, crevices and slits of molecular dimensions. Adsorptive properties of activated carbons are quietly high due to micro-porous structure, high surface area, and a high degree of surface reactivity.

Adsorption Using Natural/Agricultural Wastes

Recently natural or agricultural wastes materials have been widely used for the adsorption process due to low-cost, locally available and eco-friendly for environment.

Advantage and Disadvantage of Different Removal Technologies

Although different removal process has been applied for the treatment of industrial effluents but they have some disadvantage and limitations. Adsorption process using natural/bio-adsorbents are more applicable due to its simplicity, low cost, high adsorption capacity and wide range of adsorbents availability. Table 2 represents the advantages and disadvantage of different removal processes.

Table 2. Advantage and disadvantage of different metal ion removal process

Removal Process	Advantage	Disadvantage
Ion exchange	1. Metal selective 2. High regeneration of materials	1. Less number of metal ions removed 2. High cost
Chemical precipitation	1. Inexpensive 2. Simple method 3. Most of metals can be removed	1. Large amounts of sludge produced 2. Disposal problems
Coagulation & flocculation	1. Cheap process	1. Disposal of precipitated material possess major problem 2. Process complicated
Membrane filtration	1. Less solid waste produced 2. Less chemical consumption 3. High efficiency (>95% for single metal)	1. High initial and running cost 2. Low flow rates 3. Removal (%) decreases with the presence of other metals
Electrochemical treatment	1. No consumption of chemicals 2. Metal selective 3. Pure metals can be achieved	1. High capital cost 2. High running cost 3. Initial solution pH and current density
Cementation	1. Simple process 2. Low energy utilization 3. Recovery of very high purity metals.	1. Dissolution of the sacrificial metal into the solution.
Photocatalysis	1. Promising technology	1. Cost effective technology
Adsorption using activated carbon	1. Most of metals can be removed 2. High efficiency (>99%)	1. Cost of activated carbon 2. No regeneration 3. Performance depends upon adsorbent
Adsorption using natural/ agricultural wastes	1. Low cost 2. Local availability 3. High adsorption capacity	

ADSORPTION EXPERIMENTS

Adsorption experiments may be conducted by batch or column technique (Singha and Das, 2013; Mitra et al., 2014). Batch process applicable for small scale where as column process is used for the continuous and large scale purpose.

Batch Technique

Adsorption process is highly pH dependent. Contact time of adsorbent onto the metal ion, adsorbent dosage, initial metal ion concentration and temperature etc. play an important role for the effective adsorption process. Batch process optimizes those process parameters. Before the batch experiments, pH of the solution must be adjusted using acid or base solution. The metal ion solution of known concentration at particular pH and necessary amount of adsorbent must be added in conical flasks and content may be shaken in an electrically thermostated shaker as per require. After that the treated solution filtered using appropriate filter paper and filtrate analyzed using Atomic absorption spectrophotometer or UV visible spectrophotometer (Clescerl et al., 1998). The amount of metal ions ($q_{e,}$) retained in the adsorbent phase is estimated by the following equation,

$$q_e = \frac{\left(C_0 - C_t\right)V}{W} \tag{1}$$

The percentage of removed metal ions in the solution is calculated using following equation,

$$\%\text{ removal of metal ions} = \frac{\left(C_0 - C_t\right)}{C_0} \text{X}100\% \tag{2}$$

Where C_o is the initial metal ion concentration, C_t is the metal ion concentration at any time t, V is the volume of the solution and W is the amount of natural/agricultural wastes.

Effect of Different Parameters for Batch Process

Effect of Initial Solution pH

One of the most important parameter for the adsorption process is initial pH of the solution. Not only different metals show different optimum pH for their sorption but also may vary from one kind of biomass to the other (Tiwari et al., 1995). The pH of the solution or effluents play an important role for the determination of adsorption capacities due to its influence on the surface properties of aforesaid adsorbents and different ionic forms of the metal ions in the solutions. In general, adsorption of cation is favored at pH > pHpzc (Srivastava et al., 2006). Where pH_{PZC} is the point of zero charge of the adsorbents. This study helps in designing the appropriate pH of the effluent/wastewater to achieve maximum efficiency in the removal of toxic heavy metal ions by natural/agricultural wastes.

At very low pH, surface of the adsorbents would be surrounded by the hydronium ions. If metal ions exists as negatively charged species then at that pH adsorption capacities increases.

As for example at lower pH chromium ions exist in the form of $HCrO_4^-$, while in the increase in pH (up to pH = 6) different forms such as $Cr_2O_7^{2-}$, $HCrO_4^-$, and $Cr_3O_{10}^{2-}$, coexists, of which $HCrO_4^-$ predominates. As the pH increases equilibrium shifted form $HCrO_4^-$ to CrO_4^{2-} and $Cr_2O_7^-$. So at low pH, adsorption capacity increases. As the pH increased, the overall surface charge on the biosorbents became negative and adsorption decreased (Singha and Das, 2011). The following equilibrium may be written for the Cr(VI) anions present in aqueous solutions (Miretzkya and Cirelli, 2010).

$$H_2CrO_4 \leftrightarrow HCrO_4^- + H^+ K_1 = 1.21 \quad (3)$$

$$Cr_2O_7^{2-} + H_2O \leftrightarrow 2HCrO_4^- K_2 = 35.5 \quad (4)$$

$$HCrO_4^- \leftrightarrow CrO_4^{2-} + H^+ K_3 = 3 \times 10^{-7} \quad (5)$$

Adsorption of Cr(VI) is low at pH values more than 6 due to dual complexation of the anions (CrO_4^{2-}, $Cr_2O_7^{2-}$ and OH^-) to be adsorbed on the surface of the adsorbents, of which OH^- predominates (Mallick et al., 2006). This can also be explained by the property of point of zero charge of the adsorbents.

If metal ions exists as positively charged species then at low pH (pH 2-3) less metal ions sorption occurred. This is due to the competition between protonated surface of the adsorbents and positively charged metal ion species for occupancy of binding sites (Naiya et al., 2009a; Hasana et al., 2010; Reddy et al., 2010). As the pH increased and adsorbents surface functional groups activated, it resulting the increased metal ions adsorption. This phenomena may be explained by adsorption of Pb(II) ions. The Pb(II) ions may undergo hydrolysis and salvation in aqueous solution as follows (Naiya et al., 2009b),

$$Pb^{2+} + nH_2O = Pb(H_2O)_n^{2+} \quad (6)$$

$$Pb(H_2O)_n^{2+} = [Pb(H_2O)_{n-1}(OH)]^+ + H^+ \quad (7)$$

$$Pb^{2+} + nH_2O \rightleftarrows [Pb(H_2O)_{n-1}(OH)]^+ + H^+ \quad (8)$$

So at higher pH the removal efficiency increases for Pb(II) ions. But some difficulty arises due the precipitation of Pb(II) at pH>5. So at pH > 5, removal efficiency increased may be a combination of both adsorption and precipitation of lead hydroxide from the solution (Low et al., 2000; Saeeda et al., 2005).

Effect of Adsorbent Dosage

Generally, adsorption efficiency increases with the increasing adsorbent dosage. Initially metal ions uptake rate with the increment of adsorbent dosage increases rapidly then it became very low as the surface metal ion concentration and the solution metal ion concentration came to equilibrium with each other. With further increase in adsorbent dosage, the removal efficiency becomes almost remained constant. The variation of adsorption efficiency between the various adsorbents could be related to the type and number of the functional group responsible for the adsorption process (Naiya et al., 2009c). With increase in adsorbent dosage more surface area is available for adsorption due to increase in active sites on the adsorbent surface.

Effect of Contact Time

Adsorption of heavy metal ions is highly depended on contact time with adsorbent. It is obvious that increase in contact time the percentage removal increases significantly. Initially the adsorption rate is very high that gives away a very slow approach to equilibrium. The nature of adsorbent and its available sorption sites affected the time needed to reach equilibrium. When adsorption process reaches to equilibrium than its rate decreases significantly with contact time variation.

Effect of Initial Metal Ions Concentration

The percentage of adsorption decreases with the increase in initial metal ions concentration. At lower concentration, all the metal ions in the solution would react with the binding sites of adsorbent and thus facilitated almost complete adsorption. At higher concentration more metal ions left un-adsorbed in the solution due to the saturation of the binding sites. This indicates that the energetically less favorable sites become involved with increasing metal ions concentration in aqueous solutions (Kadirvelu and Namasivayam, 2003; Bhattacharya et al., 2006).

Kinetic Modeling Using Batch Experiment Data

The kinetics of adsorption studies describe the solute uptake rate and evidently this rate controlled the residence time of adsorbate uptake at the solid–solution interface including the diffusion process. Physical and chemical properties of adsorbents are highly responsible for the adsorption mechanism as well as on the mass transfer process. The experimental data from contact time variation experiment may be analyze to predict the adsorption mechanism. The different kinetic models are Pseudo-first order, Pseudo-second order, Intra particle diffusion, Reichenberg, Elovich model, Prediction of rate limiting step, Mass transfer analysis, Determination of diffusivity etc. The conformity between experimental data and the model predicted values are expressed by correlation coefficient and chi-square test. The high correlation coefficients and low chi-square values of any model indicate the applicability of this one.

Pseudo-First Order Kinetic Model

The Pseudo first-order kinetic model (Lagergren, 1898) is generally expressed as follows,

$$\frac{dq_t}{dt} = k_1\left(q_e - q_t\right) \tag{9}$$

After integration and applying boundary conditions

t=0 $q_t = 0$

and $t=t$ $q_t=q_t$

The integrated form of Eq. (9) became,

$$\log(q_e - q_t) = \log q_e - \frac{k_1}{2.303}t \tag{10}$$

Where k_1 is the Pseudo first-order rate constant, q_e is the amount adsorb per unit mass of the adsorbent at equilibrium, q_t is the amount adsorb per unit mass of adsorbent at any time t. The values of $\log(q_e - q_t)$ may be linearly correlated with t and the rate constants are also determined from the slope of the plot.

Pseudo-Second Order Kinetic Model

The Pseudo-second order adsorption kinetic model is expressed as (Ho et al., 2000),

$$\frac{dq_t}{dt} = k_2 (q_e - q_t)^2 \tag{11}$$

For the boundary conditions

$t=0$ $q_t = 0$

and $t=t$ $q_t=q_t$

The integrated linearized form of Eq. (11) becomes,

$$\frac{t}{q_t} = \frac{1}{k_2 q_e^{\ 2}} + \frac{1}{q_e}t \tag{12}$$

Where k_2 is the Pseudo second-order rate constant. From the plot of t/q_t vs. t, the values of q_e and k_2 will be determined. Applicability of this model may be tested using the value of correlation coefficients and chi-square.

Intra Particle Diffusion Model

If the structure of the adsorbent is highly porous then this type of structure are suitable for Intra particle diffusion. The rate constants for intra particle diffusion may be determined using the following equation given by Weber and Morris (1963).

$$q_t = k_i t^{0.5} + C \tag{13}$$

Where k_i is the Intra-particle rate constant and C is the Intra particle diffusion constant. The value of correlation coefficients and chi-square may indicate the applicability of this model.

Reichenberg Model

The rate of adsorption may also determine by applying well known equation for the diffusion and mass transfer phenomena. For the fast reaction, the adsorption may be due to film diffusion (Reichenberg, 1953) and occur within the micropore of the bioadsorbents. Reicherberg equation is expressed by the following equation,

$$F = \left(1 - \frac{6}{\pi^2}\right) e^{-Bt} \tag{14}$$

Above equation may be written as,

$$Bt = -0.4977 \ln\left[1 - F(t)\right] \tag{15}$$

Where *F(t)* is the ratio of amount of metal ion adsorbed per unit mass of adsorbent at any time *t* to that of at equilibrium time and *B* is the time dependent factor. Linearity of the plot of *Bt* versus *t* indicates that the adsorption process may controlled by film diffusion (Naiya et al., 2009a).

Elovich Model

The kinetics of the chemisorption of gases on solids is described by the Elovich model (Low, 1960). In this model, the variation in energy of chemisorption is attributed to a change in the surface coverage or to a continuous and specific range of site reactivities (Teng and Hsieh, 1999). The equation is formulated as,

$$q_t = \frac{1}{b_1} \ln(a_1 b_1) + \frac{1}{b_1} \ln\left(t + t_0\right) \tag{16}$$

Parameter a_1 gives an idea of the reaction rate constant, b_1 represents the rate of chemisorption at zero coverage, and t_0 is an adjustable parameter that makes the plot liner over the entire range. The linear correlation coefficients may indicate the adsorption process is chemical in nature (Ho and McKay, 2002).

Prediction of Rate Limiting Step

An important feature of adsorption process is the prediction of rate-limiting step.

Solid-liquid adsorption process may be described by the following categories.

1. Film or outer diffusion- It is the metal ion transport from bulk solution through liquid film to the adsorbent exterior surface.
2. Intra particle or inner diffusion- It is the metal ion transport from the adsorbent exterior surface to the pores of the adsorbent.

3. Third one is the bonding between the active group present in the inner and outer surface of the adsorbent.

The rate of adsorption is generally controlled by outer diffusion or inner diffusion or both (Kalavathy et al., 2005). The diffusion process of metal ions on the exterior and interior surface of the microspheres is described by the Fick's equation, Eq. (17) (Seeber et al., 1998; Zaki et al., 2000) as follows

$$\frac{q_t}{q_\alpha} = \frac{6}{R_a}\sqrt{\frac{D_e t}{\pi}} \tag{17}$$

Where q_∞ is the amount adsorb per unit mass of adsorbent at infinite time, R_a is the radius of the adsorbent particle and D_e is the effective diffusion coefficient of adsorbate in the adsorbent phase. Plot of $\frac{q_t}{q_\alpha}$ (q_α may replace by q_e) versus $\sqrt{t}$ may represented the different stages of adsorption process.

Mass Transfer Analysis

Mass transfer analysis for the adsorption of metal ions from aqueous solution/effluents by the natural/agricultural waste may be carried out using the following equation as proposed by Mckay et al., (1981),

$$\ln\left(\frac{C_t}{C_0} - \frac{1}{1+MK_{bq}}\right) = \ln\left|\frac{MK_{bq}}{1+MK_{bq}}\right| - \left(\frac{1+MK_{bq}}{MK_{bq}}\right)\beta S_s t \tag{18}$$

Where M is the mass of the adsorbent per unit volume, K_{bq} is the mass transfer constant, β is mass transfer coefficient and S_S is the external surface area of the adsorbent per unit volume. The plot of $\ln\left(\frac{C_t}{C_0} - \frac{1}{1+MK_{bq}}\right)$ versus t may results a straight line of slope $\left(\frac{1+MK_{bq}}{MK_{bq}}\right)\beta S_S$ from which the mass transfer coefficients (β) may be calculated. The values of mass transfer coefficients indicate whether the velocity of the metal ion transport from the bulk to the solid phase may fast or not.

Determination of Diffusivity

Contact time variation data from batch experiment may also be treated by using the Boyd model (Boyd et al., 1947). With the diffusion rate controlling adsorption process in the adsorption of metal ion onto natural/agricultural waste, the solution of the simultaneous set of differential and algebraic equation leads to,

$$F(t) = 1 - \frac{6}{\pi^2}\sum_{n_1=1}^{\alpha}\frac{1}{n_1^2}\exp\left(-\frac{D_e t \pi^2 n_1^2}{R_a^2}\right) \tag{19}$$

Where n_l is an integer. Assuming spherical shape of the adsorbents the Vermeulen's approximation (Vermeulen, 1953) of the Eq. (19) fits the whole range 0 < F(t) < 1 as,

$$F(t)=\left[1-\exp\left(\frac{-\pi^2 D_e t}{R^2{}_a}\right)\right]^{\frac{1}{2}} \quad (20)$$

This equation could be further simplified to cover most of the data points for calculating effective particle diffusivity.

$$\ln\left[\frac{1}{1-F^2(t)}\right]=\frac{\pi^2}{R_a^2}D_e t \quad (21)$$

The linear plot of $\ln\left[\frac{1}{1-F^2(t)}\right]$ versus t whose slope, $\frac{\pi^2}{R_a^2}D_e$, thus diffusion coefficient, D_e can be calculated. If the value of De falls well in between 10^{-9} to 10^{-17} m^2/s then we can say the system is chemisorption (Srivastava et al., 2009).

Isotherm Modeling Using Batch Experiment Data

Metal ions adsorption isotherm process may describe using initial metal ion concentration variation data. Several adsorption isotherm models have been published in literature but Langmuir, Freundlich, Temkin, Harkins-Jura, Halsey, Dubinin-Radushkevich etc. are the more applicable.

Langmuir Isotherm Model

Langmuir isotherm model (Langmuir, 1918) may be used to describe the monolayer adsorption onto the surface with a finite number of identical sites. The equation may use for the purpose is as follows,

$$\frac{C_e}{q_e}=\frac{1}{q_{\max}b}+\frac{C_e}{q_{\max}} \quad (22)$$

Where C_e is the metal ion concentration in solution at equilibrium, q_{max} is the maximum adsorption capacity and b is the Langmuir constant. The plot of C_e/q_e vs. C_e gives the value of q_{max} and b. Linearity of the plots indicated the applicability of the isotherm model. The feasibility of the Langmuir isotherm in terms of dimensionless constant separation factor (Webi and Chakravort, 1974) or equilibrium parameter, R_L, which is defined as,

$$R_L=\frac{1}{1+bC_0} \quad (23)$$

Where R_L is the separation factor. If the value of R_L lies between 0 and 1 then it is favorable adsorption, $R_L > 1$ represents unfavorable adsorption, $R_L = 1$ indicates linear adsorption while the adsorption process is irreversible if $R_L = 0$ (Mckay et al., 1982).

Freundlich Isotherm Model

The empirical equation may use for the adsorption on the heterogeneous surface is Freundlich isotherm model. The linear form of this adsorption isotherm can be defined by the following equation (Freundlich, 1906),

$$\log q_e = \log K_f + \frac{1}{n}\log C_e \tag{24}$$

The Freundlich isotherm constant n is an empirical parameter that varies with the degree of heterogeneity and K_f is related to adsorption capacity. The constants K_f and n are calculated from Eq. (24) and plots $\log q_e$ vs $\log C_e$. The values of n in between 1 and 10 represent favorable adsorption (Bhattacharya et al., 2008).

Temkin Isotherm Model

The interaction between adsorbing species and adsorbate may be explained by Temkin isotherm model. This model assumes,

1. The adsorbent-adsorbate interactions results the decreasing of the heat of adsorption linearly of all the molecules in the layer with coverage.
2. This model also describe the uniform distribution of binding energies, up to a maximum binding energy.

The Temkin isotherm assumes that the fall in the heat of sorption is linear rather than logarithmic, as implied in the Freundlich equation. This isotherm has commonly been applied in the following form (Temkin and Pyzhev, 1940),

$$q_e = b_T \ln K_T + b_T \ln C_e \tag{25}$$

Where b_T and K_T are the Temkin isotherm constants.

Harkins-Jura isotherm model

Harkins-Jura isotherm model (Murthy et al., 2010) may be applied for multilayer adsorption as well as heterogeneous pore distribution in the adsorbents surface as

$$\frac{1}{q_e^2} = \left[\frac{B_H}{A_H}\right] - \left[\frac{1}{A_H}\right] \log C_e \tag{26}$$

Where A_H and B_H are Harkins-Jura isotherm constants. The value of correlation coefficients and chi-square indicate the applicability of the model.

Halsey Isotherm Model

Multilayer adsorption of metal ions may also be described using Halsey adsorption isotherm model (Murthy et al., 2010) as,

$$\ln q_e = \left[\left(\frac{1}{n_H}\right) \ln K_H\right] - \left(\frac{1}{n_H}\right) \ln\left(\frac{1}{C_e}\right) \tag{27}$$

Where K_H and n_H are the Halsey isotherm model constants. These isotherm constants, correlation coefficients and chi-square values are calculated using the Eq. (27).

Dubinin-Radushkevich Isotherm Model

The Dubinin-Radushkevich (Dubinin et al., 1947) isotherm model may be used to predict the nature of adsorption process by calculating adsorption energy whether it is chemical or physical in nature. The linearized mathematical from of the model is described as,

$$\ln C_{abs} = \ln X_m - \lambda \varepsilon^2 \tag{28}$$

Where C_{abs} is the amount of metal ion adsorbed onto adsorbent surface, X_m is the maximum sorption capacity of adsorbent, λ is the constant related to sorption energy and ε is the polanyi potential. The Polanyi potential (Polanyi, 1932) which is equal to,

$$\varepsilon = RT \ln(1 + \frac{1}{C_e}) \tag{29}$$

Where R is the ideal gas constant and T is the Temperature. The plot of $\ln C_{abs}$ vs. ε^2 gives the values of λ and X_m for the adsorbents used. Using the value of λ, the mean sorption energy, *E*, may be evaluated as,

$$E = \frac{1}{\sqrt{-2\lambda}} \tag{30}$$

Where E is the mean sorption energy. The Dubinin-Radushkevich isotherm relates the heterogeneity of energies close to the adsorbent surface. If a very small sub-region of the sorption surface is considered and assumed to be approximately by the Langmuir isotherm, the quantity $\sqrt{-\lambda}$ can be related to the mean sorption energy, E. The value of sorption indicates the chemical or physical nature of the adsorption process (Helfferich, 1962; Malik et al., 2005; Sawalha et al., 2006).

1. If $E < 8$ kJ/mol: The process is physical in nature.
2. If E ranges in-between 8 to 16 kJ/mol: The process is chemical in nature.

Thermodynamics Study for Adsorption Process

The adsorption process is highly influenced by the solution temperature. The thermodynamic equilibrium constant (K_c^0) may obtained by calculating apparent equilibrium constant ($K_c^{'}$) at different temperature and initial metal ion concentration for each system and extrapolating to zero (Dakiky et al., 2002),

$$K_c^{'} = \frac{C_a}{C_e} \tag{31}$$

Where C_a is the amount of metal ion adsorbed on the adsorbent per unit volume. The change of Gibbs free energy (ΔG^0), enthalpy (ΔH^0), and entropy (ΔS^0), for the adsorption processes may be calculated using following equations.

$$\ln K_c^0 = -\frac{\Delta H^0}{RT} + \frac{\Delta S^0}{R} \tag{32}$$

And

$$\Delta G^0 = -RT \ln K_c^0 \tag{33}$$

From the slope and intercept of the plot $\ln K_c^0$ vs. *1/T*, the values of ΔH^0 and ΔS^0 may be computed, while ΔG^0 is calculated using Eq. (33). The negative value of ΔG^0 at all temperatures may indicate the spontaneous nature of the adsorption process. Positive values of the enthalpy change suggest the endothermic nature of the adsorption process and vice versa. The increased randomness at the solid/solution interface during the adsorption process may results due to the positive value of entropy change (Malkoc et al., 2007).

Measurement of Sticking Probability for Adsorption Process

The sticking probability may be applied to measure the metal ion capacity to remain adsorbed indefinitely (Horsfall and Spiff, 2005). The sticking probability in the surface coverage (θ) may be evaluated by the following equation,

$$\theta = 1 - \frac{C_e}{C_o} \tag{34}$$

The activation energy (E_a) and sticking probability (S^*) are estimated by a modified Arrhenius type equation related to surface coverage as,

$$S^* = (1-\theta)\exp\left(-\frac{E_a}{RT}\right) \tag{35}$$

Designing of Single-Stage Batch Adsorption System

The single-stage batch adsorption system can be designed by adsorption isotherm data (Kumar et al., 2011). The initial metal ions concentration is reduced from C_0 to C_t of the solution volume V. Metal ions loaded into M g of adsorbents is changed from q_0 to q_t . The mass balance equation is applied for the single-stage batch adsorption system can be applied as,

$$V(C_0 - C_t) = M(q_t - q_0) = Mq_t \tag{36}$$

When adsorption process is in equilibrium then $C_t = C_e$ and $q_t = q_e$. If the Langmuir isotherm model is the most fitted isotherm model then the above equation can be expressed as,

$$\frac{M}{V} = \frac{(C_0 - C_t)}{q_t} = \frac{(C_0 - C_t)}{q_e} = \frac{(C_0 - C_t)}{\frac{q_{\max} bC_e}{1 + bC_e}} \tag{37}$$

If Freundlich isotherm model is the best fitted isotherm model then this model is used in the batch adsorption design system. So we can write in the above equation as,

$$\frac{M}{V} = \frac{(C_0 - C_t)}{q_t} = \frac{(C_0 - C_t)}{q_e} = \frac{(C_0 - C_t)}{K_f C_e^{\frac{1}{n}}} \tag{38}$$

Using this model engineers can design the single-stage batch adsorption system

Column Technique

Continuous process is very efficient for the treatment of industrial effluents for large scale purpose. This process is also depended on the inlet metal ion concentration, influent flow rate, bed height and diameter etc. The bed of the column is prepared by appropriate adsorbents with require height. In this process, the flow rate will be controlled using peristaltic pump.

Effect of Different Parameters for Column Process

The design of continuous process is needed for the prediction of breakthrough curves and the adsorption capacities under different operating conditions such as follows.

Effect of Initial pH

Adsorption of toxic heavy metal ions onto natural/agricultural waste surface is highly pH dependent which is discussed in the batch process. The removal efficiency may be increases at lower pH for the metal ions which exists as negatively charged species. For positively charged species it is vise versa.

Effect of Flow Rate

The flow rate of influent plays a vital role in the performance of continuous adsorption process. The experiment may carried out by varying the flow rate. Generally breakthrough time decreases as the flow rate increases. This is because at lower flow rate, metal ions have more time to be interact with the surface of adsorbents compare to higher flow rate, which results in the higher removal of metal ions from solution.

Effect of Influent Metal Ion Concentration

Adsorption efficiency of natural/agricultural wastes may increases with the decrease in influent metal ion concentration and it has taken more time to reach the saturation time. So the saturation time decreases with the increasing influent metal ion concentration. This is because, the adsorption sites may be covered rapidly by the metal ion.

Effect of Bed Height

Bed height of the column is one of the important parameter for the continuous process. Generally, if bed height increases then the breakthrough time increases due to the fact that more adsorption sites may available. The equilibrium adsorption capacity decreases with the increase in the bed height. That indicates the smaller bed height, the adsorbate concentration ratio increases more rapidly than compared to the higher bed height due to the availability of less adsorbent sites.

Mathematical Modeling With Column Experiment Data

The well-known simple empirical and semi-empirical models have been used to analyze the continuous experimental data such as Bohart–Admas model, Wolborska model, Thomas model, Yoon–Nelson model etc. (Mitra et al., 2014).

Bohart–Admas model

One of the important models for continuous process is Bohart–Admas model (Bohart and Admas, 1920). The mathematical expression for this model is as follows,

$$ln\left(\frac{C_t}{C_0}\right) = K_{AB}C_0t - K_{AB}N_0X\frac{Z}{v} \quad \ln\left(\frac{C_t}{C_0}\right) = K_{AB}C_0t - K_{AB}N_0x\frac{Z}{V} \tag{39}$$

Where C_0 is the influent metal ion concentration initially, C_t is the effluent metal ion concentration at time t, K_{AB} is the kinetic constant for Bohart-Admas model, N_0 is the saturation concentration, Z is the bed depth of the column and v is the flow rate. This model assumes that the adsorption rate is proportional to both the residual capacity of adsorbent and the concentration of the metal ions. It is often used to describe the initial part of the breakthrough curve. The value of K_{AB} of this model indicates whether system kinetically control by the external mass transfer process or not (Taxier et al., 2002; Aksu and Gönen, 2004).

Wolborska Model

Wolborska model may suggest the concept of mass transfer in diffusion mechanism (Wolborska, 1989; Wolborska and Pustelnik, 1996). This model is applicable only for low metal ion concentration, negligible axial diffusion and only the external diffusion with a constant kinetic coefficient. The mathematical expression of the model is,

(40)

Where β_a is the Wolborska rate constant and q is the metal ion adsorbed.

Thomas Model

The well-known Thomas model (Thomas, 1948) is expressed as follows

$$ln\left(\frac{C_0}{C_t} - 1\right) = \frac{K_{th}q_0x}{v} - \frac{K_{th}C_0V_{eff}}{v} \tag{41}$$

Where K_{th} is the Thomas rate constant, q_0 is the maximum metal uptake per unit mass of adsorbent, x is the amount of adsorbent present in the column and V_{eff} is the effluent volume. The validity of external or internal diffusion can be judge by this model. Generally adsorption may controlled by the interphase mass transfer and also the existence of an axial dispersion.

Yoon–Nelson Model

Another important kinetic model for continuous process is Yoon-Nelson model (Yoon and Nelson, 1984). The mathematical model as follows

$$ln\left(\frac{C_t}{C_0}\right) = K_{AB}C_0t - K_{AB}N_0x\frac{Z}{v} \tag{42}$$

Where K_{YN} is the Yoon- Nelson model constant and τ is the time. This model assumed that the rate of decrease in the probability of adsorption for each adsorbate molecule is proportional to the probability of adsorbate and also the probability of adsorbate breakthrough on adsorbent (Han et al., 2006).

Fourier Transform Infrared Spectroscopy (FTIR) Analysis

Fourier Transform Infrared Spectroscopy analysis is important to identify the functional group present and responsible for the adsorption process. FTIR study may be carried out to identify the functional groups presents in the natural/agricultural waste at 4000-400 cm^{-1} range. Generally –OH, -COOH, -CHO, >C=O, $-NH_2$, $-NO_2$ etc. may responsible for biosorption process. Crystal radius is also vital measurement to fit into the binding site of the natural adsorbents and bind to several functional group present in the adsorbents simultaneously. As for example it is 0.52Å for Cr(VI) ion.

Desorption Experiment

Desorption experiments is just reverse of the adsorption experiment. This process may be carried out for reuse of the adsorbents as well as reducing the disposal problem of toxic metal loaded adsorbents in the environment. Desorption experiments may conduct using NaOH, HNO_3 or HCl solutions as appropriate for each metal ion at optimum batch condition.

Application Study Using Industrial Effluent

Toxic metal ion containing industrial effluent may collect from nearby industries. Batch experiment using suitable process parameter may be conducted to identify the applicability of the natural/agricultural waste for the particular metal ion. Result from the batch adsorption process, column experiment may be performed.

CONCLUSION

In third world countries like India, the most of the small and medium size industries located in rural areas. These industries are unable to treat their effluent containing toxic pollutants due to their financial constrain and also the unavailability of the common discharge facility. So these toxic pollutants containing waste water enter directly into the local water bodies and/or in the agricultural fields. Thus there is a possibility to enter these materials into the food chain. Natural/ agricultural waste may be viewed as useful material, considering the economic aspects of wastewater treatment. Fixed-bed column technique is the best process compare to the batch technique. Expert should analysis the different adsorption models and appropriate one may be applied for scale-up design which will be economically helpful for small and medium size industries.

REFERENCES

Aksu, Z., & Gönen, F. (2004). Biosorption of phenol by immobilized activated sludge in a continuous packed bed: Prediction of breakthrough curves. *Process Biochemistry*, *39*(5), 599–613. doi:10.1016/S0032-9592(03)00132-8

Babel, S., & Kurniawan, T. A. (2004). Cr(VI) removal from synthetic wastewater using coconut shell charcoal and commercial activated carbon modified with oxidizing agents and/or chitosan. *Chemosphere*, *54*(7), 951–967. doi:10.1016/j.chemosphere.2003.10.001 PMID:14637353

Bernardo, G.R., & Rene, R.J., & Catalina, A.L.T.M. (2009). Chromium (III) uptake by agro-waste biosorbents: Chemical characterization, sorption-desorption studies, and mechanism. *Journal of Hazardous Materials*, *170*, 845–854. doi:10.1016/j.jhazmat.2009.05.046 PMID:19523757

Bhatnagar, A., Vilar, V. J. P., Botelho, C. M. S., & Boaventura, R. A. R. (2011). A review of the use of red mud as adsorbent for the removal of toxic pollutants from water and wastewater. *Environmental Technology*, *32*(3-4), 231–249. doi:10.1080/09593330.2011.560615 PMID:21780692

Bhattacharya, A. K., Mandal, S. N., & Das, S. K. (2006). Adsorption of Zn(II) from aqueous solution by using different adsorbents. *Chemical Engineering Journal*, *123*(1-2), 43–51. doi:10.1016/j.cej.2006.06.012

Bhattacharya, A. K., Naiya, T. K., Mandal, S. N., & Das, S. K. (2008). Adsorption, kinetics and equilibrium studies on the removal of Cr(VI) from aqueous solutions using different low-cost adsorbents. *Chemical Engineering Journal*, *137*, 529–541.

Bohart, G., & Admas, E. N. (1920). Some aspects of the behavior of charcoal with respect to chlorine. *Journal of the American Chemical Society*, *42*(3), 523–544. doi:10.1021/ja01448a018

Boyd, G. E., Adamson, A. W., & Myers, L. S. Jr. (1947). The exchange adsorption of ions from aqueous solutions by organic zeolites. II. Kinetic. *Journal of the American Chemical Society*, *69*(11), 2836–2848. doi:10.1021/ja01203a066 PMID:20270838

Chowdhury, P., Elkamel, A., & Ray, A. K. (2014). Photocatalytic processes for the removal of toxic metal ions. In S. Sharma (Ed.), *Heavy Metals In Water: Presence, Removal and Safety* (pp. 25–43). Royal Society of Chemistry. doi:10.1039/9781782620174-00025

Clescerl, L. S., Greenberg, A. E., & Eaion, A. D. (1998). *Standard methods for examination of water and wastewater* (20th ed.). APHA.

Dakiky, M., Khamis, M., Manassra, A., & Mer'eb, M. (2002). Selective adsorption of Chromium(VI) in industrial wastewater using low-cost abundantly available adsorbents. *Advances in Environmental Research*, *6*(4), 533–540. doi:10.1016/S1093-0191(01)00079-X

Dubinin, M. M., Zaverina, E. D., & Radushkevich, L. V. (1947). Sorption and structure of active carbons I. Adsorption of organic vapors. *Журнал физической химии*, *21*, 1351–1362.

Eckenfelder, W., & Ford, D. L. (2009). *Industrial water quality* (4th ed.). McGraw-Hill.

EPA. (1990). *Environmental Pollution Control Alternatives: Drinking Water Treatment for Small Communities*. Retrieved from https://cfpub.epa.gov/si/si_public_record_Report.cfm?Lab=NRMRL&dirEntryId=40845&CFID=9119927&CFTOKEN=70327879

Freundlich, H. (1906). Adsorption in solution. *Zeitschrift für Physikalische Chemie*, *57*, 384–410.

Han, R., Zhang, J., Xiao, W. Z. H., Shi, J., & Liu, H. (2006). Biosorption of copper(II) and lead(II) from aqueous solution by chaff in a fixed-bed column. *Journal of Hazardous Materials*, *133*(1-3), 262–268. doi:10.1016/j.jhazmat.2005.10.019 PMID:16298055

Hasana, S. H., Srivastavaa, P., & Talatb, M. (2010). Biosorption of lead using immobilized *Aeromonas hydrophila* biomass in up flow column system: Factorial design for process optimization. *Journal of Hazardous Materials*, *177*(1-3), 312–322. doi:10.1016/j.jhazmat.2009.12.034 PMID:20036460

Helfferich, F. (1962). *Ion Exchange*. McGraw-Hill.

Ho, Y. S., & McKay, G. (2002). Application of kinetic models to the sorption of copper(II) onto peat. *Adsorption Science and Technology*, *20*(8), 797–815. doi:10.1260/026361702321104282

Ho, Y. S., Mckay, G., Wase, D. A. J., & Foster, C. F. (2000). Study of the sorption of divalent metal on to peat. *Adsorption Science and Technology*, *18*(7), 639–650. doi:10.1260/0263617001493693

Horsfall, M. Jnr, & Spiff, A. (2005). Effects of temperature on the sorption of Pb^{2+} and Cd^{2+} from aqueous solution by *Caladium bicolor* (Wild Cocoyam) biomass. *Electronic Journal of Biotechnology*, *8*(2), 162–169. doi:10.2225/vol8-issue2-fulltext-4

IS 10500. (1992). *Drinking water specification (reaffirmed 1993)*. Retrieved from https://rwua.org.in/download/DRINKING%20WATER%20SPECIFICATION%20IS%2010500,%201992.pdf

Kadirvelu, K., & Namasivayam, C. (2003). Activated carbon from coconut coir pitch as metal adsorbent: Adsorption of Cd(II) from aqueous solution. *Advances in Environmental Research*, *7*(2), 471–478. doi:10.1016/S1093-0191(02)00018-7

Kalavathy, M. H., Karthikeyan, T., Rajgopal, S., & Miranda, L. R. (2005). Kinetic and isotherm studies of Cu(II) adsorption onto H_3PO_4-activated rubber wood sawdust. *Journal of Colloid and Interface Science*, *292*(2), 354–362. doi:10.1016/j.jcis.2005.05.087 PMID:16040040

Kumar, P. S., Ramalingam, S., Kirupha, S. D., Murugesan, A., Vidhyadevi, T., & Sivanesan, S. (2011). Adsorption behavior of nickel(II) onto cashew nut shell: Equilibrium, thermodynamics, kinetics, mechanism and process design. *Chemical Engineering Journal*, *167*(1), 122–131. doi:10.1016/j.cej.2010.12.010

Lagergren, S. (1898). Zur theorie der sogenannten adsorption geloster stoffe. Kungliga Svenska Vetenskapsakademiens. *Handlingar*, *24*, 1–39.

Langmuir, I. (1918). The adsorption of gases on plane surfaces of glass, mica and platinum. *Journal of the American Chemical Society*, *40*(9), 1361–1403. doi:10.1021/ja02242a004

Low, K. S., Lee, C. K., & Liew, S. C. (2000). Sorption of cadmium and lead from aqueous solutions by spent grain. *Process Biochemistry*, *36*(1-2), 59–64. doi:10.1016/S0032-9592(00)00177-1

Low, M. J. D. (1960). Kinetics of chemisorption of gases on solids. *Chemical Reviews*, *60*(3), 267–312. doi:10.1021/cr60205a003

Malik, U. R., Hasany, S. M., & Subhani, M. S. (2005). Sorptive potential of sunflower stem for Cr(III) ions from aqueous solutions and its kinetic and thermodynamic profile. *Talanta*, *66*(1), 166–173. doi:10.1016/j.talanta.2004.11.013 PMID:18969977

Malkoc, E., & Nuhoglu, Y. (2007). Determination of kinetic and equilibrium parameters of the batch adsorption of Cr(VI) onto waste acorn of *Quercus ithaburensis. Chemical Engineering and Processing*, *46*(10), 1020–1029. doi:10.1016/j.cep.2007.05.007

Mallick, S., Dash, S. S., & Parida, K. M. (2006). Adsorption of Hexavalent Chromium on Manganese Nodule Leached Residue Obtained from NH_3-SO_3 Leaching. *Journal of Colloid and Interface Science*, *297*(2), 419–425. doi:10.1016/j.jcis.2005.11.001 PMID:16330035

Mckay, G., Blair, H. S., & Gardener, J. R. (1982). Adsorption of dyes on chitin. I. Equilibrium studies. *Journal of Applied Polymer Science*, *27*(8), 3043–3057. doi:10.1002/app.1982.070270827

Mckay, G., Otterburn, M. S., & Sweeney, A. G. (1981). Surface mass transfer processes during colour removal from effluent using silica. *Water Research*, *15*(3), 327–331. doi:10.1016/0043-1354(81)90036-1

Miretzkya, P., & Cirelli, A. F. (2010). Cr(VI) and Cr(III) Removal from Aqueous Solutions by Raw and Modified Lignocellulosic Materials: A Review. *Journal of Hazardous Materials*, *180*(1-3), 1–19. doi:10.1016/j.jhazmat.2010.04.060 PMID:20451320

Mitra, T., Singha, B., Bar, N., & Das, S. K. (2014). Removal of Pb(II) ions from aqueous solution using water hyacinth root by fixed-bed column and ANN modeling. *Journal of Hazardous Materials*, *273*, 94–103. doi:10.1016/j.jhazmat.2014.03.025 PMID:24727010

Murthy, Z.V.P., Prabhakar, T., & Kumar, R. (2010). Isotherm and Kinetics Modeling of Fluoride Removal from Industrial Effluent by Alumina. *Chemical Product and Process Modeling, 5*.

Naiya, T. K., Bhattacharya, A. K., & Das, S. K. (2009a). Adsorption of Cd(II) and Pb(II) from aqueous solutions on activated alumina. *Journal of Colloid and Interface Science*, *333*(1), 14–26. doi:10.1016/j.jcis.2009.01.003 PMID:19211112

Naiya, T. K., Bhattacharya, A. K., Mandal, S. N., & Das, S. K. (2009b). The sorption of lead(II) ions on rice husk ash. *Journal of Hazardous Materials*, *163*(2-3), 1254–1264. doi:10.1016/j.jhazmat.2008.07.119 PMID:18783880

Naiya, T. K., Chowdhuary, P., Bhattacharya, A. K., & Das, S. K. (2009c). Saw dust and neem bark as low-cost natural biosorbent for adsorptive removal of Zn(II) and Cd(II) ions from aqueous solutions. *Chemical Engineering Journal*, *148*(1), 68–79. doi:10.1016/j.cej.2008.08.002

Patterson, S. C., & Jancuk, W. A. (1977). Cementation treatment of copper in wastewater. In *Proceedings of 32nd Industrial Waste Conference* (pp. 85-865), Perdue University.

Polanyi, M. (1932). Theories of the adsorption of gases. A general survey and some additional remarks. *Transactions of the Faraday Society*, *28*(0), 316–333. doi:10.1039/TF9322800316

Reddy, D. H. K., Harinatha, Y., Seshaiah, K., & Reddy, A. V. R. (2010). Biosorption of Pb(II) from aqueous solutions using chemically modified *Moringa oleifera* tree leaves. *Chemical Engineering Journal*, *162*(2), 626–634. doi:10.1016/j.cej.2010.06.010

Reichenberg, D. (1953). Properties of ion-exchange resins in relation to their structure. III. Kinetics of exchange. *Journal of the American Chemical Society*, *75*(3), 589–597. doi:10.1021/ja01099a022

Saeeda, A., Iqbala, M., & Akhtarb, M. W. (2005). Removal and recovery of lead(II) from single and multimetal (Cd, Cu, Ni, Zn) solutions by crop milling waste (black gram husk). *Journal of Hazardous Materials*, *B117*(1), 65–73. doi:10.1016/j.jhazmat.2004.09.008 PMID:15621354

Sawalha, M. F., Peralta-Videa, J. R., Romero-Gonzalez, J., & Gardea-Torresdey, J. L. (2006). Biosorption of Cd(II), Cr(III), and Cr(VI) by saltbush (*Atriplex canescens*) biomass: Thermodynamic and isotherm studies. *Journal of Colloid and Interface Science*, *300*(1), 100–104. doi:10.1016/j.jcis.2006.03.029 PMID:16600278

Seeber, G., Buchmeiser, M. R., Bonn, G. K., & Bertsch, T. (1998). Determination of airborne, volatile amines from polyurethane foams by sorption onto a high-capacity cation-exchange resin based on poly (succinic acid). *Journal of Chromatography. A*, *809*(1-2), 121–129. doi:10.1016/S0021-9673(98)00146-0 PMID:9677714

Singha, B., & Das, S. K. (2011). Biosorption of Cr(VI) ions from aqueous solutions: Kinetics, equilibrium, thermodynamics and desorption studies. *Colloids and Surfaces. B, Biointerfaces*, *84*(1), 221–232. doi:10.1016/j.colsurfb.2011.01.004 PMID:21282045

Singha, B., & Das, S. K. (2013). Adsorptive removal of Cu(II) from aqueous solution and industrial effluent using natural/agricultural wastes. *Colloids and Surfaces. B, Biointerfaces*, *107*, 97–106. doi:10.1016/j.colsurfb.2013.01.060 PMID:23466548

Srivastava, V. C., Mall, I. D., & Mishra, I. M. (2006). Characterization of mesoporous rice husk ash (RHA) and adsorption kinetics of metal ions from aqueous solution onto RHA. *Journal of Hazardous Materials*, *B134*(1-3), 257–267. doi:10.1016/j.jhazmat.2005.11.052 PMID:16386363

Srivastava, V. C., Mall, I. D., & Mishra, I. M. (2009). Competitive adsorption of cadmium(II) and nickel(II) ions from aqueous solution onto rice husk ash. *Chemical Engineering and Processing*, *48*(1), 370–379. doi:10.1016/j.cep.2008.05.001

Taxier, A. C., Andrès, Y., Brasquet, C. F., & Cloirec, P. I. (2002). Fixed-bed study for anthanide (La, Eu, Yb) in removal from aqueous solution by immobilized Pseudomonas aeruginosa: Experiental data and modelization. *Chemosphere*, *47*(3), 333–342. doi:10.1016/S0045-6535(01)00244-2 PMID:11996155

Temkin, M. J., & Pyzhev, V. (1940). Recent modifications of Langmuir isotherms. *Acta Physiochim URSS*, *12*, 217–222.

Teng, H., & Hsieh, C. T. (1999). Activation energy for oxygen chemisorptions on carbon at low temperature. *Industrial & Engineering Chemistry Research*, *38*(1), 292–297. doi:10.1021/ie980107j

Thomas, H. G. (1948). Chromatography: A problem in kinetics. *Annals of the New York Academy of Sciences*, *49*(2), 161–182. doi:10.1111/j.1749-6632.1948.tb35248.x PMID:18911539

Tiwari, D. P., Singh, D. K., & Saksena, D. N. (1995). Hg(II) adsorption from aqueous solutions using rice-husk ash. *Journal of Environmental Engineering*, *121*(6), 479–481. doi:10.1061/(ASCE)0733-9372(1995)121:6(479)

Tong, K. S., Kassim, M. J., & Azraa, A. (2011). Adsorption of copper ion from its aqueous solution by a novel biosorbent *Uncaria gambir*: Equilibrium, kinetics, and thermodynamic studies. *Chemical Engineering Journal*, *170*(1), 145–153. doi:10.1016/j.cej.2011.03.044

Vermeulen, T. (1953). Theory for irreversible and constant pattern solid diffusion. *Industrial & Engineering Chemistry*, *45*(8), 1664–1670. doi:10.1021/ie50524a025

Weber, W. J., & Morris, J. C. (1963). Kinetics of adsorption on carbon from solution. *Journal of the Sanitary Engineering Division*, *89*, 31–60.

Webi, T. W., & Chakravort, R. K. (1974). Pore and solid diffusion models for fixed bed absorbers. *AIChE Journal. American Institute of Chemical Engineers*, *20*(2), 228–238. doi:10.1002/aic.690200204

WHO. (1993). *Guidelines for drinking water quality*. Retrieved from https://www.who.int/water_sanitation_health/publications/gdwq2v1/en/

Wolborska, A. (1989). Adsorption on activated carbon of p-nitrophenol from aqueous solution. *Water Research*, *23*(1), 85–91. doi:10.1016/0043-1354(89)90066-3

Wolborska, A., & Pustelnik, P. (1996). A simplified method for determination of breakthrough time of an adsorbent layer. *Water Research*, *30*(11), 2643–2650. doi:10.1016/S0043-1354(96)00166-2

Yoon, Y. H., & Nelson, J. H. (1984). Application of gas adsorption kinetics II. A theoretical model for respirator cartridge service life. *American Industrial Hygiene Association Journal*, *45*(8), 509–516. doi:10.1080/15298668491400197 PMID:6475759

Zaki, A. B., El-Sheikh, M. Y., Evans, J., & El-Safty, S. A. (2000). Kinetics and mechanism of the sorption of some aromatic amines onto amberlite IRA-904 anion-exchange resin. *Journal of Colloid and Interface Science*, *221*(1), 58–63. doi:10.1006/jcis.1999.6553 PMID:10623452

ADDITIONAL READING

Banerjee, M., Basu, R. K., & Das, S. K. (2018). Cr(VI) adsorption by a green adsorbent walnut shell: Adsorption studies, regeneration studies, scale-up design and economic feasibility. *Process Safety and Environmental Protection*, *116*, 693–702. doi:10.1016/j.psep.2018.03.037

Das, A., Bar, N., & Das, S. K. (2020). Pb(II) adsorption from aqueous solution by nutshells, green adsorbent: Adsorption studies, regeneration studies, scale-up design, its effect on biological indicator and MLR modeling. *Journal of Colloid and Interface Science*, *580*, 245–255. doi:10.1016/j.jcis.2020.07.017 PMID:32683121

Farooq, U., Kozinski, J. A., Khan, M. A., & Athar, M. (2010). Biosorption of heavy metal ions using wheat based biosorbents – A review of the recent literature. *Bioresource Technology*, *101*(14), 5043–5053. doi:10.1016/j.biortech.2010.02.030 PMID:20223652

Hossain, M. A., Ngo, H. H., Guo, W. S., & Setiadi, T. (2012). Adsorption and desorption of copper(II) ions onto garden grass. *Bioresource Technology*, *121*, 386–395. doi:10.1016/j.biortech.2012.06.119 PMID:22864175

Labidi, A., Salaberria, A. M., Fernandes, S. C. M., & Abderrabba, J. M. (2016). Adsorption of copper on chitin-based materials: Kinetic and thermodynamic studies. *Journal of the Taiwan Institute of Chemical Engineers*, *65*, 140–148. doi:10.1016/j.jtice.2016.04.030

Nag, S., Bar, N., & Das, S. K. (2019). Sustainable bioremadiation of Cd(II) in fixed bed column using green adsorbents: Application of Kinetic models and GA-ANN technique. *Environmental Technology & Innovation*, *13*, 130–145. doi:10.1016/j.eti.2018.11.007

Sellaoui, L., Kehili, M., Lima, E. C., Thue, P. S., Petriciolet, A. B., Lamine, A. B., Dotto, G. L., & Erto, A. (2019). Adsorption of phenol on microwave-assisted activated carbons: Modeling and interpretation. *Journal of Molecular Liquids*, *274*, 309–314. doi:10.1016/j.molliq.2018.10.098

Tang, Y., Chen, L., Wei, X., Yao, Q., & Li, T. (2013). Removal of lead ions from aqueous solution by the dried aquatic plant, *Lemna perpusilla* Torr. *Journal of Molecular Liquids*, *244-245*, 603–612. PMID:23182246

Wu, Q., Chen, J., Clark, M., & Yu, Y. (2014). Adsorption of copper to different biogenic oyster shell structures. *Applied Surface Science*, *311*, 264–272. doi:10.1016/j.apsusc.2014.05.054

Xinyuan, X., Runkang, D., Youqun, P., Yan, B., Wenjie, Z., & Yanhui, Z. (2017). Adsorption of copper(II) by sulfur microparticles. *Chemical Engineering Journal*, *314*, 434–442. doi:10.1016/j.cej.2016.11.163

KEY TERMS AND DEFINITIONS

Activation Energy: Minimum amount of energy required to start of a chemical reaction.

Adsorbate: The substance which adsorbed is known as adsorbate.

Adsorbent: The surface on which adsorption occur is called adsorbent.

Adsorption: Adsorption is a phenomena that occur at the surfaces not in the bulk of material. In adsorption, the atoms, ions, or molecules of a substance is attached on the surface of a liquid film or on a solid surface.

Adsorption Isotherm: A curve which relates between adsorbate and adsorbent in a constant-temperature for adsorption process.

Breakthrough Curve: It is the way of the effluent adsorptive concentration at the outlet of fixed bed column during the continuous adsorption process.

Diffusion: Diffusion is the movement of atom, ions, or molecules from a region of higher concentration to a region of lower concentration.

Enthalpy: It is the summation of internal energy of a system and pressure volume work.

Entropy: It is the randomness of system.

Gibbs Free Energy: The maximum amount of energy available to a system during a process that can be converted into useful work.

pH: It is the negative logarithm of hydrogen ion concentration. Acidic solutions have a pH lower than 7, while basic solutions have pH higher than 7. At room temperature (25°C or 77°F), pure water is neither acidic nor basic and has a pH of 7.

Point of Zero Charge: The point of zero charge is the pH at which the net charge of absorbent's surface is equal to zero.

Sticking Probability: It is the probability of atom, ion or molecules that are trapped on the surfaces of adsorbent for adsorption chemically.

Chapter 16
Wastewater Treatment Using Microalgae:
Nutrient Removal

Soumya Saswat Priyadarshi
College of Engineering and Technology, Bhubaneswar, India

ABSTRACT

Six different concentrations of Chlorella vulgaris microalgae culture are incorporated with raw wastewater to improve treatment efficiency in three phases. In phase one of study, 30% and 40% microalgae were incorporated with raw wastewater with 24 hrs retention period for 18 days in a batch mode and these conditions are made same for all phases. During phase two and phase three of study, 20%, 25% and 35%, 45% micro-algae concentrations used to treat raw wastewater respectively. Amongst all the six concentrations, 30% microalgae behaves most efficiently and is able to reduce 34 to 81% COD, 43 to 75% nitrate, and 36 to 99% of phosphorous. Overall ammonia nitrogen reduction varies from 80% to 99%. No attempt was made to control temperature or evaporation losses. These attempts can definitely increase the growth process of microalgae and increase the removal efficiency of nutrients in wastewater.

INTRODUCTION

As the population increases demand for freshwater also increases and most of the water is fit for drinking purposes. Conventional waste water treatment processes are less efficient in removing Nutrient in form of Nitrogen and phosphorous (Migliaceio et al., 2009). Discharge of wastewater with elevated quantities of nitrate and phosphates contributes to eutrophication in waterbodies. It has been observed as significant water pollution (Tredici et al., 1992). Costal eutrophication causes higher mortality rates in aquatic species, and enhanced organic particle sedimentation (Driscoll et al., 2003). In this process, microalgae consumes nutrients and carbon dioxide, generates oxygen for aerobic microorganisms. Microalgae biomass can be utilized for bio-fuel extraction because of its rich fatty acid content (Sheehan et al., 1998).Microalgae survives easily in water enriched with nutrients this ability of microalgae signifies its importance in wastewater treatment and can be a incorporated with conventional wastewater treatment

DOI: 10.4018/978-1-7998-4921-6.ch016

systems (Ruiz et al., 2011).Microalgae can utilizes organic carbonaceous matter for cell synthesis (Wang et al., 2010).Various kinds of algal species starting from single-cell to progressively complex multi-cellular structures, amongst which microalgae are widely available in various situations. Photosynthesis and growth rates of microalgae are comparatively higher than other plants present on land surface (Li et al., 2008). The most widely accepted microalgae strains used for treating wastewater are Chlorella and Scenedesmus (Garcia et al., 2000). Chlorella species and tolerant flexible and can be harvested physically and bio-mechanically under different wastewater conditions. Algal wastewater treatment method helps to remove nutrient and provide oxygen as a by-product in the system which is not only reduce the aeration demand also provide oxygen to aerobic microbes

Figure 1.

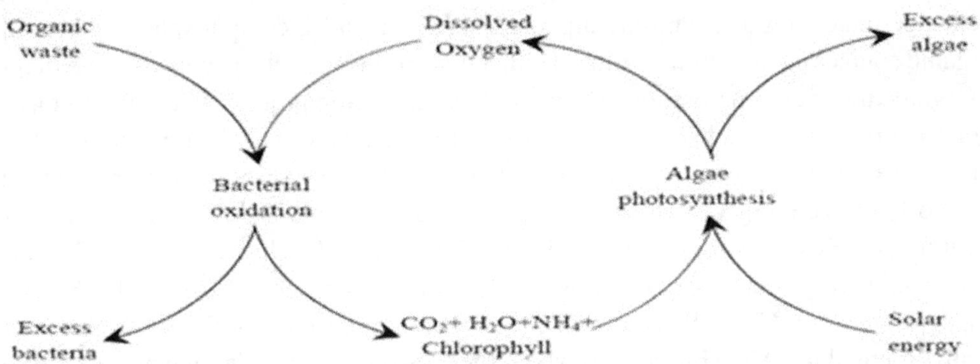

BACKGROUND

Microalgae got substantial interest of research now a days due to their nutrient uptake from pollutants in addition to their capability to produce valuable end product which can be utilized in various industrial applications like food processing, pharmaceuticals, organic fertilizer, animal feed, bio-fuel and biogas. In wastewater treatment applications, microalgae absorbs nutrients, uptake carbon di-oxide and provide oxygen for aerobic bacteria,. Global warming and deficiency in petroleum derivatives have been the main thrusts for people to research wellsprings of sustainable power source that could give a superior option in contrast to non-renewable energy sources (Barbara, 2007). Microalgae biomass is an alluring choice because of its high oil content contrasted with earthbound yields which are utilized for bio-fuel a extraction (Sheehan et al., 1998). Also, handling of green growth biomass to blend biofuel is all the more naturally cordial since it utilizes less synthetics and radiates less CO_2. Microalgae require nutrients which are found in wastewater to develop. In this way, Nutrient enriched wastewater possibly be utilized as a feed material to develop microalgae, could possibly accomplish the expulsion of nutrients and different toxinsand generation of microalgae biomass simultaneously. Subsequently, microalgae are environment frienly microorganisms that can help with wastewater treatment a supportable manner (Pienkos and Darzins, 2009). Microalgae treatment additionally empowers the reuse of wastewater and resource recovery such as the recovery of nutrients (Noue and Pauw, 1988).Another attractive benefit of using microalgae is the assimilation of CO_2. Recently, microalgae have been used for carbon sequestration due to its ability to uptake CO_2 from the environment through photosynthesis process (Hende et al., 2011). Disposal of wastewater which is full of nutrients can be a reason for eutrophication which is a serious issue particularly in enclosed water bodies. Eutrophication causes due to presence of algae growth in contaminated water due to presence of very high concentrations of Nutrients in form of plant nutrients, It reduces the amount of oxygen in dissolve form in aquatic system to less than the natural condition which causes a shortage in Dissolve oxygen, required for survival of aquatic organism .microalgae can grow quickly wastewater that is rich in nutrients due to this reason it can be used in wastewater treatment and it will helps to reduce the aeration demand.. Nutrient expulsion by algae follows different metabolic mechanism. Minerals which are essential for a development of algae are commonly Carbon, nitrogen, phosphorous and sulfur. Metals, for example, potassium, sodium, calcium, iron, and magnesium. fill in as micronutrients for the development of algae. Ammonical nitrogen and Nitrate is the major component in biological activity of microalgae and substances like enzymes, peptides, proteins, chlorophylls helps in energy exchange molecules as adenosine diphosphate and adenosine-triphosphate (Cai et al., 2013) .various inorganic sources such as nitrate, nitric acid, ammonium, and nitrogen gas are the major source from which Organic nitrogen can possibly collected . In assimilation process Microalgae converts inorganic nitrogen to nitrite, nitrate and ammonical nitrogen to which can easily be utilized in cell synthesis of microalgae. Algae which consists of eukaryotic cell only can follow assimilation of inorganic nitrogen. Microalgae majorly utilize inorganic carbon source. Carbon dioxide is the primary decision for microalgae and bicarbonate (HCO_3^-) is the second decision without CO_2. Be that as it may, HCO_3^-) needs the protein carbonic anhydrase to change over it into CO_2 (Borowitzka, 1998; Oswald, 1988). A group of microalgae species, name as Chlorella can work distinctively in presence of light . Phototrophic conditions essentially required for Algal biomass in which they absorbs carbon dioxide as the carbon source. Likewise, some species can develop by absorbing organic carbon following heterotrophic conditions.

MATERIALS AND METHODOLOGY

The microalgae strain used for the study *Chlorella vulgaris* in this study. The microalgae culture was provided by an environmental consultant of Ahmedabad. During the beginning of study for the survival of algal biomass culture, Nutrient were added to the bottle and kept in sunlight. The Fresh wastewater sample was collected from Umra pumping station near SVNIT SURAT on daily basis. Chlorella vulgaris was a microalgae strain used in this research. Chlorella vulgaris is a green unicellular organism obtaining food through photosynthesis. The Microalgae culture is provided by an environmental consultant of Ahmedabad, Gujrat. Initially during the project for the survival of algal bio-mass, Nutrient were added and kept in sunlight. Four type of Nutrient were used for the preservation of microalgae culture provided by consultancy i.e. - Urea, Biotech-A, Biotech-Plus, SSP. Six different concentrations of microalgae culture are incorporated with raw wastewater to improve treatment efficiency in three phases. In phase one of study, 30% and 40% microalgae were incorporated with raw wastewater with 24 hrs retention period for 18 days in a batch mode and these conditions are made same for all phases. During phase two and phase three of study, 20%, 25% and 35%, 45% micro-algae concentrations used to treat raw wastewater respectively. For the above mentioned phases microalgae culture was added to wastewater once at the beginning of Experiment on basis of (volume/volume) and allowed a acclimatization period of 48 hrs. All experiments were carried out in batch-scale reactors of capacity 2liter providing a constant aeration of 24 hours. No provision of artificial light was made. Before collection of effluent from the reactor a significant amount of time (1hour for all concentration) for settlement of algal bio mass was given and supernatant was collected. Each of samples was filtered through filter paper and both the filter and non-filter sample was collected and following tests are conducted on daily basis pH, TS, TDS, TSS, Turbidity, COD, NO_3^{2-}, PO_4^{3-}, NH^{3}-N, and D.O. All measurements were conducted with standard procedure according to APHA,

Issues, Controversies, Problems

Conventional wastewater treatment process is inefficient in removal of Nutrients which booms other problems like eutrophication. This problem can potentially impact the Ecosystem of effluent receiving water bodies. Microalgae can be a possible solutions in term of nutrient removal as it is incorporated with wastewater and provide oxygen which helps to reduce aeration demand. Phosphorus is extracted in form of fatty acids and proteins This assumes a vital job in cellular synthesis of microalgae. At the time of photosynthesis, phosphorus majorly converted to $H_2PO_4^{-}$ and HPO_4^{2-} is joined into natural mixes through a procedure called phosphorylation. This process requires energy. Process begins from the oxidation of substrates, and electron transport action of the mitochondria, or from incident photo energy. Time of ATP from ADP occurs during phosphorylation and microalgae can ingest and store phosphorus in overabundance for its cellular development which not only helps to overcome Eutrophication but also helps to treat wastewater.. These stores will be accessible during the development cycle without phosphorus in the media. The development rate of algae may in this way not react without a moments delay to changes in the outer convergence of phosphorus, instead of the prompt reactions to temperature and light (Larsdotter et al., 2006). Phosphorus is the second most fundamental supplement for microalgal growth.. It is the most significant component that involves 1.5% of the cellular dry mass (Brown and Shilton, 2014). Be that as it may, microalgae can take-up more than what they need through their capacity to integrate and collect polyphosphate in their cells. Removal of phosphorous by assimilation of these nutrients by microalgae cells considered it to be a direct process whereas; indirect nutrient removal involves removal of phosphorous by precipitation and phosphorous stripping at a elevated pH (below 9) level caused by microalgal metabolisim. Thanks to this activity, prolonged retention times helps in nutrient removal (Cai et al., 2013). Most important factor that is associated with nutrient removal in wastewater treatment using microalgae is the hydraulic retention time (Larsdotter, 2006 ;). It can be concluded that for optimal removal of nutrient adequate retention time should be required. Phosphorous concentration in wastewater can be reduced by treatment with microalgae by direct removal and indirect removal. Some filtration process also helps to remove phosphorous because of absorption and adsorption as well.

RESULTS AND DISCUSSION

COD Uptake

COD reductions for samples 20% Non-Filter (NF), 25% NF, 30% NF, 35% NF, 40%NF, and 45% NF were analyzed on basis of their reduction potential.

Figure 2.

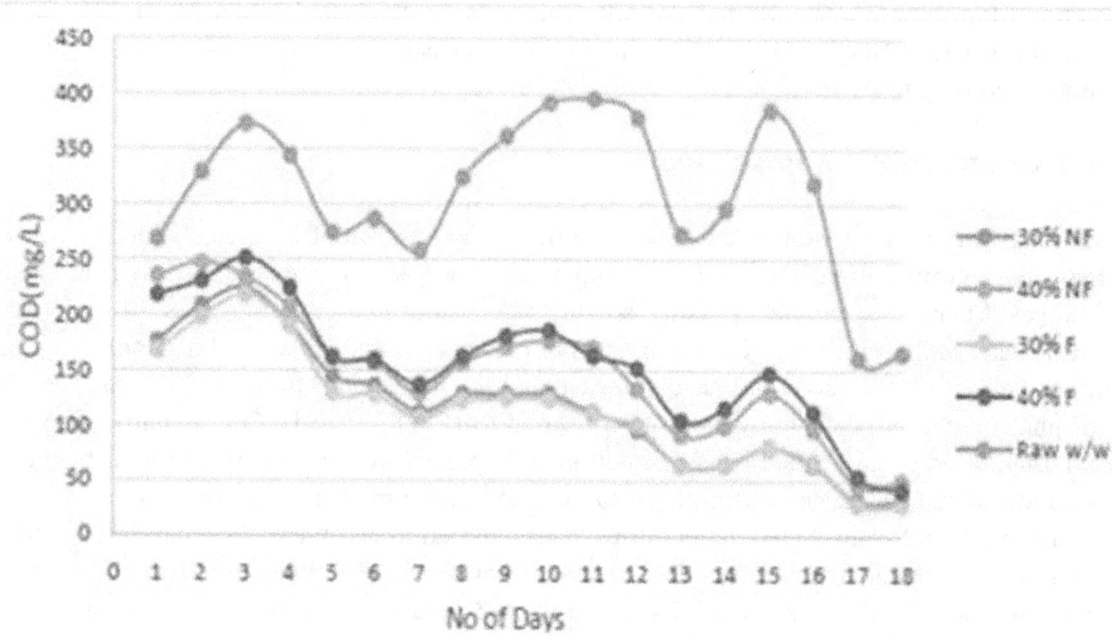

30% non-filter sample has maximum cod removal potential of 80.65% amongst the entire above sample. After that COD reduction potential of 25%. Microalgae is comparatively higher than other concentration. Among all the six concentration least COD reduction potential of microalgae concentration

Figure 3.

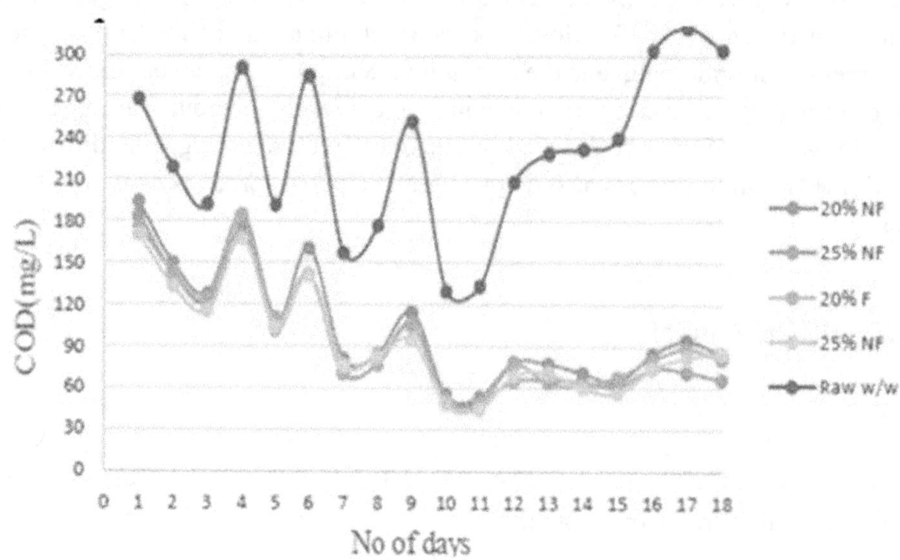

Figure 4.

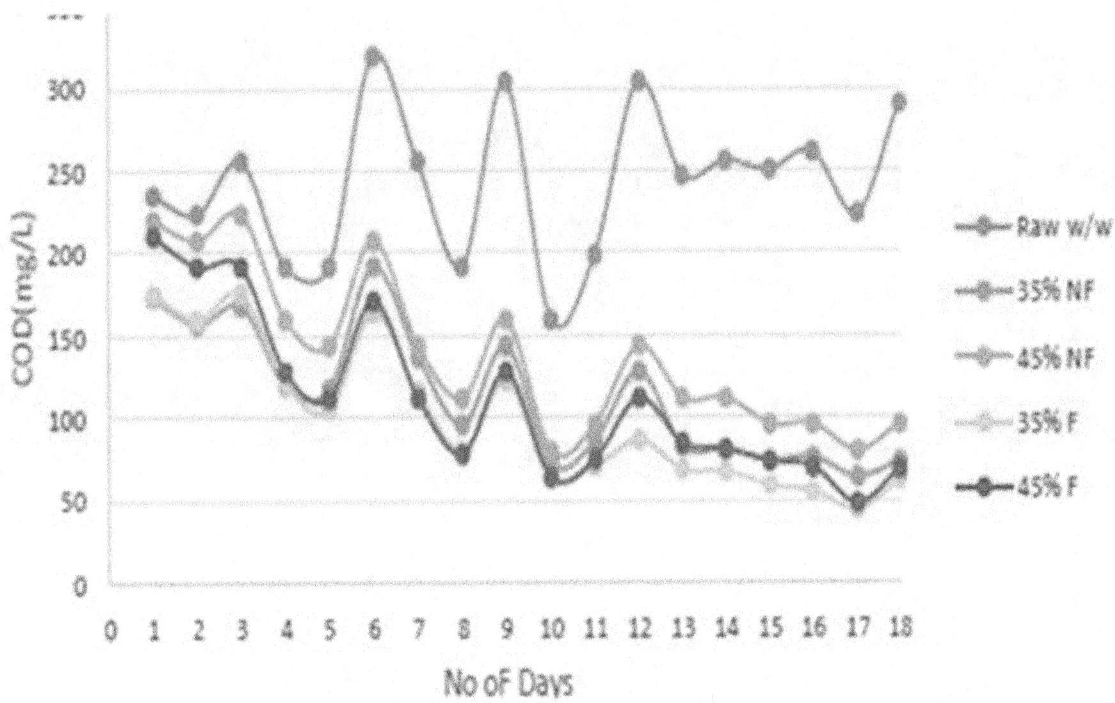

Figure 5.

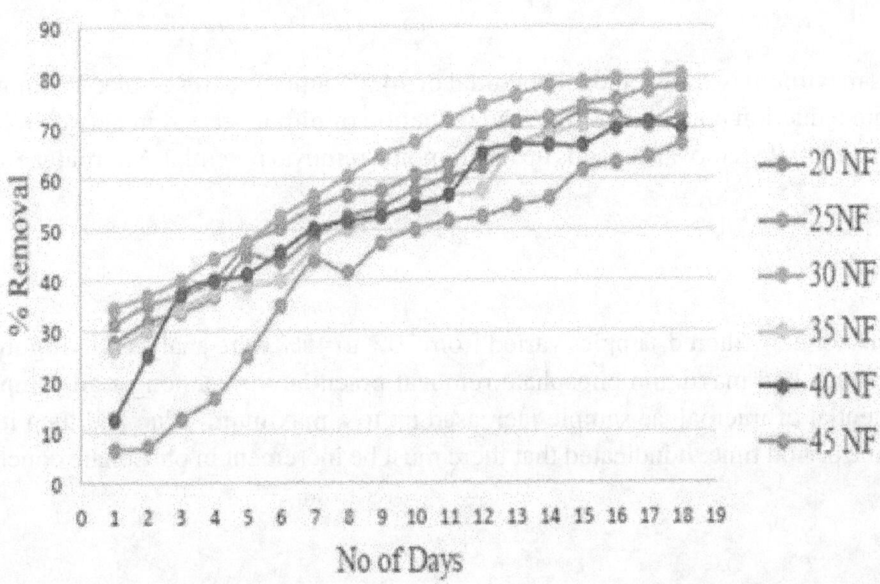

Figure 6.

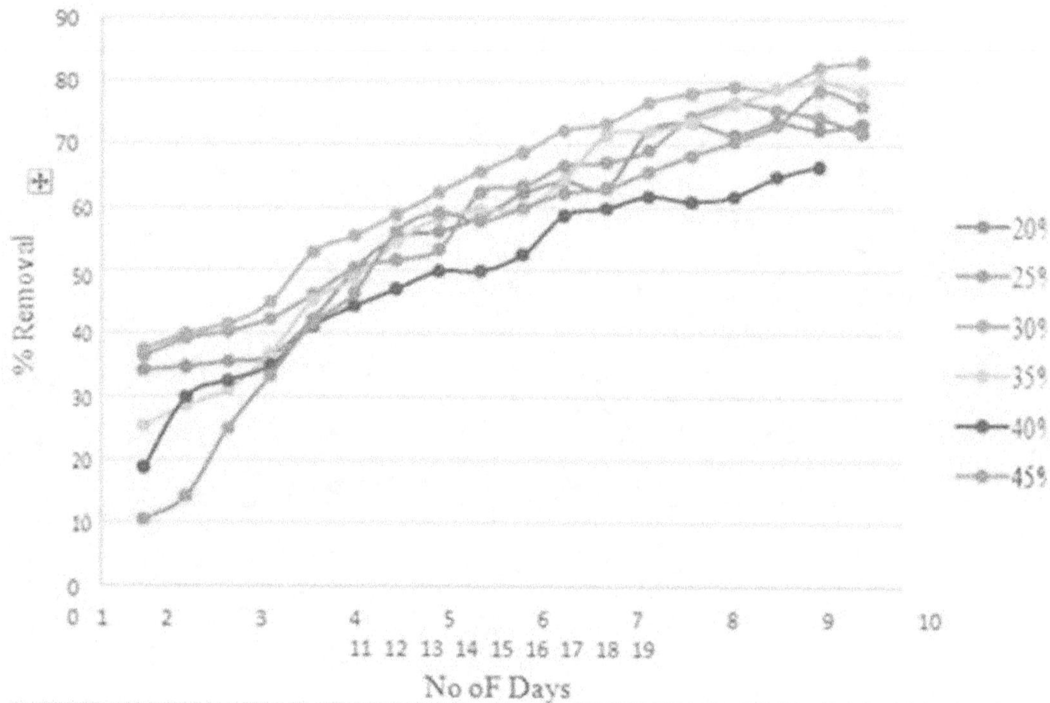

is 45%. Similarly COD reduction potential of Filtered sample were analyzed in Figure 2.4.Amongst all the concentration 30% filtered sample has maximum cod removal potential of 83.23% . 25% filtered and 35% filtered microalgae culture are effective in removal of COD

Nitrate Uptake

30% NF sample has maximum Nitrate removal potential of 74.5% amongst all sample. 25% non-filtered has maximum nitrate reduction potential of 70%.The reduction of nitrate started in between 7 to 8 days of experiment, because nitrate removal depends up on ammonia removal potential. Microalgae assimilate ammonia first.

Phosphate Uptake

Phosphate reductions for non-filtered samples varied from 20% to 45% were analyzed in FIgureure4.78. 30% non-filtered sample had maximum phosphate removal potential 99% amongst all sample. Phosphate reduction potential of microalgae sample increased up to a maximum value and then it's started decreasing after some period time. It indicated that there must be increment in phosphate concentration.

Figure 7.

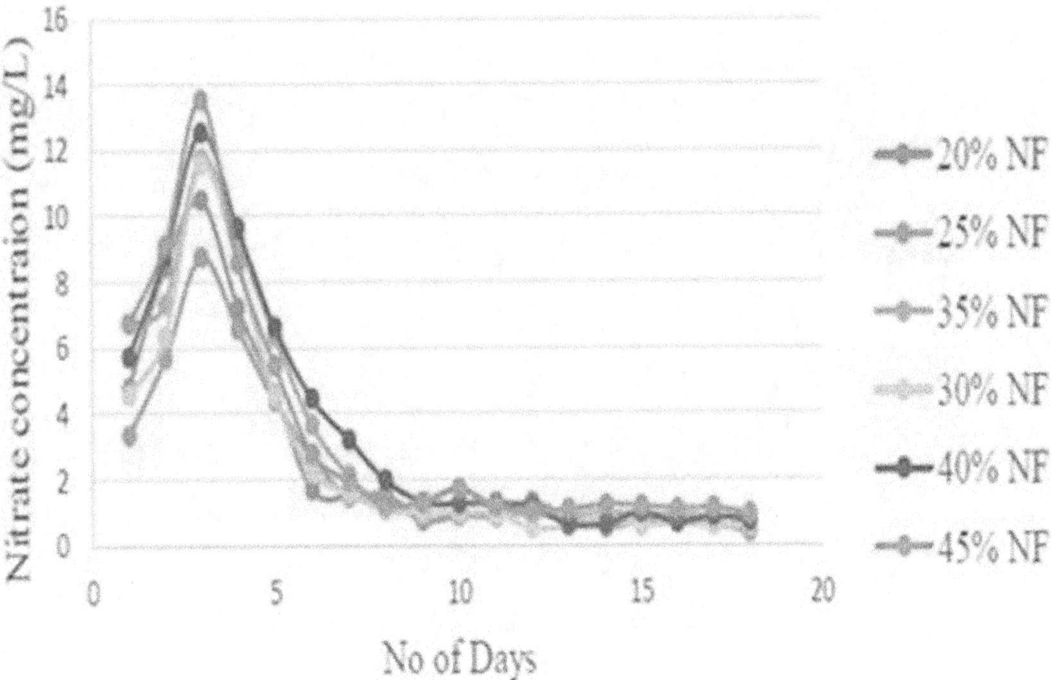

Turbidity Removal

In this study 30% and 25% micro algae had maximum turbidity removal potential however after 12 days all the samples turbidity reduction potential increases drastically. From the above plot 30% and 40% micro algae had maximum turbidity removal potential after filtration however after 10 days all the samples turbidity reduction potential increases drastically. Filtered sample have less suspended solids and very low Turbidity compared to Non- filtered sample.

Ammonia Uptake

Ammonia Nitrogen reductions for all six samples of both filter and Non-filter micro algae concentration varied from 20 to 45%. Initially 30% microalgae sample has maximum ammonia Nitrogen removal potential. After 12 days ammonia Nitrogen removing potential for other sample also increased. Ammonia reduction for all samples was above 95%.. Stripping ammonia due to high pH was not the main removal mechanism because ammonia started to decrease at neutral pH. There was also an increase in nitrate concentration associated with the decrease in ammonia. It was noted that nitrate uptake began when ammonia was depleted in the wastewater. Microalgae primarily use inorganic nitrogen as a source of nutrients for their cell synthesis, and ammonia is the first preferable type of inorganic nitrogen for microalgae

Figure 8.

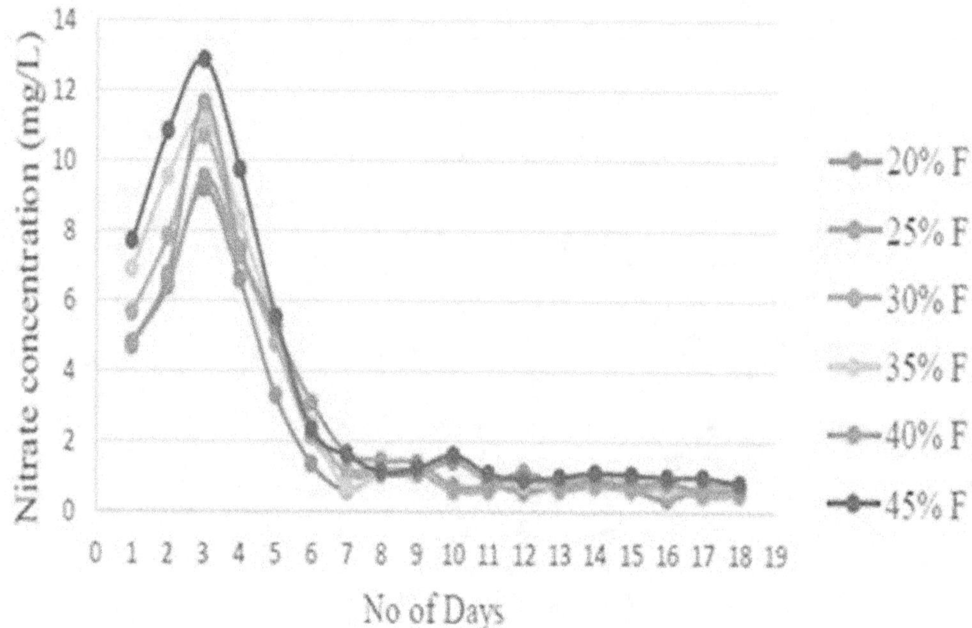

FUTURE RESEARCH DIRECTIONS

Microalgae can incorporate with various conventional wastewater treatment processes to increases the nutrient removal efficiency. It can also help to remove carbon emission by reducing aeration demand. It can possibly be an alternative towards achieving sustainability. The major problem with microalgal wastewater treatment is the concentration of Total solids, Total dissolve solids and Total suspended solids is quite high. Any technique which is efficient enough to remove these parameters can be used for the further experiment. Various studies can be performed with the effluent of Primary and secondary process of wastewater treatment to know the efficiency of microalgae in nutrient removal aspect. Microalgae are photosynthetic microorganisms that can be used as oxygen providers for aerobic bacteria during the decomposition of organic matter. Thus, further research is suggested to better utilize this advantage and reduce the overall aeration requirements for biological treatment.

CONCLUSION

In phase one of study, 30% and 40% microalgae were incorporated with raw wastewater amongst which 30% microalgae performed most efficiently on basis of percentage removal. Maximum Reduction potential of COD, Nitrate, ammonia, Phosphate varied as 83.23%, 74.56%, 99.06% and 99%. According to the GPCB Disposal Limits of wastewater COD value (117mg/L) is slightly higher. pH and Total suspended solids exceeded the limiting value. All other parameters total dissolve solids (TDS), ammonia, phosphate,

Figure 9.

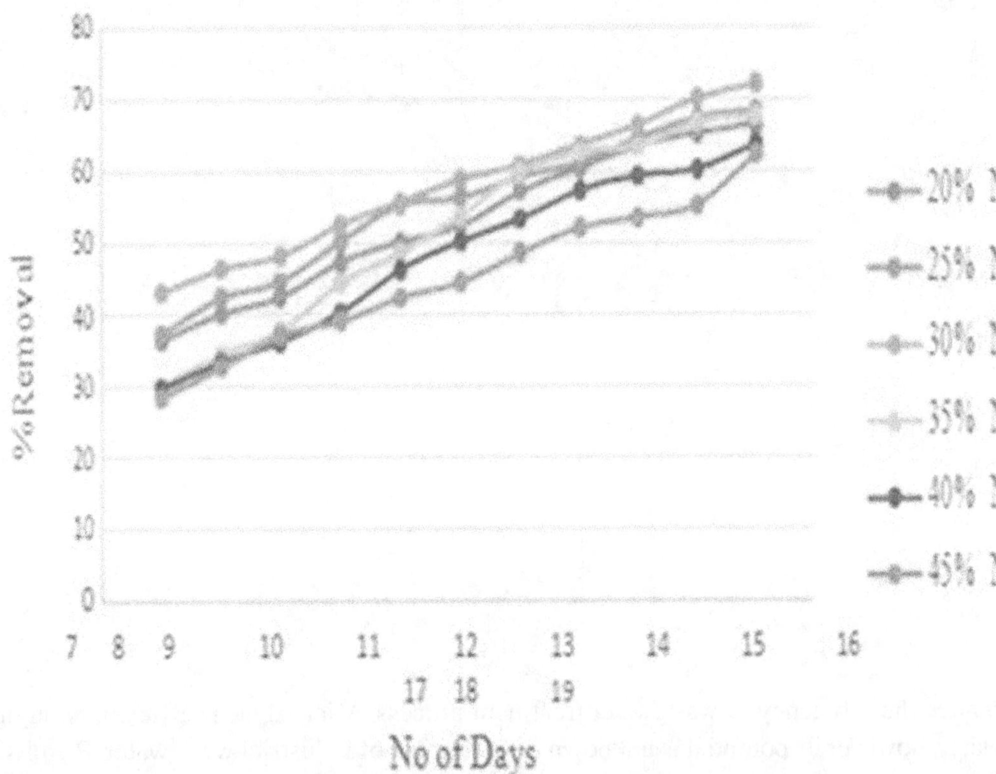

and Nitrate were within the limits. In phase two of study, 20% and 25% concentration of microalgae were incorporated with raw wastewater amongst which 25% microalgae performed most efficiently on basis of percentage removal. Maximum Reduction potential of COD, Nitrate, ammonia, Phosphate varied as 73.23%, 72%, 98% and 99%. Refering to the Disposal Limits of wastewater, pH and total suspended solids exceeded the limiting value. All other parameters i.e. COD, total dissolve solids (TDS), ammonia, phosphate, and Nitrate were within the permissible limits. In phase three of study, 35% and 45% microalgae were used to treat raw wastewater. Amongst which 35% microalgae performed most efficiently on basis of percentage removal. Maximum Reduction potential of COD, Nitrate, ammonia, Phosphate varied as 74.67%, 67.6%, 98.5% and 97.86%. According to the GPCB Disposal Limits of wastewater COD value is slightly higher. pH and Total suspended solids exceeded the limiting value. All other parameters total dissolve solids (TDS), ammonia, phosphate, and Nitrate were within the limits. According to Percentage Removal of COD, nitrate, ammonia and Phosphate 30% behaves most efficiently amongst all six concentrations. While treating raw wastewater using microalgae measures to control pH and filtration

Figure 10.

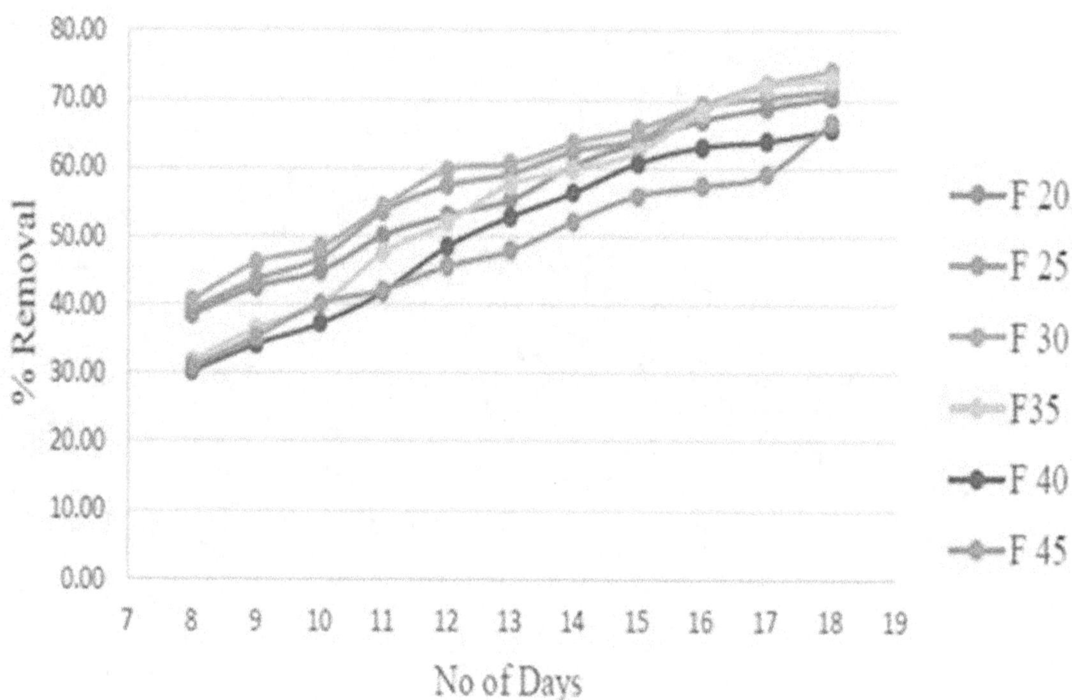

can enhance the efficiency of wastewater treatment process. Microalgae is efficient enough to treat raw wastewater, however its potential is unknown for treatment of industrial wastewater. Results showed that incorporating microalgae to conventional wastewater treatment can improve the treatment performance.

Figure 11.

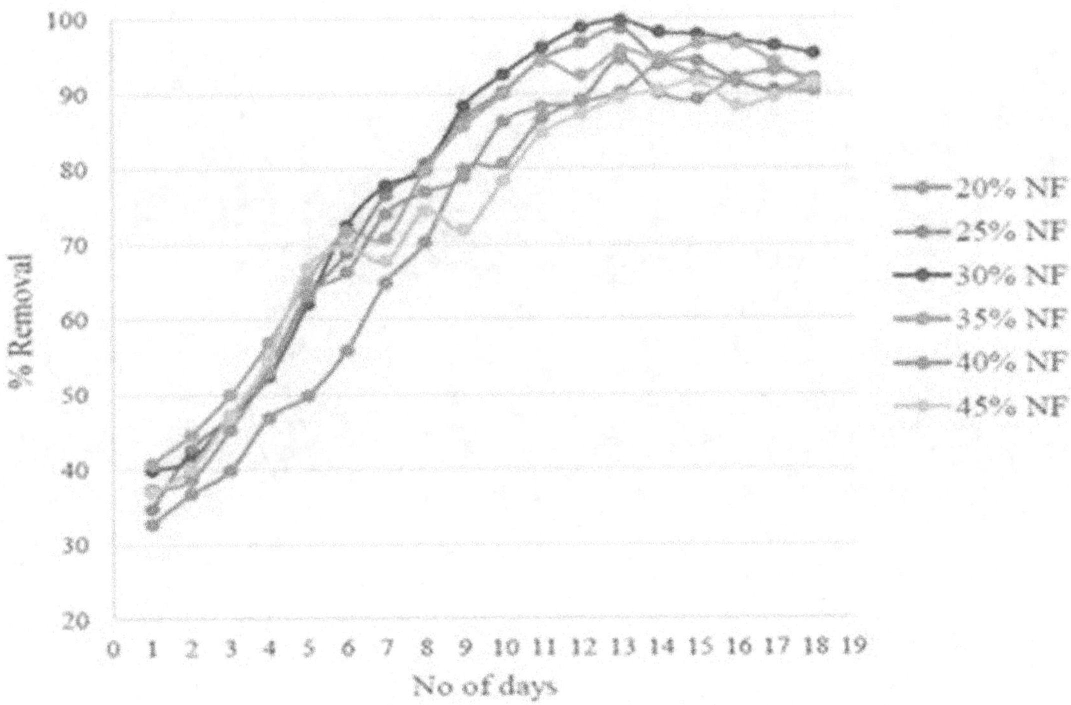

Figure 12.

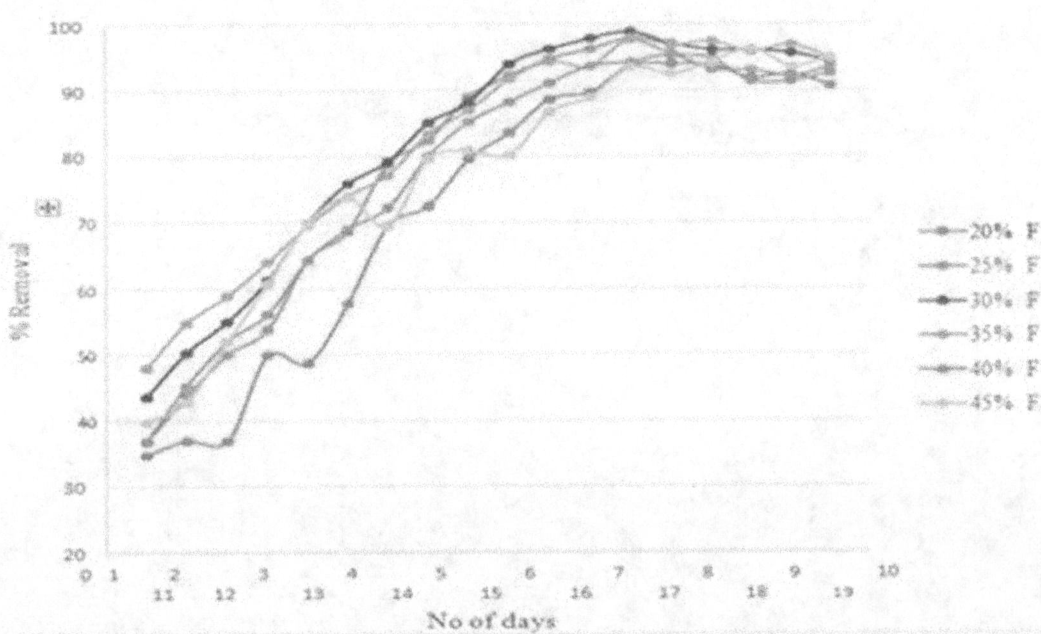

Figure 13.

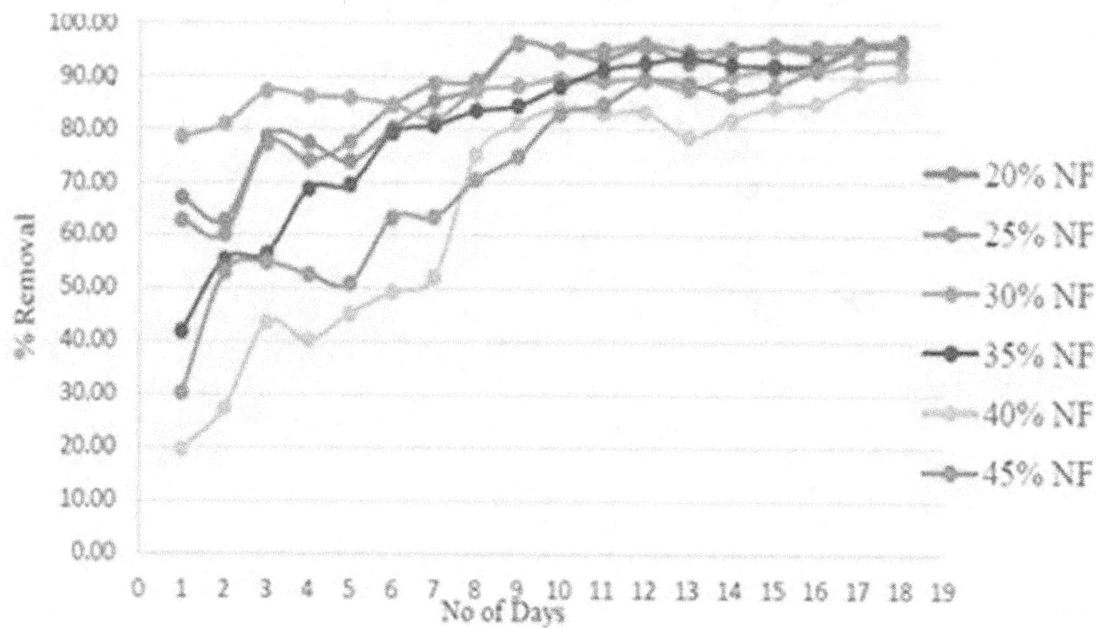

Figure 14.

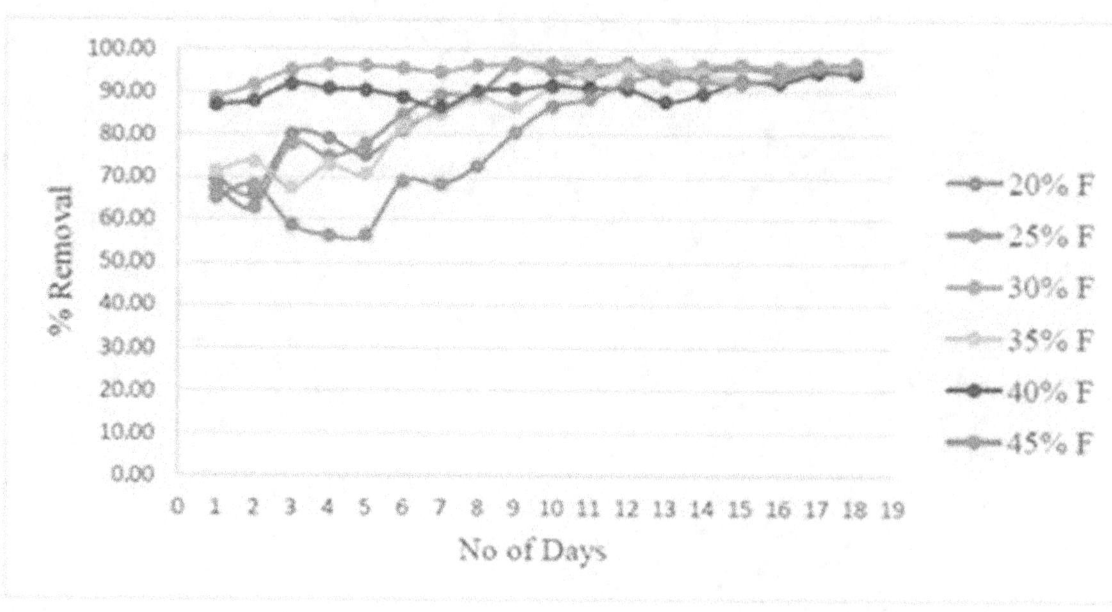

Figure 15.

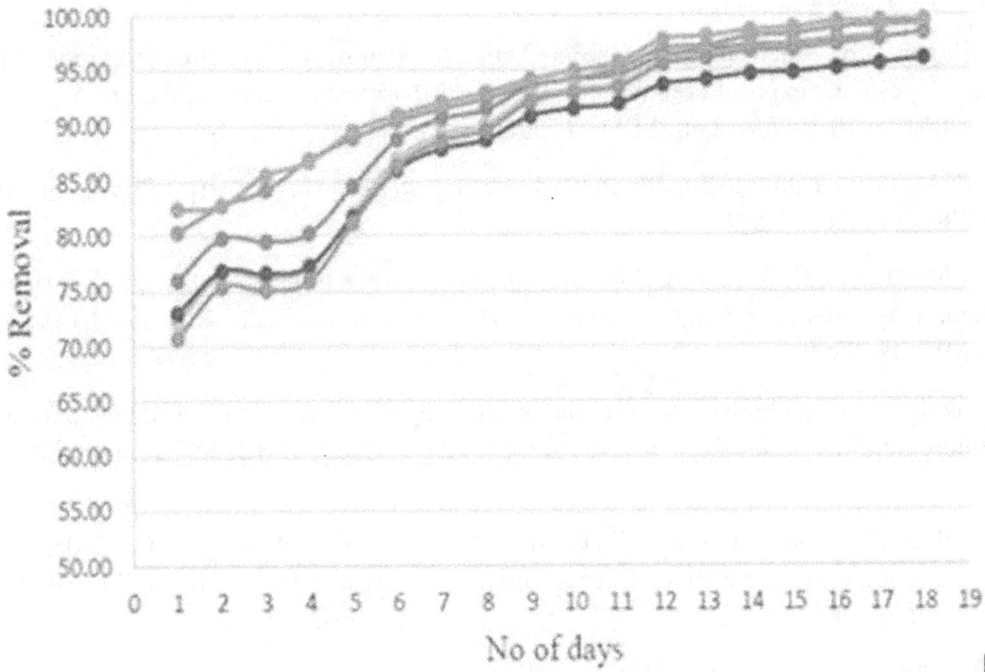

Figure 16.

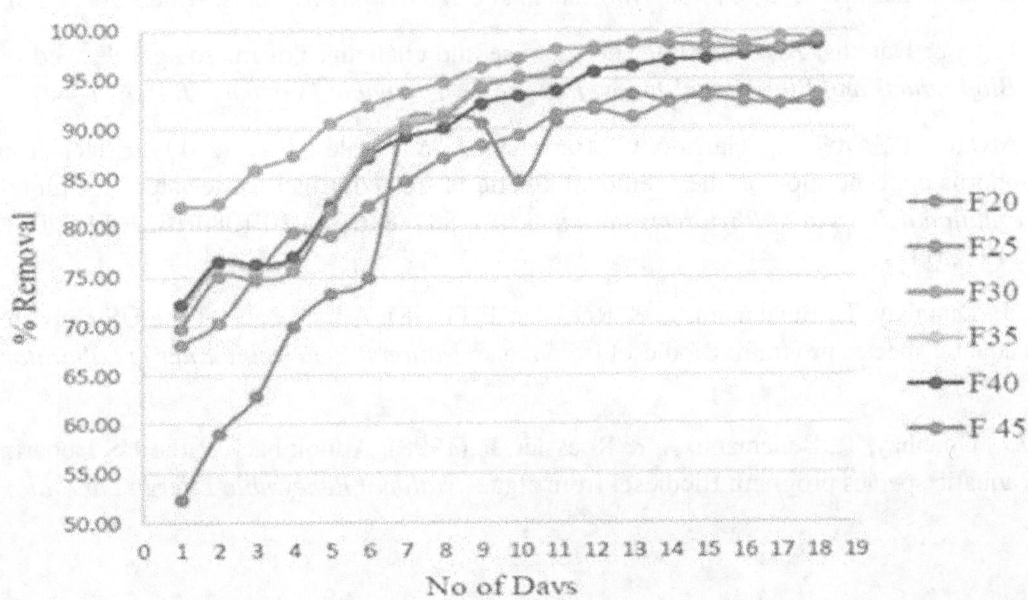

REFERENCES

Barbara, J. S. (2007). *The fales promise of biofuels*. The International Forum on Globalization and the Institute for Policy Studies.

Betarbet, R., Sherer, T. B., MacKenzie, G., Garcia-Osuna, M., Panov, A. V., & Greenamyre, J. T. (2000). Chronic systemic pesticide exposure reproduces features of Parkinson's disease. *Nature Neuroscience*, *3*(12), 1301–1306. doi:10.1038/81834 PMID:11100151

Borowitzka, M. A. (1998). Limits to growth. In *Wastewater treatment with algae* (pp. 203–226). Springer. doi:10.1007/978-3-662-10863-5_12

Carey, R. O., & Migliaccio, K. W. (2009). Contribution of wastewater treatment plant effluents to nutrient dynamics in aquatic systems: A review. *Environmental Management*, *44*(2), 205–217. doi:10.100700267-009-9309-5 PMID:19458999

de la Noue, J., & de Pauw, N. (1988). The potential of microalgal biotechnology: A review of production and uses of microalgae. *Biotechnology Advances*, *6*(4), 725–770. doi:10.1016/0734-9750(88)91921-0 PMID:14550037

Driscoll, C. T., Whitall, D., Aber, J., Boyer, E., Castro, M., Cronan, C., ... Lawrence, G. (2003). Nitrogen pollution in the northeastern United States: Sources, effects, and management options. *Bioscience*, *53*(4), 357–374. doi:10.1641/0006-3568(2003)053[0357:NPITNU]2.0.CO;2

Larsdotter, K. (2006). Wastewater treatment with microalgae-a literature review. *Vatten*, *62*(1), 31.

Li, Y., Horsman, M., Wu, N., Lan, C. Q., & Dubois-Calero, N. (2008). Biofuels from microalgae. *Biotechnology Progress*, *24*(4), 815–820. PMID:18335954

Oswald, W., & Gotass, H. (1957). *Photosynthesis in sewage treatment*. Trans. Amer. Soc. Civil Engrs.

Oswald, W., & Gotass, H. (1957). *Photosynthesis in sewage treatment*. Trans. Amer. Soc. Civil Engrs.

Pienkos, P. T., & Darzins, A. L. (2009). The promise and challenges of microalgal-derived biofuels. *Biofuels, Bioproducts and Biorefining: Innovation for a Sustainable Economy, 3*(4), 431-440.

Ruiz, J., Álvarez, P., Arbib, Z., Garrido, C., Barragán, J., & Perales, J. A. (2011). Effect of nitrogen and phosphorus concentration on their removal kinetic in treated urban wastewater by Chlorella vulgaris. *International Journal of Phytoremediation*, *13*(9), 884–896. doi:10.1080/15226514.2011.573823 PMID:21972511

Sheehan, J., Dunahay, T., Benemann, J., & Roessler, P. (1998). A look back at the US Department of Energy's aquatic species program: Biodiesel from algae. *National Renewable Energy Laboratory*, *328*, 1–294.

Sheehan, J., Dunahay, T., Benemann, J., & Roessler, P. (1998). A look back at the US Department of Energy's aquatic species program: Biodiesel from algae. *National Renewable Energy Laboratory*, *328*, 1–294.

Tredici, M. R., Margheri, M. C., Zittelli, G. C., Biagiolini, S., Capolino, E., & Natali, M. (1992). Nitrogen and phosphorus reclamation from municipal wastewater through an artificial food-chain system. *Bioresource Technology*, *42*(3), 247–253. doi:10.1016/0960-8524(92)90028-V

Van Den Hende, S., Vervaeren, H., Desmet, S., & Boon, N. (2011). Bioflocculation of microalgae and bacteria combined with flue gas to improve sewage treatment. *New Biotechnology*, *29*(1), 23–31. doi:10.1016/j.nbt.2011.04.009 PMID:21565287

Way, D. A., & Oren, R. (2010). Differential responses to changes in growth temperature between trees from different functional groups and biomes: A review and synthesis of data. *Tree Physiology*, *30*(6), 669–688. doi:10.1093/treephys/tpq015 PMID:20368338

KEY TERMS AND DEFINITIONS

Chemical Oxygen Demand: It is the amount of oxygen required to consume all biodegradable and non-biodegradable organic matter.

Nutrients: Essentially required for growth of any living organisms. Plants required 11 types of compound for the growth process. Majorly 3 compound N-P-K (nitrate, phosphate, potassium) are essential for plant growth. These are commonly known as plant nutrients.

Photosynthesis: It is a process followed by plants in which light energy converted to chemical energy which supports the plant growth. It is a natural process.

Turbidity: It is the degree to which transparent liquid scatters light. Turbidity causes due to very small particle which are normally not visible to naked eye.

Chapter 17
Safe Use of Wastewater in Agriculture Through Bioremediation Processes

Rachana Dubey
ICAR Research Complex for Eastern Region, India

Arbind Kumar Choudhary
ICAR Research Complex for Eastern Region, India

Janki Sharan Mishra
ICAR Research Complex for Eastern Region, India

Ashutosh Upadhyaya
ICAR Research Complex for Eastern Region, India

Sharad Kumar Dwivedi
ICAR Research Complex for Eastern Region, India

Surajit Mondal
ICAR Research Complex for Eastern Region, India

Karnena Koteswara Rao
ICAR Research Complex for Eastern Region, India

Abhishek Kumar Dubey
ICAR Research Complex for Eastern Region, India

ABSTRACT

Population growth, industrialisation, urbanisation, and climate change have created huge pressure on freshwater resources to fulfil the demand. Approx. 70-80% of the freshwater supply returns as wastewater, which is difficult to tackle and manage. We need to tackle the freshwater demand from different sectors like domestic, industrial, and agriculture. Most important is how to use the wastewater safely

DOI: 10.4018/978-1-7998-4921-6.ch017

in agriculture. Therefore, it is an apt time to refocus on ways to recycle water especially in sectors like agriculture and for ecosystem services. The major concern in using wastewater in agriculture is its quality as the wastewater may carry pathogens, heavy metals, and many other pollutants, which might reach to human beings and animals via food chain. A solution to wastewater reuse is through bioremediation techniques. Bioremediation should be considered as a feasible and futuristic technology for safe use of wastewater in agriculture as it will reduce the burden on centralised water treatment system as well as it being economic and eco-friendly.

BACKGROUND

A major concern now a days internationally and nationally is of quantity and quality of irrigation water. There is a conflicting problem as on one hand there is an increasing scarcity of freshwater availability and on the other side wastewater generation has enormously increased. Wastewater generation is linked with increasing urbanisation, population pressure, industrialisation and all these have created a huge pressure on managing this wastewater which otherwise will create socio-economic and environmental problem (Chaidez et al., 2014). Before going into the details of wastewater, let us be clear about wastewater term.

Wastewater is any water which is not intended to be used directly for human consumption. In general wastewater is 99.9% water and rest are pollutants which we intend to remove or which makes it a wastewater. This 0.1% contains organic matter, inorganic compounds, nutrients, pathogens and toxic metals. Generally, municipal wastewater is a mix of domestic wastewater, storm water and industrial wastewater. Domestic wastewater is the water which is discharged from households, schools and other institutions of a place. While industrial wastewater refers to wastewater generated from the industries. The composition of industrial wastewater differs with the industry type (Hussain et al., 2002). As demand for water is increasing, the per capita availability of water has declined and day by day it is widening which is posing threat for human existence. Around 36% of global population lives in water scarce regions and as per the estimate by World Bank (2019), this figure will reach to more than 50% by 2050. Therefore researchers globally are focussing on measures for conserving water and at the same time it is appropriate time now to find ways to use wastewater as a resource and reuse it in agriculture (Hussain et al., 2002) as it will promote water security and resilience to climate variability in long term. In many regions of the world especially in water scarce areas falling under arid and semi-arid regions as well as peri urban and peri-industrial area where competition for water is very high among different sectors like industries, agriculture, domestic use, landscape etc. agriculture is the largest sufferer. Since agriculture uses more than 80% of freshwater use of wastewater for irrigation purpose is common. Figure 1 depicts use of wastewater for irrigation by different countries. As there is no official figure of wastewater use for irrigation in India therefore the figure 1 did not have mention of India. As per one estimate by Raychaudhuri et al. (2017), 62000 million litres per day of wastewater is produced in India of which only 27% is treated and rest is untreated. This untreated wastewater is being used by farmers in peri urban and peri industrial areas for their crop irrigation under water scarce situation. More than 73000 ha area is being irrigated with wastewater (Raychaudhuri et al. 2017).

International Water Management Institute estimated irrigation potential with wastewater in class I and II cities of India. Volume of treated wastewater and untreated wastewater generated from these cit-

Figure 1.

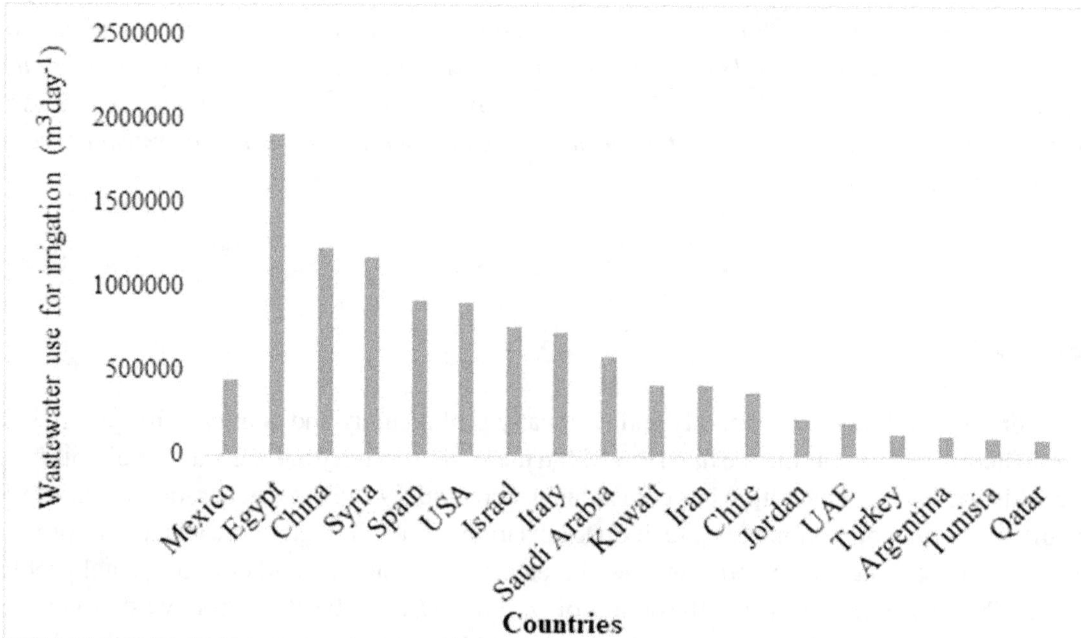

ies are 11787 mld and 326467 mld respectively which has the potential to irrigate approximately 70722 and 1032213 ha of area respectively.

Current Status of Wastewater Use in Agriculture

Although proper documentation related to use of wastewater is lacking but we cannot deny its use of irrigation especially in peri-urban and peri-industrial area. There are many benefits for which farmers prefer to use wastewater like round the year availability, nutrient present in it, which reduces fertilizer cost and overall production cost (Banerjee, 2016; Rachana and Subhasis, 2019). In spite of all these benefits still wastewater use has many limitations. In most of the developing and under developed nation's use of untreated sewage is prevalent which is associated with many health hazards like bacterial, viral and protozoan diseases. Many types of parasitic worms like hookworm, roundworm etc. are prevalent. Additionally, depending on wastewater source heavy metals, persistent pollutants, and emerging pollutants are present which not only poses threat to soil but through the food chain and cycle reaches to human being and animals (Mateo-Sagasta et al., 2013). Many instances in the past has been found where persistent pollutants have created more havoc to the animals/ plants/human beings of higher hierarchy in the food chain via bio-magnification. Therefore globally the focus is on safe use of wastewater in agriculture. Treatment of wastewater is the most appropriate option where wastewater is converted into bilge water which can be discharged back into the environment. But operation and maintenance cost of treatment plants is very high because of which most of the treatment plants are not functional even after establishment. For example, India only treats 37% of its total sewage water generated and rest goes directly on land or nearby water bodies creating risk for its users (CPCB, 2016). Keeping this in view

in 2017, United Nations emphasized on wastewater with the theme 'Why Waste Water'- which was about reducing the use of water and reusing wastewater for sustainable development. Thus out of many ways of reusing wastewater most economical and sustainable option is bioremediation which has been discussed in detail in below section.

Bioremediation Process

Bioremediation refers to use of bio or living entities to degrade, alter or remove toxic/undesirable compounds or pollutants from the wastewater. Here organisms play a key role and rate of bioremediation depends on the catabolic activities of the microorganism which use wastewater as a source of food for their energy requirement. These microorganisms can either be native to contaminated site or it can be brought and adjusted to the new site Bioremediation provides ample scope to use wastewater irrigation safely in agriculture.

Bioremediation process has benefits as it can be done directly on contaminated site reducing its cost and space. Although bioremediation seems simple in application, the process is not that easy as it involves organism which needs to adapt to a site, moreover they should use the contaminant as their energy source in the process of degradation. In other words, efficiency of bioremediation process depends on contaminant properties as well as microorganism ability (Mora-Ravelo et al., 2017). Bioremediation process is mainly classified in two broad categories i.e. *ex-situ and in-situ* (figure 2). *In-situ* bioremediation refers to the process where contaminants biologically degrades to simple products like carbon dioxide and water or other transformed product at the site only. This is a preferred process since it involves low cost and maintenance, additionally it is environmentally sustainable (Megharaj et al., 2011). Ex-situ refers to remediating contaminated site after taking the pollutants from one place to the site of treatment (Azubuike et al., 2016). All the bioremediation process can be categorised into three types depending on the process involved. The first one is Bio-attenuation: also known as natural attenuation where naturally the remediation process goes on at the contaminated site. Second one is bio-stimulation: here the natural process of bioremediation is stimulated externally by supplying either water, energy, electron donors or acceptors which enhances the remediation process by increasing the metabolic process of microorganism. Third process is bio-augmentation: here according to the dominant contaminant to be removed from a site, well known microbes are added externally which have the capability to degrade these contaminants (Ying, 2018). The type of bioremediation process to be used will depend on environmental plus microbial condition like site property, climate, quality and quantity of pollutants present etc. (Ying, 2018).

Use of Bioremediation for Wastewater Treatment

Although there are numerous bioremediation techniques but most common ones being practised in wastewater remediation is phytoremediation, chelation, methylation and precipitation (Martínez-Prado et al., 2011).

Phytoremediation

Phytoremediation is made of two words i.e. phyto- means plants and remediation- to rectify. As per United Nations, phytoremediation is defined as the use of living green plants for in-situ removal/ degradation/ reduction of contaminants in the affected area of soil, water or groundwater. Phytoremediation can be

Figure 2. Types of bioremediation processes

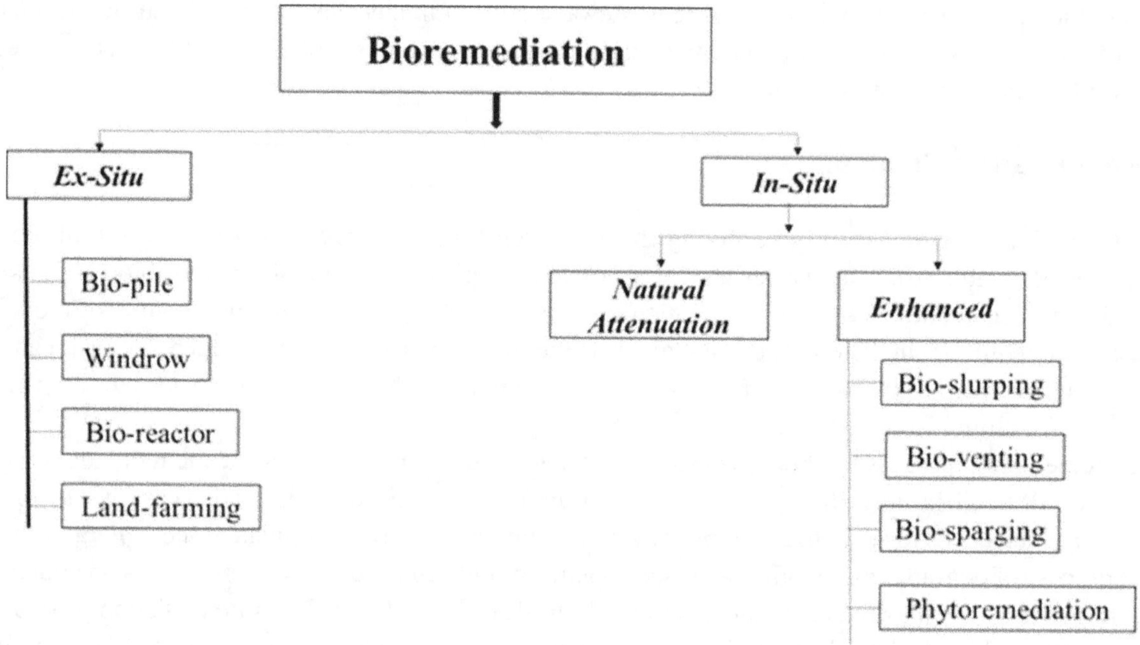

done to remediate metals like lead, zinc, cadmium, copper, nickel, mercury, nickel etc., metalloids like arsenic, inorganic compounds like nitrate, phosphate, ammonium; radioactive compounds, petroleum compounds, pesticides, herbicides, chlorinated compounds (Favas et al., 2014).

Phytoremediation process is broadly divided into six types namely phytoextraction, phytodegradation, phytovolatilisation, rhizofiltration, phytostabilisation and rhizodegradation. Detail are being given below:

1. *Phytoextraction*: Phyto-extraction refers to extraction of metals or organic compounds by plants from contaminated sites. In this process, pollutants are absorbed by plants and are translocated to above ground biomass. Plants are harvested from the contaminated site. This process is suitable where contamination is of low to moderate levels.mn. Plants selected for phytoextraction should be tolerant type which are able to tolerate high concentration of toxic metals or compounds in their body. The plants must be fast grower with high biomass production (Farraji, 2014). Phytoextraction can be continuous or induced process. Continuous phytoextraction refers to the process where plants accumulate high level of toxic metals/compounds in their parts during the entire life cycle of plants. Induced phytoextraction means the extraction process is enhanced via use of chelators at any particular plant stage at which the plants extraction capacity is highest (Farraji, 2014). For example chelators like EDTA or low molecular weight acids have the ability to enhance the efficiency of phytoextraction. In a study done by Mirza et al. (2010) where species *Arundo donax* was tested for arsenic from wastewater under greenhouse condition. Arsenic concentration of upto 600 $\mu g\ L^{-1}$ was extracted from *A. donax* spp. and above that concentration (1000 $\mu g\ L^{-1}$), plant started oxidative stress. Therefore *A. donax* was found suitable for phyto-extracting arsenic from wastewater. Similarly, Monferran et al. (2012) showed *Potamogeteon pusillus* leaves and roots tolerant towards Cr and Cu which can be used for phytoextraction of these metals from contaminated water.

2. Phytodegradation: Phytodegradation or phytotransformtion is breaking of organic compounds by plants. Breakdown of organic compounds or contaminants is done either by metabolic activities within the plants or via metabolites released by plants. These plants via metabolites such as nitroreductases, dehalogenase, oxygenase etc. degrade organic compounds present in plants (Farraji, 2014). Plans photodegrades contaminants either by uptake, metabolism or translocation to different plat parts (Ferraji, 2014). Nutrient load reduction in wastewater through microalgae is one of the emerging option (Akpor et al., 2014) as these microalgae can assimilate nutrients in their cells. These microalgae are variedly used to degrade chlorinated solvents, insecticides and pesticides and many other inorganic and organic nutrients (Akpor et al., 2014). Plants ability to metabolise these nutrients determines the efficacy of the process. When a toxic compound is taken by plants the first stage is activation of compounds by the action of different metabolites like flavin, monooxygenase etc. then its conjugation in plant parts and lastly the plant compartmentalise it in different parts preferably vacuoles or apoplast which ensures safe disposal of these toxic compounds (Akpor et al., 2014).
3. Phytovolatilsation: As the name suggest, phytovolatilsation refers to volatilisation of contaminants via plant parts. Here plant takes up contaminants from the affected site transforms them into volatile form which then vaporises in the air via plant parts (Farraji, 2014). This process is used for remediation of volatile organic compounds either through stem and leaves (known as direct volatilisation) or from soil due to root activity of plant (indirect volatilisation) (Limmer & Burken, 2016). A classic example of volatilisation is removal of Selenium. Selenium as a toxic element is found in most of the arid and semi-arid regions of the world with seleniferous soils originating from marine sediments. Selenite can leach from soil to groundwater or nearby surface water which through bioaccumulation can reach to human and plants. Plants are effective in removing selenium from contaminated site(Adel et al., 1999). When selenium is absorbed by plant roots and translocated to shoots it is volatilised into dimethyl selenide which gets volatilised from the stem or leaves. Hyperaccumulators of Se are mostly from *Fabaceae* group including *Brassica juncea, Brassica napus,* Stanleya, Astragalus spp. etc. (Adel et al., 1999).

Another example of phytovolatilsation is constructed wetlands used for remediation of wastewater. It can be used to remediate industrial, domestic, municipal or agricultural wastewater. In this process helipohytes takes contaminants from the wastewater which are then taken up by plants with the assistance of microbial activities and are transformed to different secondary products and released from the plants (Stottmeister et al., 2003).

4. Rhizofiltration: Rhizofiltration in simple terms refers to filtration by roots. It is the adsorption or absorption into plant roots of contaminants present in soil solution. It is a preferred technique to remediate groundwater. Plants are grown hydroponically in contaminated water where they take up contaminants alongwith water. When the roots gets saturated with the contaminants they are taken off from the site and fresh plants are again grown in these contaminated water (Abdullahi, 2015; Dushenkov et al., 1995). Very popular example of it is use of Sunflower plants for remdiation of groundwater from radioactive contaminants in Ukraine (Abdullahi, 2015). In another study by Elias et al. (2014) where water hyacinth was used to treat ceramic industry wastewater containing organic pollutants and heavy metals like Cd, Cr, Cu, Mn, Zn, Fe and B. Studies showed that most of the metal and organic pollutant absorption was through roots of the plants with an efficiency of

99%. Rhizofiltration ability of *Carex pendula* (wetland plant) was tested against lead in wastewater. C.pendula showed tolerance level of 1600mg/kg when water concentration in wastewater was 10 ppm. Results showed that C. pendula is an effective plant for rhizofiltration of lead from wastewater (Yadav et al., 2011)

Rhizofiltration is mostly used for wastewater remediation especially for heavy metals as it is cost-effective and environmental friendly. Moreover the best part is that accumulated metals can be recovered from root system which after extraction can be used as bio-ore. As this technique is in-situ so it does not disturbs the ecological balance of the site as well as it is aesthetically pleasing which makes it more sustainable option for metal recovery of heavy metals (Yadav et al., 2011)

5. Phytostabilisation: Phytostabilisation refers to immobilisation of contaminants in the soil/water via accumulation and adsorption by the plants roots or rhizospheric zone. In this process movement of contaminant is restricted by plant roots so that they are spread by erosion or water movement (OB and Muchie, 2010).

In a study done at urban stream sediments of the Nakivubo in Kampala, Uganda found that *Commelina benghalensis* (L.) and *Cynodon dactylon* (L.) bioaccumulation factor was more than 1 compared to translocation factor with respect to Pb, Cd and Zn which showed that these plant spp. can be utilised for phytostabilisation of heavy metals in these area (Sekabira et al., 2011)

6. Rhizodegradation: Rhizodegradation means degradation of compounds (mostly organic contaminants) by rhizospheric microorganism for nutrient use in plants. This process is also known as phytostimulation. Under this process microbial activity is enhanced in rhizospheric zone via plant

Table 1. Different phytoremediation techniques with some examples

Phytoremediation type	Process description	Medium	Type of contaminants	Plants involved
Rhizodegradation/ Phytodegradation	Degradation of contaminants in rhizosphere assisted by microorganism	Soil, Sludge	Aromatic, aliphatic and petroleum hydrocarbons, pesticides, chlorinated solvents	Grasses Alfalfa Hybrid poplar Cassia Brassica spp.
Phytostabilisation	Stabilisation of contaminants by plants through complexation or absorption.	Sediments, soil, sludge	Mostly metals like Pb, Fe, Cu, Cr, As, Ni radionuclides etc.	Sunflower Chenopodium
Phytoextraction	Extraction of contaminants by roots and accumulation in upper parts like shoots or vacuoles.	Sediments, soil, sludge	Mostly metals like Pb, Fe, Cu, Cr, As, Ni radionuclides etc.	Sunflower Brassica Alyssum
Rhizofiltration	Removal of contaminants by roots	Wastewater Groundwater Surface water	Metals, Radionuclides	Lemna Eichhornia
Phytovolatilisation	Contaminants are volatilised from leaves or other exposed surfaces of plants	Soil Sediments	Organic and inorganic contaminants including metals	Poplar Scirpus

(Source: Dhir, 2013)

exudates such as carboxylic acid, carbohydrates or proteins are released from plant which enhances microbial activity and hence degrades organic contaminats present in soil or water (Ferraji, 2014). Some examples of rhizodegradation by plants are- *Hibiscus cannabinus* for used lubricating oil, *Plantago major* for Imidachlorpid insecticide, *Sorghum bicolor* for crude oil, willow spp. for perchloeate removal from soil and water (Ferraji, 2014). Table 1. describes the phytoremediation techniques in brief citing some examples.

Merits and Limitations of Phytoremediation Process

There are many advantages associated with phytoremediation techniques (Koptsik, 2014) some of them are:

1. It can be used to remediate a wide range of contaminants including organic, inorganic and radionuclides.
2. As *in-situ* remediation occurs, it reduces the quantum of waste which enters into landfills or dump sites
3. Again as it is *in-situ*, it reduces the soil disturbance and spread of pollutants.
4. As compared to other chemical or traditional methods, it is less costly or cheaper technique.
5. The technique of phytoremediation is simple which is easy to use and maintain. At the same time it is renewable one which can be easily adopted in field with least technical skill requirement.
6. It is environmentally safe, aesthetically pleasant and socially acceptable method.
7. It is less noisy method.

Limitations (Koptsik, 2014)

The benefits of phytoremediation are many but still this technique has some limitations which has not made it very popular amongst industrialist to adopt at wider scale:

1. This technique is restricted within the rhizospheric zone and contaminants removal depends on solubility and availability of pollutants so that plants can take them up.
2. It is time consuming process, can take several years (2-15 years) to remediate a contaminated site.
3. Remediation via phytoremediation can be successful only if the contaminats level is low-medium.
4. After phytoextraction of contaminants mostly heavy metals, proper disposal is a concern and challenge to tackle. Although bio-ores have come into role as one of the option for removal of these contaminants but still its technology is not very sound to be used at commercial scale.
5. Plants can grow under certain agro-climatic condition and any change in it may affect their remediating capacity.
6. Sometimes introduction of a plant species in a certain area for remediation purpose can cause menace as it may be a threat to biodiversity of that place.
7. Variation in agronomic practices may render ineffectiveness in phytoremediation capacity.

In one publication by Carolyn Beans in PNAS (2017) titled 'Phytoremediation advances in the lab but lags in the field'. The author stated that despite phytoremediation being a successful technology still it is mostly limited to lab studies or pilot study. Major lacuna is time intensive where plants years to remediate which discouraged industrialist to invest in this technique. Secondly, there are more than 1000 publications on phytoremediation but they all are limited to lab studies and very limited number of publication showed long term field studies in this regard which may precisely determine the total time and cost involved in remediating a contaminated site considering all the factors in a given area. Therefore long term studies on phytoremediation is utmost important to make it a successful technology at commercial level.

Depending on contaminant type the whole phytoremediation process can be categorised into three broad categories

1. Remediation of inorganic contaminants site.
2. Remediation of organic contaminants site.
3. Remediation of both inorganic and organic contaminants site.

With increasing pressure on freshwater resources with burgeoning population and in turn increased demand for food, energy and other resources have put tremendous pressure on our water resources. With agriculture being the largest user of freshwater resources the brunt of impact of pollution and climate change is bear by agriculture focusing on irrigation water around the world. Pollution by saltwater intrusion, algal blooms, soil erosion, improper sanitation, detergents, pesticides, chemicals etc. have posed serious concern (Falconer & Humpage, 2005)

Although understanding of remediation after categorising it into organic or inorganic pollutants is easy to understand but in in real we find places or wastewater contaminated with organic as well as inorganic pollutants called as co-contaminated sites. In these areas, first of all analysis of the wastewater is required and then depending on the major type of pollutant whose concentration is high action is taken upon. In these areas, phytoremediation activity is affected by interaction of different substances which affects their form and availability. In these areas, different pollutant affects plant-microbe interaction and therefore the micro-environment changes within the wastewater or a particular site. Therefore engineered microorganism or plants are best suited for remediation of this type of affected wastewater.

In an attempt to remove 4-chlorophenol from wastewater both photocatalysis and biological treatment was done. Pre-treatment of Titanium dioxide as photocatlayst. Integrating biological treatment with photochemical showed greater remediating efficiency to 4-chlorophenol. A concentration of 400 ppm of 4-chlorophenol in wastewater took 264 hr when only biodegradation was done while the same was done in 96 hr when both photolysis and biological treatment was applied (Goel et al., 2010). An attempt was made by Gupta and Ali (2004) where lead and chromium present in wastewater was removed through bagasse fly ash obtained abundantly from local sugar industry in India. The mechanism involved was particle diffusion and reaction was exothermic in nature. With column flow rate of 0.5 ml min^{-1} removal efficiency through adsorption was 95-98%. In another study phenol was removed from wastewater using palm seed coated carbon through film diffusion method. For phenol concentration of 25 ppm palm seed coated carbon showed removal efficiency twice as effective compared to commercially available activated carbon in the pH range of 4-9 (Rengaraj et al., 2002). Trovaslet et al. (2007) tested the efficacy of white rot fungi for treating the dye in wastewater. From fungal strain *Pycnoporus sanguineus* MUCL 41582 laccases was extracted and isoenzyme LAC-1 was tested for textile industry effluents. As textile industry effluents mostly contains either sodium sulphate or sodium hydroxide and LAC-1 showed promising result as the affinity was very high. Algae consortia are also great agents for bioremediation. In an attempt by Mahapatra et al. (2014), algae was grown in reactors and was tested for filtered wastewater. Results showed that 86% total organic carbon, 90% total nitrogen, 89% ammoniacal nitrogen, 70% total phosphorous and 76% ortho phosphate were removed from the wastewater. It showed that algae consortia if replicated in sewage receiving ponds or lakes can help in remediation of wastewater that too in decentralised manner. Not only this, algae are good source of energy so at one place we will be able to meet some energy demand with reduction in ghg emission.

Algae mainly removes nutrients or contaminants by assimilation in their body as in general different algal species have different nutrient removal capability under certain optimum environmental condition. From livestock wastewater algae species *Chroococcus* removed 80% of ammoniacal and nitrate nitrogen and total phosphorous (Prajapati et al., 2014). In another study of effluent of anaerobic digestion *Arthrospira maxima and Tetradesmus obliquus* removed 98-99% of nitrogen while *Phaeodactylum tricornutum and Botryococcus braunii* removed 80-89% of phosphorous. One of the nutrient removal system is High rate algae pond (HRAP). Many studies have been done which removes nutrients and other contaminants. HRAP removed 51-57% of P, 80-85% of ammoniacal N and 50-60% of total nitrogen from anaerobic digested wastewater (Nhat et al., 2014). Dominant algal species in *HRAP* were *Cyanobacteria, Planctomycetes Alpha, beta, gamma epsilon, delta-proteobacteria, Firmicutes, and Bacteroidetes* (Nhat et al., 2018). Although algae offers many benefits but its production at commercial level handles many challenges which makes its economic production unfeasible. Most important one is harvesting and dewatering of microbial cells, stabilising nutrient level in water. Therefore existing techniques does not seems practically feasible and economically profitable (Sharma and Malaviya, 2016). Fungus aided algal association is one technique which is getting attention as harvesting efficiency gets enhanced. Therefore algal assisted fungal biomass is a renewable and sustainable association for remediation of wastewater (Muradov et al., 2015). Because of the filamentous nature, fungi offers many benefits over algae or bacteria as it can tolerate high flow pressure, can be handled and cultured easily. Fungi have high tolerance and uptake ability towards metals (Salvadori et al., 2014).

As we know that waste from industries are more toxic compared to domestic or municipal wastewater. Not only we need to reduced chemical oxygen demand, colour or nutrients from it but more concern is dyes used in the industries. As dyes are colourful, therefore they impact light diffusion in water impacting the organism activities (Salvadori et al., 2014). Sharma and Malaviya, (2016) tested for chromium resistant ascomycetes fungi *(Cladosporeum perangustum and Penicillium commune)* isolated from tannery wastewater. These fungi were used for remediation of tannery wastewater. *Paecilomyces lilacinus* and *Fusarium equiseti* were used for the remediation purpose. Nylon mesh was used to immobilise fungal consortia in a bioreactor where results were promising. These fungi consortia were able to remove 82% of chemical oxygen demand, 86% colour, and 100% chromium (VI) and 92% total lead in 5 days. The treated wastewater was then tried in wheat crop which showed around 50% germination as compared to untreated wastewater.

Industries particularly electroplating, leather, tannery effluents contains large amount of heavy metal. Fungi can remove these from wastewater through bioaccumulation and biosorption. Joshi et al. (2011) isolated some fungi namely *Phanerochaete chrysosporium, Aspegillus awamori, Aspergillus flavus, Trichoderma virid* from sewage and industrial effluents. Most of these fungi were able to tolerate 400 ppm of heavy metals namely lead, chromium, cadmium and nickel. Results showed that lead, cadmium, chromium and nickel showed bioaccumulation of 60 ppm, 16ppm,0.6 ppm and 0.6 ppm respectively by Pb3 (*Aspergillus terreus*), *Trichoderma viride*, *Cr8 (Trichoderma longibrachiatum)*, and *isolate Ni27 (A. niger)* respectively, which proves the bioremediation capacity of these fungal strains in removal of heavy metals from industrial effluents.

Macrophytes are another group of aquatic system which have the potential for remediation of wastewater. Now a days with increasing load of nutrients growth of macrophytes is massive in the water system. These plants helps in stabilising surface of water, provides physical filtration for water. It also provides allows oxygen transfer which enhances degradation of organics in wastewater (Lone et al., 2014). In this regard to test the efficacy of three macrophytes viz. water lettuce, water hyacinth and duckweed were used which were grown locally in the Taxila area of Pakistan (Shah et al., 2014). These were evaluated in terms of biological oxygen demand, chemical oxygen demand, nitrogen and phosphorous. After 10 days of retention of wastewater with these macrophytes it was tested for their removal efficiency. Water hyacinth showed 50%, 47%, 41% and 19% removal of BOD, COD, N and P respectively. Duckweed removed 33% BOD, 27% COD, 18% N and 15% P. Similarly water lettuce was able to reduce BOD,COD, N and P by 33%, 26%, 18% and 15% respectively (Shah et al., 2014).This shows that macrophytes are not only weeds for aquatic ecosystem but at the same time they can do appreciable remediation from wastewater as well. Thus we can say that not only plants but different other microorganism also play a crucial in bioremediation of wastewater which needs to be exploited at commercial scale.

Wastewater Reuse in Agriculture

When we talk of wastewater reuse it can be (i) direct or indirect use- Direct use refers to the use of wastewater from source (municipal/ sewer/ industrial effluents) directly. Indirect use refers to the use of water in which wastewater is being discharged like river or lake water. In the latter case wastewater which is being used in diluted form where farmers may do not know about the use of wastewater in its irrigation (ii) Treatment level of wastewater: It determines the condition/ quality of wastewater. Now this treatment level can be either untreated, treated or reclaimed wastewater (requires additional treatment to meet the desired standard of wastewater for use) and (iii) its intended or unintended use: whether farmers use the wastewater knowingly or unknowingly in agriculture. (Jeong et al., 2016). The above methods for irrigation is being followed in developed countries where the regulatory measures are followed strictly. In India scenario is different, as infrastructure for collection and treatment is inadequate, most of the wastewater flows into water system directly. As discussed earlier use of this water is round the year with some nutritional benefits farmers use this in irrigation (Saldias et al., 2015). Although this is beneficial in terms of water availability but it poses health risk to the farmers, users, soil and overall ecosystem of that place. Therefore one thing is institutional focus on this issue and essential regulatory framework to deal with wastewater and use this wastewater in agriculture in safer ways. As we all accept that wastewater irrigation can no longer be ignored or stopped with increasing water scarcity. Therefore going for bioremediation measures can help to lessen the burden of wastewater treatment system and can decentralise it (Saldias et al., 2015).

CONCLUSION

The agricultural sector is the major user of freshwater resources and in future with increasing water scarcity major chunk of pressure will fall upon this sector only. Therefore we need to take precautions and a step forward to use the wastewater as irrigation source. It is desirable to utilise the wastewater safely to minimise any ill health effect on soil as well as humans. Therefore bioremediation should be considered as a feasible and futuristic technology for safe use of wastewater in agriculture. Irrigation of non-food crops is a much safer option which is already in practice at many places and its wings are extending in horticultural crops as well. However further research is required to validate the advantages of bioremediation process for cleaning and make water safe for irrigation with international standards. It's high time that we turn this practice in planned and controlled manner for sustainable agriculture with sustainable ecosystem.

REFERENCES

Abdullahi, M. S. (2015). Soil contamination, remediation and plants: Prospects and challenges. Soil Remediation and Plants.

Akpor, O. B., Otohinoyi, D. A., Olaolu, D. T., & Aderiye, B. I. (2014). Pollutants in wastewater effluents: impacts and remediation processes. *International Journal of Environmental Research and Earth Science, 3*(3), 50-59.

Azubuike, C. C., Chikere, C. B., & Okpokwasili, G. C. (2016). Bioremediation techniques–classification based on site of application: Principles, advantages, limitations and prospects. *World Journal of Microbiology & Biotechnology, 32*(11), 180. doi:10.100711274-016-2137-x PMID:27638318

Banerjee, A. (2016). Wastewater use for agriculture in India: A background review. *Wastewater Use for Agriculture in India*, 1-12.

Beans, C. (2017). Core Concept: Phytoremediation advances in the lab but lags in the field. *Proceedings of the National Academy of Sciences of the United States of America, 114*(29), 7475–7477. doi:10.1073/pnas.1707883114 PMID:28720737

Bishnoi, N. R., & Garima. (2005). Fungus-an alternative for bioremediation of heavy metal containing wastewater: A review. *Journal of Scientific and Industrial Research, 64*, 93–100.

Chaidez, C., Soto, M., & Jiménez, M. (2014). Water: waste, recycling and irrigation in fresh produce processing. In Global Safety of Fresh Produce (pp. 119-132). Woodhead Publishing. doi:10.1533/9781782420279.2.119

Dhir, B. (2013). *Phytoremediation: Role of Aquatic Plants in Environmental Clean-Up*. Springer. doi:10.1007/978-81-322-1307-9

Dubey & Swain. (2019). Wastewater use in agriculture: issues and prospects. *Agriculture & Food, 1*(10).

Dushenkov, V., Kumar, P. N., Motto, H., & Raskin, I. (1995). Rhizofiltration: The use of plants to remove heavy metals from aqueous streams. *Environmental Science & Technology, 29*(5), 1239–1245. doi:10.1021/es00005a015 PMID:22192017

Elias, S. H., Mohamed, M., Nor-Anuar, A., Muda, K., Hassan, M. A. H. M., Othman, M. N., & Chelliapan, S. (2014). Ceramic industry wastewater treatment by rhizofiltration system-application of water hyacinth bioremediation. *The IIOAB Journal, 5*(1), 6.

Falconer, I. R., & Humpage, A. R. (2005). Health risk assessment of cyanobacterial (blue-green algal) toxins in drinking water. *International Journal of Environmental Research and Public Health*, *2*(1), 43–50. doi:10.3390/ijerph2005010043 PMID:16705800

Farraji, H. (2014). Wastewater treatment by phytoremediation methods. *Wastewater Engineering: Advanced Wastewater Treatment Systems, 194*.

Favas, P. J., Pratas, J., Varun, M., D'Souza, R., & Paul, M. S. (2014). Phytoremediation of soils contaminated with metals and metalloids at mining areas: potential of native flora. *Environmental Risk Assessment of Soil Contamination, 3*, 485-516.

Goel, M., Chovelon, J. M., Ferronato, C., Bayard, R., & Sreekrishnan, T. R. (2010). The remediation of wastewater containing 4-chlorophenol using integrated photocatalytic and biological treatment. *Journal of Photochemistry and Photobiology. B, Biology*, *98*(1), 1–6. doi:10.1016/j.jphotobiol.2009.09.006 PMID:19914843

Gupta, V. K., & Ali, I. (2004). Removal of lead and chromium from wastewater using bagasse fly ash—A sugar industry waste. *Journal of Colloid and Interface Science*, *271*(2), 321–328. doi:10.1016/j.jcis.2003.11.007 PMID:14972608

Hussain, I., Raschid, L., Hanjra, M. A., Marikar, F., & Van Der Hoek, W. (2002). *Wastewater use in agriculture: Review of impacts and methodological issues in valuing impacts* (Vol. 37). IWMI.

Jeong, H., Seong, C., Jang, T., & Park, S. (2016). Classification of wastewater reuse for agriculture: A case study in South Korea. *Irrigation and Drainage*, *65*, 76–85. doi:10.1002/ird.2053

Joshi, P. K., Swarup, A., Maheshwari, S., Kumar, R., & Singh, N. (2011). Bioremediation of heavy metals in liquid media through fungi isolated from contaminated sources. *Indian Journal of Microbiology*, *51*(4), 482–487. doi:10.100712088-011-0110-9 PMID:23024411

Koptsik, G. N. (2014). Problems and prospects concerning the phytoremediation of heavy metal polluted soils: A review. *Eurasian Soil Science*, *47*(9), 923–939. doi:10.1134/S1064229314090075

Limmer, M., & Burken, J. (2016). Phytovolatilization of organic contaminants. *Environmental Science & Technology*, *50*(13), 6632–6643. doi:10.1021/acs.est.5b04113 PMID:27249664

Lone, P. A., Bhardwaj, A. K., & Shah, K. W. (2014). Macrophytes as powerful natural tools for water quality improvement. *Journal of Botany (Faisalabad)*, *9*(2), 24–30. doi:10.3923/rjb.2014.24.30

Mahapatra, D. M., Chanakya, H. N., & Ramachandra, T. V. (2014). Bioremediation and lipid synthesis through mixotrophic algal consortia in municipal wastewater. *Bioresource Technology*, *168*, 142–150. doi:10.1016/j.biortech.2014.03.130 PMID:24791711

Martínez-Prado, A., Pérez-López, M., Pinto-Espinoza, J., Gurrola-Nevarez, B., & Osorio-Rodríguez, A. (2011). Biorremediación de suelo contaminado con hidrocarburos empleando lodos residuales como fuente alterna de nutrientes. –. *Revista Internacional de Contaminación Ambiental*, *27*(3), 241–252.

Mateo-Sagasta, J., Medlicott, K., Qadir, M., Raschid-Sally, L., & Drechsel, P. (2013). *Proceedings of the UN-water project on the safe use of wastewater in agriculture*. Academic Press.

Megharaj, M., Ramakrishnan, B., Venkateswarlu, K., Sethunathan, N., & Naidu, R. (2011). Bioremediation approaches for organic pollutants: A critical perspective. *Environment International*, *37*(8), 1362–1375. doi:10.1016/j.envint.2011.06.003 PMID:21722961

Mirza, N., Mahmood, Q., Pervez, A., Ahmad, R., Farooq, R., Shah, M. M., & Azim, M. R. (2010). Phytoremediation potential of Arundo donax in arsenic-contaminated synthetic wastewater. *Bioresource Technology*, *101*(15), 5815–5819. doi:10.1016/j.biortech.2010.03.012 PMID:20363125

Monferrán, M. V., Pignata, M. L., & Wunderlin, D. A. (2012). Enhanced phytoextraction of chromium by the aquatic macrophyte Potamogeton pusillus in presence of copper. *Environmental Pollution*, *161*, 15–22. doi:10.1016/j.envpol.2011.09.032 PMID:22230062

Mora-Ravelo, S. G., Alarcon, A., Rocandio-Rodriguez, M., & Vanoye-Eligio, V. (2017). Bioremediation of wastewater for reutilization in agricultural systems: A review. *Applied Ecology and Environmental Research*, *15*(1), 33–50. doi:10.15666/aeer/1501_033050

Muradov, N., Taha, M., Miranda, A. F., Wrede, D., Kadali, K., Gujar, A., Stevenson, T., Ball, A. S., & Mouradov, A. (2015). Fungal-assisted algal flocculation: Application in wastewater treatment and biofuel production. *Biotechnology for Biofuels*, *8*(1), 24. doi:10.118613068-015-0210-6 PMID:25763102

Nhat, P. V. H., Ngo, H. H., Guo, W. S., Chang, S. W., Nguyen, D. D., Nguyen, P. D., Bui, X. D., Zhang, X. B., & Guo, J. B. (2018). Can algae-based technologies be an affordable green process for biofuel production and wastewater remediation? *Bioresource Technology*, *256*, 491–501. doi:10.1016/j.biortech.2018.02.031 PMID:29472123

Ob, A., & Muchie, M. (2010). *Remediation of heavy metals in drinking water and wastewater treatment systems: Processes and applications*. Academic Press.

Raychaudhuri, S., Raychaudhuri, M., Rautaray, S. K., Parida, S., & Kumar, A. (2017). *Urban wastewater–A potential irrigation source for summer paddy (Oryza sativa L.) in India*. Academic Press.

Rengaraj, S., Moon, S. H., Sivabalan, R., Arabindoo, B., & Murugesan, V. (2002). Agricultural solid waste for the removal of organics: adsorption of phenol from water and wastewater by palm seed coat activated carbon. *Waste Management, 22*(5), 543-548.

Saldías, C., Speelman, S., Amerasinghe, P., & Van Huylenbroeck, G. (2015). Institutional and policy analysis of wastewater (re) use for agriculture: Case study Hyderabad, India. *Water Science and Technology*, *72*(2), 322–331. doi:10.2166/wst.2015.220 PMID:26177416

Salvadori, M. R., Ando, R. A., Oller Do Nascimento, C. A., & Correa, B. (2014). Bioremediation from wastewater and extracellular synthesis of copper nanoparticles by the fungus Trichoderma koningiopsis. *Journal of Environmental Science and Health. Part A, Toxic/Hazardous Substances & Environmental Engineering*, *49*(11), 1286–1295. doi:10.1080/10934529.2014.910067 PMID:24967562

Sekabira, K., Oryem–Origa, H., Mutumba, G., Kakudidi, E., & Basamba, T. A. (2011). *Heavy metal phytoremediation by Commelina benghalensis (L) and Cynodon dactylon (L) growing in urban stream sediments*. Academic Press.

Shah, M., Hashmi, H. N., Ali, A., & Ghumman, A. R. (2014). Performance assessment of aquatic macrophytes for treatment of municipal wastewater. *Journal of Environmental Health Science & Engineering*, *12*(1), 106. doi:10.1186/2052-336X-12-106 PMID:25089203

Sharma, S., & Malaviya, P. (2016). Bioremediation of tannery wastewater by chromium resistant novel fungal consortium. *Ecological Engineering*, *91*, 419–425. doi:10.1016/j.ecoleng.2016.03.005

Stottmeister, U., Wießner, A., Kuschk, P., Kappelmeyer, U., Kästner, M., Bederski, O., Müller, R. A., & Moormann, H. (2003). Effects of plants and microorganisms in constructed wetlands for wastewater treatment. *Biotechnology Advances*, *22*(1-2), 93–117. doi:10.1016/j.biotechadv.2003.08.010 PMID:14623046

Trovaslet, M., Enaud, E., Guiavarc'h, Y., Corbisier, A. M., & Vanhulle, S. (2007). Potential of a Pycnoporus sanguineus laccase in bioremediation of wastewater and kinetic activation in the presence of an anthraquinonic acid dye. *Enzyme and Microbial Technology*, *41*(3), 368–376. doi:10.1016/j.enzmictec.2007.03.007

World Bank. (2019). *From Waste to Resource: Shifting paradigms for smarter wastewater interventions in Latin America and the Caribbean*. World Bank Group.

Yadav, B. K., Siebel, M. A., & van Bruggen, J. J. (2011). Rhizofiltration of a heavy metal (lead) containing wastewater using the wetland plant Carex pendula. CLEAN–Soil, Air. *Water (Basel)*, *39*(5), 467–474.

Ying, G. G. (2018). Ecological risk assessment of pesticides used in agriculture. In *Integrated Analytical Approaches for Pesticide Management* (pp. 67–79). Academic Press. doi:10.1016/B978-0-12-816155-5.00005-1

Zayed, A., Pilon-Smits, E., deSouza, M., Lin, Z.-Q., & Terry, N. (1999). *Remediation of Selenium-Polluted Soils and Waters by Phytovolatilization. In Phytoremediation of contaminated soil and water*. CRC Press.

Chapter 18
Wood Waste Characterization and Reuse Possibilities

Eraldo Antonio Bonfatti Júnior
https://orcid.org/0000-0002-2730-7681
Universidade Federal do Paraná, Brazil

Thiago Campos Monteiro
Federal University of Paraná, Brazil

Elaine Cristina Lengowski
Federal University of Mato Grosso, Brazil

ABSTRACT

Wood is the most versatile natural material. It can be used to obtain sawn wood, wood for civil construction, naval purpose, packaging, furniture, wood flooring, composites, bioenergy, pulp and paper. More recently, nanotechnology and the biorefinery processes create new possibilities for wood use; however, wood processing still generates a large amount of waste. Wood can be processed mechanically or chemically. The mechanical processing of wood has residues like sawdust, cuttings, and wood chips. From the chemical processing of wood, the main residue is black liquor.

BACKGROUND

The production of wood waste in the world exceeded 234.6 million m^3 in 2018 (FAO, 2020). China is the greatest producer of this waste, with 44.3% of the total, followed by Brazil (8.2%) and the United States (6.1%). Different forest-based industries generate this waste due to different factors, such as i) economic: low-cost raw material obtained from reforestation and with defects such as knots, cracks, and growth stresses; ii) legal: laws, regulatory norms or certification that may require the maximum use of native species, as in the Brazilian Amazon; iii) social: needy communities close to sawmills can take advantage of wood waste to produce objects, often with high-added value, or in regions with furniture hubs where a huge number of micro and small companies optimize the use of forest resources; and iv)

DOI: 10.4018/978-1-7998-4921-6.ch018

environmental: strict laws on the destination of liquid, solid and gaseous waste, as well as pressure to value rare native species. Different ways of classifying the timber industries are used, and basically, the wood processing can be divided into mechanical, thermal and chemical.

The main mechanical wood processing industries are sawmills, veneer sheets factories, and wood processing industries such as furniture, molding, and others aiming at high-added value products. The percentage of use of the raw material in these industries, in this case, the logs, boards or veneer sheets, vary according to the final product; in veneer sheet production or log yield, it varies between 35 and 50% (Santos et al., 2015, Silva et al., 2015). On the one hand, sawmills that work with wood from native forests, such as the Amazon, generally use logs above 35%, in many cases reaching 50% of their initial volume (Mendonza et al. 2019). Both the environmental pressure to improve the use of native logs and legal issues contribute to the good use of this wood. On the other, these same companies when using reforestation logs, such as those of the genus *Pinus* or *Eucalyptus,* generally use logs close to 30%. In small companies with low automation, these values can be greater than 40% (Monteiro et al. 2017), but in companies with high automation, they can reach values below 20% for products with high-added value. In these industries, large production often prevails at the expense of better use of raw material. In addition, the cost and quality of the logs of these species are inferior when compared to logs of native species.

The residues generated in the sawmills are produced in the different stages of the mechanical processing of wood. These residues have different dimensions and anatomical, chemical and physical composition. In large part, the heterogeneity of the raw material is due to the heterogeneity of the wood present in the tree, which varies between trees and within the same individual, both in the longitudinal and in the radial axes. There is a wide range of use of these residues depending on the location of the sawmill. Countries like Canada and Finland integrate companies from different segments, and sawmill waste is directed to industries such as cellulose, composites, and energy production. They are also commonly used in the sawmill itself, as an energy source for drying boards. In emerging countries, such as Brazil, its use as an energy source is widely adopted in the sawmills themselves or the sale of co-products to ceramic industries or boilers used in agribusiness, such as for drying grains. The energy potential of these residues is high (Monteiro et al., 2017) and there are companies with great potential to be self-sufficient in energy. This fact is very important in countries with a petroleum-based energy matrix (Dowaki & Mori, 2005), mainly in those with policies to reduce the emission of greenhouse gases. The possibility of transforming lignocellulosic waste into new products or even its use as an energy source makes wood advantageous considering the environmental aspects, when compared with other substitute materials such as concrete, steel, aluminum, and plastic, mainly using the life cycle assessment of materials (Ferguson et al., 1996).

One example of thermal processing of wood is pyrolysis, the main technique used to produce charcoal. The raw materials for its production are planted forests and native forests, and the equipment employed to produce coal uses simple techniques such as hot tail ovens or rectangular ovens. They are simple ovens, with varied dimensions, and generally present a yield close to 30% in relation to the wood that starts the process (Fortaleza et al., 2019). Most of this wood is released in the form of gases in the atmosphere, impacting the emission of greenhouse gases. Some companies are using gas burners to reduce this emission and generate electricity (Pereira et al., 2016).

The main form of chemical processing of wood is the conversion of this material into cellulosic pulp through the Kraft pulping process (Bonfatti Júnior et al., 2019a). There are two segments of pulp production: short fiber and long fiber. The first is used to produce writing, printing and sanitary paper; the second, for packaging production. In the pulp mill, there is waste such as bark and wood knot, but the most relevant one is the black liquor, which contains fifty percent of alkali solubilized wood. This waste is rich in lignin, degraded carbohydrates, wood extractives, and kraft pulping reagents. Although it is part of a closed recovery cycle, the use of a small portion of black liquor to manufacture high-added value products, especially lignin derivatives, is increasingly common in the pulp segment. Lignin has the potential to produce carbon fiber adhesive, additives to produce energy briquettes, and other bioproducts (Lourençon et al., 2015).

Vegetal biomass is formed by three different feedstocks, carbohydrates like starch; cellulose and hemicelluloses; lignin and extractives (Waldron, 2014). These three feedstocks together can limit the use of wood waste to energy use; however, by fractioning the vegetal biomass, the lignin, hemicellulose, cellulose, and extracts can be isolated. With these separated chemical components, products that were originally produced from petroleum can be developed from biomass (Ramos, 2003).

WASTE FROM MECHANICAL PROCESSING OF WOOD

Lumber Production

Sawmills are companies that transform logs obtained from native or reforestation forests into smaller pieces such as planks, boards, and slats. They generally employ a significant amount of workforce, mostly with low qualification, and have a huge variety of layouts, as well as different equipment and deployment methods. Lumber is the main product of this industry. This material is renewable, biodegradable, has immobilized carbon, and has been widely used by man for thousands of years. The parts and residues produced in this industry have a high moisture content and great heterogeneity in their anatomical, chemical and physical properties (Kollmann & Côté Junior, 1968). In the last decades, changes in the destination of these residues have occurred. In the past, it was common to incinerate waste by burning it outdoors or even to discard it for example in rivers of the Amazon rainforest, a rare reality today, but still present in the complex industry of mechanical processing of wood in Brazil.

The industrialization of wood in sawmills differs greatly, depending for example on the raw material; whether the logs come from native forests or reforestation; the dimensions of the logs, mainly diameter; the company's automation, whether it is fixed or mobile; among other classifications. The production of sawn wood can use a variety of techniques and equipment; generally, one of the first steps in this industry is to detect metals in logs. This step is essential to maintain the cutting tools and to avoid wasting time with the maintenance of the saws, as well as financial losses due to equipment damage. For this purpose, metal detectors with different incorporated technologies are generally used. It is important to notice that the metal is often found inside the log, incorporated with the secondary growth of the tree over the years. Depending on the capacity of the metal detector, nails or screws are not detected for example. When the detection of these metals occurs many times, the logs are completely discarded and can be used as firewood, or generally crushed for chip generation and later energy generation.

On the one hand, low-quality logs containing many knots, high curving, presence of *sapopemas,* and degraded heartwood or bifurcations are usually discarded in automated sawmills. On the other, small producers of sawn wood, artisans, architects, and designers have been using these same logs to make high-added value pieces such as dining or living room tables, counters, cabinets, doors, among other objects. These uses for defective logs add a lot of value to the waste and immobilize tons of carbon; besides, there is a demand from society for these parts.

At sawmills, logs can be debarked before they are split. The bark, depending on the species, has high energy power; when processed mechanically with the log, it can reduce the useful life of the cutting tools. The peelers are of several models, but the most common one is disc or drum. The bark volume depends on the species processed and on the granulometry after its removal. The bark generally represents between 10 and 21% of the volume of *Eucalyptus* logs and 9 to 15% of the volume of *Pinus* logs (Oliveira et al., 1999, Leite & Pereira, 2017). It is important to notice that after its removal, the decompression process occurs and its volume considerably increases, for example, in the proportion of 1 m^3 of bark present in the log to 3 m^3 of bark pile. Many companies are interested in the bark due to its potential for generating energy or even for sale (Seixas et al., 2005). Removing the bark of the logs, the residues generated in the breakdown are wood commercially called clean residues. When the removal of the bark does not occur, the residues may present a mixture of wood and bark, being usually called dirty residues.

Some sawmills control the length of the logs. In this stage, systems with cutting chains are generally used, such as sabers or disks. This step generates wooden discs that are usually transported to chippers. Few places use this technique, which is generally adopted in companies with a high degree of automation. The sawdust generated in this process can be used to generate energy.

After these steps, the primary breakdown of the log begins to produce lumber at the sawmill. The first residues generated are wood slab, which is obtained after its introduction in equipment such as band or circular saws. These residues have high dimensions in comparison to the other sawmill residues and have a flat and a curved face, close to the shape of a wedge. Its dimensions depend on the quality of the log, for example, taper, and on the breakdown methods used. Approximately 25% of the log is converted into slab wood (Monteiro et al., 2013). These residues have the potential for making small wooden objects (Vieira et al., 2010), ranging from toys to cutlery handles and decorative pieces.

In small and medium sawmills, it is common to reuse these residues and convert them into small parts and to generate chips as residues, which are like the slab wood ones, but with very small dimensions. Chip formation occurs at all cutting stages, usually with high moisture content. In many companies, the wood slab is sent to chippers and transformed into chips for power generation. In modern sawmills, it is common to use chipper canter equipment for instant conversion of the wood slab into chips generally sized for use in the pulp or wood panel industry.

After standardizing the width and thickness of the board, even in the secondary split, the length of the sawn wood can be controlled. Uncapping machines are used for this purpose, usually with the actuation of a circular saw. The residues of this stage are sawdust and fragments of the ends of the boards, with a short length. These residues are often taken to chippers and converted into chips for energy generation.

At the end of the production line, the sawn wood is classified and prepared for drying. In this stage, there is a very large demand for energy, mainly for steam generation, and a large part of the sawmill waste is consumed.

Production of Veneer Sheets

Wood laminates are generally used for making engineered floors, door faces or plywood panels. China, Russia, and Brazil are major producers of laminates and also wood waste (FAO, 2020). The production of wooden blades is performed by a piece of equipment called cutlery or lathe, the most used one. The lamination consists of obtaining a continuous wooden sheet called laminate, with thickness varying between 0.13 and 6.35 mm, produced through the rotation of the log fixed in the lathe system and the knife/pressure bar set (Marra, 1992, Iwakiri, 2005).

In this process, the residues generated are mainly pieces of blades and the central part of the log, called rest roller. The use of the log to generate the blades is usually greater than 50% of the initial volume of the logs (Santos et al., 2015; Silva et al., 2015). Defective blades are commonly generated due to taper and flattening of the logs. The production of the laminate starts when the log is perfectly cylindrical. Losses considering the bark and blades with defects represent approximately 35% of the initial volume of the log (Santos et al., 2015). The rest roll is the residue located in the central region of the log and represents 6.8% of the initial volume of the log (Santos et al., 2015). These residues have a high moisture content and are generally used in making other panels or generating energy.

Wood Processing Industry

Dry sawn wood and panel made of lignocellulosic materials are the basis of the wood processing industry (Thoemen et al., 2010). Their industries are more complex when compared to sawmills and many have already been inserted in the context of Industry 4.0, commonly with high automation of processes. The wood and equipment used in these companies have smaller dimensions when compared to those used in sawmills.

Wood processing in industries depends on the final product, such as furniture, household utensils, frames, and other products with high-added value. The residues generated in the beneficiation industry are small, often with exhaust systems for their storage, low moisture content, and high calorific value. The shape and size of the waste vary depending on the machining operation performed in the wood.

Generally, wood processing starts by standardizing the thickness of the boards through planing. The residues from this process are shavings that have a small thickness, great length, and low moisture content. Planers generally have an exhaust system for suction and storage of these residues. Its use can be as an energy source, a raw material for composites, bedding for animals such as chickens, among others. The mechanical sawing process usually occurs with circular or band saws, which are used to standardize the length or width of the pieces. The residues generated in these operations are sawdust, particles much smaller than wood shavings. Sawdust is usually burned to generate steam or electricity, but it can also be pressed to make pellets or briquettes.

The optimizers (Automatic Wood Cross Cut-off Saw Machine) are equipment that uses discs as a cutting tool and are widely used in molding industries to eliminate defects such as knots. They are a piece of equipment usually associated with sensors and automation systems and high speed of feeding and cutting. This equipment generates a large amount of waste such as sawdust and also defective parts, mainly with the presence of knots. These residues are usually directed to chippers and used for burning in boilers or for producing pellets and briquettes. In the last few years, the largest residues with the presence of knots have been reprocessed and amended to form structures such as beams and pillars. This is a form of adding value to wood residues and producing unique pieces. The Asian market, mainly China, is the main consumer of these products.

Finally, at the end of the production line of wood processing, molders that use milling cutters as tools add a lot of value to the products. A variety of formats are made in this equipment and consequently reflect in a wide variety of waste. Molding planers can generate larger waste similar to shavings while milling cutters produce finger-joint waste similar to sawdust. The destination of this waste is comparable to shavings and sawdust as they can be used to generate steam and energy or can be pressed to produce pellets and briquettes. Some companies may send it for the production of composites or even for use in animal litter, such as chickens.

Biomass Energy Industry

The thermal processing of wood through its direct burning has been very important for centuries in agriculture, cooking and especially in the Industrial Revolution. After this period, wood was gradually replaced by fossil fuels, but even today the direct burning of wood is important for cooking food in regions with low human development index (HDI). In these regions, such as in Africa, charcoal production is used for cooking food and as biochar for improving soil quality and consequently for food production (Häring et al., 2017). Brazil is the largest consumer and producer of charcoal, and most of that coal is used in the steel industry as a reducing agent in steel production (FAO, 2020). However, the African continent is the largest producer of this input, mainly in countries such as Nigeria and Ethiopia, for residential use such as cooking food.

Pyrolysis is the main technique used to produce charcoal, some companies add in this process gas burners to reduce this emission and generate electricity (Pereira et al., 2016); in this process, it is also possible to separate the condensed gases. One of the products contained in this gas is pyroligneous liquor, which is composed of several different chemical compounds such as sugars, alcohols, acetic acid, ethers, phenols, and others. The liquor is used as an antibacterial and antifungal agent (Araújo et al., 2017), adding value to the residue.

In addition to this use, the wood waste can be employed as an energy source, either by burning wood directly, using it as charcoal, using waste from the exploitation or using bio-oils from pyrolysis. Despite the energy potential, the use of these residues directly in energy generation can present problems related to the high moisture content of biomass, low energy density, varied granulometry, and high hygroscopicity (Hansted et al., 2016), in addition to the low density hinder transportation and storage, which would make transportation and logistic activities more expensive. These problems are common in the employment of direct burning of forest biomass, for example, for the drying of soybeans and corn in Brazil. For decades, cooperatives have used forest biomass in the form of logs, with large dimensions and high moisture content. Nowadays, this sector has begun to use forest biomass in the form of chips and with lower moisture content in the wood. Anyway, regardless of the thermal processing used in the wood, there are still many possibilities for better use of the residues, mainly of the gases emitted in these industries.

WASTE FROM CHEMICAL PROCESSING OF WOOD

Pulping of wood can occur by processes that use different energy sources such as mechanical, thermal and chemical or a combination of these three. Despite the great diversity of pulping processes, those that use chemical energy at alkaline pH are the most common. The most famous and used one is the kraft process developed by the German chemist Carl F. Dahl in 1879 (Hu, Zhang & Lee, 2018) and patented in 1884 (Smook, 2016); today it is responsible for the production of 90% of all cellulosic pulp in the world (Oliveira, Mateus & Santos, 2018). This process is the result of the evolution of the soda process, whose reactants are sodium hydroxide (NaOH) and sodium sulfide (Na_2S). It aims to dissolve the middle lamella by removing the lignin, causing the individualization of the fibers (Lengowski et al., 2019). For that, wood chips are processed in pressurized digesters and heated with cooking liquor.

The kraft process was established worldwide in the pulp and paper sector for being adaptable to different sources of lignocellulosic fibers, for producing cellulosic pulps of high quality and whiteness (Lengowski et al. 2019), and mainly for being a closed cycle of chemical and energy recovery (Bajpai,

2018). However, this process has some disadvantages such as low efficiency in converting wood into cellulosic pulp when compared to mechanical and thermomechanical processes (Smook, 2016), high costs of implantation, maintenance and operation of an industrial plant (Diesen, 2007), and strong unpleasant odor caused by emissions of mercaptans (Chagas et al., 2019), which are organosulfur compounds containing a –SH group attached to a carbon atom (–C – S – H or R – S – H, where R represents any group of atoms containing carbon such as alkanes and alkenes) (Cremlyn, 1996).

The pulping process has two main products: cellulosic pulp and black liquor, which is the main effluent in the industry. This black liquor is composed of residual process chemicals such as sodium carbonate (Na_2CO_3), sodium hydroxide (NaOH), sodium sulfide (Na_2S), sodium sulfate (Na_2SO_4), silica (SiO_2), iron oxide (Fe_2O_3), aluminum oxide (Al_2O_3), sodium chloride (NaCl), potassium hydroxide (KOH), among others, and of compounds from the solubilization of wood. Proportionally, the inorganic fraction of the black liquor varies between 18 to 25% while the organic one ranges from 75 to 82% (Wallberg et al., 2006).

The performance of the pulping process is often evaluated by its yield, which is the relationship between the cellulosic pulp mass after the process and the mass of chips used in pulping in percentage. When looking for kappa number 18 ± 0.5 in kraft pulping of several genetic materials other than the genus *Eucalyptus* spp., Gomide et al. (2005) found yield values that ranged from 49.3 to 57.6%, whereas Bonfatti Júnior et al. (2019b) found a 49.4% yield for the production of cellulosic pulp, with a kappa number of 55 ± 2, from *Pinus taeda* wood. Therefore, it can be considered that around fifty percent of the wood is transformed into a cellulosic pulp and the other fifty percent is dissolved in the black liquor.

The cellulosic pulp can be bleached and used for the production of tissue papers for absorbent purposes, printing papers, and writing papers, or it can be converted without being bleached into paper for packaging production. The black liquor is destined for the energy and chemical recovery of the industrial plant (Gavrilescu, 2008), represented in Figure 1. This system is responsible for the economic viability of the pulping kraft process, as it manages to recover about 90% of the chemical reagents present in the black liquor in addition to converting the energy released by burning the substances in condensed water vapor or electricity (Diesen, 2007). For a long time, the recovery system was the only destination for black liquor; however, the possibilities of extracting the organic fraction (lignin, polysaccharides, and extracts) and converting it into products with high-added value drove the development of techniques known as biorefinery.

In addition to the black liquor, the cellulosic pulp industry has tall oil, bark, branches, leaves and non-standard wood by-products for the pulping process, among others (Bajpai, 2018). Figure 2 shows the layout of a fiber line from a cellulose industry based on the kraft process, the different biorefinery possibilities applicable to this type of industrial plant and the products with the highest added value that can be obtained.

Biorefinery can be defined as a process or a combination of processes for converting biomass into several other products with high-added value, including fuels, energy, chemical compounds, adhesives, drugs, and materials (González & Bustillos, 2019). It is the application of the oil refinery concept that produces gasoline, diesel oil, lubricating oils, LPG, asphalt products, kerosene, polymers, and phenolic compounds, but starting from biomass (Santos et al., 2013).

The term biomass is defined by IUPAC (International Union of Pure and Applied Chemistry) as all material produced by the growth of microorganisms, plants or animals. Within these options, the most economical interest to implement biorefinery is the biomass of plant origin or lignocellulosic biomass (Wertz et al., 2019). One of the prerequisites for the successful implementation of the biorefinery concept is the development of chemical fractionation methods that enable the recovery of the main constituents

Figure 1. Kraft recovery process

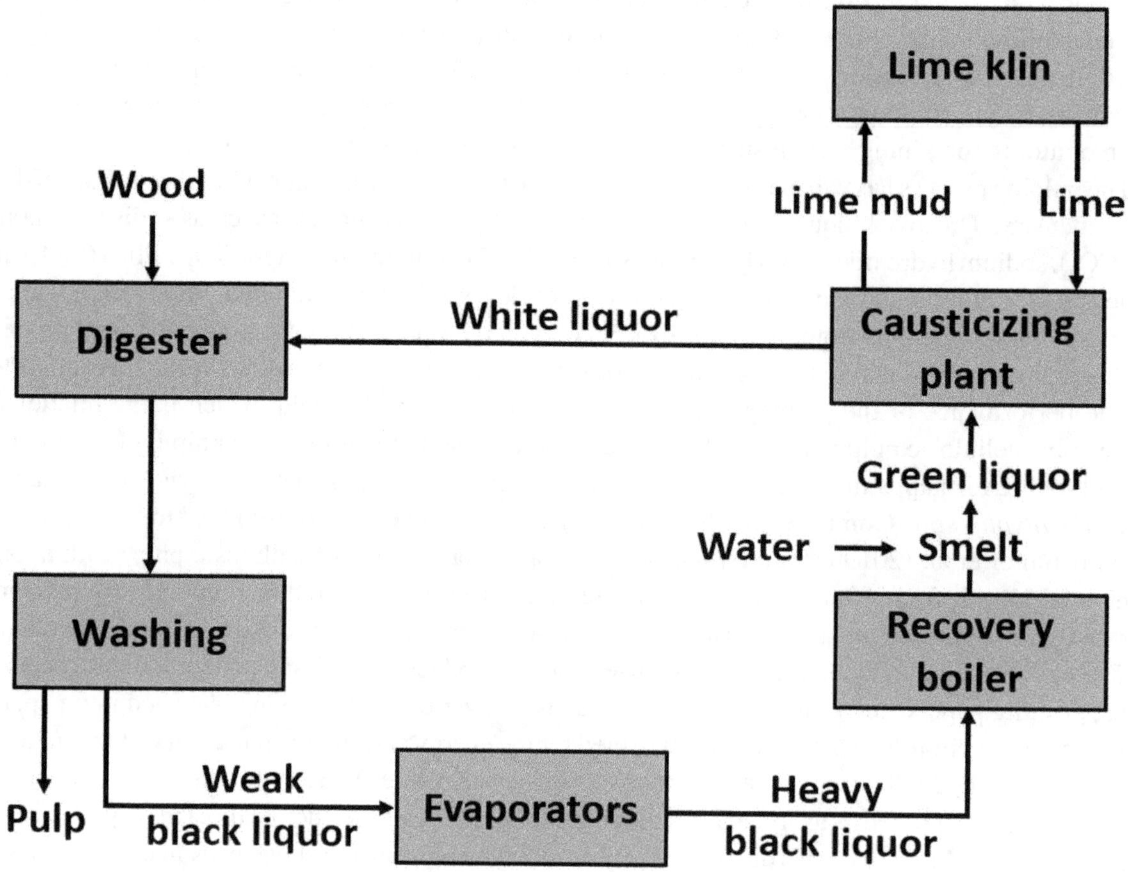

of lignocellulosic biomass, such as cellulose, lignin, and hemicelluloses, in a sufficiently pure form (Yanqiao et al., 2008; Wertz et al., 2019). This fractionation can be performed by physical or chemical treatments or even a sequence of them; these processes enable the complete use of the waste from the chemical processing of wood.

Hemicelluloses are a heterogeneous group of polysaccharides present in wood, occurring to a lesser degree of polymerization than cellulose and of lower molecular weight. Various sugars of 5 or 6 carbon atoms may appear doomed, such as pentoses (β-D-xylose and a-L-arabinose) and hexoses (α-D-Galactose, β-D-glucose and β-D-mannose), in addition to uronic acids and deoxyhexoses (Wastowski, 2018). The fraction of hemicelluloses can be extracted by the hydrothermal method that includes the submission of lignocellulosic materials to water and steam at temperatures between 150 to 230°C with the addition of an acid catalyst (Garrote et al., 2006). The hydrothermal treatment has many applications, including the removal of hemicelluloses from the wood before the pulping process to produce soluble cellulose (Bajpai, 2018).

Hemicelluloses have a great chemical potential for use due to the presence of side chains in their structure, a characteristic that makes it more reactive than cellulose. From the hydrolysis of hemicelluloses, sugars can be obtained to produce ethanol, and by steam exploitation, furfural is obtained,

Figure 2. Possibilities of biorefinery in a line of kraft fibers

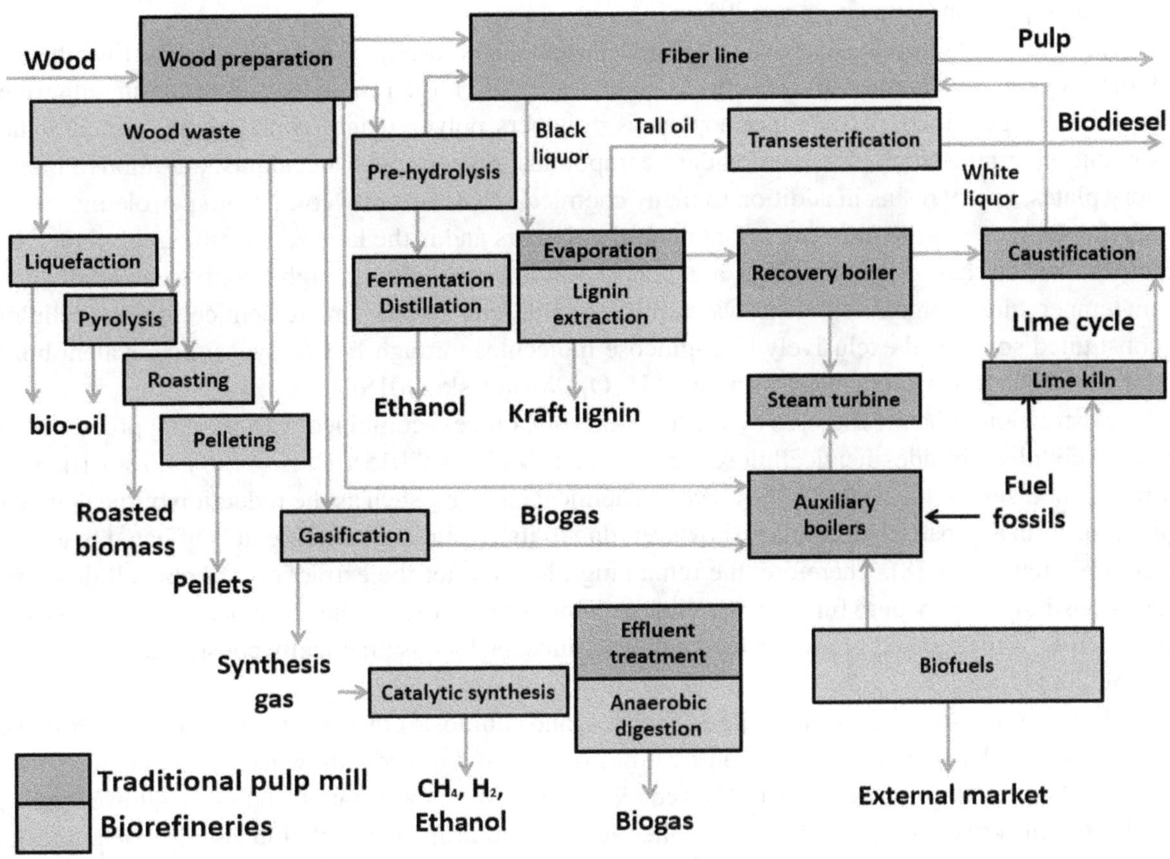

which can be used to produce resins. Besides, it is possible to produce xylitol and mannitol by catalytic hydrogenation of hemicelluloses.

The term "lignin" is used to describe phenolic polymers that together with cellulose and hemicelluloses form the cellular structure of vegetables (Bruijnincx et al., 2015). In the plant biomass, it represents between 15 to 40% of all dry mass (Abreu et al., 1999). Among the phenolic polymers, lignin is the most abundant in the world and its basic unit is phenyl propane (C6-C3 or C9). It has as precursors three types of subunits of hydroxy-cinnamic alcohols (monolignols): p-coumaryl alcohol (H), coniferyl alcohol (G), and synaphyl alcohol (S), with structural differences in the extension of the methyl radical in the position 3 ¢ or 3¢ – 5 ¢ of the phenolic rings (Tian et al., 2016). These precursor molecules are linked together covalently in different ways, which provides lignin with great structural complexity (Solomons et al., 2016).

The extraction of lignin can be performed by the organosolv process or by pulping processes, such as kraft, soda, and sulfite ones (Santos et al., 2013). The lignin extracted by the organosolv process is chemically very close to the natural lignin, present in the original material, whereas the lignins extracted by the pulping processes are quite modified and are solubilized in the black liquor (Cha, 2020). The most common form to extract lignin kraft is to acidify the black liquor; the protonation of the hydroxyl groups causes the phenolic structures to be little active nucleophilic centers by neutralizing the free hydroxyls

(Öhman, 2006). Then, the forces of attraction between the molecules intensify so that they aggregate causing precipitation (Gouvêa, et al., 2018).

Lignin represents a great source of renewable carbon, and biorefinery aims to enhance this phenolic polymer beyond the simple use of energy (Gillet et al., 2017), the products used in lignin, adhesives, composites, carbon fibers, plastics, thermoplastics, polymers, polyurethane foams, polymeric membranes, absorbent materials, emulsifiers, antioxidant compounds, dispersants, flocculants, automotive brakes, battery plates, epoxy resins in addition to many chemicals that currently come from petroleum.

Cellulose is one of the most important natural polymers and is the largest constituent of plants. It is a polysaccharide that presents itself as a condensation polymer with enough length to be insoluble at room temperature in organic solvents, water, diluted acids, and alkalis. Unlike hemicelluloses, cellulose is constituted solely and exclusively by β-glucose molecules through B-1,4-glycosidic covalent bonds and is represented by the chemical formula $C_6H_{10}O_5$ (Wastowski, 2018).

The extraction of hemicelluloses by the hydrothermal process combined with pulping processes can separate cellulose besides hemicelluloses and lignins (Vallejos, 2015). During the hydrothermal and pulping processes, cellulose undergoes several chemical changes, such as the reduction of the degree of polymerization and partial dissolution; however, due to its crystalline structure, it is affected to a lesser extent (Wastowski, 2018). Therefore, the remaining residue after the extraction of hemicelluloses and lignin constitutes a very pure fraction of cellulose. This combination of the aforementioned processes is the principle of the method of producing soluble cellulose, which is practically constituted by cellulose (Bajpai, 2018).

Cellulose is used to produce paper, its derivatives, and cellulosic ethanol through hydrolysis followed by fermentation. However, chemical modifications of cellulose already allow its use in diagnostics and biomedical areas in the form of fibers (Trivedi & Fardim, 2019), and nanotechnology allowed the use of cellulose in active food packaging, controlled release and drug encapsulation, as additives in resins, adhesives, and in the manufacture of paper as well as in the production of foams, hydrogels, and aerogel.

Extractives are chemical compounds present in the cell wall of the wood but do not have a structural function. Generally formed from greases, fatty acids, fine alcohols, phenols, terpenes, steroids, acidic resins, resins, and waxes, they can exist in the form of monomers, dimers or polymers (Wastowski, 2018). Wood extracts are substances of low molecular weight, very soluble in organic solvents and low degree of polymerization; in general, softwoods have more extracts than hardwoods. In addition to the quantity, the softwoods have some fats that, after the kraft process, form tall oil in the black liquor, which can be separated into different fractions and used to produce biodiesel through the transesterification reaction.

Table 1 shows some examples of wood waste biorefineries from pyrolysis to produce bio-oil, gasification to produce synthesis gas or fuel gas to replace natural gas, lignin extraction, and lastly the case of UPM Biorefinery Finland, an example of an industry that produces biodiesel from tall oil.

Over time, technological advances have revolutionized the production and development of materials, with nanotechnology being a remarkable tool for this purpose (Sparks, 2012). The application of nanotechnology in the wood-based industry will become increasingly common in the coming years (Wegner et al., 2010). Research on the use of biomass has evolved in recent decades; within the cellulose and paper sector, the study on nanocellulose has stood out as this material can be produced by different routes, fibrillation or breakdown by chemical or enzymatic hydrolysis, increasingly showing potential for use. Because it is a renewable material, which is more advantageous to the environment when compared to materials of fossil origin, products containing nanocellulose have a wide range of commercialization

Table 1. Cases of wood waste biorefineries

Process	Company	Location
Pyrolysis	Dynamotive Energy Systems	Canada
	BTG	Malaysia and the Netherlands
	Fortum (Valmet/VTT/UPM)	Finland
	Ensyn	Canada and the US
Gasification	Lahti	Finland
	Vassa	Finland
	Södra Cell	Sweden
	Göteborg Energi	Sweden
	Metsä Fibre	Finland
	APP	Indonesia
Lignin extraction	Nordic Paper - Bäckhammar	Sweden
	Domtar	US
	Stora Enso	Finland
	West Frase	Canada
	Suzano	Brazil
Transesterification	UPM	Finland

and opportunities since the demands for biodegradable products made from renewable and sustainable resources and low environmental impact are increasing.

Among the products derived from wood or vegetable fibers, nanocellulose is one of those with the highest added value and greatest applicability; therefore, the production of nanocellulose fits perfectly into the concept of biorefinery (Lengowski et al., 2019). When using wood waste for the production of nanocellulose, there is yet another opportunity to increase the gains in wood processing, whether using chemical or mechanical processing (Rajinipriya et al., 2018).

The term “nanocellulose” refers to cellulosic materials having at least one of their dimensions on a nanometric scale (Lengowski et al., 2019). Nanocelluloses can be produced by different methods from various lignocellulosic sources (Eichhorn et al., 2010; Abdul Khalil et al., 2014) and can be divided into three types: nanocrystalline cellulose (CNC), microfibrillated cellulose (MFC), and nanofibrillated cellulose (NFC) (Fujisawa et al., 2011). In dimensional terms, MFCs are fibers with a diameter between 25 and 100 nm whereas NFCs have diameters between 5 and 30 nm and length between 2 and 10 μm (Sehaqui et al., 2011; Rojas et al., 2015), both having amorphous and crystalline zones. CNCs are crystals with diameters between 2 and 20 nm and length between 100 and 600 nm (Julkapli & Bagheri, 2016), without the presence of amorphous zones.

From nanocellulose, it is possible to produce food packaging, foams, aerogels, electronic components, additives for the production of papers, resins, and polymeric composites, as well as means of encapsulation and delivery of drugs in the pharmaceutical field. Nanocellulose applications in civil construction and the automatic, naval, aerospace, and war industries are already present in the market, where this new material is used to improve important characteristics of goods.

CONCLUDING REMARKS

The residues of the wood industry present high heterogeneity in their anatomical, chemical, and physical composition. These properties change according to the type of wood processing; for example, in sawmills, they have high dimension parts and moisture content while in the processing industries, they have smaller and more homogeneous parts with low moisture content.

In industries of mechanical wood processing, historically one of the main destinations for waste was burning to generate steam, electrical energy or even to incinerate the material. In the last decades, the reuse of waste for making objects and even structural parts with high-added value is a reality.

Direct wood burning or pyrolysis has been very important for the development of industries in the past centuries. Nowadays, the thermal processing of wood is still a reality, especially in developing countries. This segment makes very little use of waste, especially gases. Thus, the potential for using these gaseous residues exists, mainly in the chemical industry.

Wood is a natural complex formed by polymeric compounds of saccharide and phenolic nature, and its maximum commercial gain comes from the separation of these components for the individual use of each one. Both lignin and extracts, as well as cellulose and hemicelluloses, are used as raw materials for products with high-added value. When it comes to the production of cellulosic pulp, the fractionation of biomass is already facilitated since the pulping processes aim to individualize the cellulose fibers.

In addition to the use of lignin and hemicelluloses for the production of chemical products with several uses in the most variable industrial fields, nanotechnology can be applied for the development of nanocelluloses, which brings great application opportunities. Environmentally, the use of products of lignocellulosic origin is milder than those of fossil origin due to the renewable characteristic of this material, and therefore the biorefinery is essential for a cleaner and more sustainable industry.

ACKNOWLEDGMENT

We are thankful to the Pulp, Paper and Biorefinery Laboratory team from the Federal University of Mato Grosso (Brazil) for the support of this research.

This study was financed in part by the Coordination of Improvement of Higher Education Personnel - Brazil (CAPES) - Finance Code 001.

REFERENCES

Abdul, K. (2014). Production and modification of nanofibrillated celulose using various mechanical processes: A review. *Carbohydrate Polymers*, *99*, 649–665. doi:10.1016/j.carbpol.2013.08.069 PMID:24274556

Abreu, H. S., Nascimento, A. M., & Maria, M. A. (1999). Lignin structure and wood properties. *Wood and Fiber Science*, *31*(4), 426–433.

Araujo, E., Pimenta, A. S., Feijo, F., Castro, R. V. O., Lima, K., Lima, M., & Monteiro, T. C. (2017). Antibacterial and antifungal activity activity of pyroligneous acid from wood of and. *Journal of Applied Microbiology*, *124*(1), 85–96. doi:10.1111/jam.13626 PMID:29095556

Bajapai, P. (2018). *Biermann's handbook of pulp and paper*. Elsevier.

Bonfatti Júnior, E. A., Raia, R. R., Bila, N. F., Lopes, M. S., Klock, U., Andrade, A. S., & Vivian, M. A. (2019b). Kraft punlping and papermaking of *Cryptomeria japonica. Scientia Forestalis, 47*(124), 811–822. doi:10.18671cifor.v47n124.22

Bruijnincx, P. C. A., Rinald, R., & Weckhuysen, R. M. (2015). Unlocking the potential of a sleeping giant: Lignins as sustainable raw materials for renewable fuels, chemicals and materials. *Green Chemistry, 15*(11), 4860–4861. doi:10.1039/C5GC90055G

Cha, Y. L., Alam, A. M., Park, S. M., Moon, Y. H., Kim, K. S., Lee, J. E., Kwon, D. E., & Kang, Y. G. (2020). Hydrothermal-process-based direct extraction of polydisperse lignin microspheres from black liquor and their physicochemical characterization. *Bioresource Technology, 297*, e122399. doi:10.1016/j.biortech.2019.122399 PMID:31759855

Cremlyn, R. J. (1996). *An introduction to organosulfur chemistry*. Wiley.

Dieden, M. (2007). Economics of pulp and paper industry (2nd ed.). Espoo: Puunjalostusinsinöörit ry.

Dowaki, K., & Mori, S. (2005). Biomass energy used in a sawmill. *Applied Energy, 80*(3), 327–339. doi:10.1016/j.apenergy.2004.04.006

Eichhorn, S. J., Dufresne, A., Aranguren, M., Marcovich, N. E., Capadona, J. R., Rowan, S. J., Werder, C., Thielemans, W., Roman, M., Renneckar, S., Gindl, W., Veigel, S., Keckes, S., Yano, H., & Abe, K. (2010). Review: Current international research into cellulose nanofibers and nanocomposites. *Journal of Materials Science, 45*(1), 1–33. doi:10.100710853-009-3874-0

FAO. (2020). *FAOSTAT: Forestry Production and Trade*. Rome: UN.

Felix, A. L., Narciso, C. R. P., Lima, F. S., Mendes, R. F., Mendes, L. M., & Scolforo, J. R. S. (2016). Use of waste wood for particleboard production. *Key Engineering Materials, 668*, 375–380. doi:10.4028/www.scientific.net/KEM.668.375

Ferguson, I., La Fontarie, B., Vinden, P., Bren, L., Hateley, R., & Hermesec, B. (1996). *Environmental Properties of Timber*. Forest & Wood Products Research & Development Corporation.

Fortaleza, A. P., Filho, J. J. P. N., Silva, R. P. S., Barros, D. S., & Silva, S. S. (2019). Biomassa de espécies florestais para produção de carvão vegetal. *Ciência Florestal, 29*(3), 1436–1451. doi:10.5902/1980509831639

Fujisawa, S., Okita, Y., Fukuzumi, H., Saito, T., & Isogai, A. (2011). Preparation and characterization of TEMPO-oxidized cellulose nanofibril films with free carboxyl groups. *Carbohydrate Polymers, 84*(1), 579–583. doi:10.1016/j.carbpol.2010.12.029

Garrote, G., Domínguez, H., & Parajó, J. C. (1999). Hydrothermal processing of lignocellulosic materials. *Holz als Roh- und Werkstoff, 57*(3), 191–202. doi:10.1007001070050039

Gavrilescu, D. (2008). Energy from biomass in pulp and paper mills. *Environmental Engineering and Management Journal, 7*(5), 537–546. doi:10.30638/eemj.2008.077

Gillet, S., Aguedo, M., Petitjen, L., Morais, A. R. C., Lopes, A. M. C., Łukasik, R. M., & Anastas, P. T. (2017). Lignin transformations for high value applications: Towards targeted modifications using green chemistry. *Green Chemistry, 19*(18), 4200–4233. doi:10.1039/C7GC01479A

Gomide, J. L., Neto, H. F., & Regazzi, A. J. (2005, February). Caracterização tecnológica, para a produção de celulose, da nova geração de clones de *Eucalyptus* do Brasil. *Revista Árvore, 29*(1), 129–137. doi:10.1590/S0100-67622005000100014

Gonzáles, A. F. G. & Bustillos, L. G. T. (2019). *Biorrefinerias y Economia Circular*. Ciudad de México: UAM.

Gouvêa, A. F. G., Carvalho, A. M. M. L., & Carneiro, A. C. O. (2018). *Extração da lignina kraft de eucalipto e uso para energia*. Appris.

Hansted, A. L. S., Nakashima, G. T., Martins, M. P., Yamamoto, H., & Yamaji, F. M. (2016). Comparative analyses of fast-growing species in different moisture content for high quality solid fuel production. *Fuel, 184*, 180–184. doi:10.1016/j.fuel.2016.06.071

Häring, V., Manka'abusi, D., Akoto-Danso, E. K., Werner, S., Atiah, K., Steiner, C., Lompo, D. J. P., Adiku, S., Buerkert, A., & Marschner, B. (2017). Effects of biochar, wastewater irrigation and fertilization on soil properties in West African urban agriculture. *Scientific Reports, 7*(1), 10738. doi:10.103841598-017-10718-y PMID:28878251

Hu, J., Zhang, Q., & Lee, D.-J. (2018). Kraft lignin biorefinery: A perspective. *Bioresource Technology, 247*, 1181–1183. doi:10.1016/j.biortech.2017.08.169 PMID:28899675

Iwakiri, S. (2005). *Painéis de Madeira reconstituída*. FUPEF.

Julkapli, N. W., & Bagheri, S. (2016). Developments in nano-additives for paper industry. *Journal of Wood Science, 62*(2), 117–130. doi:10.100710086-015-1532-5

Júnior, B. (2019a). Briquetagem dos resíduos dos processamentos mecânico e químico de *Pinus* spp. *Cadernos de Ciência & Tecnologia, 36*(3), e26522. doi:10.35977/0104-1096.cct2019.v36.26522

Kollmann, F. F. P., & Côté, W. A. (1968). *Principles of wood science and technology*. Springer-Verlag. doi:10.1007/978-3-642-87928-9

Leite, C., & Pereira, H. (2005). Cork – Containing barks – A Review. *Frontiers in Materials, 3*, e63.

Lengowski, E. C., Bonfatti Júnior, E. A., Kumode, M. M. N., Carneiro, M. E., & Satyanarayna, K. G. Nanocellulose in the paper making. In Sustainable Polymer Composites and Nanocomposites (pp. 1027–1066). Cham: Springer. doi:10.1007/978-3-030-05399-4_36

Lourençon, T. V., Hansel, F. A., Silva, T. A., Ramos, L. P., Muñiz, G. I. B., & Magalhães, W. L. E. (2015). Hardwood and softwood kraft lignins fractionation by simple sequential acid precipitation. *Separation and Purification Technology, 154*(5), 82–88. doi:10.1016/j.seppur.2015.09.015

Marra, A. A. (1992). *Technology of wood bonding: principles and practice*. Van Nostrand Reinhold.

Mendes, L. M., Mendes, R. F., Tonoli, G. H. D., Bufalino, L., Mori, F. A., & Guimarães, J. B. Junior. (2012). Lignocellulosic composites made from agricultural and forestry wastes in brazil. *Key Engineering Materials, 517*, 556–563. doi:10.4028/www.scientific.net/KEM.517.556

Mendoza, Z. M. S. H., Borges, P. H. M., Morais, P. H. M., & Elias, M. P. S. (2019). Use of beta regression to estimate the volumetric yield coefficient in logs of native species of the legal Amazon. *Nativa (Sinop), 7*(3), 323–329. doi:10.31413/nativa.v7i3.6834

Monteiro, T. C., Lima, J. T., Silva, J. R. M., Trugilho, P. F., & Andrade, B. C. L. (2013). Avaliação do desdobro de toras de Eucalyptus para a obtenção de peças estruturais. *Cerne, 19*(3), 357–364. doi:10.1590/S0104-77602013000300001

Monteiro, T. C., Lima, J. T., Silva, J. R. M., Trugilho, P. F., & Barauna, E. E. P. (2017). Energy balance in sawing *Eucalyptus grandis* logs. *BioResources, 12*(3), 5790–5800. doi:10.15376/biores.12.3.5790-5800

Öhman, F. (2006). *Precipitation and separation of lignin from kraft black liquor* (PhD Thesis). Chalmers University of Technology, Gothenburg, Sweden.

Oliveira, J. T. S., Hellmeister, J. C., Simões, J. W., & Tomazello Filho, M. (1999). Caracterização da madeira de sete espécies de eucaliptos para a construção civil: 1- avaliações dendrométricas das árvores. *Scientia Forestalis, 56*, 113–124.

Oliveira, R. C. P., Mateus, M., & Santos, D. M. F. (2018). Chronoamperometric and chronopotentiometric investigation of kraft black liquor. *International Journal of Hydrogen Energy, 43*(35), 16817–16823. doi:10.1016/j.ijhydene.2018.01.046

Pereira, E. G., Martins, M. A., Pecenka, R., & Carneiro, A. C. O. (2016). Pyrolysis gases burners: Sustainability for integrated production of charcoal, heat and electricity. *Renewable & Sustainable Energy Reviews, 75*, 592–600. doi:10.1016/j.rser.2016.11.028

Rajinipriya, M., Nagalakshmaiah, M., Robert, M., & Elkoun, S. (2018). Importance of Agricultural and Industrial Waste in the Field of Nanocellulose and Recent Industrial Developments of Wood Based Nanocellulose: A Review. *ACS Sustainable Chemistry & Engineering, 6*(3), 2807–2828. doi:10.1021/acssuschemeng.7b03437

Ramos, L. P. (2003). The chemistry involved in the steam treatment of lignocellulosic materials. *Quimica Nova, 26*(6), 863–871. doi:10.1590/S0100-40422003000600015

Rojas, J., Bedoya, M., & Ciro, Y. Current Trends in the Production of Cellulose Nanoparticles and Nanocomposites for Biomedical Applications. In M. Poletto (Ed.), *Cellulose – Fundamental aspects and current trends* (pp. 193–228). IntechOpen. doi:10.5772/61334

Santos, A. F., Guimarães, J. B. Junior, Martins, E. H., Lima, J. V., & Protásio, T. P. (2015). Rendimento efetivo em laminação de madeira de *Pinus oocarpa* cultivada no estado de Minas Gerais. *Enciclopédia Biosfera, 11*(21), 1662–1670.

Santos, F., Queiroz, J. H. & Queiroz, J. H. (2013). *Bioenergia & Biorrefinaria: Cana-de-Açúcar & Espécies Florestais*. Viçosa: UFV.

Sehaqui, H., Alliais, M., Zhou, Q., & Berglund, L. A. (2011). Wood cellulose biocomposites with fibrous structures at micro and nanoscale. *Composites Science and Technology, 71*(3), 382–387. doi:10.1016/j.compscitech.2010.12.007

Seixas, F., Bauch, S. C., & Oliveira Júnior, E. D. (2005). Energetic balance and economical study of two debarking alternatives for *Eucalypts* wood. *Scientia Forestalis, 67*, 37–43.

Silva, G. F., Mendonça, A. R., Hoffmann, R. G., Zaneti, L. Z., Chichorro, J. F. & Ferreira, R. L. C. (2015). *Rendimento em laminação de madeira de parica na região de Paragominas*. Academic Press.

Smook, G. (2016). *Handbook for pulp and paper technologists* (4th ed.). TAPPI Press.

Solomons, T. W. G., Fryhle, C. B., & Snyder, S. A. (2016). *Organic Chemistry* (12th ed.). Wilwy.

Sparks, S. (2012). *Nanotechnology: business applications and commercialization*. CRC Press.

Thoemen, H., Irle, M., & Sernek, M. (2010). *Wood-based panels: an introduction for specialists*. Brunel University Press.

Tian, X., Fang, Z., Smith, R. L. Junior, Wu, Z., & Liu, M. (2016). Properties, Chemical Characteristics and Application of Lignin and Its Derivatives. In Z. Fang & R. L. Smith Junior., (Eds.), *Production of Biofuels and Chemicals from Lignin* (pp. 3–34). Springer. doi:10.1007/978-981-10-1965-4_1

Trivedi, P., & Fardim, P. (2019). Recent advances in cellulose chemistry and potential applications. In Z. Fang, J. R. Smith, & X. F. Tian (Eds.), *Production of Materials from Sustainable Biomass Resources. Biofuels and Biorefineries* (pp. 99–115). Springer. doi:10.1007/978-981-13-3768-0_4

Vallejos, M. E., Zambon, M. D., Area, M. C., & Curvelo, A. A. S. (2015). Low liquid-solid ratio fractionation of sugarcane bagasse by hot water autohydrolysis and organosolv delignification. *Industrial Crops and Products, 65*, 349–353. doi:10.1016/j.indcrop.2014.11.018

Vieira, R. S., Lima, J. T., Silva, J. R. M., Hein, P. R. G., Bailleres, H., & Baraúna, E. E. P. (2010). Small wooden objects using eucalypt sawmill wood waste. *BioResources, 5*, 1463–1472.

Waldron, K. (2014). *Advances in Biorefineries: Biomass and Waste Supply Chain Exploitation*. Cambridge: Woodhead.

Wallberg, O., Linde, M., & Jonsson, A. (2006). Extraction of lignin and hemiceluloses from black liquor. *Desalination, 199*(1-3), 413–414. doi:10.1016/j.desal.2006.03.094

Wastowski, A. D. (2018). *Química da madeira*. Interciência.

Wegner, T., Skog, K. E., Ince, P. J., & Michler, C. J. (2010). Uses and desirable properties of wood in the 21st century. *Journal of Forestry, 108*(4), 165–173.

Wertz, J. L., Deleu, M., Coppée, S., & Richel, A. H. (2019). *Hemicelluloses and lignin in biorefinerires*. CRC Press.

Yanqiao, P., Zhang, D., Singh, P., & Ragauskas, A. (2008). The new forestry biofuels sector. *Biofuels, Bioproducts & Biorefining, 2*(1), 58–73. doi:10.1002/bbb.48

ADDITIONAL READING

Bajpai, P. (2017). *Carbon fiber from lignin*. Springer. doi:10.1007/978-981-10-4229-4

Bhaskar, T., Pandey, A., Mohan, S. V., Lee, D. J., & Khanal, S. K. (2018). *Waste Biorefinery: Potential and Perspectives*. Elsevier.

Inamuddin, Boddula, R. & Asiri, A. M. (2020). *Advanced Applications of Bio-degradable Green Composites*. Millersville: Material. *Research Forum*.

Kazi, S. N. (2018). *Pulp and Paper Processing*. IntechOpen. doi:10.5772/intechopen.68843

Lengowski, E. C., Bonfatti Júnior, E. A., Kumode, M. M. N., Carneiro, M. E., & Satyanarayna, K. G. Nanocellulose-Reinforced Adhesives for Wood-Based Panels. In. Inamuddin, S. Thomas & R. K. Asiri (Eds.), Sustainable Polymer Composites and Nanocomposites, (pp. 1001–1025). Chan: Springer. doi:10.1007/978-3-030-05399-4_35

Nolasco, A. M., & Uliana, L. R. (2014). *Waste management in the wooden floor industry*. ANPM.

Poletto, M. (2015). *Cellulose: Fundamental Aspects and Current Trends*. IntechOpen. doi:10.5772/59889

Servos, M. R., Munkittrick, K. R., Carey, J. H., & Van der Kraak, G. (1996). *Environmental Fate and Effects of Pulp and Paper: Mill Effluents*. CRC Press.

Van Heiningen, A. (2006). Converting a kraft pulp mill into an integrated forest biorefinery. *Pulp & Paper Canada*, *107*, 141–146.

Walker, J. C. F. (2006). *Primary Wood Processing* (2nd ed.). Springer Netherlands.

KEY TERMS AND DEFINITIONS

Biomass: Renewable carbon source with the potential to be processed in a biorefinery in a similar way to petroleum in conventional refineries, to produce fuels, chemicals, and high-added value products.

Biorefinery: Industrial process, or set of processes, that converts biomass into fuels, chemicals, electricity, heat, refined derivatives, resin, membranes, among other products with high-added value.

Black Liquor: Main by-product of kraft pulping process used in the manufacture of cellulose pulp.

Green Chemistry: Area of chemistry responsible for developing products through processes that eliminate substances harmful to health and the environment.

Kraft Pulping Process: Cellulose pulp production process that uses sodium hydroxide ($NaOH$) and sodium sulfide (Na_2S) as reagents.

Veneer Sheets: Thin slices of wood, usually thinner than 3mm, that are usually glued to make wood panels like plywood and LVL.

Wood Pyrolysis: Process of obtaining charcoal through the action of heat that eliminates most of the volatile components of wood.

Wood Slab: A piece initially sawn from a log face when squaring up the log. It has no potential to be made into lumber with two flat faces.

Compilation of References

Abbas, S., Ismail, I., Mostafa, T., & Abbas, S. (2014). Biosorption of heavy metals [review]. *J ChemSciTechnol, 3*, 74–102.

Abd Rahman, N. I.-G. G. for H. T. (2013). Driving SMEs towards sustainable future. *Malaysian Biomass Industry Action Plan 2020, 1*(Driving SMEs Towards Sustainable Future), 1–80.

Abdel-Shafy, H., & Mansour, M. (2018). Solid waste issue: Sources, composition, disposal, recycling and valorisation. *Egyptian Journal of Petroleum, 27*(4), 1275–1290. doi:10.1016/j.ejpe.2018.07.003

Abderrahmane, Y., Okkacha, Y., & Hassibaa, B. (2014). Sustainable solid waste management in the city of Mecheria (Western Algeria). *Energy Procedia, 50*, 953–959. doi:10.1016/j.egypro.2014.06.114

Abdoli, M., Samieifard, R., & Jalili, G. M. (2008). *Rural solid waste management.* Academic Press.

Abdul, K. (2014). Production and modification of nanofibrillated celulose using various mechanical processes: A review. *Carbohydrate Polymers, 99*, 649–665. doi:10.1016/j.carbpol.2013.08.069 PMID:24274556

Abdullah, N., & Town, G. (2013). *The oil palm wastes in Malaysia.* doi:10.5772/55302

Abdullahi, M. S. (2015). Soil contamination, remediation and plants: Prospects and challenges. Soil Remediation and Plants.

Abdullah, N., & Chin, N. L. (2010). Simplex-centroid mixture formulation for optimised composting of kitchen waste. *Bioresource Technology, 101*(21), 8205–8210. doi:10.1016/j.biortech.2010.05.068 PMID:20624604

Abdullah, N., Sulaiman, F., & Gerhauser, H. (2011). Characterisation of oil palm empty fruit bunches for fuel application. *The Journal of Physiological Sciences; JPS, 22*(1), 1–24.

Abou-Zeid, D., Müller, R., & Deckwer, W. (2001). Degradation of natural and synthetic polyesters under anaerobic conditions. *Journal of Biotechnology, 86*(2), 113–126. doi:10.1016/S0168-1656(00)00406-5 PMID:11245900

Abreu, H. S., Nascimento, A. M., & Maria, M. A. (1999). Lignin structure and wood properties. *Wood and Fiber Science, 31*(4), 426–433.

Acién, F. G., Gómez-Serrano, C., Morales-Amaral, M. D. M., Fernández-Sevilla, J. M., & Molina-Grima, E. (2016). Wastewater treatment using microalgae: How realistic a contribution might it be to significant urban wastewater treatment? *Applied Microbiology and Biotechnology, 100*(21), 9013–9022. doi:10.100700253-016-7835-7 PMID:27645298

Acién, F. G., Molina, E., Reis, A., Torzillo, G., Zittelli, G. C., Sepúlveda, C., & Masojídek, J. (2017). Photobioreactors for the production of microalgae. In *Microalgae-based biofuels and bioproducts* (pp. 1–44). Woodhead Publishing.

Adachi, O., Moonmangmee, D., Toyama, H., Yamada, M., Shinagawa, E., & Matsushita, K. (2003). New developments in oxidative fermentation. *Applied Microbiology and Biotechnology, 60*(6), 643–653. doi:10.100700253-002-1155-9 PMID:12664142

Adams, R. P. (2007). *Identification of Essential Oil Components by Gas Chromatography/Mass Spectrometry* (4th ed.). Allured Publishing.

Addaney, M., & Oppong, R. (2015). Critical issues of municipal solid waste management in Ghana. *Journal of Energy and Natural Resource Management*, *2*, 30–36.

Adefemi, E., & Awokunmi, E. (2009). The impact of municipal solid waste disposal in Ado-Ekiti metropolis, Ekiti-state, Nigeria. *African Journal of Environmental Science and Technology*, *3*(8), 186–189. doi:10.5897/AJEST09.075

Adriano, D. C. (2001). *Trace Elements in Terrestrial Environments; Biochemistry, Bioavailability and Risks of Metals*. Springer-Verlag. doi:10.1007/978-0-387-21510-5

Agamuthu, P., Khidzir, K. M., & Hamid, F. S. (2009). Drivers of sustainable waste management in Asia. *Waste Management & Research*, *27*(7), 625–633. doi:10.1177/0734242X09103191 PMID:19470545

Agarwal, N. (2015). Recent Growth Trends of Automobile Industry in India. *Indian Journal of Applied Research*, *5*(7), 376–378.

Agdag, O. N., & Sponza, D. T. (2007). Co-digestion of mixed industrial sludge with municipal solid wastes in anaerobic simulated landfilling bioreactors. *Journal of Hazardous Materials*, *140*(1-2), 75–85. doi:10.1016/j.jhazmat.2006.06.059 PMID:16884847

Agrawal, A., Pathak, P., Mishra, D., & Sahu, K. K. (2012). Solvent mediated interactions for the selective recovery of Cadmium from Ni–Cd battery waste. *Journal of Molecular Liquids*, *173*, 77–84. doi:10.1016/j.molliq.2012.06.017

Ahmad, A. L., Ismail, S., & Bhatia, S. (2005). Membrane treatment for palm oil mill effluent: Effect of transmembrane pressure and crossflow velocity. *Desalination*, *179*(1-3), 245–255. doi:10.1016/j.desal.2004.11.071

Ahmad, A. L., Sumathi, S., & Hameed, B. H. (2005). Adsorption of residue oil from palm oil mill effluent using powder and flake chitosan: Equilibrium and kinetic studies. *Water Research*, *39*(12), 2483–2494. doi:10.1016/j.watres.2005.03.035 PMID:15985277

Ahmad, A. L., Sumathi, S., & Hameed, B. H. (2006). Coagulation of residue oil and suspended solid in palm oil mill effluent by chitosan, alum and PAC. *Chemical Engineering Journal*, *118*(1–2), 99–105. doi:10.1016/j.cej.2006.02.001

Ahmed, T., Shahid, M., Azeem, F., Rasul, I., Shah, A., Noman, M., Hameed, A., Manzoor, N., Manzoor, I., & Muhammad, S. (2018). Biodegradation of plastics: Current scenario and future prospects. *Environmental Science and Pollution Research International*, *25*(1), 1–12. doi:10.100711356-018-1234-9 PMID:29332271

Ahn, H. K., Mulbry, W., White, J. W., & Kondrad, S. L. (2011). Pile mixing increases greenhouse gas emissions during composting of dairy manure. *Bioresource Technology*, *102*(3), 2904–2909. doi:10.1016/j.biortech.2010.10.142 PMID:21111610

Akhtar, M. S., Chali, B., & Azam, T. (2013). Bioremediation of Arsenic and Lead by Plants and Microbes from Contaminated Soil. *Res. Plant Sciences.*, *1*(3), 68–73.

Akkawuttiwanich, P., & Yenradee, P. (2018). Fuzzy QFD approach for managing SCOR performance indicators. *Computers & Industrial Engineering*, *122*(May), 189–201. doi:10.1016/j.cie.2018.05.044

Akpor, O. B., Otohinoyi, D. A., Olaolu, D. T., & Aderiye, B. I. (2014). Pollutants in wastewater effluents: impacts and remediation processes. *International Journal of Environmental Research and Earth Science, 3*(3), 50-59.

Aksu, Z., & Gönen, F. (2004). Biosorption of phenol by immobilized activated sludge in a continuous packed bed: Prediction of breakthrough curves. *Process Biochemistry*, *39*(5), 599–613. doi:10.1016/S0032-9592(03)00132-8

Albertsson, A. C. (1980). The shape of the biodegradation curve for low- and high-density polyethylene's in prolonged series of experiments. *European Polymer Journal*, *16*(7), 623–630. doi:10.1016/0014-3057(80)90100-7

Albertsson, A. C., Andersson, S. O., & Karlsson, S. (1987). The mechanism of biodegradation of polyethylene. *Polymer Degradation & Stability*, *18*(1), 73–87. doi:10.1016/0141-3910(87)90084-X

Albertsson, A. C., Barenstedt, C., & Karlsson, S. (1994). Abiotic degradation products from enhanced environmentally degradable polyethylene. *Acta Polymerica*, *45*(2), 97–103. doi:10.1002/actp.1994.010450207

Albertsson, A. C., & Karlsson, S. (1990). The influence of biotic and abiotic environments on the degradation of polyethylene. *Progress in Polymer Science*, *15*(2), 177–192. doi:10.1016/0079-6700(90)90027-X

Alcántara, C., Posadas, E., Guieysse, B., & Muñoz, R. (2015). Microalgae-based wastewater treatment. In *Handbook of Marine Microalgae* (pp. 439–455). Academic Press. doi:10.1016/B978-0-12-800776-1.00029-7

Alegbeleye, O. O., Opeolu, B. O., & Jackson, V. A. (2017). Polycyclic aromatic hydrocarbons: A critical review of environmental occurrence and bioremediation. *Environmental Management*, *60*(4), 758–783. doi:10.100700267-017-0896-2 PMID:28573478

Aleksandrovich, S., Nicolaevich, E., & Ivanovich, E. (1998). *Method of processing of products based on ahalcogenides of base metals containing metals of platinum group and gold.* Russian Patent, RU2112064, C22B, 11(02).

Ali, A. (2010). Wasting time on solid waste in developing countries. *Waste Management (New York, N.Y.)*, *30*(8-9), 1437–1438. doi:10.1016/j.wasman.2010.04.025 PMID:20493678

Alkhatib, M. F., Mamun, A. A., & Akbar, I. (2014). *Application of response surface methodology (RSM) for optimization of color removal from POME by granular activated carbon.* doi:10.100713762-014-0504-4

Alpizar, F., Carlsson, F., Lanza, G., Carney, B., Daniels, R., Jaime, M., Ho, T., Nie, Z., Salazar, C., Tibesigwa, B., & Wahdera, S. (2020). A framework for selecting and designing policies to reduce marine plastic pollution in developing countries. *Environmental Science & Policy*, *109*, 25–35. doi:10.1016/j.envsci.2020.04.007

Alsaed, O., & Jalham, I. S. (2012). Polyvinyl Butyral. (PVB) and Ethyl Vinyl Acetate (EVA) as a Binding Material for Laminated Glass. *Jordan Journal of Mechanical and Industrial Engineering*, *6*(2), 127–133.

Al-Salem, S., Antelava, A., Constantinou, A., Manos, G., & Dutta, A. (2017). A review on thermal and catalytic pyrolysis of plastic solid waste (PSW). *Journal of Environmental Management*, *197*, 177–198. doi:10.1016/j.jenvman.2017.03.084 PMID:28384612

Al-Salem, S., Lettieri, P., & Baeyens, J. (2009). Recycling and recovery routes of plastic solid waste (PSW): A review. *Waste Management (New York, N.Y.)*, *29*(10), 2625–2643. doi:10.1016/j.wasman.2009.06.004 PMID:19577459

Alshehrei, F. (2017). Biodegradation of synthetic and natural plastic by microorganisms. *Journal of Applied & Environmental Microbiology*, *5*(1), 8–19.

Amat, N. A. A., Tan, Y. H., Lau, W. J., Lai, G. S., Ong, C. S., Mokhtar, N. M., ... Lai, S. O. (2015). Journal of Water Process Engineering Tackling colour issue of anaerobically-treated palm oil mill effluent using membrane technology. *Journal of Water Process Engineering*, *8*, 221–226. doi:10.1016/j.jwpe.2015.10.010

American Public Health Association. (2001). *5210 Biochemical Oxygen Demand.* BOD.

American Society of Mechanical Engineers (ASME). (2014). *Waste to Energy, Summary Report.* Available Online: http://www.asme.org

Amoah, S. T., & Kosoe, E. A. (2014). Solid waste management in urban areas of Ghana: Issues and experiences. *Wa Journal of Environmental Pollution and Human Health, 2*, 110–117.

Anbalagan, A., Toledo-Cervantes, A., Posadas, E., Rojo, E. M., Lebrero, R., González-Sánchez, A., Nehrenheim, E., & Muñoz, R. (2017). Continuous photosynthetic abatement of CO_2 and volatile organic compounds from exhaust gas coupled to wastewater treatment: Evaluation of tubular algal-bacterial photobioreactor. *Journal of CO_2 Utilization, 21*, 353-359.

Andalib, M., Elbeshbishy, E., Mustafa, N., Hafez, H., Nakhla, G., & Zhu, J. (2014). Performance of an anaerobic fluidized bed bioreactor (AnFBR) for digestion of primary municipal wastewater treatment biosolids and bioethanol thin stillage. *Renewable Energy, 71*, 276–285. doi:10.1016/j.renene.2014.05.039

Anderson, J., & Shive, M. (2012). Biodegradation and biocompatibility of PLA and PLGA microspheres. *Advanced Drug Delivery Reviews, 64*, 72–82. doi:10.1016/j.addr.2012.09.004

Andrady, A. L. (2011). Microplastics in the marine environment. *Marine Pollution Bulletin, 62*(8), 1596–1605. doi:10.1016/j.marpolbul.2011.05.030 PMID:21742351

Andreola, F., Barbieri, L., Lancellotti, I., Leonelli, C., & Manfredini, T. (2016). Recycling of industrial wastes in ceramic manufacturing: State of art and glass case studies. *Ceramics International, 42*(12), 13333–13338. doi:10.1016/j.ceramint.2016.05.205

Ang, W. L., Mohammad, A. W., Benamor, A., & Hilal, N. (2016). Chitosan as Natural Coagulant in Hybrid Coagulation-Nanofiltration Membrane Process for Water Treatment. *Biochemical Pharmacology*. Advance online publication. doi:10.1016/j.jece.2016.03.029

Ang, W. L., Mohammad, A. W., Teow, Y. H., Benamor, A., & Hilal, N. (2015). Hybrid chitosan/FeCl3 coagulation–membrane processes: Performance evaluation and membrane fouling study in removing natural organic matter. *Separation and Purification Technology, 152*, 23–31. doi:10.1016/j.seppur.2015.07.053

Annepu, R. R. K. (n.d.). *Sustainable solid waste management in India, Waste-to-Energy Research and Technology Council (WTERT)*. Columbia University. Retrieved from http://www.seas.columbia.edu/earth/wtert/sofos/Sustainable%20 Solid%20Waste%20Management%20 in%20India_Final.pdf

Anschutz, J., Ijgosse, J., & Scheinberg, A. (2004). *Putting integrated solid waste management to practice*. Waste.

Ant Spirits Sdn Bhd. (2018). *Low speed mini granulator*. Author.

Anupama, Rao, Saxena, & Rayabag. (2015). Challenges in manufacturing and end-of-life recycling or disposal of solar PV panels. *Journal of Electrical and Electronics Engineering (Oradea), 10*(4), 81–87.

Anwar, F., Sulman, M., Hussain, A. I., Saari, N., Iqbal, C., & Rashid, U. (2011). Physicochemical composition of hydro-distilled essential oil from coriander (*Coriandrum sativum* L.) seeds cultivated in Pakistan. *Journal of Medicinal Plants Research, 5*(15), 3537–3544.

Aparicio, M. A., Eiroa, M., Kennes, C., & Veiga, M. C. (2007). *Combined post-ozonation and biological treatment of recalcitrant wastewater from a resin-producing factory*. doi:10.1016/j.jhazmat.2006.09.025

Aracil, C., Haro, P., Fuentes-Cano, D., & Gomez-Barea, A. (2018). Implementation of waste-to-energy options in landfill-dominated countries: Economic evaluation and GHG impact. *Waste Management (New York, N.Y.), 76*, 443–456. doi:10.1016/j.wasman.2018.03.039 PMID:29610061

Araujo, E., Pimenta, A. S., Feijo, F., Castro, R. V. O., Lima, K., Lima, M., & Monteiro, T. C. (2017). Antibacterial and antifungal activity activity of pyroligneous acid from wood of and. *Journal of Applied Microbiology, 124*(1), 85–96. doi:10.1111/jam.13626 PMID:29095556

Arendse, L., & Godfrey, L. (2010). *Waste management indicators for national state of environment reporting.* Available Online: http://www.unep.or.jp/ietc/kms/data/2010.pdf

Arifuzzaman, Md., Zahid Hasan, Md., Badier Rahaman, S. M., & Kamruzzaman Pramanik, Md. (2014). Isolation and characterization of *Aectobacter and Gluconobacter spp* from sugarcane and rotten fruits. *Research and Reviews in Bio Sciences, 8*(9), 359–365.

Ariunbaatar, J., Panico, A., Esposito, G., Pirozzi, F., & Lens, P. N. L. (2014). Pre-treatment methods to enhance anaerobic digestion of organic solid waste. *Applied Energy, 123*, 143–156. doi:10.1016/j.apenergy.2014.02.035

Asnani, P. U. (2005). *United States Asia Environmental Partnership Report, United States Agency for International Development, Centre for Environmental Planning and Technology, Ahmedabad; 2005. Technical Committee Report, West Bengal SWM Mission 2005*. Government of West Bengal.

Association of Plastics Manufacturers in Europe (APME). (2004). 1995. Plastics—a material of choice for the electrical and electronic industry—plastics consumption and recovery. In *Western Europe* (Vol. 1). APME.

Astals, S., Batstone, D. J., Mata-Alvarez, J., & Jensen, P. D. (2014). Identification of synergistic impacts during anaerobic co-digestion of organic wastes. *Bioresource Technology, 169*, 421–427. doi:10.1016/j.biortech.2014.07.024 PMID:25079207

ASTM. (2004). 6400-04 Standard specification for compostable plastics. *Journal of ASTM International.*

Atalia, K., Buha, D., Bhavsar, K., & Shah, N. (2015). A review on composting of municipal solid waste. *Journal of Environmental Science. Toxicology and Food Technology, 9*(5), 20–29.

Augusto, G. (2017). *Shaft design*. Academic Press.

Awaja, F., & Pavel, D. (2005). Recycling of PET. *European Polymer Journal, 41*(7), 1453–1477. doi:10.1016/j.eurpolymj.2005.02.005

Awasthi, S. K. (2000). Prevention of Food Adulteration Act no 37 of 1954. Central and State Rules as Amended for 1999, 303-327.

Awasthi, A. K., Zeng, X., & Li, J. (2016). Environmental pollution of electronic waste recycling in India: A critical review. *Environmental Pollution, 211*, 259–270. doi:10.1016/j.envpol.2015.11.027 PMID:26774773

Ayodele, T., Ogunjuyigbe, A., & Alao, M. (2017). Life cycle assessment of waste-to-energy (WtE) technologies for electricity generation using municipal solid waste in Nigeria. *Applied Energy, 201*, 200–218. doi:10.1016/j.apenergy.2017.05.097

Ayuso, M., Hernandez, T., Garcia, C., & Pascual, J. A. (1996). Biochemical and chemical structural characterization of different organic materials used as manures. *Bioresource Technology, 57*(2), 201–207. doi:10.1016/0960-8524(96)00070-3

Azhar, D. D. P. (2009). *Palm oil industry in Malaysia skills & knowledge for sustained development origin of palm oil palm oil value chain & applications*. Academic Press.

Aziz, D. A. (2015). MIDA Biomass Conference. In *Driving National Biomass Agenda*. MPOB.

Azmi, N. A., Yunos, K. F. M., & Zakaria, R. (2012). Application ofsandwich membrane for the treatment of palm oil mill effluent (POME) for water reuse. Procedia Engineering, 44, 1980–1981. doi:10.1016/j.proeng.2012.09.014

Azmi, N. S., & Yunos, K. F. M. (2014). Wastewater Treatment of Palm Oil Mill Effluent (POME) by Ultrafiltration Membrane Separation Technique Coupled with Adsorption Treatment as Pre-treatment. *Agriculture and Agricultural Science Procedia, 2*, 257–264. doi:10.1016/j.aaspro.2014.11.037

Azubuike, C. C., Chikere, C. B., & Okpokwasili, G. C. (2016). Bioremediation techniques–classification based on site of application: Principles, advantages, limitations and prospects. *World Journal of Microbiology & Biotechnology, 32*(11), 180. doi:10.100711274-016-2137-x PMID:27638318

Babayemi, J., Nnorom, I., Osibanjo, O., & Weber, R. (2019). Ensuring sustainability in plastics use in Africa: Consumption, waste generations and projections. *Environmental Sciences Europe, 31*(1), 60. doi:10.118612302-019-0254-5

Babel, S., & Kurniawan, T. A. (2004). Cr(VI) removal from synthetic wastewater using coconut shell charcoal and commercial activated carbon modified with oxidizing agents and/or chitosan. *Chemosphere, 54*(7), 951–967. doi:10.1016/j.chemosphere.2003.10.001 PMID:14637353

Bajapai, P. (2018). *Biermann's handbook of pulp and paper*. Elsevier.

Bakar, N. A. B. U., May, C. Y., Lim, J., Lim, S., Chin, E., Loh, S. O. H. K., & Soon, L. I. M. W. (2015). *Improved process for the production of low-ash empty fruit bunch pellet*. Academic Press.

Balde, C. P., Kuehr, R., Blumenthal, K., Fondeur Gill, S., Kern, M., Micheli, P., . . . Huisman, J. (2015). E-waste statistics-Guidelines on classification, reporting and indicators. United Nations University, IAS-SCYCLE.

Balde, C. P., Forti, V., Gray, V., Kuehr, R., & Stegmann, P. (2017). *The Global E-Waste Monitor—2017. United Nations University (UNU), International Telecommunication Union (ITU) & International Solid Waste Association*. ISWA.

Banerjee, A. (2016). Wastewater use for agriculture in India: A background review. *Wastewater Use for Agriculture in India*, 1-12.

Baratta, M. T., Dorman, H. J. D., Deans, S. G., Biondi, D. M., & Ruberto, G. (1998). Chemical composition, antimicrobial and antioxidative activity of laurel, sage, rosemary, oregano and coriander essential oils. *The Journal of Essential Oil Research, 10*(6), 618–627. doi:10.1080/10412905.1998.9700989

Barbara, J. S. (2007). *The fales promise of biofuels*. The International Forum on Globalization and the Institute for Policy Studies.

Barber, W. P., & Stuckey, D. C. (1999). The use of the anaerobic baffled reactor (ABR) for wastewater treatment: A review. *Water Research, 33*(7), 1559–1578. doi:10.1016/S0043-1354(98)00371-6

Barker, A. V. (1997). Composition and uses of compost. In *Agricultural uses of by-products and wastes*. ACS Symposium Series 668. 10.1021/bk-1997-0668.ch010

Barnes, D., & Milner, P. (2005). Drifting plastic and its consequences for sessile organism dispersal in the Atlantic Ocean. *Marine Biology, 146*(4), 815–825. doi:10.100700227-004-1474-8

Barrington, S., Choiniere, D., Trigui, M., & Knight, W. (2003). Compost convective airflow under passive aeration. *Bioresource Technology, 86*(3), 259–266. doi:10.1016/S0960-8524(02)00155-4 PMID:12688469

Bartone, C., & Leitman, J. (1994). *Towards environmental strategies for cities, urban infrastructure notes, urban No. UE-11*. World Bank.

Batista, A. P., Ambrosano, L., Graça, S., Sousa, C., Marques, P. A., Ribeiro, B., Botrel, E. P., Neto, P. C., & Gouveia, L. (2015). Combining urban wastewater treatment with biohydrogen production–an integrated microalgae-based approach. *Bioresource Technology, 184*, 230–235. doi:10.1016/j.biortech.2014.10.064 PMID:25453433

Baud, I., Grafakos, S., Hordijk, M., & Post, J. (2001). Quality of life and alliances in solid waste management contributions to urban sustainable development. *Cities (London, England), 18*(1), 3–12. doi:10.1016/S0264-2751(00)00049-4

Beans, C. (2017). Core Concept: Phytoremediation advances in the lab but lags in the field. *Proceedings of the National Academy of Sciences of the United States of America*, *114*(29), 7475–7477. doi:10.1073/pnas.1707883114 PMID:28720737

Behera, S. K., Park, J. M., Kim, K. H., & Park, H. (2010). Methane production from food waste leachate in laboratory-scale simulated landfill. *Waste Management (New York, N.Y.)*, *30*(8-9), 1502–1508. doi:10.1016/j.wasman.2010.02.028 PMID:20227867

Bello, I., Ismail, M., & Kabbashi, N. (2016). Solid waste management in Africa: Integrated development plan review: A city growing with you, Eastern Cape, SA. A review. *International Journal of Waste Resources*, *6*, 1–4.

Bello, M. M., & Abdul Raman, A. A. (2017). Trend and current practices of palm oil mill effluent polishing: Application of advanced oxidation processes and their future perspectives. *Journal of Environmental Management*, *198*, 170–182. doi:10.1016/j.jenvman.2017.04.050 PMID:28460324

Bell, R., & Russell, C. (2002). Environmental policy for developing countries. *Issues in Science and Technology*, *18*, 63–70.

Benson, H. J. (1990). *Microbiological Application – A Laboratory Manual in General Microbiology. C.* Brown Publishers.

Bernardo, G.R., & Rene, R.J., & Catalina, A.L.T.M. (2009). Chromium (III) uptake by agro-waste biosorbents: Chemical characterization, sorption-desorption studies, and mechanism. *Journal of Hazardous Materials*, *170*, 845–854. doi:10.1016/j.jhazmat.2009.05.046 PMID:19523757

Bertoldi, M. (1983). The biology of composting: A review. *Waste Management & Research*, *1*(1), 157–176. doi:10.1177/0734242X8300100118

Betarbet, R., Sherer, T. B., MacKenzie, G., Garcia-Osuna, M., Panov, A. V., & Greenamyre, J. T. (2000). Chronic systemic pesticide exposure reproduces features of Parkinson's disease. *Nature Neuroscience*, *3*(12), 1301–1306. doi:10.1038/81834 PMID:11100151

Bharadwaj, K. K. R. (1995). Improvements in microbial compost technology: A special reference to microbiology of composting. In S. Khanna & K. Mohan (Eds.), *Wealth from waste* (pp. 115–135). Tata Energy Research Institute.

Bharti, Sharma, Singh, Singh, Arora, Gogoi, Rai, Meenu, Tyagi, Tyagi, & Kaicker. (2019). *Waste Utilization and product development.* Demonstration of products Technology in exhibition showcase section of 'Ecothon-2019: Solutions and sustainability - conference & exhibition' organized by Vprospurs Singapore & Green Dream Foundation India, IGNCA, New Delhi, India.

Bharti, P. K. (2019). *Product Development and Waste Recycling*. Discovery Publishing House.

Bhatia, S., Othman, Z., & Ahmad, A. L. (2007). *Coagulation – flocculation process for POME treatment using Moringa oleifera seeds extract : Optimization studies.* doi:10.1016/j.cej.2007.01.034

Bhatnagar, A., Vilar, V. J. P., Botelho, C. M. S., & Boaventura, R. A. R. (2011). A review of the use of red mud as adsorbent for the removal of toxic pollutants from water and wastewater. *Environmental Technology*, *32*(3-4), 231–249. doi:10.1080/09593330.2011.560615 PMID:21780692

Bhattacharya, A. K., Mandal, S. N., & Das, S. K. (2006). Adsorption of Zn(II) from aqueous solution by using different adsorbents. *Chemical Engineering Journal*, *123*(1-2), 43–51. doi:10.1016/j.cej.2006.06.012

Bhattacharya, A. K., Naiya, T. K., Mandal, S. N., & Das, S. K. (2008). Adsorption, kinetics and equilibrium studies on the removal of Cr(VI) from aqueous solutions using different low-cost adsorbents. *Chemical Engineering Journal*, *137*, 529–541.

Bhavana, Sonashree, Rashmi, Rao, & Praveena. (2019). A review on the preparation of vinegar from fruit peels. *World Journal of Pharmacy and Pharmaceutical Sciences, 8*(9), 227-233.

Bhuie, A. K., Ogunseitan, O. A., Saphores, J. D., & Shapiro, A. A. (2004, May). Environmental and economic trade-offs in consumer electronic products recycling: a case study of cell phones and computers. In *IEEE International Symposium on Electronics and the Environment, 2004. Conference Record. 2004* (pp. 74-79). IEEE. 10.1109/ISEE.2004.1299691

Biedenbach, D. J., Kazmierczak, K., Bouchillon, S. K., Sahm, D. F., & Bradford, P. A. (2015). In vitro activity of aztreonam-avibactam against a global collection of Gram-negative pathogens from 2012 and 2013. *Antimicrobial Agents and Chemotherapy, 59*(7), 4239–4248. doi:10.1128/AAC.00206-15 PMID:25963984

Bishnoi, N. R., & Garima. (2005). Fungus-an alternative for bioremediation of heavy metal containing wastewater: A review. *Journal of Scientific and Industrial Research, 64*, 93–100.

Bisschop, L. (2016). How e-waste challenges environmental governance. In *Hazardous Waste and Pollution* (pp. 27–43). Springer. doi:10.1007/978-3-319-18081-6_3

Blaauw, P. (2017). Informal employment in South Africa: Still missing pieces in the vulnerability puzzle. *Southern African Business Review, 21*, 339–360.

Bodkhe, S. (2008). Development of an improved anaerobic filter for municipal wastewater treatment. *Bioresource Technology, 99*(1), 222–2226. doi:10.1016/j.biortech.2006.11.026 PMID:17207998

Boe, K., & Angelidaki, I. (2009). Serial CSTR digester configuration for improving biogas production from manure. *Water Research, 43*(1), 166–172. doi:10.1016/j.watres.2008.09.041 PMID:18976792

Bohart, G., & Admas, E. N. (1920). Some aspects of the behavior of charcoal with respect to chlorine. *Journal of the American Chemical Society, 42*(3), 523–544. doi:10.1021/ja01448a018

Bolzonella, D., Fatone, F., Pavan, P., & Cecchi, F. (2005). Anaerobic fermentation of organic municipal solid wastes for the production of soluble organic compounds. *Industrial & Engineering Chemistry Research, 44*(10), 3412–3418. doi:10.1021/ie048937m

Bonfatti Júnior, E. A., Raia, R. R., Bila, N. F., Lopes, M. S., Klock, U., Andrade, A. S., & Vivian, M. A. (2019b). Kraft punlping and papermaking of *Cryptomeria japonica. Scientia Forestalis, 47*(124), 811–822. doi:10.18671cifor.v47n124.22

Bonhomme, S., Cuer, A., Delort, A. M., Lemaire, J., Sancelme, M., & Scott, G. (2003). Environmental biodegradation of polyethylene. *Polymer Degradation & Stability, 81*(3), 441–452. doi:10.1016/S0141-3910(03)00129-0

Boopathy, R. (2000). Factors Limiting Bioremediation Technologies. *Bioresource Technology, 74*(1), 63–67. doi:10.1016/S0960-8524(99)00144-3

Borowitzka, M. A. (1998). Limits to growth. In *Wastewater treatment with algae* (pp. 203–226). Springer. doi:10.1007/978-3-662-10863-5_12

Borowitzka, M. A. (2013). High-value products from microalgae—Their development and commercialisation. *Journal of Applied Phycology, 25*(3), 743–756. doi:10.100710811-013-9983-9

Boucher, J., & Friot, D. (2017). *Primary Microplastics in the Oceans: a Global Evaluation of Sources*. International Union for Conservation of Nature and Natural Resources. doi:10.2305/IUCN.CH.2017.01.en

Bougrier, C., Delgenès, J. P., & Carrère, H. (2007). Impacts of thermal pre-treatments on the semi-continuous anaerobic digestion of waste activated sludge. *Biochemical Engineering Journal, 34*(1), 20–27. doi:10.1016/j.bej.2006.11.013

Boyandin, A., Prudnikova, S., Karpov, V., Ivonin, V., Đo, N., Nguyen, T., Lê, T. M. H., Filichev, N. L., Levin, A. L., Filipenko, M. L., Volova, T. G., & Gitelson, I. I. (2013). Microbial degradation of polyhydroxyalkanoates in tropical soils. *International Biodeterioration & Biodegradation*, *83*, 77–84. doi:10.1016/j.ibiod.2013.04.014

Boyd, G. E., Adamson, A. W., & Myers, L. S. Jr. (1947). The exchange adsorption of ions from aqueous solutions by organic zeolites. II. Kinetic. *Journal of the American Chemical Society*, *69*(11), 2836–2848. doi:10.1021/ja01203a066 PMID:20270838

Bradley, P., & Yee, S. (2015). *Using the DPSIR framework to develop a conceptual model: Technical support document.* USEPA.

Braun, R. (2002). *Potential of Co-digestion.* https://www.novaenergie.ch/iea-bioenergytask37/Dokumente/final.PDF

Brennan, L., & Owende, P. (2010). Biofuels from microalgae—A review of technologies for production, processing, and extractions of biofuels and co-products. *Renewable & Sustainable Energy Reviews*, *14*(2), 557–577. doi:10.1016/j.rser.2009.10.009

Briski, F., Vukovic, M., Papa, K., Gomzi, Z., & Domanovac, T. (2007). Modelling of compositing of food waste in a column reactor. *Chemical Papers*, *61*(1), 24–29. doi:10.247811696-006-0090-0

Brooks, R. R. (1998). *Plants that hyperaccumulate heavy metals.* Cambridge University Press.

Brown & Caldwell. (2001). Description of Commonly Considered Water Quality Constituents. *Watershed Protection Plan Development Guidebook*, 1–11.

Brown, D., & Li, Y. (2013). Solid state anaerobic co-digestion of yard waste and food waste for biogas production. *Bioresource Technology*, *127*, 275–280. doi:10.1016/j.biortech.2012.09.081 PMID:23131652

Bruijnincx, P. C. A., Rinald, R., & Weckhuysen, R. M. (2015). Unlocking the potential of a sleeping giant: Lignins as sustainable raw materials for renewable fuels, chemicals and materials. *Green Chemistry*, *15*(11), 4860–4861. doi:10.1039/C5GC90055G

Brunner, P. H., & Rechberger, H. (2015). Waste to energy – key element for sustainable waste management. *Waste Management (New York, N.Y.)*, *37*, 3–12. doi:10.1016/j.wasman.2014.02.003 PMID:24630214

Brunner, P., & Fellner, J. (2007). Setting priorities for waste management strategies in developing countries. *Waste Management & Research*, *25*(3), 234–240. doi:10.1177/0734242X07078296 PMID:17612323

Brusselaers, J., Hageluken, C., Mark, F., Mayne, N., & Tange, L. (2005). An eco-efficient solution for plastics-metals-mixtures from electronic waste: the integrated metals smelter. *5th Identiplast.*

Budak, N. H., Aykin, E., Seydim, A. C., Greene, A. K., & Guzel-Seydim, Z. B. (2014). Functional Properties of Vinegar. *Journal of Food Science*, *79*(5), 757–765. doi:10.1111/1750-3841.12434 PMID:24811350

Buenrostro, O., Bocco, G., & Cram, S. (2001). Classification of sources of municipal solid wastes in developing countries. *Resources, Conservation and Recycling*, *32*(1), 29–41. doi:10.1016/S0921-3449(00)00094-X

Buffalo City Metropolitan Municipality (BCMM). (2014). *Metro waste management forum to be launched.* Available Online: http://www.bcmmnews.co.za/blog.php?id=8

Buffalo City Metropolitan Municipality. (2011). *Buffalo City Metropolitan Integrated Development Five-Year Plan: 2011 to 2016. Adopted by Council on 30th June 2011.* Buffalo City Metropolitan Municipality.

Buffalo City Metropolitan Municipality. (2018). *Solid waste management department.* Available Online: https://www.buffalocitymetro.gov.za/Municipality/MSWaste

Bulan, R., Mandang, T., & Hermawan, W. (2015). *Physical and mechanical properties of palm frond for the development of palm oil waste chopper and pressing machine design*. Academic Press.

Burken, J., Vroblesky, D., & Balouet, J. C. (2011). Phytoforensics, dendrochemistry and phytoscreening: New green tools for delineating contaminants from past and present. *Environmental Science & Technology*, *45*(15), 6218–6226. doi:10.1021/es2005286 PMID:21749088

Burnley, S. J. (2007). A review of municipal solid waste composition in the United Kingdom. *Waste Management (New York, N.Y.)*, *27*(10), 1274–1285. doi:10.1016/j.wasman.2006.06.018 PMID:17011771

Busselle, L. D., Moore, T. A., Shoemaker, J. M., & Allred, R. E. (1999, May). Separation processes and economic evaluation of tertiary recycling of electronic scrap. In *Proceedings of the 1999 IEEE International Symposium on Electronics and the Environment* (Cat. No. 99CH36357) (pp. 192-197). IEEE. 10.1109/ISEE.1999.765874

Buthiyappan, A., Raman, A., Aziz, A., Mohd, W., & Wan, A. (2016). *Recent advances and prospects of catalytic advanced oxidation process in treating textile effluents*. doi:10.1515/revce-2015-0034

Calabia, B., & Tokiwa, Y. (2006). A novel PHB depolymerase from a thermophilic *Streptomyces sp. Biotechnology Letters*, *28*(6), 383–388. doi:10.100710529-005-6063-5 PMID:16614903

Cao, H. (2005). An initial study on DPSIR model. *Environmental Science & Technology*, *28*, 110–111.

Carey, R. O., & Migliaccio, K. W. (2009). Contribution of wastewater treatment plant effluents to nutrient dynamics in aquatic systems: A review. *Environmental Management*, *44*(2), 205–217. doi:10.100700267-009-9309-5 PMID:19458999

Carlsson, M., Lagerkvist, A., & Morgan-Sagastume, F. (2012). The effects of substrate pre-treatment on anaerobic digestion systems: A review. *Waste Management (New York, N.Y.)*, *32*(9), 1634–1650. doi:10.1016/j.wasman.2012.04.016 PMID:22633466

Caruso, G. (2015). Plastic degrading microorganisms as a tool for bioremediations of plastic contamination in aquatic environments. *Journal of Pollution Effects and Control*, *3*(3), e112. doi:10.4172/2375-4397.1000e112

Carvajal-zarrabal, O., Nolasco-hipólito, C., Barradas-dermitz, D. M., Hayward-jones, P. M., Aguilar-uscanga, M. G., & Bujang, K. (2012). Treatment of vinasse from tequila production using polyglutamic acid. *Journal of Environmental Management*, *95*, S66–S70. doi:10.1016/j.jenvman.2011.05.001 PMID:21600690

Casetti, F., Bartelke, S., Biehler, K., Augustin, M., Schempp, C. M., & Frank, U. (2012). Antimicrobial activity against bacteria with dermatological relevance and skin tolerance of the essential oil from *Coriandrum sativum* L. fruits. *Phytotherapy Research*, *26*(3), 420–424. doi:10.1002/ptr.3571 PMID:21815228

Castillo, E. F. M., Cristancho, D. E., & Arellano, V. A. (2006). Study of the operational conditions for anaerobic digestion of urban solid wastes. *Waste Management (New York, N.Y.)*, *26*(5), 546–556. doi:10.1016/j.wasman.2005.06.003 PMID:16321515

Centre for Science and Environment. (2015). *Recommendations to Address the Issues of Informal Sector Involved in E-Waste Handling*. http://www.cseindia.org/ userfiles/moradabad-e-waste.pdf

Cerda, A., Artola, A., Font, X., Barrena, R., Gea, T., & Sánchez, A. (2018). Composting of food wastes: Status and challenges. *Bioresource Technology*, *248*, 57–67. doi:10.1016/j.biortech.2017.06.133 PMID:28693949

Cesaro, A., & Belgiorno, V. (2013). Sonolysis and ozonation as pre-treatment for anaerobic digestion of solid organic waste. *Ultrasonics Sonochemistry*, *20*(3), 931–936. doi:10.1016/j.ultsonch.2012.10.017 PMID:23231941

Cesaro, A., & Belgiorno, V. (2014). Pre-treatment methods to improve anaerobic biodegradability of organic municipal solid waste fractions. *Chemical Engineering Journal*, *240*, 24–37. doi:10.1016/j.cej.2013.11.055

Chaidez, C., Soto, M., & Jiménez, M. (2014). Water: waste, recycling and irrigation in fresh produce processing. In Global Safety of Fresh Produce (pp. 119-132). Woodhead Publishing. doi:10.1533/9781782420279.2.119

Chandegara, V., Cholera, S., Nandasana, J., Kumpavat, M., & Pate, K. (2015). Plastic Packaging Waste Impact on Climate Change and its Mitigation. In R. Subbaiah & G. Prajapati (Eds.), *Water management and climate smart agriculture. Adaptation of Climatic Resilient Water Management and Agriculture*. Gyan Publishing House.

Chang, J. S., Show, P. L., Ling, T. C., Chen, C. Y., Ho, S. H., Tan, C. H., Nagarajan, D., & Phong, W. N. (2017). Photobioreactors. In *Current Developments in Biotechnology and Bioengineering* (pp. 313–352). Elsevier. doi:10.1016/B978-0-444-63663-8.00011-2

Chattopadhyay, S., Singh, S., Pramanik, N., Niyogi, U., Khandal, R., Uppaluri, R., & Ghoshal, A. (2011). Biodegradability studies on natural fibers reinforced polypropylene composites. *Journal of Applied Polymer Science*, *121*(4), 2226–2232. doi:10.1002/app.33828

Chaudhary, R., Dikshit, A. K., & Tong, Y. W. (2018). Carbon-dioxide biofixation and phycoremediation of municipal wastewater using Chlorella vulgaris and Scenedesmus obliquus. *Environmental Science and Pollution Research International*, *25*(21), 20399–20406. doi:10.100711356-017-9575-3 PMID:28656576

Cha, Y. L., Alam, A. M., Park, S. M., Moon, Y. H., Kim, K. S., Lee, J. E., Kwon, D. E., & Kang, Y. G. (2020). Hydrothermal-process-based direct extraction of polydisperse lignin microspheres from black liquor and their physicochemical characterization. *Bioresource Technology*, *297*, e122399. doi:10.1016/j.biortech.2019.122399 PMID:31759855

Cheah, W. Y., Show, P. L., Chang, J. S., Ling, T. C., & Juan, J. C. (2015). Biosequestration of atmospheric CO_2 and flue gas-containing CO_2 by microalgae. *Bioresource Technology*, *184*, 190–201. doi:10.1016/j.biortech.2014.11.026 PMID:25497054

Chee, J. W., Amirul, A. A., Muhammad, T. T., Majid, M. I. A., & Mansor, S. M. (2008). The influence of copolymer ratio and drug loading level on the biocompatibility of P (3HB-co-4HB) synthesized by Cupriavidus sp.(USMAA2-4). *Biochemical Engineering Journal*, *38*(3), 314–318. doi:10.1016/j.bej.2007.07.018

Chen, C. Y., Chang, C. Y., Lo, Y. C., Ho, S. H., & Chang, J. S. (2014). Enhancing biohydrogen production from *Chlorella vulgaris* FSP-E under mixotrophic cultivation conditions. *Energy Procedia*, *61*, 870–873. doi:10.1016/j.egypro.2014.11.984

Chen, C. Y., Chen, Y. C., Huang, H. C., Huang, C. C., Lee, W. L., & Chang, J. S. (2013a). Engineering strategies for enhancing the production of eicosapentaenoic acid (EPA) from an isolated microalga *Nannochloropsis oceanica* CY2. *Bioresource Technology*, *147*, 160–167. doi:10.1016/j.biortech.2013.08.051 PMID:23994697

Chen, C. Y., Kao, P. C., Tsai, C. J., Lee, D. J., & Chang, J. S. (2013b). Engineering strategies for simultaneous enhancement of C-phycocyanin production and CO_2 fixation with *Spirulina platensis*. *Bioresource Technology*, *145*, 307–312. doi:10.1016/j.biortech.2013.01.054 PMID:23664178

Chen, M., Xu, P., Zeng, G., Yang, C., Huang, D., & Zhang, J. (2015). Bioremediation of soils contaminated with polycyclic aromatic hydrocarbons, petroleum, pesticides, chlorophenols and heavy metals by composting: Applications, microbes and future research needs. *Biotechnology Advances*, *33*(6), 745–755. doi:10.1016/j.biotechadv.2015.05.003 PMID:26008965

Chen, Y., Cheng, J. J., & Creamer, K. S. (2008). Inhibition of anaerobic digestion process: A review. *Bioresource Technology*, *99*(10), 4044–4064. doi:10.1016/j.biortech.2007.01.057 PMID:17399981

Chmielewski, A. G., Urbañski, T. S., & Migdał, W. (1997). Separation technologies for metals recovery from industrial wastes. *Hydrometallurgy*, *45*(3), 333–344. doi:10.1016/S0304-386X(96)00090-4

Choe, C., & Fraser, I. (1998). The economics of household waste management: A review. *The Australian Journal of Agricultural and Resource Economics*, *42*(3), 269–302. doi:10.1111/1467-8489.00052

Chojnacka, K. (2010). Biosorption and bioaccumulation – the prospects for practical. *Appl. Environ.*, *36*, 299–307. PMID:20051290

Chong, S., Sen, T. K., Kayaalp, A., & Ang, H. M. (2012). The performance enhancements of upflow anaerobic sludge blanket (UASB) reactors for domestic sludge treatment—A state-of-the-art review. *Water Research*, *46*(11), 3434–3470. doi:10.1016/j.watres.2012.03.066 PMID:22560620

Chowdhury, A. H., Mohammad, N., Haque, M. R. U., & Hossain, T. (2014). Developing 3Rs (Reduce, Reuse and Recycle) strategy for waste management in the urban areas of Bangladesh: Socioeconomic and climate adoption mitigation option. *IOSR Journal of Environmental Science, Toxicology and Food Technology (IOSR-JESTFT)*, *8*(5), 9-18.

Chowdhury, P., Elkamel, A., & Ray, A. K. (2014). Photocatalytic processes for the removal of toxic metal ions. In S. Sharma (Ed.), *Heavy Metals In Water: Presence, Removal and Safety* (pp. 25–43). Royal Society of Chemistry. doi:10.1039/9781782620174-00025

Christenson, L., & Sims, R. (2011). Production and harvesting of microalgae for wastewater treatment, biofuels, and bioproducts. *Biotechnology Advances*, *29*(6), 686–702. doi:10.1016/j.biotechadv.2011.05.015 PMID:21664266

Chu, F., Labadi, N., & Prins, C. (2006). A scatter search for the periodic capacitated arc routing problem. *European Journal of Operational Research*, *169*(2), 586–605. doi:10.1016/j.ejor.2004.08.017

Chu, L., Yan, S.-T., Xing, X. H., Yu, A.-F., Sun, X.-L., & Jurcik, B. (2008). Enhanced sludge solubilisation by microbubble ozonation. *Chemosphere*, *72*(2), 205–212. doi:10.1016/j.chemosphere.2008.01.054 PMID:18328535

Chung, S. Y., Yettella, R. R., Kim, J. S., Kwon, K., Kim, M. C., & Min, D. B. (2011). Effects of grilling and roasting on the levels of polycyclic aromatic hydrocarbons in beef and pork. *Food Chemistry*, *129*(4), 1420–1426. doi:10.1016/j.foodchem.2011.05.092

Chynoweth, D., Turick, C., Owens, J., Jerger, D. E., & Peck, M. W. (1993). Biochemical methane potential of biomass and waste feedstocks. *Biomass and Bioenergy*, *5*(1), 95–111. doi:10.1016/0961-9534(93)90010-2

Cioanca, O., Hritcu, L., Mihasan, M., & Hancianu, M. (2013). Cognitive-enhancing and antioxidant activities of inhaled coriander volatile oil in amyloid β (1-42) rat model of Alzheimer's disease. *Physiology & Behavior*, *120*, 193–202. doi:10.1016/j.physbeh.2013.08.006 PMID:23958472

Claudio, & Estibaliz, Cerezo, Torija, & Albert. (2010). Production of fruit vinegars: Technological process for persimmon and strawberry vinegars. *Int.J. of Wine Research*, *2*, 55–61.

Cleghorn, W. L. (2016). *Mechanics of machines*. Academic Press.

Clescerl, L. S., Greenberg, A. E., & Eaion, A. D. (1998). *Standard methods for examination of water and wastewater* (20th ed.). APHA.

Cobbinah, P. (2015). Local attitudes towards natural resources management in rural Ghana. *Management of Environmental Quality*, *26*(3), 423–436. doi:10.1108/MEQ-04-2014-0061

Cobbinah, P., Addaney, M., & Agyeman, K. (2017). Locating the role of urbanites in solid waste management in Ghana. *Environmental Development*, *24*, 9–12. doi:10.1016/j.envdev.2017.06.004

Cobbinah, P., Erdiaw-Kwasie, M., & Amoateng, P. (2015). Africa's urbanisation: Implications for sustainable development. *Cities (London, England)*, *47*, 62–72. doi:10.1016/j.cities.2015.03.013

Congeevaram, S., Dhanarani, S., Park, J., Dexilin, M., & Thamaraiselvi, K. (2007). Biosorption of chromium and nickel by heavy metal resistant fungal and bacterial isolates. *Journal of Hazardous Materials*, *146*, 270–277. PMID:17218056

Contreras-Lisperguer, R., Batuecas, E., Mayo, C., Díaz, R., Pérez, F., & Springer, C. (2018). Sustainability assessment of electricity cogeneration from sugarcane bagasse in Jamaica. *Journal of Cleaner Production*, *200*, 390–401. doi:10.1016/j.jclepro.2018.07.322

Cooper, T. (2000). WEEE, WEEE, WEEE, WEEE, all the way home? An evaluation of proposed electrical and electronic waste legislation. *European Environment*, *10*(3), 121–130. doi:10.1002/1099-0976(200005/06)10:3<121::AID-EET226>3.0.CO;2-N

Corral-Verdugo, V. (1997). Dual 'realities' of conservation behavior: self-reports vs observations of re-use and recycling behavior. *Journal of Environmental Psychology*, *17*(2), 135–145. doi:10.1006/jevp.1997.0048

CPCB. (2000). *Status of solid waste generation, collection, treatment and disposal in metero cities* (Series:CUPS/46/1999-2000). CPCB.

Craggs, R. J., Heubeck, S., Lundquist, T. J., & Benemann, J. R. (2011). Algal biofuels from wastewater treatment high rate algal ponds. *Water Science and Technology*, *63*(4), 660–665. doi:10.2166/wst.2011.100 PMID:21330711

Cremlyn, R. J. (1996). *An introduction to organosulfur chemistry*. Wiley.

Crini, G., & Badot, P. M. (2008). Application of chitosan, a natural aminopolysaccharide, for dye removal from aqueous solutions by adsorption processes using batch studies: A review of recent literature. *Progress in Polymer Science (Oxford)*, *33*(4), 399–447. doi:10.1016/j.progpolymsci.2007.11.001

Cruz-Pinto, J. J. C., Carvalho, M. E. S., & Ferreira, J. F. A. (1994). The kinetics and mechanism of polyethylene photo-oxidation. *Die Angewandte Makromolekulare Chemie: Applied Macromolecular Chemistry and Physics*, *216*(1), 113–133. doi:10.1002/apmc.1994.052160108

Cui, J., & Forssberg, E. (2003). Mechanical recycling of waste electric and electronic equipment: A review. *Journal of Hazardous Materials*, *99*(3), 243–263. doi:10.1016/S0304-3894(03)00061-X PMID:12758010

Cui, J., & Forssberg, E. (2007). Characterization of shredded television scrap and implications for materials recovery. *Waste Management (New York, N.Y.)*, *27*(3), 415–424. doi:10.1016/j.wasman.2006.02.003 PMID:16624540

Cui, Y. J., Zhu, Y. G., Zhai, R. H., Chen, D. Y., Huang, Y. Z., Qui, Y., & Liang, J. Z. (2004). Transfer of metals from near a smelter in Nanning, China. *Environment International*, *30*, 785–791. doi:10.1016/j.envint.2004.01.003 PMID:15120196

Dacera, D. D. M., Babel, S., & Parkpian, P. (2009). Potential for land application of contaminated sewage sludge treated with fermented liquid from pineapple wastes. *Journal of Hazardous Materials*, *167*(1-3), 866–872. doi:10.1016/j.jhazmat.2009.01.064 PMID:19232826

Dai, C., Duan, N., Dong, B., & Dai, L. (2013). High-solids anaerobic co-digestion of sewage sludge and food waste in comparison with mono-digestions: Stability and performance. *Waste Management (New York, N.Y.)*, *33*(2), 308–316. doi:10.1016/j.wasman.2012.10.018 PMID:23177568

Dakiky, M., Khamis, M., Manassra, A., & Mer'eb, M. (2002). Selective adsorption of Chromium(VI) in industrial wastewater using low-cost abundantly available adsorbents. *Advances in Environmental Research*, *6*(4), 533–540. doi:10.1016/S1093-0191(01)00079-X

Danko, A., Luo, M., Bagwell, C., Brigmon, R., & Freedman, D. (2004). Involvement of linear plasmids in aerobic biodegradation of vinyl chloride. *Applied and Environmental Microbiology*, *70*(10), 6092–6097. doi:10.1128/AEM.70.10.6092-6097.2004 PMID:15466555

Daoji, L., & Daler, D. (2004). Ocean pollution from land-based sources: East China Sea, China. *Ambio*, *33*(1), 107–113. doi:10.1579/0044-7447-33.1.107 PMID:15083656

Dasgupta, A. M., David, D. J., & Misra, A. (1992). Synthesis and Characterization of Ionomeric Poly(Vinyl Butyral). *Journal of Applied Polymer Science*, *44*(7), 1213–1221. doi:10.1002/app.1992.070440711

Daskalopoulos, E., Badr, O., & Probert, S. (1998). An integrated approach to municipal solid waste management. *Resources, Conservation and Recycling*, *24*(1), 33–50. doi:10.1016/S0921-3449(98)00031-7

Day, J. G. (1984). *U.S. Patent No. 4,427,442*. Washington, DC: U.S. Patent and Trademark Office.

De Freitas Borghi, D., Guirardello, R., & Cardozo Filho, L. (2009). Storage logistics of fruits and vegetables: Effect of temperature. *Chemical Engineering Transactions*, *17*, 951–956.

de la Noue, J., & de Pauw, N. (1988). The potential of microalgal biotechnology: A review of production and uses of microalgae. *Biotechnology Advances*, *6*(4), 725–770. doi:10.1016/0734-9750(88)91921-0 PMID:14550037

De Ley, J., Gillis, M., & Swings, J. (1984). *Baltimore: Williams and Wilkins Baltimore*. Bergey's Manual of Systematic Bacteriology.

DEA. (2019). *National waste management strategy*. Available online: https://www.environment.gov.za/documents/strategicdocuments/wastemanagement

DEAT. (1999). *National waste management strategy*. Available Online: http://www.deat.gov.za/ProjProg/WasteMgmt/waste.html

DEAT. (2007). Assessment of the status of waste service delivery and capacity at the local government level. DEAT: General waste management directorate and Palmer Development group, Pretoria, SA.

del Mar Morales-Amaral, M., Gómez-Serrano, C., Acién, F. G., Fernández-Sevilla, J. M., & Molina-Grima, E. (2015). Outdoor production of Scenedesmus sp. in thin-layer and raceway reactors using centrate from anaerobic digestion as the sole nutrient source. *Algal Research*, *12*, 99–108. doi:10.1016/j.algal.2015.08.020

Delgenès, J. P., Penaud, V., & Moletta, R. (2003). Pretreatments for the enhancement of anaerobic digestion of solid waste. In J. Mata-Alvarez (Ed.), Bio-methanization of the Organic Fraction of Municipal Solid Waste. IWA Publishing. doi:10.1002/chin.200313271

Demirbas, A., & Demirbas, M. F. (2011). Importance of algae oil as a source of biodiesel. *Energy Conversion and Management*, *52*(1), 163–170. doi:10.1016/j.enconman.2010.06.055

Department of Economic Development and Environmental Affairs (DEDEA). (2009). *Draft integrated general waste management plan*. GIBB Engineering, Province of the Eastern Cape.

Department of Economic Development, Environmental Affairs and Tourism (DEDEAT). (2013). *2013-14 Policy Statement: Honourable MEC Mcebisi Jonas, Eastern Cape Provincial Government.* Available Online: http://www.dedea.gov.za/Speeches/Policy%20Speech%202013%20-%20MEC%20Jonas.pdf

Department of Environmental Affairs (DEA). (2016). *Report on the Determination of the Extent and Role of Waste Picking in SA; DEA: Pretoria, SA, 2016*. Available online: http://sawic.environment.gov.za/documents/5413.pdf

Department of Environmental Affairs and Technology (DEAT). (2008). The capacity audit and needs analysis survey for environmental impact administrators. Department of Environmental Affairs and Tourism. Undertaken by Regenesys, Pretoria, SA.

Department of Statistics Malaysia. (2017). *Small and medium enterprises gross domestic product (SMEs GDP) 2016*. Author.

Derraik, J. (2002). The pollution of the marine environment by plastic debris: A review. *Marine Pollution Bulletin*, *44*(9), 842–852. doi:10.1016/S0025-326X(02)00220-5 PMID:12405208

De, S., & Debnath, B. (2016). Prevalence of health hazards associated with solid waste disposal-A case study of kolkata, India. *Procedia Environmental Sciences*, *35*, 201–208. doi:10.1016/j.proenv.2016.07.081

Devendra, C., & Leng, R. A. (2011). Feed resources for animals in Asia: Issues, strategies for use, intensification and integration for increased productivity. *Asian-Australasian Journal of Animal Sciences*, *24*(3), 303–321. doi:10.5713/ajas.2011.r.05

Dewil, R., Appels, L., Baeyens, J., & Degrève, J. (2007). Peroxidation enhances the biogas production in the anaerobic digestion of biosolids. *Journal of Hazardous Materials*, *146*(3), 577–581. doi:10.1016/j.jhazmat.2007.04.059 PMID:17532118

Dhir, B. (2013). *Phytoremediation: Role of Aquatic Plants in Environmental Clean-Up*. Springer. doi:10.1007/978-81-322-1307-9

Dieden, M. (2007). Economics of pulp and paper industry (2nd ed.). Espoo: Puunjalostusinsinöörit ry.

Dixit, R., Wasiullah Malaviya, D., Pandiyan, K., Singh, U. B., Sahu, A., Shukla, R., Singh, B. P., Rai, J. P., Sharma, P. K., Lade, H., & Paul, D. (2015). Bioremediation of heavy metals from soil and aquatic environment: An overview of principles and criteria of fundamental processes. *Sustainability*, *7*, 2189–2212.

Dlamini, B., Rampedi, I., & Ifegbesan, A. (2017). Community resident's opinions and perceptions on the effectiveness of waste management and recycling potential in the Umkhanyakude and Zululand district municipalities in the KwaZulu-Natal province of SA. *Sustainability*, *9*(10), 1–19. doi:10.3390u9101835

Dlamini, S., & Simatele, D. (2016). Unrecognized informal solid waste recycling in an emerging African megacity: A study of Johannesburg, SA. *WIT Transactions on Ecology and the Environment*, *202*, 1–3.

Dlamini, S., Simatele, M., & Kubanza, N. (2019). Municipal solid waste management in South Africa: From waste to energy recovery through waste-to-energy technologies in Johannesburg. *Local Environment*, *24*(3), 249–257. doi:10.1080/13549839.2018.1561656

Do Nascimento, C. W. A., & Xing, B. S. (2006). Phytoextraction: A review on enhanced metal availability and plant accumulation. *Scientia Agrícola*, *63*(3), 299–311. doi:10.1590/S0103-90162006000300014

Dong, S., Zhang, X., He, F., Dong, S., & Wang, B. (2014). *Visible-light photocatalytic degradation of methyl orange over spherical activated carbon-supported and Er 3 + : YAlO 3 -doped TiO 2 in a fluidized bed*. doi:10.1002/jctb.4391

Donoso-Bravo, A., Ortega-Martinez, E., & Ruiz-Filippi, G. (2016). Impact of milling, enzyme addition, and steam explosion on the solid waste biomethanation of an olive oil production plant. *Bioprocess and Biosystems Engineering*, *39*(2), 331–340. doi:10.100700449-015-1519-z PMID:26670779

Dookhi, A. (2016). Engineering compliance- East London regional waste disposal site- A case study. *Proceedings of the 23rd WasteCon Conference*. Available Online: http://www.infrastructurene.ws/wp-content/uploads/sites/4/2017/10/17.-Dookhi-AS.pdf

Dookhi, N., Jewskiewitz, S., & Jewaskiewitz, B. (2015). East London Regional (Roundhill) waste disposal site- wasting no time on the road to compliance. *Civil Engineering (New York, N.Y.)*, *23*, 35–41.

Doran, M. D., Dee, P. E., & Techknowledge, M. D. (2008). *Wastewater and Wastewater Treatment Very Basics*. Academic Press.

Dowaki, K., & Mori, S. (2005). Biomass energy used in a sawmill. *Applied Energy*, *80*(3), 327–339. doi:10.1016/j.apenergy.2004.04.006

Driscoll, C. T., Whitall, D., Aber, J., Boyer, E., Castro, M., Cronan, C., ... Lawrence, G. (2003). Nitrogen pollution in the northeastern United States: Sources, effects, and management options. *Bioscience*, *53*(4), 357–374. doi:10.1641/0006-3568(2003)053[0357:NPITNU]2.0.CO;2

Drzyzga, O., & Prieto, A. (2018). Plastic waste management, a matter for the community. *Microbial Biotechnology*, *0*(0), 1–3. PMID:30411497

Dubey & Swain. (2019). Wastewater use in agriculture: issues and prospects. *Agriculture & Food, 1*(10).

Dubinin, M. M., Zaverina, E. D., & Radushkevich, L. V. (1947). Sorption and structure of active carbons I. Adsorption of organic vapors. *Журнал физической химии*, *21*, 1351–1362.

Dunn, W. E., Jr., Carda, D. D., & Storbeck, T. A. (1991). U.*S. Patent No. 5,004,500*. Washington, DC: U.S. Patent and Trademark Office.

Dushenkov, V., Kumar, P. N., Motto, H., & Raskin, I. (1995). Rhizofiltration: The use of plants to remove heavy metals from aqueous streams. *Environmental Science & Technology*, *29*(5), 1239–1245. doi:10.1021/es00005a015 PMID:22192017

Dutta, D., & Goel, S. (2017). Electronic Waste (E-Waste) Generation and Management. In *Advances in Solid and Hazardous Waste Management* (pp. 249–266). Springer. doi:10.1007/978-3-319-57076-1_12

Eckenfelder, W., & Ford, D. L. (2009). *Industrial water quality* (4th ed.). McGraw-Hill.

Edewor, J. O. (2000). *A Comparison of Treatment Methods for Palm Oil Mill Effluent (POME) Wastes*. Academic Press.

Eichhorn, S. J., Dufresne, A., Aranguren, M., Marcovich, N. E., Capadona, J. R., Rowan, S. J., Werder, C., Thielemans, W., Roman, M., Renneckar, S., Gindl, W., Veigel, S., Keckes, S., Yano, H., & Abe, K. (2010). Review: Current international research into cellulose nanofibers and nanocomposites. *Journal of Materials Science*, *45*(1), 1–33. doi:10.100710853-009-3874-0

El-Gendy, N. S., Hamdy, A., Fatthallah, N. A., & Omran, B. A. (2016). Recycling of some domestic wastes to produce green corrosion inhibitors and biocides for sulfate reducing bacteria. *Energy Sources. Part A, Recovery, Utilization, and Environmental Effects*, *38*(24), 3722–3732. doi:10.1080/15567036.2016.1179364

Elias, S. H., Mohamed, M., Nor-Anuar, A., Muda, K., Hassan, M. A. H. M., Othman, M. N., & Chelliapan, S. (2014). Ceramic industry wastewater treatment by rhizofiltration system-application of water hyacinth bioremediation. *The IIOAB Journal*, *5*(1), 6.

Emmanuel, E., Perrodin, Y., Keck, G., Blanchard, J.-M., & Vermande, P. (2002). Effects of hospital wastewater on aquatic ecosystem. *Proceedings of the XXVIII Congreso Interamericano de Ingenieria Sanitaria y Ambiental*.

EP1639041 A1. (2004). *Polyvinylbutyral compositions and blends having enhanced surface properties and articles made therefrom*. European Patent Office Application.

EP2308919 A1. (2008). *Polyvinyl butyral recycling method*. European Patent Office Application.

EP2308919A1. (2011). *Polyvinyl Butyral Recycling Method.* WO2009/118426 (01.10.2009 Gazette 2009/40).

EPA. (1990). *Environmental Pollution Control Alternatives: Drinking Water Treatment for Small Communities.* Retrieved from https://cfpub.epa.gov/si/si_public_record_Report.cfm?Lab=NRMRL&dirEntryId=40845&CFID=9119927&CFTOKEN=70327879

Epstein, E. (1997). *The science of composting*. Technomic Publishing.

Eriksen, M., Lebreton, L., Carson, H., Thiel, M., Moore, C., Borerro, J., Galgani, F., Ryan, P. G., & Reisser, J. (2014). Plastic pollution in the world's oceans: More than 5 trillion plastic pieces weighing over 250,000 tons afloat at sea. *PLoS One*, *10*(12), e111913. doi:10.1371/journal.pone.0111913 PMID:25494041

Esenduran, G., Atasu, A., & Van Wassenhove, L. N. (2019). Valuable e-waste: Implications for extended producer responsibility. *IISE Transactions*, *51*(4), 382–396. doi:10.1080/24725854.2018.1515515

EU (European Union). (2008). Directive 2008/98/EC of the European Parliament and of the Council of 19 November 2008 on Waste and Repealing Certain Directives.

EU RoHS Directive 2002/95/EC. (2003). https://eur-lex.europa.eu/homepage.html

E-Waste Rajyasabha Report. (2011). https://rajyasabha.nic.in/rsnew/publication_electronic/EWaste_in_india.pdf/

E-Waste Recycling Facts and Figures. (2018). https://www.thebalance.com/e-waste-recycling-factsand-figures-2878189

Falconer, I. R., & Humpage, A. R. (2005). Health risk assessment of cyanobacterial (blue-green algal) toxins in drinking water. *International Journal of Environmental Research and Public Health*, *2*(1), 43–50. doi:10.3390/ijerph2005010043 PMID:16705800

Fantola, A. (1997). *Introduction to Solid Waste Management Engineering*. Bibis Press.

FAO. (2020). *FAOSTAT: Forestry Production and Trade*. Rome: UN.

Farraji, H. (2014). Wastewater treatment by phytoremediation methods. *Wastewater Engineering: Advanced Wastewater Treatment Systems, 194*.

Faryal, R., Sultan, A., Tahir, F., Ahmed, S., & Hameed, A. (2007). Biosorption of lead by indigenous fungal strains. *Pakistan Journal of Botany*, *39*, 615–622.

Farzana, R. R., Rajarao, R., & Sahajwalla, V. (2016). Characteristics of waste automotive glasses as silica resource in ferrosilicon synthesis. *Waste Management & Research*, *34*(2), 113–121. doi:10.1177/0734242X15617010 PMID:26628051

Farzana, R., & Sahajwalla, V. J. (2015). Novel Recycling to Transform Automotive Waste Glass and Plastics into SiC-Bearing Resource by Silica Reduction. *Journal of Sustainable Metallurgy*, *1*(1), 65–74. doi:10.100740831-014-0004-2

Favas, P. J., Pratas, J., Varun, M., D'Souza, R., & Paul, M. S. (2014). Phytoremediation of soils contaminated with metals and metalloids at mining areas: potential of native flora. *Environmental Risk Assessment of Soil Contamination, 3*, 485-516.

Favas, J., Monteiro, E., & Rouboa, A. (2017). Hydrogen production using plasma gasification with steam injection. *International Journal of Hydrogen Energy*, *42*(16), 10997–11005. doi:10.1016/j.ijhydene.2017.03.109

Fdéz-Güelfo, A., Álvarez-Gallego, C., & Sales-Márquez, D., & Romero, G. L. I. (2011). Biological pre-treatment applied to industrial organic fraction of municipal solid waste (OFMSW): Effect on anaerobic digestion. *Chemical Engineering Journal*, *172*, 321–325. doi:10.1016/j.cej.2011.06.010

Fdez-Güelfo, A., Álvarez-Gallego, C., Sales-Márquez, D., & Romero García, L. I. (2012). New parameters to determine the optimum pre-treatment for improving the biomethanization performances. *Chemical Engineering Journal, 198–199*, 81–86. doi:10.1016/j.cej.2012.05.077

Fdez-Güelfo, L. A., Álvarez-Gallego, C., Sales Márquez, D., & Romero García, L. I. (2011). The effect of different pretreatments on biomethanation kinetics of industrial Organic Fraction of Municipal Solid Waste (OFMSW). *Chemical Engineering Journal, 171*, 411–417. doi:10.1016/j.cej.2011.03.095

Felix, A. L., Narciso, C. R. P., Lima, F. S., Mendes, R. F., Mendes, L. M., & Scolforo, J. R. S. (2016). Use of waste wood for particleboard production. *Key Engineering Materials, 668*, 375–380. doi:10.4028/www.scientific.net/KEM.668.375

Ferguson, I., La Fontarie, B., Vinden, P., Bren, L., Hateley, R., & Hermesec, B. (1996). *Environmental Properties of Timber*. Forest & Wood Products Research & Development Corporation.

Fernández-Sevilla, J. M., Fernández, F. A., & Grima, E. M. (2010). Biotechnological production of lutein and its applications. *Applied Microbiology and Biotechnology, 86*(1), 27–40. doi:10.100700253-009-2420-y PMID:20091305

Ferreira, A., Ribeiro, B., Marques, P. A., Ferreira, A. F., Dias, A. P., Pinheiro, H. M., Reis, A., & Gouveia, L. (2017). Scenedesmus obliquus mediated brewery wastewater remediation and CO_2 biofixation for green energy purposes. *Journal of Cleaner Production, 165*, 1316–1327. doi:10.1016/j.jclepro.2017.07.232

Ferrer, I., Ponsá, S., Vázquez, F., & Font, X. (2008). Increasing biogas production by thermal (70°C) sludge pre-treatment prior to thermophilic anaerobic digestion. *Biochemical Engineering Journal, 42*(2), 186–192. doi:10.1016/j.bej.2008.06.020

Fesseha, H. & Abebe, F. (2019). Degradation of plastic materials using microorganisms: A review. *Public Health Open Journal, 4*(2), 57-63.

Fiehn, H. & Ball, J. (2005). *Integrated waste management: background research paper produced for the SA Environment Outlook report on behalf of the Department of Environmental Affairs and Tourism*. Jarrod Ball and Associates.

Fobil, J., Armah, N., Hogarh, J., & Carboo, D. (2008). The influence of institutions and organizations on urban waste collection systems: An analysis of waste collection system in Accra, Ghana. *Journal of Environmental Management, 86*(1), 262–271. doi:10.1016/j.jenvman.2006.12.038 PMID:17291677

Fong, M. (2014). *Treatment and decolorization of biologically treated Palm Oil Mill Ef fl uent (POME) using banana peel as novel biosorbent*. doi:10.1016/j.jenvman.2013.11.031

Fortaleza, A. P., Filho, J. J. P. N., Silva, R. P. S., Barros, D. S., & Silva, S. S. (2019). Biomassa de espécies florestais para produção de carvão vegetal. *Ciência Florestal, 29*(3), 1436–1451. doi:10.5902/1980509831639

Fountoulakis, M. S., Drakopoulou, S., Terzakis, S., Georgaki, E., & Manios, T. (2008). Potential for methane production from typical Mediterranean agro-industrial by-products. *Biomass and Bioenergy, 32*(2), 155–161. doi:10.1016/j.biombioe.2007.09.002

Francois-Heude, A., Richard, E., Desnoux, E., & Colin, X. (2014). Influence of temperature, UV-light wavelength and intensity of polypropylene photo-thermal oxidation. *Polymer Degradation and Stabilization, 100*, 10–20. doi:10.1016/j.polymdegradstab.2013.12.038

Freundlich, H. (1906). Adsorption in solution. *Zeitschrift für Physikalische Chemie, 57*, 384–410.

Fu, B., Jiang, Q., Liu, H., & Liu, H. (2014). Occurrence and reactivation of viable but non-culturable E. coli in sewage sludge after mesophilic and thermophilic anaerobic digestion. *Biotechnology Letters, 36*(2), 273–279. doi:10.100710529-013-1361-9 PMID:24101245

Fu, F., & Wang, Q. (2011). Removal of heavy metal ions from wastewaters : A review. *Journal of Environmental Management, 92*(3), 407–418. doi:10.1016/j.jenvman.2010.11.011 PMID:21138785

Fujisawa, S., Okita, Y., Fukuzumi, H., Saito, T., & Isogai, A. (2011). Preparation and characterization of TEMPO-oxidized cellulose nanofibril films with free carboxyl groups. *Carbohydrate Polymers, 84*(1), 579–583. doi:10.1016/j.carbpol.2010.12.029

Fulke, A. B., Mudliar, S. N., Yadav, R., Shekh, A., Srinivasan, N., Ramanan, R., Krishnamurthi, K., Devi, S. S., & Chakrabarti, T. (2010). Bio-mitigation of CO_2, calcite formation and simultaneous biodiesel precursors production using *Chlorella* sp. *Bioresource Technology, 101*(21), 8473–8476. doi:10.1016/j.biortech.2010.06.012 PMID:20580227

Future Works Sustainability Consulting (FWSC). (2014). *Review of the Buffalo City Metropolitan Municipality integrated environmental management plan and coastal zone management plan: Draft state of the environment report.* FWSC.

Gajalakshmi, S., & Abbasi, S. A. (2008). Solid Waste Management by Composting: State of the Art. *Critical Reviews in Environmental Science and Technology, 38*(5), 311–400. doi:10.1080/10643380701413633

Gao, D.-W., Hu, Q., Yao, C., & Ren, N.-Q. (2014). Treatment of domestic wastewater by an integrated anaerobic fluidized-bed membrane bioreactor under moderate to low temperature conditions. *Bioresource Technology, 159*, 193–198. doi:10.1016/j.biortech.2014.02.086 PMID:24650533

Garcia, C., Hernandez, T., Costa, F., & Ayusho, M. (1992). Evaluation of maturity of municipal waste compost using simple chemical parameters. *Communications in Soil Science and Plant Analysis, 23*(13-14), 1501–1512. doi:10.1080/00103629209368683

Garlapati, V. K. (2016). E-waste in India and developed countries: Management, recycling, business and biotechnological initiatives. *Renewable & Sustainable Energy Reviews, 54*, 874–881. doi:10.1016/j.rser.2015.10.106

Gar, M., Tawfik, A., & Ookawara, S. (2015). Journal of Water Process Engineering Comparison of solar TiO 2 photocatalysis and solar photo-Fenton for treatment of pesticides industry wastewater : Operational conditions, kinetics, and costs. *Journal of Water Process Engineering, 8*, 55–63. doi:10.1016/j.jwpe.2015.09.007

Garrote, G., Domínguez, H., & Parajó, J. C. (1999). Hydrothermal processing of lignocellulosic materials. *Holz als Roh- und Werkstoff, 57*(3), 191–202. doi:10.1007001070050039

Gavrilescu, D. (2008). Energy from biomass in pulp and paper mills. *Environmental Engineering and Management Journal, 7*(5), 537–546. doi:10.30638/eemj.2008.077

Geyer, R., Jambeck, J., & Law, K. (2017). Production, use and fate of all plastics ever made. *Science Advances, 3*(7), 1–5. doi:10.1126ciadv.1700782 PMID:28776036

Ghorbanpour, M., Asgari Lajayer, H., & Hadian, J. (2016). Influence of copper and zinc on growth, metal accumulation and chemical composition of essential oils in sweet Basil (*Ocimum basilicum* L.). *Faslnamah-i Giyahan-i Daruyi, 15*(59), 132–144.

Ghosh, S., Pal, S., & Ray, S. (2013). Study of microbes having potentiality for biodegradation of plastics. *Environmental Science and Pollution Research International, 20*(7), 4339–4355. doi:10.100711356-013-1706-x PMID:23613206

Gibson, R. W., Goodman, P. D., Holt, L., Dalrymple, I. M., & Fray, D. J. (2003). *U.S. Patent No. 6,641,712*. Washington, DC: U.S. Patent and Trademark Office.

Gillet, S., Aguedo, M., Petitjen, L., Morais, A. R. C., Lopes, A. M. C., Łukasik, R. M., & Anastas, P. T. (2017). Lignin transformations for high value applications: Towards targeted modifications using green chemistry. *Green Chemistry, 19*(18), 4200–4233. doi:10.1039/C7GC01479A

Glass, J. E., & Swift, G. (1989). Agricultural and synthetic polymers, biodegradation and utilization. In ACS symposium series (Vol. 433, pp. 9-64). Academic Press.

Glavac, N. K., Djogo, S., Razic, S., Kreft, S., & Veber, M. (2017). Accumulation of heavy metals from soil in medicinal plants. *Arhiv za Higijenu Rada i Toksikologiju*, *68*(3), 236–244. doi:10.1515/aiht-2017-68-2990 PMID:28976884

Godfrey, L. (2019). Waste plastic, the challenge facing developing countries-ban it, change it, collect it? *Recycling*, *4*(3), 1–6. doi:10.3390/recycling4010003

Godfrey, L., & Oelofse, S. (2017). Historical Review of Waste Management and Recycling in South Africa. *Resources*, *6*(4), 57. doi:10.3390/resources6040057

Goel, M., Chovelon, J. M., Ferronato, C., Bayard, R., & Sreekrishnan, T. R. (2010). The remediation of wastewater containing 4-chlorophenol using integrated photocatalytic and biological treatment. *Journal of Photochemistry and Photobiology. B, Biology*, *98*(1), 1–6. doi:10.1016/j.jphotobiol.2009.09.006 PMID:19914843

Gogoi, B. K., Dutta, N. N., Goswami, P., & Krishna Mohen, T. R. (2002). A Case Study of Petroleum hydrocarbon Contaminated Soil at a Crude Oil Site. *Advances in Environmental Research*, *7*(4), 767–782.

Goh, C. S., Tan, K. T., Lee, K. T., & Bhatia, S. (2010). Bioresource technology bio-ethanol from lignocellulose : Status, perspectives and challenges in Malaysia. *Bioresource Technology*, *101*(13), 4834–4841. doi:10.1016/j.biortech.2009.08.080 PMID:19762229

Goldblatt, M. (2009). *Research paper: Macro-economic trends and economic instruments*. Department of Environmental Affairs, Pretoria.

Gomide, J. L., Neto, H. F., & Regazzi, A. J. (2005, February). Caracterização tecnológica, para a produção de celulose, da nova geração de clones de *Eucalyptus* do Brasil. *Revista Árvore*, *29*(1), 129–137. doi:10.1590/S0100-67622005000100014

Gonzáles, A. F. G. & Bustillos, L. G. T. (2019). *Biorrefinerias y Economia Circular*. Ciudad de México: UAM.

Gouvêa, A. F. G., Carvalho, A. M. M. L., & Carneiro, A. C. O. (2018). *Extração da lignina kraft de eucalipto e uso para energia*. Appris.

Grand View Research. (2015). *Synthetic Fibers Market Analysis, By Type (Acrylics, Polyester, Nylon, Polyolefin), By Application (Clothing, Home Furnishing, Automotive, Filtration), By Region, And Segment Forecasts, 2014 – 2025*. https://www.grandviewresearch.com/industry-analysis/synthetic-fibers-market

Greene, J. (2007). Biodegradation of compostable plastics in green yard-waste compost environment. *Journal of Polymers and the Environment*, *15*(4), 269–273. doi:10.100710924-007-0068-1

Gregory, M. R. (2009). Environmental implications of plastic debris in marine settings—Entanglement, ingestion, smothering, hangers-on, hitch-hiking and alien invasions. *Philosophical Transactions of the Royal Society of London. Series B, Biological Sciences*, *364*(1526), 2013–2025. doi:10.1098/rstb.2008.0265 PMID:19528053

Gu, J. D., Ford, T. E., Mitton, D. B., & Mitchell, R. 2000. Microbial degradation and deterioration of polymeric materials. In The Uhlig Corrosion Handbook (2nd ed.). New York: Wiley.

Guerrero, M., Re, M., Kazimierski, L. D., Menéndez, Á. N., & Ugarelli, R. (2013). Effect of climate change on navigation channel dredging of the Parana River. *International Journal of River Basin Management, 11*(4), 439-448.

Guidi, W., Palm, E., Mancuso, S., & Azzarello, E. (2018). Trace element phytoextraction from contaminated soil. a case study under Mediterranean climate. *Environmental Science and Pollution Research International*, *18*, 1197–1205.

Gumbi, S. (2016). *Current waste management and minimisation patterns and practices: an exploratory study on the Ekurhuleni metropolitan municipality in South Africa* (MSc Thesis). University of South Africa, RSA.

Gumbo, T., & Simelane, T. (2015). Innovations in municipal solid waste management: Experiences from eThekwini municipality, SA. In T. Simelane & R. Mohee (Eds.), *Future directions of municipal solid waste management in Africa* (pp. 202–217). Africa Institute of SA. doi:10.2307/j.ctvh8r2sj.18

Gunaseelan, V. N. (1997). Anaerobic digestion of biomass for methane production: A review. *Bio Bioener, 1997*(13), 83–114. doi:10.1016/S0961-9534(97)00020-2

Gupta, P. L., Lee, S. M., & Choi, H. J. (2015). A mini review: Photobioreactors for large scale algal cultivation. *World Journal of Microbiology & Biotechnology, 31*(9), 1409–1417. doi:10.100711274-015-1892-4 PMID:26085485

Gupta, S. M., & McLean, C. R. (1996). Disassembly of products. *Computers & Industrial Engineering, 31*(1-2), 225–228. doi:10.1016/0360-8352(96)00146-5

Gupta, V. K., & Ali, I. (2004). Removal of lead and chromium from wastewater using bagasse fly ash—A sugar industry waste. *Journal of Colloid and Interface Science, 271*(2), 321–328. doi:10.1016/j.jcis.2003.11.007 PMID:14972608

Gutberlet, J. (2018). *Waste in the city: Challenges and opportunities for urban agglomerations, urban agglomeration Mustafa Ergen*. IntechOpen. Retrieved 09 July 2020 from: https://www.intechopen.com/books/urban-agglomera

Gwada, B., Ogendi, G., & Trott, S. (2018). Composition of plastic waste discarded by household and its management approaches. *Global Journal of Environmental Science and Management, 5*(1), 83–94.

Hadad, D., Geresh, S., & Sivan, A. (2005). Biodegradation of polyethylene by the thermophilic bacterium Brevibacillus borstelensis. *Journal of Applied Microbiology, 98*(5), 1093–1100. doi:10.1111/j.1365-2672.2005.02553.x PMID:15836478

Hageluken, C. (2006, May). Improving metal returns and eco-efficiency in electronics recycling-a holistic approach for interface optimisation between pre-processing and integrated metals smelting and refining. In *Proceedings of the 2006 IEEE International Symposium on Electronics and the Environment*, 2006 (pp. 218-223). IEEE. 10.1109/ISEE.2006.1650064

Hageluken, C., & Art, S. (2007). *Recycling of e-scrap in a global environment: opportunities and challenges. Tackling e-waste towards efficient management techniques*. TERI Press.

Hailu, S., Admassu, S., & Jha, Y. K. (2012). *Vinegar Production Technology–An Overview*. Beverage & Food World.

Halden, R. (2010). Plastics and health risks. *Annual Review of Public Health, 31*(1), 179–194. doi:10.1146/annurev.publhealth.012809.103714 PMID:20070188

Hamlet, C., Matte, T. & Mehta, S. (2018). *Combating plastic air pollution on earth's day. Vital strategies environmental health division*. Academic Press.

Ha, N. N., Agusa, T., Ramu, K., Tu, N. P. C., Murata, S., Bulbule, K. A., Parthasaraty, P., Takahashi, S., Subramanian, A., & Tanabe, S. (2009). Contamination by trace elements at e-waste recycling sites in Bangalore, India. *Chemosphere, 76*(1), 9–15. doi:10.1016/j.chemosphere.2009.02.056 PMID:19345395

Hankin, L. H., Poincelot, R. P., & Anagnostakis, S. L. (1976). Microorganisms from composting leaves: Ability to produce extracellular enzymes. *Microbial Ecology, 2*(4), 296–308. doi:10.1007/BF02011649 PMID:24241392

Han, R., Zhang, J., Xiao, W. Z. H., Shi, J., & Liu, H. (2006). Biosorption of copper(II) and lead(II) from aqueous solution by chaff in a fixed-bed column. *Journal of Hazardous Materials, 133*(1-3), 262–268. doi:10.1016/j.jhazmat.2005.10.019 PMID:16298055

Hansted, A. L. S., Nakashima, G. T., Martins, M. P., Yamamoto, H., & Yamaji, F. M. (2016). Comparative analyses of fast-growing species in different moisture content for high quality solid fuel production. *Fuel*, *184*, 180–184. doi:10.1016/j.fuel.2016.06.071

Häring, V., Manka'abusi, D., Akoto-Danso, E. K., Werner, S., Atiah, K., Steiner, C., Lompo, D. J. P., Adiku, S., Buerkert, A., & Marschner, B. (2017). Effects of biochar, wastewater irrigation and fertilization on soil properties in West African urban agriculture. *Scientific Reports*, *7*(1), 10738. doi:10.103841598-017-10718-y PMID:28878251

Hasana, S. H., Srivastavaa, P., & Talatb, M. (2010). Biosorption of lead using immobilized *Aeromonas hydrophila* biomass in up flow column system: Factorial design for process optimization. *Journal of Hazardous Materials*, *177*(1-3), 312–322. doi:10.1016/j.jhazmat.2009.12.034 PMID:20036460

Hashemi, V. H., Ghanadi, A., & Sharif, B. (2003). Anti-inflammatory and analgesic effects of *Coriandrum sativum* L. in animal models. *Shahrekord University of Medical Sciences Journal*, *5*(2), 8–15.

Hashim, T. A., Abbas, H. H., Farid, I. M., El-Husseiny, O. H. M., & Abbas, M. H. H. (2017). Accumulation of some heavy metals in plants and soils adjacent to Cario-Alexandria agricultural highway. *Egyptian. Journal of Soil Science*, *57*(2), 215–232.

Hassan, M. A., Yacob, S., Shirai, Y., & Hung, Y. (2005). Treatment of Palm Oil Wastewaters. *Waste Treatment in the Food Processing Industry*, 101–117. doi:10.1201/9781420037128.ch4

Hassan, M. A., & Abd-aziz, S. (2011). *Waste and Environmental Management in the Malaysian Palm Oil Industry. Palm Oil: Production, Processing, Characterization, and Uses*. AOCS Press. doi:10.1016/B978-0-9818936-9-3.50026-5

Hassan, M. N., Rahman, R. A., Chong, T. L., Zakaria, Z., & Awang, M. (2000). Waste recycling in Malaysia: Problems and prospects. *Waste Management & Research*, *18*(4), 320–328. doi:10.1034/j.1399-3070.2000.00133.x

Hassan, S. A., Jameel, N. G. M., & Şekeroğlu, B. (2016). Smart solid waste monitoring and collection system. *International Journal (Toronto, Ont.)*, *6*(10).

Hassan, S. H., Abd el Nasser, A. Z., & Kassim, R. M. (2019). Electricity generation from sugarcane molasses using microbial fuel cell technologies. *Energy*, *178*, 538–543. doi:10.1016/j.energy.2019.04.087

Haug, R. T., Stuckey, D. C., Gossett, J. M., & Mac Carty, P. L. (1978). Effect of thermal pre-treatment on digestibility and dewaterability of organic sludges. *Journal - Water Pollution Control Federation*, *23*, 73–85.

Hawash, S. I., Farah, J. Y., & El-Diwani, G. (2017). Pyrolysis of agriculture wastes for bio-oil and char production. *Journal of Analytical and Applied Pyrolysis*, *124*, 369–372. doi:10.1016/j.jaap.2016.12.021

Hazra, T., & Goel, S. (2009). Solid waste management in Kolkata, India: Practices and challenges. *Waste Management (New York, N.Y.)*, *29*(1), 470–478. doi:10.1016/j.wasman.2008.01.023 PMID:18434129

He, L., Huang, G., Zeng, G., & Hu, H. (2009). Identifying optimal regional solid waste management strategies through an inexact integer programming model containing infinite objectives and constraints. *Journal of Waste Management*, *29*(1), 21–31. doi:10.1016/j.wasman.2008.02.003 PMID:18406594

Held, T., & Don, H. (2000). *In Situ* Remediation. *Biotechnology: An Indian Journal*, *11b*, 350–370.

Helfferich, F. (1962). *Ion Exchange*. McGraw-Hill.

Hellmann, B., Zelles, L., Palojarvi, A., & Bai, Q. (1997). Emission of climate-relevant trace gases and succession of microbial communities during open-windrow composting. *Applied and Environmental Microbiology*, *63*(3), 1011–1018. doi:10.1128/AEM.63.3.1011-1018.1997 PMID:16535535

Hemalatha, R., & Anbuselvi, S. (2013). Physicochemical constituents of pineapple pulp and waste. *Journal of Chemical and Pharmaceutical Research*, *5*(2), 240–242.

Hettiarachchi, H., Meegoda, J., & Ryu, S. (2018). Organic waste buyback as a viable method to enhance sustainable municipal solid waste management in developing countries. *International Journal of Environmental Research and Public Health*, *15*(11), 1–15. doi:10.3390/ijerph15112483 PMID:30405058

Hidalgo, C., Vegas, E., Mateo, W., Tesfaye, A. B., Cerezo, R. M., Callejón, M., Poblet, J. M., Guillamón, A., Mas, M. J., & Torija, M. J. (2010). Effect of barrel design and the inoculation of Acetobacter pasteurianus in wine vinegar production. *International Journal of Food Microbiology*, *141*(1-2), 56–62. doi:10.1016/j.ijfoodmicro.2010.04.018 PMID:20478638

Hirai, H., Takada, H., Ogata, Y., Yamashita, R., Mizukawa, K., Saha, M., Kwan, C., Moore, C., Gray, H., Laursen, D., Zettler, E. R., Farrington, J. W., Reddy, C. M., Peacock, E. E., & Ward, M. W. (2011). Organic micropollutants in marine plastic debris from the open ocean and remote and urban beaches. *Marine Pollution Bulletin*, *62*(8), 1683–1692. doi:10.1016/j.marpolbul.2011.06.004 PMID:21719036

Hlihor, R. M., Gavrilescu, M., Tavares, T., Fvier, L., & Olivieri, G. (2017). Editorial bioremediation: An overview of current practices, advances, and new perspectives in environmental pollution treatment Hindawi. *BioMed Research International*, *2*, 3–5.

Hoang, K., Tseng, M., & Shu, W. (2007). Degradation of polyethylene succinate (PES) by a new thermophilic Micro-bispora strain. *Biodegradation*, *18*(3), 333–342. doi:10.100710532-006-9067-5 PMID:17109189

Hoffmann, J. E. (1992). Recovering precious metals from electronic scrap. *JOM*, *44*(7), 43–48. doi:10.1007/BF03222275

Hohenthal, C., Leon, J., Dobon, A., Kujanpää, M., Meinl, G., Ringman, J., & Forsström, U. (2019). The ISO 14067 approach to open-loop recycling of paper products: Making it operational. *Journal of Cleaner Production*, *224*, 264–274. doi:10.1016/j.jclepro.2019.03.179

Honda, R., Teraoka, Y., Noguchi, M., & Yang, S. (2017). Optimization of hydraulic retention time and biomass concentration in microalgae biomass production from treated sewage with a membrane photobioreactor. *Journal of Water and Environment Technology*, *15*(1), 1–11. doi:10.2965/jwet.15-085

Horsfall, M. Jnr, & Spiff, A. (2005). Effects of temperature on the sorption of Pb^{2+} and Cd^{2+} from aqueous solution by *Caladium bicolor* (Wild Cocoyam) biomass. *Electronic Journal of Biotechnology*, *8*(2), 162–169. doi:10.2225/vol8-issue2-fulltext-4

Ho, S. H., Chan, M. C., Liu, C. C., Chen, C. Y., Lee, W. L., Lee, D. J., & Chang, J. S. (2014). Enhancing lutein productivity of an indigenous microalga *Scenedesmus obliquus* FSP-3 using light-related strategies. *Bioresource Technology*, *152*, 275–282. doi:10.1016/j.biortech.2013.11.031 PMID:24296122

Ho, S. H., Chen, C. Y., & Chang, J. S. (2012). Effect of light intensity and nitrogen starvation on CO_2 fixation and lipid/carbohydrate production of an indigenous microalga *Scenedesmus obliquus* CNW-N. *Bioresource Technology*, *113*, 244–252. doi:10.1016/j.biortech.2011.11.133 PMID:22209130

Ho, S. H., Chen, C. Y., Lee, D. J., & Chang, J. S. (2011). Perspectives on microalgal CO2-emission mitigation systems—A review. *Biotechnology Advances*, *29*(2), 189–198. doi:10.1016/j.biotechadv.2010.11.001 PMID:21094248

Ho, S. H., Li, P. J., Liu, C. C., & Chang, J. S. (2013). Bioprocess development on microalgae-based CO_2 fixation and bioethanol production using Scenedesmus obliquus CNW-N. *Bioresource Technology*, *145*, 142–149. doi:10.1016/j.biortech.2013.02.119 PMID:23566474

Hosseini Koupaie, E., Barrantes Leiva, M., Eskicioglu, C., & Dutil, C. (2014). Mesophilic batch anaerobic co-digestion of fruit-juice industrial waste and municipal waste sludge: Process and cost-benefit analysis. *Bioresource Technology, 152*, 66–73. doi:10.1016/j.biortech.2013.10.072 PMID:24280084

Howard, M. (2018). Plastic pollution of the world's seas and oceans as a contemporary challenge in ocean governance. *Nature Communications, 9*(1), 667. doi:10.103841467-018-03104-3 PMID:29445166

Ho, Y. S., & McKay, G. (2002). Application of kinetic models to the sorption of copper(II) onto peat. *Adsorption Science and Technology, 20*(8), 797–815. doi:10.1260/026361702321104282

Ho, Y. S., Mckay, G., Wase, D. A. J., & Foster, C. F. (2000). Study of the sorption of divalent metal on to peat. *Adsorption Science and Technology, 18*(7), 639–650. doi:10.1260/0263617001493693

Hrebicek, J., Kalina, J., Soukopova, J., Prasek, J., & Valta, J. (2017). *The forecasting waste generation model based on linked open data and the DPSIR framework: Case study concerning municipal waste in the Czech Republic.* Available Online: https://www.muni.cz/en/research/publications/1348999

Huang, J. C., Shetty, A. S., & Wang, M. S. (1990). Biodegradable plastics: A review. *Advances in Polymer Technology, 10*(1), 23–30. doi:10.1002/adv.1990.060100103

Huang, S. J., Roby, M. S., Macri, C. A., & Cameron, J. A. (1992). The effects of structure and morphology on the degradation of polymers with multiple groups. *Special Publication - Royal Society of Chemistry, 109*(1), 149–149.

Hu, J., Huang, Y., Yao, Y., Pan, G., Sun, J., Zeng, X., & Wong, C.-P. (2017). Polymer composite with improved thermal conductivity by constructing a hierarchically ordered three-dimensional interconnected network of BN. *ACS Applied Materials & Interfaces, 9*(15), 13544–13553. doi:10.1021/acsami.7b02410 PMID:28362080

Hu, J., Zhang, Q., & Lee, D.-J. (2018). Kraft lignin biorefinery: A perspective. *Bioresource Technology, 247*, 1181–1183. doi:10.1016/j.biortech.2017.08.169 PMID:28899675

Hussain, I., Raschid, L., Hanjra, M. A., Marikar, F., & Van Der Hoek, W. (2002). *Wastewater use in agriculture: Review of impacts and methodological issues in valuing impacts* (Vol. 37). IWMI.

Hussein, Al-Mayaly, & Khudeir. (2015). Isolation, Screening and Identification of Low-Density Polyethylene (LDPE) degrading bacteria from contaminated soil with plastic wastes. *Journal, 1*(4), 1–14.

Idris, A., Inanc, B., & Hassan, M. N. (2004). Overview of waste disposal and landfills/dumps in Asian countries. *Journal of Material Cycles and Waste Management, 6*(2), 104–110. doi:10.100710163-004-0117-y

Idris, N. A., Loh, S. K., Lau, H. L. N., Mustafa, E. M., Vello, V., Tan, C. Y., & Phang, S. M. (2017). Cultivation of microalgae in medium containing palm oil mill effluent and its conversion into biofuel. *Journal of Oil Palm Research, 29*(2), 291–299. doi:10.21894/jopr.2017.2902.13

Ihedioha, J. N., Ukoha, P. O., & Ekere, N. R. (2017). Ecological and human health risk assessment of heavy metal contamination in soil of a municipal solid waste dump in Uyo, Nigeria. *Environmental Geochemistry and Health, 39*(3), 497–515. doi:10.100710653-016-9830-4 PMID:27170184

Ilankoon, I. M. S. K., Ghorbani, Y., Chong, M. N., Herath, G., Moyo, T., & Petersen, J. (2018). E-waste in the international context–A review of trade flows, regulations, hazards, waste management strategies and technologies for value recovery. *Waste Management (New York, N.Y.), 82*, 258–275. doi:10.1016/j.wasman.2018.10.018 PMID:30509588

Ilyas, S., Anwar, M. A., Niazi, S. B., & Ghauri, M. A. (2007). Bioleaching of metals from electronic scrap by moderately thermophilic acidophilic bacteria. *Hydrometallurgy, 88*(1-4), 180–188. doi:10.1016/j.hydromet.2007.04.007

Infrastructure News. (2016). *Roundhill rehabilitation reaches new heights*. Available Online: http://www.infrastructurene.ws/2016/08/02/roundhill-rehabilitation-reaches-new-heights/

Inter-American Development Bank. (2003). *Economic instruments for solid waste management: global review and applications for Latin America and the Caribbean*. D.C. Inter-American Development Bank.

IPCC. (1996). *Intergovernmental panel on climate change*. http:/www.ipcc-nggip-.org.jp.

IPCC. (2007). *Intergovernmental panel on climate change*. https://www.ipcc.ch/pdf/assessment-report/ar4/wg1/ar4-wg1-spm.pdf

Iranzo, M., Sainz-Padro, I., Boluda, R., Sanchez, J., & Mormeneo, S. (2001). The Use of Microorganisms in Environmental Engineering. *Annals of Microbiology*, *51*, 135–143.

Iranzo, M., Sainz-Pardo, I., Boluda, R., Sanchez, J., & Mormeneo, S. (2001). The use of microorganisms in environmental remediation. *Annals of Microbiology*, *51*(2), 135–144.

IS 10500. (1992). *Drinking water specification (reaffirmed 1993)*. Retrieved from https://rwua.org.in/download/DRINKING%20WATER%20SPECIFICATION%20IS%2010500,%201992.pdf

Işıldar, A., Rene, E. R., van Hullebusch, E. D., & Lens, P. N. L. (2018). Electronic waste as a secondary source of critical metals: Management and recovery technologies. *Resources, Conservation and Recycling*, *135*, 296–312. doi:10.1016/j.resconrec.2017.07.031

Islam, M. T., & Huda, N. (2020). Assessing the recycling potential of "unregulated" e-waste in Australia. *Resources, Conservation and Recycling*, *152*, 104526. doi:10.1016/j.resconrec.2019.104526

Itelima, J., Onwuliri, F., Onwuliri, E., Onyimba, I., & Oforji, S. (2013). Bio-Ethanol production from banana, plantain and pineapple peels by simultaneous saccharification and fermentation process. *International Journal of Environmental Sciences and Development*, *4*, 213–216. doi:10.7763/IJESD.2013.V4.337

Iwakiri, S. (2005). *Painéis de Madeira reconstituída*. FUPEF.

Iwata, T. (2015). Biodegradable and bio-based polymers: Future prospects of eco-friendly plastics. *Angewandte Chemie International Edition*, *54*(11), 3210–3215. doi:10.1002/anie.201410770 PMID:25583677

Izumi, K., Okishio, Y., Nagao, N., Niwa, C., Yamamoto, S., & Toda, T. (2010). Effects of particle size on anaerobic digestion of food waste. *Int. Biodeterior. Biodegr*, *64*(7), 601–608. doi:10.1016/j.ibiod.2010.06.013

Jackowiak, D., Sfrigon, J. C., Ribeiro, T., Pauss, A., & Guiot, S. (2011). Enhancing solubilisation and methane production kinetic of switchgrass by microwave pre-treatment. *Bioresource Technology*, *102*(3), 3535–3540. doi:10.1016/j.biortech.2010.11.069 PMID:21147525

Jackson, M. L. (1958). *Soil Chemical Analysis*. Prentice Hall.

Jacob-Lopes, E., Scoparo, C. H. G., Queiroz, M. I., & Franco, T. T. (2010). Biotransformations of carbon dioxide in photobioreactors. *Energy Conversion and Management*, *51*(5), 894–900. doi:10.1016/j.enconman.2009.11.027

Jain, R., Ahmad, P., Gadgil, K., & Sarma, S. (2012). Heavy metal toxicity: Effect of plant growth, Biochemical parameters and metal accumulation by Brasica juncea. *L. Int.J. Plant Production.*, *3*(3), 65–75.

Jain, S., Jain, S., Wolf, I. T., Lee, J., & Tong, Y. W. (2015). A comprehensive review on operating parameters and different pre-treatment methodologies for anaerobic digestion of municipal solid waste. *Renewable & Sustainable Energy Reviews*, *52*, 142–154. doi:10.1016/j.rser.2015.07.091

Jambeck, J., Geyer, R., Wilcox, C., Siegler, T., Perryman, M., Andrady, A., Narayan, R., & Law, K. L. (2015). Plastic waste inputs from land into the ocean. *Science*, *347*(6223), 768–771. doi:10.1126cience.1260352 PMID:25678662

Jarvis, P., Jefferson, B., & Parsons, S. A. (2006). Floc structural characteristics using conventional coagulation for a high doc, low alkalinity surface water source. *Water Research*, *40*(14), 2727–2737. doi:10.1016/j.watres.2006.04.024 PMID:16765408

Jasmine Praveena, R., & And Estherlydia, D. (2014). Comparative Study of Photochemical Screening and Antioxidant Capacities of Vinegar Made from Peel and Fruit of Pineapple (Ananas Comosus L.). *Int. Pharm. Bio. Sci*, *5*(4), 394–403.

Jeong, H., Seong, C., Jang, T., & Park, S. (2016). Classification of wastewater reuse for agriculture: A case study in South Korea. *Irrigation and Drainage*, *65*, 76–85. doi:10.1002/ird.2053

Jewaskiewitz, S. & Dookhi, N. (2017). Roundhill restoration. *Sustainable Landfills,* 16-19.

Jiya, A., Ijah, U. J., Galadima, M., & Akpan, U. (2019). Isolation of Microorganisms Associated with Biodegradation of Household Domestic Wastes for Biogas Production in Niger State, Nigeria. *Journal of BioMedical Research and Clinical Practice*, *2*(1), 70–76. doi:10.46912/jbrcp.82

Jody, B. J., Daniels, E. J., Duranceau, C. M., Pomykala, J. A., & Spangenberger, J. S. (2010). End-of-Life Vehicle Recycling: State of the Art of Resource Recovery from Shredder Residue. Energy Systems Division, Argonne National Laboratory.

Jones, C. A., Langer, H. W., Anderson, K., McDermott, T. R., & Inskeep, W. P. (2000). Rates of microbially mediated arsenate reduction and solubilization. *Soil Science Society of America Journal*, *64*, 600–608.

Joshi, P. K., Swarup, A., Maheshwari, S., Kumar, R., & Singh, N. (2011). Bioremediation of heavy metals in liquid media through fungi isolated from contaminated sources. *Indian Journal of Microbiology*, *51*(4), 482–487. doi:10.100712088-011-0110-9 PMID:23024411

Joshi, R., & Ahmed, S. (2016). Status and challenges of municipal solid waste management in India: A review. *Cogent Environmental Science*, *2*(1), 1139434. doi:10.1080/23311843.2016.1139434

Julkapli, N. W., & Bagheri, S. (2016). Developments in nano-additives for paper industry. *Journal of Wood Science*, *62*(2), 117–130. doi:10.100710086-015-1532-5

Juneja, A., Ceballos, R. M., & Murthy, G. S. (2013). Effects of environmental factors and nutrient availability on the biochemical composition of algae for biofuels production: A review. *Energies*, *6*(9), 4607–4638. doi:10.3390/en6094607

Júnior, B. (2019a). Briquetagem dos resíduos dos processamentos mecânico e químico de *Pinus* spp. *Cadernos de Ciência & Tecnologia*, *36*(3), e26522. doi:10.35977/0104-1096.cct2019.v36.26522

Kabata-Pendias, A. (2011). *Trace Element in Soils and Plants* (4th ed.). Boka Raton, Taylor and Francis Group.

Kadere, T. T., Miamoto, T., Oniang'o, R. K., Kutima, P. M., & Njoroge, S. M. (2008). Isolation and identification of genera *Acetobacter* and *Gluconobacter* in coconut toddy (mnazi). *African Journal of Biotechnology*, *7*, 2963–2971.

Kadirvelu, K., & Namasivayam, C. (2003). Activated carbon from coconut coir pitch as metal adsorbent: Adsorption of Cd(II) from aqueous solution. *Advances in Environmental Research*, *7*(2), 471–478. doi:10.1016/S1093-0191(02)00018-7

Kalavathy, M. H., Karthikeyan, T., Rajgopal, S., & Miranda, L. R. (2005). Kinetic and isotherm studies of Cu(II) adsorption onto H_3PO_4-activated rubber wood sawdust. *Journal of Colloid and Interface Science*, *292*(2), 354–362. doi:10.1016/j.jcis.2005.05.087 PMID:16040040

Kale, G., Kijchavengkul, T., Auras, R., Rubino, M., Selke, S., & Singh, S. (2007). Compostability of bioplastic packaging materials: An overview. *Macromolecular Bioscience*, *7*(3), 255–277. doi:10.1002/mabi.200600168 PMID:17370278

Kalule, S. (n.d.). *A case study for developing a framework for ecological risk assessment for landfill sites in SA*. Available Online: https://www.academia.edu/4558844/EGM_826_assignment_4_Manuscript_Kitumba_Kalule

Kamath, R., Rentz, J. A., Schnoor, J. L., & Alvarez, P. J. J. (2007). *Phytoremediation of Hydrocarboncontaminated Soils; Principles and Applications*. USA University of Iowa.

Kao, C. Y., Chen, T. Y., Chang, Y. B., Chiu, T. W., Lin, H. Y., Chen, C. D., Chang, J. S., & Lin, C. S. (2014). Utilization of carbon dioxide in industrial flue gases for the cultivation of microalga Chlorella sp. *Bioresource Technology*, *166*, 485–493. doi:10.1016/j.biortech.2014.05.094 PMID:24950094

Kapetanois, E. G., Loizidu, M., & Vakanas, G. (1993). Compost production from Greek domestic refuse. *Bioresource Technology*, *43*(1), 13–16. doi:10.1016/0960-8524(93)90201-L

Karakus, K. (2016). Polycaprolactone (PCL) based polymer composites filled wheat straw flour. *Kastamonu University Journal of Forestry Faculty*, *16*(1), 264–268. doi:10.17475/kujff.03251

Karami, R., Hosseini, M., Mohammadpour, T., Ghorbani, A., Sadeghnia, H. R., Rakhshandeh, H., Vafaee, F., & Esmaeilizadeh, M. (2015). Effects of hydroalcoholic extract of *Coriandrum sativum* on oxidative damage in pentylenetetrazole-induced seizures in rats. *Iranian Journal of Neurology*, *14*(2), 59–66. PMID:26056549

Kargupta, W., Ganesh, A., & Mukherji, S. (2015). Estimation of carbon dioxide sequestration potential of microalgae grown in a batch photobioreactor. *Bioresource Technology*, *180*, 370–375. doi:10.1016/j.biortech.2015.01.017 PMID:25616748

Kasirajan, S., & Ngouajio, M. (2012). Polyethylene and biodegradable mulches for agricultural applications: A review. *Agronomy for Sustainable Development*, *32*(2), 501–529. doi:10.100713593-011-0068-3

Katisiimeh, M. W., Burger, K., & Mol, A. P. (2013). Informal waste collection and its co-existence with the formal waste sector: The case of Kampala, Uganda. *Habitat International*, *38*, 1–9. doi:10.1016/j.habitatint.2012.09.002

Kazuva, E., Zhang, J., Tong, Z., Si, A., & Na, L. (2018). The DPSIR model for environmental risk assessment of municipal solid waste in Dar es Salaam city, Tanzania. *International Journal of Environmental Research and Public Health*, *15*(8), 1–30. doi:10.3390/ijerph15081692 PMID:30096829

Ke'er, C. (2010). *Controlling municipal solid waste (MSW) over-generation with a "waste disposal fee" based on an efficient stakeholder partnership* (Msc Thesis). Lund University.

Keeley, J., Smith, A. D., Judd, S. J., & Jarvis, P. (2014). Reuse of recovered coagulants in water treatment : An investigation on the effect coagulant purity has on treatment performance. *Separation and Purification Technology*, *131*, 69–78. doi:10.1016/j.seppur.2014.04.033

Keller, F. A., Hamilton, J., & Nguyen, Q. A. (2003). Microbial pre-treatment of biomass. *Appl. Biotechnol. Bioeng*, *105*, 27–41.

Kernbaum, S. (2014). Industrial scale purification of PVB of windshield waste. *Eco-Innovation*. https://ec.europa.eu/environment/eco-innovation/projects

Khajuria, A. (2010). *Application of the DPSIR framework for municipal solid waste management in south Asian developing countries* (PhD Dissertation). Osaka University.

Khalekuzzaman, M., Alamgir, M., Islam, M. B., & Hasan, M. (2019). A simplistic approach of algal biofuels production from wastewater using a Hybrid Anaerobic Baffled Reactor and Photobioreactor (HABR-PBR) System. *PLoS One*, *14*(12), e0225458. doi:10.1371/journal.pone.0225458 PMID:31805078

Khalid, A., Arshad, M., Anjum, M., Mahmood, T., & Dawson, L. (2011). The anaerobic digestion of solid organic waste. *Waste Management (New York, N.Y.), 31*(8), 1737–1744. doi:10.1016/j.wasman.2011.03.021 PMID:21530224

Khalil, Z. I., Asker, M. M., El-Sayed, S., & Kobbia, I. A. (2010). Effect of pH on growth and biochemical responses of *Dunaliella bardawil* and *Chlorella ellipsoidea. World Journal of Microbiology & Biotechnology, 26*(7), 1225–1231. doi:10.100711274-009-0292-z PMID:24026927

Khan, M. S., Zaidi, A. A., Wani, A. P. A., & Mohammad, A. (2009). Role of plant growth promoting rhizobacteria in the remediation of metal contaminated soils. *Environmental Chemistry Letters, 7*, 1–19.

Khope, P. B., & Modak, J. P. (2013). *Design of experimental set-up for establishing empirical relationship for Chaff cutter energized by human powered flywheel motor*. Academic Press.

Khumalo, S. (2016). *Environmental impact of household solid waste disposal practices on plant growth in rural areas of KwaZulu-Natal: A case study of UThukela district municipality* (Msc. Dissertation). University of SA.

Kiddee, P., Naidu, R., & Wong, M. H. (2013). Electronic waste management approaches: An overview. *Waste Management, 33*(5), 1237-1250.

Kiely, G., Tayfur, G., Dolan, C., & Tanji, K. (1997). Physical and mathematical modelling of anaerobic digestion of organic wastes. *Water Research, 31*(3), 534–540. doi:10.1016/S0043-1354(96)00175-3

Kilic, S., & Kilic, M. (2017). Effects of cadmium-induced stress on essential oil production, morphology and physiology of Lemon balm (Melissa officinalis L., Lamiaceae). *Applied Ecology and Environmental Research, 15*(3), 1653–1669. doi:10.15666/aeer/1503_16531669

Kilmer, V. J., & Alexander, L. T. (1949). Soil texture methods of making mechanical analyses of soils. *Soil Science, 68*(1), 15–24. doi:10.1097/00010694-194907000-00003

Kim, D. H., & Oh, S. E. (2011). Continuous high-solids anaerobic co-digestion of organic solid wastes under mesophilic conditions. *Waste Management (New York, N.Y.), 31*(9-10), 1943–1948. doi:10.1016/j.wasman.2011.05.007 PMID:21684733

Kim, E. Y., Lee, J. C., Kim, B. S., Kim, M. S., & Jeong, J. (2007). Leaching behavior of nickel from waste multi-layer ceramic capacitors. *Hydrometallurgy, 86*(1-2), 89–95. doi:10.1016/j.hydromet.2006.11.007

Kim, H. W., Nam, J. Y., & Shin, H. S. (2011). A comparison study on the high-rate co-digestion of sewage sludge and food waste using a temperature-phased anaerobic sequencing batch reactor system. *BioresTechn, 102*(15), 7272–7279. doi:10.1016/j.biortech.2011.04.088 PMID:21600764

Kim, J., Kim, K., Ye, H., Lee, E., Shin, C., McCarty, P. L., & Bae, J. (2010). Anaerobic fluidized bed membrane bioreactor for wastewater treatment. *Environmental Science & Technology, 45*(2), 576–581. doi:10.1021/es1027103 PMID:21158433

Kim, J., Park, C., Kim, T. H., Lee, M., Kim, S., Kim, S. W., & Lee, J. (2003). Effect ofvarious pretreatmentsforenhancedanaerobicdigestionwithwasteactivatedsludge. *Journal of Bioscience and Bioengineering, 95*(3), 271–275. doi:10.1016/S1389-1723(03)80028-2 PMID:16233404

Kim, Y., Yeo, S., Kum, J., Song, H. G., & Choi, H. T. (2005). Cloning of a manganese peroxidase cDNA gene repressed by manganese in Trametes versicolor. *Journal of Microbiology (Seoul, Korea), 43*(6), 569–571. PMID:16410775

Klawpiyapamornkun, T., Bovonsombut, S., & Bovonsombut, S. (2015). Isolation and Characterization of Acetic acid Bacteria from Fruits and Fermented fruit juices for Vinegar Production. *Food and Applied Bioscience Journal, 3*(1), 30–38.

Kleijntjens, R. H., & Luyben, K. C. A. M. (2000). Bioremediation. *Biotechnology: An Indian Journal, 11b*, 329–347.

Klein, J. (2000). Possibilities, Limits and Future Developments of Soil Bioremediation. *Biotechnology: An Indian Journal, 11b*, 466–476.

Kliopova, I., Staniškis, J. K., Stunžėnas, E., & Jurovickaja, E. (2019). Bio-nutrient recycling with a novel integrated biodegradable waste management system for catering companies. *Journal of Cleaner Production, 209*, 116–125. doi:10.1016/j.jclepro.2018.10.185

Kogan, V. (2006). *Process for the recovery of precious metals from electronic scrap by hydrometallurgical technique.* International Patent, WO/2006/013568 (C22B 11/00), WIP Organization.

Koji, N., Mizutani, K., Zhang, R., Sugimoto, K., & Ogata, N. (2012). Poly(Vinyl Butyral)-Zirconia Hybrid Films Formed by Sol-Gel Process. *World Journal of Engineering, 9*(3), 233–238. doi:10.1260/1708-5284.9.3.233

Kollmann, F. F. P., & Côté, W. A. (1968). *Principles of wood science and technology.* Springer-Verlag. doi:10.1007/978-3-642-87928-9

Koning, M., Hupe, K., & Stegmann, R. (2000). Thermal Processes, Scrubbing/Extraction, Bioremediation and Disposal. *Biotechnology: An Indian Journal, 11b*, 306–317.

Konopka, A., Zakharova, T., Bischoff, M., Oliver, L., Nakatsu, C., & Turco, R. F. (1999). Microbial biomass and activity in lead-contaminated soil. *Applied and Environmental Microbiology, 65*, 2256–2259. PMID:10224032

Koptsik, G. N. (2014). Problems and prospects concerning the phytoremediation of heavy metal polluted soils: A review. *Eurasian Soil Science, 47*(9), 923–939. doi:10.1134/S1064229314090075

Kothari, R., Tyagi, V. V., & Pathak, A. (2010). Waste-to-energy: A way from renewable energy sources to sustainable development. *Renewable & Sustainable Energy Reviews, 14*(9), 3164–3170. doi:10.1016/j.rser.2010.05.005

Koushal, V., Sharma, R., Sharma, M., Sharma, R., & Sharma, V. (2014). Plastics: Issues Challenges and Remediation. *International Journal of Waste Resources, 4*, 134.

Kowalczyk, A., Chyc, M., Ryszka, P., & Latowski, D. (2016). *Achromobacter xylosoxidans* as a new microorganism strain colonizing high-density polyethylene as a key step to its biodegradation. *Environmental Science and Pollution Research International, 23*(11), 11349–11356. doi:10.100711356-016-6563-y PMID:27072033

Kowser, J., Aziz, M. G., & Uddin, M. B. (2015). Isolation and characterization of Acetobacter aceti from rotten papaya. *Bangladesh Agril.Univ., 13*(2), 299–306. doi:10.3329/jbau.v13i2.28802

Krishna, A. K., & Govil, P. K. (2007). Soil contamination due to heavy metals from an industrial area of Surat, Gujarat, Western India. *Environmental Monitoring and Assessment, 124*(1-3), 263–275. doi:10.100710661-006-9224-7 PMID:17058016

Krishnaveni, M., & Vidya, R. (2015). Growth of Indian automobile industry. *International Journal of Current Research and Academic Review, 3*(2), 110–118.

Kubanza, S. (2010). *Perception and issues of solid waste management in SA, Johannesburg case study* (Msc Thesis). University of Witwatersrand.

Kubo, I., Fujita, K. I., Kubo, A., Nihei, K. I., & Ogura, T. (2004). Antibacterial activity of coriander volatile compounds against *Salmonella choleraesuis. Journal of Agricultural and Food Chemistry, 52*(11), 3329–3332. doi:10.1021/jf0354186 PMID:15161192

Kubowicz, S., & Booth, A. (2017). Biodegradability of plastics: Challenges and misconceptions. *Environmental Science & Technology, 51*(21), 12058–12060. doi:10.1021/acs.est.7b04051 PMID:29022342

Kumar, I. M. S. (2015). *Design and development Of agricultural waste shredder machine*. Academic Press.

Kumar, A. (2019). Exploring young adults'e-waste recycling behaviour using an extended theory of planned behaviour model: A cross-cultural study. *Resources, Conservation and Recycling*, *141*, 378–389. doi:10.1016/j.resconrec.2018.10.013

Kumar, A., Holuszko, M., & Espinosa, D. C. R. (2017). E-waste: An overview on generation, collection, legislation and recycling practices. *Resources, Conservation and Recycling*, *122*, 32–42. doi:10.1016/j.resconrec.2017.01.018

Kumar, A., Samadder, S. R., Kumar, N., & Singh, C. (2018). Estimation of the generation rate of different types of plastic wastes and possible revenue recovery from informal recycling. *Waste Management (New York, N.Y.)*, *79*, 781–790. doi:10.1016/j.wasman.2018.08.045 PMID:30343811

Kumar, K., Banerjee, D., & Das, D. (2014). Carbon dioxide sequestration from industrial flue gas by *Chlorella sorokiniana*. *Bioresource Technology*, *152*, 225–233. doi:10.1016/j.biortech.2013.10.098 PMID:24292202

Kumar, K., Dasgupta, C. N., Nayak, B., Lindblad, P., & Das, D. (2011). Development of suitable photobioreactors for CO_2 sequestration addressing global warming using green algae and cyanobacteria. *Bioresource Technology*, *102*(8), 4945–4953. doi:10.1016/j.biortech.2011.01.054 PMID:21334885

Kumar, M., Ou, Y. L., & Lin, J. G. (2010). Co-composting of green waste and food waste at low C/N ratio. *Waste Management (New York, N.Y.)*, *30*(4), 602–609. doi:10.1016/j.wasman.2009.11.023 PMID:20034778

Kumar, P. S., Ramalingam, S., Kirupha, S. D., Murugesan, A., Vidhyadevi, T., & Sivanesan, S. (2011). Adsorption behavior of nickel(II) onto cashew nut shell: Equilibrium, thermodynamics, kinetics, mechanism and process design. *Chemical Engineering Journal*, *167*(1), 122–131. doi:10.1016/j.cej.2010.12.010

Kusch, S., Schäfer, W., & Kranert, M. (2011). Dry digestion of organic residues. *Integr Waste Manag*, *I*, 115–134.

Kyriakopoulos, G., Doulia, D., & Hourdakis, A. (n.d.). *International Journal of Environmental Effect of ionic strength and pH on the adsorption of selected herbicides on Amberlite*. doi:10.1080/03067310500247678

Kyrikou, I., & Briassoulis, D. (2007). Biodegradation of agricultural plastic films: A critical review. *Journal of Polymers and the Environment*, *15*(2), 125–150. doi:10.100710924-007-0053-8

Kyung-Min, K., Park, H.-W., Shim, G.-S., Jang, S.-W., Kim, H.-J., Chae, G.-S., Shin, S., & Lee, J.-H. (2020). Mechanical properties and decomposition performance of peelable coating containing UiO-66 catalyst and waterborne silane-terminated polyurethane dispersions. *Journal of Materials Science*, *55*(6), 2604–2617. doi:10.100710853-019-04184-2

Lagergren, S. (1898). Zur theorie der sogenannten adsorption geloster stoffe. Kungliga Svenska Vetenskapsakademiens. *Handlingar*, *24*, 1–39.

Lagha-Benamrouche, S., & Madani, K. (2013). Phenolic contents and antioxidant activity of orange varieties (*Citrus sinensis* L. and *Citrus aurantium* L.) cultivated in Algeria: Peels and leaves. *Industrial Crops and Products*, *55*, 723–730. doi:10.1016/j.indcrop.2013.07.048

Lai, S. M., & Lin, Y. C. (2006). Mechanical Properties of Nylon 6,6/Polyvinyl Butyral Blends. *Journal Polymer-Plastics Technology and Engineering*, *45*(3), 421–428. doi:10.1080/03602550600553978

Lakshmi, S., & Raj, A. (2017). A Review Study of E-Waste Management in India. *Asian Journal of Applied Science and Technology*, *1*, 33–36.

Lal, R. M., Nagpure, A. S., Luo, L., Tripathi, S. N., Ramaswami, A., Bergin, M. H., & Russell, A. G. (2016). Municipal solid waste and dung cake burning discoloring the Taj Mahal and human health impacts in Agra. *Environmental Research Letters*, *11*(10), 104009. doi:10.1088/1748-9326/11/10/104009

Lam, M. K., & Lee, K. T. (2011). Renewable and sustainable bioenergies production from palm oil mill effluent (POME): Win-win strategies toward better environmental protection. *Biotechnology Advances*, *29*(1), 124–141. doi:10.1016/j.biotechadv.2010.10.001 PMID:20940036

Langmuir, I. (1918). The adsorption of gases on plane surfaces of glass, mica and platinum. *Journal of the American Chemical Society*, *40*(9), 1361–1403. doi:10.1021/ja02242a004

Larsdotter, K. (2006). Wastewater treatment with microalgae-a literature review. *Vatten*, *62*(1), 31.

Latifahmad, A., Ismail, S., & Bhatia, S. (2003). *Water recycling from palm oil mill effluent (POME) using membrane technology*. Academic Press.

Lebreton, L., & Andrady, A. (2019). Future scenarios of global plastic waste generation and disposal. *Palgrave Communications*, *5*(6), 1–11. doi:10.105741599-018-0212-7

Lee, D. H., Behera, S. K., Kim, J., & Park, H. S. (2009). Methane production potential of leachate generated from Korean food waste recycling facilities: A lab scale study. *Waste Management (New York, N.Y.)*, *29*(2), 876–882. doi:10.1016/j.wasman.2008.06.033 PMID:18796348

Lees, E. J., Noble, B., Hewitt, R., & Parsons, S. A. (2001). *Respiration rate and sludge characteristics of an activated microbial biomass*. Academic Press.

Legarth, J. B., Alting, L., & Baldo, G. L. (1995, May). Sustainability issues in circuit board recycling. In *Proceedings of the 1995 IEEE International Symposium on Electronics and the Environment ISEE* (Cat. No. 95CH35718) (pp. 126-131). IEEE. 10.1109/ISEE.1995.514963

Lehner, T. (2003, May). E&HS aspects on metal recovery from electronic scrap. In *IEEE International Symposium on Electronics and the Environment*, 2003 (pp. 318-322). IEEE. 10.1109/ISEE.2003.1208097

Leite, C., & Pereira, H. (2005). Cork – Containing barks – A Review. *Frontiers in Materials*, *3*, e63.

Lengowski, E. C., Bonfatti Júnior, E. A., Kumode, M. M. N., Carneiro, M. E., & Satyanarayna, K. G. Nanocellulose in the paper making. In Sustainable Polymer Composites and Nanocomposites (pp. 1027–1066). Cham: Springer. doi:10.1007/978-3-030-05399-4_36

Lettinga, G. (1995). Anaerobic digestion and wastewater treatment systems. *Antonie van Leeuwenhoek*, *67*(1), 3–28. doi:10.1007/BF00872193 PMID:7741528

Liao, H. T., & Wu, C. S. (2009). Preparation and characterization of ternary blends composed of polylactide, poly (ε-caprolactone) and starch. *Materials Science and Engineering A*, *515*(1-2), 207–214. doi:10.1016/j.msea.2009.03.003

Liew, W. L., Kassim, M. A., Muda, K., Loh, S. K., & Affam, A. C. (2014). Conventional methods and emerging wastewater polishing technologies for palm oil mill effluent treatment: A review. *Journal of Environmental Management*, *149*, 222–235. doi:10.1016/j.jenvman.2014.10.016 PMID:25463585

Li, H., Oraby, E., & Eksteen, J. (2020). Extraction of copper and the co-leaching behaviour of other metals from waste printed circuit boards using alkaline glycine solutions. *Resources, Conservation and Recycling*, *154*, 104624. doi:10.1016/j.resconrec.2019.104624

Li, K., Yang, F., Zhang, G., Song, S., Li, Y., Ren, D., & Song, C.-P. (2017). AIK1, a mitogen-activated protein kinase, modulates abscisic acid responses through the MKK5-MPK6 kinase cascade. *Plant Physiology*, *173*(2), 1391–1408. doi:10.1104/pp.16.01386 PMID:27913741

Lim Meng Hon, J. (2010). *A case study on palm empty fruit bunch as energy feedstock*. Academic Press.

Limmer, M., & Burken, J. (2016). Phytovolatilization of organic contaminants. *Environmental Science & Technology*, *50*(13), 6632–6643. doi:10.1021/acs.est.5b04113 PMID:27249664

Lina, L., Xu, F., Ge, X., & Li, Y. (2018). Improving the sustainability of organic waste management practices in the food-energy-water nexus: A comparative review of anaerobic digestion and composting. *Renewable & Sustainable Energy Reviews*, *89*, 151–167. doi:10.1016/j.rser.2018.03.025

Lindsay, W. L., & Norvell, W. A. (1978). Development of a DTPA soil test for zinc, iron, manganese and copper. *Soil Science Society of America Journal*, *42*(3), 421–428. doi:10.2136ssaj1978.03615995004200030009x

Lin, L., Yang, L., Xu, F., Michel, F. C. Jr, & Li, Y. (2014). Comparison of solid-state anaerobic digestion and composting of yard trimmings with effluent from liquid anaerobic digestion. *Bioresource Technology*, *169*, 439–446. doi:10.1016/j.biortech.2014.07.007 PMID:25079209

Linster, M., & Fletcher, L. (2012). *Using the pressure-state-response model to develop indicators of sustainability, USA.* Available Online: http://documentacion.ideam.gov.co/openbiblio/bvirtual/017931/DocumentosIndicadores/Temasvarios/Docum26.pdf

Linzner, R., & Lange, U. (2013). Role and size of informal sector in waste management—A review. *Water Resources Management*, *166*, 69–83.

Liu, Y., Hao, M., Guo, H., Gao, W., & Xiang, N. (2014). Establishment of assessment indicator system for environment-friendly society based on DPSIR model. *Environmental Protection Science*, *16*, 14–21.

Li, Y., Horsman, M., Wu, N., Lan, C. Q., & Dubois-Calero, N. (2008). Biofuels from microalgae. *Biotechnology Progress*, *24*(4), 815–820. PMID:18335954

Llewellyn, C. A., & Airs, R. L. (2010). Distribution and abundance of MAAs in 33 species of microalgae across 13 classes. *Marine Drugs*, *8*(4), 1273–1291. doi:10.3390/md8041273 PMID:20479978

Local Government Handbook of SA (LGHSA). (2017). *Buffalo City Metropolitan Municipality. 117 Oxford Street, Cnr North and Oxford Streets*. Trust Centre.

Loh, S. K. (2016). The potential of the Malaysian oil palm biomass as a renewable energy source. *Energy Conversion and Management*. Advance online publication. doi:10.1016/j.enconman.2016.08.081

Lokeshwari, H., & Chandrappa, G. T. (2006). Impact of heavy metal contamination of Bellandur Lake on soil and cultivated vegetation. *Current Science*, *91*(5), 622–627.

Lone, P. A., Bhardwaj, A. K., & Shah, K. W. (2014). Macrophytes as powerful natural tools for water quality improvement. *Journal of Botany (Faisalabad)*, *9*(2), 24–30. doi:10.3923/rjb.2014.24.30

Lopez Torres, M., & Espinosa Lloréns, M. (2008). Effect of alkaline pretreatment on anaerobic digestion of solid wastes. *Waste Management (New York, N.Y.)*, *28*(11), 2229–2234. doi:10.1016/j.wasman.2007.10.006 PMID:18068345

Lourençon, T. V., Hansel, F. A., Silva, T. A., Ramos, L. P., Muñiz, G. I. B., & Magalhães, W. L. E. (2015). Hardwood and softwood kraft lignins fractionation by simple sequential acid precipitation. *Separation and Purification Technology*, *154*(5), 82–88. doi:10.1016/j.seppur.2015.09.015

Low, K. S., Lee, C. K., & Liew, S. C. (2000). Sorption of cadmium and lead from aqueous solutions by spent grain. *Process Biochemistry*, *36*(1-2), 59–64. doi:10.1016/S0032-9592(00)00177-1

Low, M. J. D. (1960). Kinetics of chemisorption of gases on solids. *Chemical Reviews*, *60*(3), 267–312. doi:10.1021/cr60205a003

Lucas, N., Bienaime, C., Belloy, C., Queneudec, M., Silvestre, F., & Nava-Saucedo, J.-E. (2008). Polymer biodegradation: Mechanism and estimation techniques. *Chemosphere*, *73*(4), 429–442. doi:10.1016/j.chemosphere.2008.06.064 PMID:18723204

Luckachan, G. E., & Pillai, C. K. S. (2011). Biodegradable polymers-a review on recent trends and emerging perspectives. *Journal of Polymers and the Environment*, *19*(3), 637–676. doi:10.100710924-011-0317-1

Maal, K. B., Shafiei, R., & Kabiri, N. (2010). Production of Apricot vinegar using an Isolated acetobacter strain from Iranian Apricot. *World Academy of Science, Engineering and Technology*, *71*, 177–180.

Macaskie, L. E., Creamer, N. J., Essa, A. M. M., & Brown, N. L. (2007). A new approach for the recovery of precious metals from solution and from leachates derived from electronic scrap. *Biotechnology and Bioengineering*, *96*(4), 631–639. doi:10.1002/bit.21108 PMID:16917944

Mahapatra, D. M., Chanakya, H. N., & Ramachandra, T. V. (2014). Bioremediation and lipid synthesis through mixotrophic algal consortia in municipal wastewater. *Bioresource Technology*, *168*, 142–150. doi:10.1016/j.biortech.2014.03.130 PMID:24791711

Mahdi, M., Ameen, R., & Ibrahim, H. (2016). Study on degradation of nylon 6 by thermophilic bacteria *Anoxybacillus rupiensis* Ir3 (JQ912241). *International Journal of Advanced Research in Biological Sciences*, *3*(9), 200–209. doi:10.22192/ijarbs.2016.03.09.027

Ma, L. Q., Komar, K. M., Tu, C., Zhang, W. A., Cai, Y., & Kennelley, E. D. (2001). A fern that hyperaccumulates arsenic. *Nature*, *409*(6820), 579. doi:10.1038/35054664 PMID:11214308

Malaysian-German Chamber of Commerce and Industry (MGCC). (2017). *Oil palm biomass & biogas in Malaysia,2017*. Retrieved from www.malaysia.ahk.de

Malik, U. R., Hasany, S. M., & Subhani, M. S. (2005). Sorptive potential of sunflower stem for Cr(III) ions from aqueous solutions and its kinetic and thermodynamic profile. *Talanta*, *66*(1), 166–173. doi:10.1016/j.talanta.2004.11.013 PMID:18969977

Malkoc, E., & Nuhoglu, Y. (2007). Determination of kinetic and equilibrium parameters of the batch adsorption of Cr(VI) onto waste acorn of *Quercus ithaburensis*. *Chemical Engineering and Processing*, *46*(10), 1020–1029. doi:10.1016/j.cep.2007.05.007

Mallick, S., Dash, S. S., & Parida, K. M. (2006). Adsorption of Hexavalent Chromium on Manganese Nodule Leached Residue Obtained from NH_3-SO_3 Leaching. *Journal of Colloid and Interface Science*, *297*(2), 419–425. doi:10.1016/j.jcis.2005.11.001 PMID:16330035

Manaf, L., Samah, M., & Zukki, N. (2009). Municipal solid waste management in Malaysia: Practices and challenges. *Journal of Waste Management*, *29*(11), 2902–2906. doi:10.1016/j.wasman.2008.07.015 PMID:19540745

Manickam, S., Zainal Abidin, N. B., Parthasarathy, S., Alzorqi, I., Ng, E. H., Tiong, T. J., Gomes, R. L., & Ali, A. (2014). Role of H2O2in the fluctuating patterns of COD (chemical oxygen demand) during the treatment of palm oil mill effluent (POME) using pilot scale triple frequency ultrasound cavitation reactor. *Ultrasonics Sonochemistry*, *21*(4), 1519–1526. doi:10.1016/j.ultsonch.2014.01.002 PMID:24485395

Mao, C., Feng, Y., Wang, X., & Ren, G. (2015). Review on research achievements of biogas from anaerobic digestion. *Renewable & Sustainable Energy Reviews*, *45*, 540–555. doi:10.1016/j.rser.2015.02.032

Mark, F. E., & Lehner, T. (1999). *Plastics recovery from waste electrical and electronic equipment in non-ferrous processes*. Academic Press.

Market. (2015). *Polyvinyl Butyral (PVB) Market Poised to Surge from USD 3.0 Billion in 2015 to USD 3.9 Billion, Globally By 2021*. Retrieved from www.marketresearchstore.com

Marra, A. A. (1992). *Technology of wood bonding: principles and practice*. Van Nostrand Reinhold.

Martial-Didier, A. K., Hubert, K. K., Parfait, K. E. J., & Kablan, T. (2017). Phytochemical properties and proximate composition of papaya (Carica papaya L. var solo 8) peels. *Turkish Journal of Agriculture-Food Science and Technology*, *5*(6), 676–680. doi:10.24925/turjaf.v5i6.676-680.1154

Martínez-Prado, A., Pérez-López, M., Pinto-Espinoza, J., Gurrola-Nevarez, B., & Osorio-Rodríguez, A. (2011). Biorremediación de suelo contaminado con hidrocarburos empleando lodos residuales como fuente alterna de nutrientes. –. *Revista Internacional de Contaminación Ambiental*, *27*(3), 241–252.

Masebinu, S. O., Akinlabi, E. T., Muzenda, E., & Aboyade, A. O. (2019). A review of biochar properties and their roles in mitigating challenges with anaerobic digestion. *Renewable & Sustainable Energy Reviews*, *103*, 291–307. doi:10.1016/j.rser.2018.12.048

Masse, L., Masse, D. I., & Kennedy, K. J. (2003). Effect of enzymatic pre-treatment on fat degradation during anaerobic digestion of slaughterhouse wastewater. *Process Biochemistry (Barking, London, England)*, *38*(9), 1365–1372. doi:10.1016/S0032-9592(03)00020-7

Masud, M. H., Akram, W., Ahmed, A., Ananno, A. A., Mourshed, M., Hasan, M., & Joardder, M. U. H. (2019). Towards the effective E-waste management in Bangladesh: A review. *Environmental Science and Pollution Research International*, *26*(2), 1250–1276. doi:10.100711356-018-3626-2 PMID:30456610

Mata-Alvarez, J., Dosta, J., Romero-Güiza, M. S., Fonoll, X., Peces, M., & Astals, S. (2014). Romero-G€uiza, M.S., Fonoll, X., Peces, M., Astals, S., (2014). A critical review on anaerobic co-digestion achievements between 2010 and 2013. *Renewable & Sustainable Energy Reviews*, *36*, 412–427. doi:10.1016/j.rser.2014.04.039

Mata-Alvarez, J., Mace, S., & Llabres, P. (2000). Anaerobic digestion of organic solid wastes. An overview of research achievements and perspectives. *Bioresource Technology*, *74*(1), 3–16. doi:10.1016/S0960-8524(00)00023-7

Mata, T. M., Martins, A. A., & Caetano, N. S. (2010). Microalgae for biodiesel production and other applications: A review. *Renewable & Sustainable Energy Reviews*, *14*(1), 217–232. doi:10.1016/j.rser.2009.07.020

Mateo-Sagasta, J., Medlicott, K., Qadir, M., Raschid-Sally, L., & Drechsel, P. (2013). *Proceedings of the UN-water project on the safe use of wastewater in agriculture*. Academic Press.

Mato, Y., Isobe, T., Takada, H., Kanehiro, H., Ohtake, C., & Kaminuma, T. (2001). Plastic resin pellets as a transport medium for toxic chemical in the marine environment. *Environmental Science & Technology*, *35*(2), 318–324. doi:10.1021/es0010498 PMID:11347604

Matsushita, K., Toyama, H., & Adachi, O. (1994). Respiratory Chains and Bioenergetics of Acetic Acid Bacteria. In Advances in Microbial Physiology. Academic Press. doi:10.1016/S0065-2911(08)60181-2

Mazhar, J., & Mazumder, A. (2013). Evaluation of antidiabetic activity of methanolic leaf extract of *Coriandrum sativum* in alloxan induced diabetic rats. *Research Journal of Pharmaceutical, Biological and Chemical Sciences*, *3*(4), 500–507.

Mckay, G., Blair, H. S., & Gardener, J. R. (1982). Adsorption of dyes on chitin. I. Equilibrium studies. *Journal of Applied Polymer Science*, *27*(8), 3043–3057. doi:10.1002/app.1982.070270827

Mckay, G., Otterburn, M. S., & Sweeney, A. G. (1981). Surface mass transfer processes during colour removal from effluent using silica. *Water Research*, *15*(3), 327–331. doi:10.1016/0043-1354(81)90036-1

Mecucci, A., & Scott, K. (2002). Leaching and electrochemical recovery of copper, lead and tin from scrap printed circuit boards. Journal of Chemical Technology & Biotechnology: International Research in Process. *Environmental & Clean Technology*, *77*(4), 449–457.

Megharaj, M., Ramakrishnan, B., Venkateswarlu, K., Sethunathan, N., & Naidu, R. (2011). Bioremediation approaches for organic pollutants: A critical perspective. *Environment International*, *37*(8), 1362–1375. doi:10.1016/j.envint.2011.06.003 PMID:21722961

Mena, C., Adenso-Diaz, B., & Yurt, O. (2011). The causes of food waste in the supplier–retailer interface: Evidences from the UK and Spain. *Resources, Conservation and Recycling*, *55*(6), 648–658. doi:10.1016/j.resconrec.2010.09.006

Mendes, L. M., Mendes, R. F., Tonoli, G. H. D., Bufalino, L., Mori, F. A., & Guimarães, J. B. Junior. (2012). Lignocellulosic composites made from agricultural and forestry wastes in brazil. *Key Engineering Materials*, *517*, 556–563. doi:10.4028/www.scientific.net/KEM.517.556

Mendoza, F. J. C., Altabella, J. E., & Izquierdo, A. G. (2017). Application of inert wastes in the construction, operation and closure of landfills: Calculation tool. *Waste Management (New York, N.Y.)*, *59*, 276–285. doi:10.1016/j.wasman.2016.10.041 PMID:27816471

Mendoza, Z. M. S. H., Borges, P. H. M., Morais, P. H. M., & Elias, M. P. S. (2019). Use of beta regression to estimate the volumetric yield coefficient in logs of native species of the legal Amazon. *Nativa (Sinop)*, *7*(3), 323–329. doi:10.31413/nativa.v7i3.6834

Mengel, K., & Kirkby, E. A. (1982). *Principles of Plant Nutrition*. International Potash Institute, Bern.

Messerle, V. E., Ustimenko, A. B., & Lavrichshev, O. A. (2016). Comparative study of coal plasma gasification: Simulation and experiment. *Fuel*, *164*, 172–179. doi:10.1016/j.fuel.2015.09.095

Mian, M. M., Zeng, X., Nasry, A. A. N. B., & Al-Hamadani, S. M. (2017). Municipal solid waste management in China: A comparative analysis. *Journal of Material Cycles and Waste Management*, *19*(3), 1127–1135. doi:10.100710163-016-0509-9

Michael, T., Mokrejas, P., Merinska, D., & Peter, S. (2013). Windshield Recycling Focused on Effective Separation of PVB Sheet. *Journal of Applied Polymer Science*, *131*, 39879.

Mierzwa-Hersztek, M., Gondek, K., & Kopeć, M. (2019). Degradation of Polyethylene and Biocomponent-Derived Polymer Materials: An Overview. *Journal of Polymers and the Environment*, *27*(3), 600–611. doi:10.100710924-019-01368-4

Minghua, Z., Xiumin, F., Rovetta, A., Qichang, H., Vicentini, F., Bingkai, L., Giusti, A., & Yi, L. (2009). Municipal solid waste management in Pudong New area, China. *Waste Management (New York, N.Y.)*, *29*(3), 1227–1233. doi:10.1016/j.wasman.2008.07.016 PMID:18951780

Minh, N. P. (2014). Utilization of Ripen Star Fruit for Vinegar Fermentation. *International Journal of Multidisciplinary Research and Development*, *1*(4), 82–93.

Miretzkya, P., & Cirelli, A. F. (2010). Cr(VI) and Cr(III) Removal from Aqueous Solutions by Raw and Modified Lignocellulosic Materials: A Review. *Journal of Hazardous Materials*, *180*(1-3), 1–19. doi:10.1016/j.jhazmat.2010.04.060 PMID:20451320

Mirza, N., Mahmood, Q., Pervez, A., Ahmad, R., Farooq, R., Shah, M. M., & Azim, M. R. (2010). Phytoremediation potential of Arundo donax in arsenic-contaminated synthetic wastewater. *Bioresource Technology*, *101*(15), 5815–5819. doi:10.1016/j.biortech.2010.03.012 PMID:20363125

Mitra, T., Singha, B., Bar, N., & Das, S. K. (2014). Removal of Pb(II) ions from aqueous solution using water hyacinth root by fixed-bed column and ANN modeling. *Journal of Hazardous Materials*, *273*, 94–103. doi:10.1016/j.jhazmat.2014.03.025 PMID:24727010

Mohammadshirazi, A., Akram, A., Rafiee, S., & Bagheri Kalhor, E. (2014). Energy and cost analyses of biodiesel production from waste cooking oil. *Renewable & Sustainable Energy Reviews*, *33*, 44–49. doi:10.1016/j.rser.2014.01.067

Mohammed, R. R., Ketabachi, M. R., & Mckay, G. (2014). Combined magnetic field and adsorption process for treatment of biologically treated palm oil mill effluent (pome). *Chemical Engineering Journal*, *243*, 31–42. Advance online publication. doi:10.1016/j.cej.2013.12.084

Monferrán, M. V., Pignata, M. L., & Wunderlin, D. A. (2012). Enhanced phytoextraction of chromium by the aquatic macrophyte Potamogeton pusillus in presence of copper. *Environmental Pollution*, *161*, 15–22. doi:10.1016/j.envpol.2011.09.032 PMID:22230062

Monlau, F., Sambusiti, C., Antoniou, N., Barakat, A., & Zabaniotou, A. (2015). A new concept for enhancing energy recovery from agricultural residues by coupling anaerobic digestion and pyrolysis process. *Applied Energy*, *148*, 32–38. doi:10.1016/j.apenergy.2015.03.024

Monteiro, T. C., Lima, J. T., Silva, J. R. M., Trugilho, P. F., & Andrade, B. C. L. (2013). Avaliação do desdobro de toras de Eucalyptus para a obtenção de peças estruturais. *Cerne*, *19*(3), 357–364. doi:10.1590/S0104-77602013000300001

Monteiro, T. C., Lima, J. T., Silva, J. R. M., Trugilho, P. F., & Barauna, E. E. P. (2017). Energy balance in sawing *Eucalyptus grandis* logs. *BioResources*, *12*(3), 5790–5800. doi:10.15376/biores.12.3.5790-5800

Mora-Ravelo, S. G., Alarcon, A., Rocandio-Rodriguez, M., & Vanoye-Eligio, V. (2017). Bioremediation of wastewater for reutilization in agricultural systems: A review. *Applied Ecology and Environmental Research*, *15*(1), 33–50. doi:10.15666/aeer/1501_033050

Morten, W., Ryberg, A., & Michael, H. (2018). *Mapping of global plastics value chain and plastics losses to the environment*. United Nations Environmental Program.

Mostafa, N. A., Farag, A. A., Abo-dief, H. M., & Tayeb, A. M. (2018). Production of biodegradable plastic from agricultural wastes. *Arabian Journal of Chemistry*, *11*(4), 546–553. doi:10.1016/j.arabjc.2015.04.008

Mourao, M. C., Nunes, A. C., & Prins, C. (2009). Heuristic methods for the sectoring arc routing problem. *European Journal of Operational Research*, *196*(3), 856–868. doi:10.1016/j.ejor.2008.04.025

Mshandete, A., Bjornsson, L., Kivaisi, A. K., Rubindamayugi, M. S. T., & Mattiasson, B. (2006). Effect of particle size on biogas yield from sisal fiber waste. *Renewable Energy*, *31*(14), 2385–2392. doi:10.1016/j.renene.2005.10.015

Mudau, A. (2012). *A laboratory investigation of the effects of water content and waste composition on leachate and gas generation from simulated MSW* (MSc Thesis). University of the Witwatersrand, SA.

Muhamad, W., Othman, R., Shaharuddin, R., & Irani, M. (2015). Microorganism as plastic biodegradation agent towards sustainable environment. *Advances in Environmental Biology*, *9*, 8–14.

Municipalities of SA. (2018). *Buffalo City Metropolitan Municipality*. Available Online: https://municipalities.co.za/map/7/buffalo-city-metropolitan-municipality

Muradov, N., Taha, M., Miranda, A. F., Wrede, D., Kadali, K., Gujar, A., Stevenson, T., Ball, A. S., & Mouradov, A. (2015). Fungal-assisted algal flocculation: Application in wastewater treatment and biofuel production. *Biotechnology for Biofuels*, *8*(1), 24. doi:10.118613068-015-0210-6 PMID:25763102

Murthy, Z.V.P., Prabhakar, T., & Kumar, R. (2010). Isotherm and Kinetics Modeling of Fluoride Removal from Industrial Effluent by Alumina. *Chemical Product and Process Modeling, 5*.

Musonge, P. (2014). *Influence of Effluent Type on the Performance of Chitosan as a Coagulant*. Academic Press.

Mustafa, M. F., Liu, Y., Duan, Z., Guo, H., Xu, S., Wang, H., & Lu, W. (2017). Volatile compounds emission and health risk assessment during composting of organic fraction of municipal solid waste. *Journal of Hazardous Materials, 327*, 35–43. doi:10.1016/j.jhazmat.2016.11.046 PMID:28038430

Muzenda, E., Ntuli, F., & Pilusa, T. J. (2012). Waste management strategies and situation in SA. *Engineering and Technology, 68*, 309–312.

Nabegu, A. (2010). An Analysis of Municipal Solid Waste in Kano Metropolis, Nigeria. *Journal of Human Ecology (Delhi, India), 31*(2), 111–119. doi:10.1080/09709274.2010.11906301

Nahman, A., & Godfrey, L. (2010). Economic instruments for solid waste management in SA: Opportunities and constraints. *Resources, Conservation and Recycling, 54*(8), 521–531. doi:10.1016/j.resconrec.2009.10.009

Naiya, T. K., Bhattacharya, A. K., & Das, S. K. (2009a). Adsorption of Cd(II) and Pb(II) from aqueous solutions on activated alumina. *Journal of Colloid and Interface Science, 333*(1), 14–26. doi:10.1016/j.jcis.2009.01.003 PMID:19211112

Naiya, T. K., Bhattacharya, A. K., Mandal, S. N., & Das, S. K. (2009b). The sorption of lead(II) ions on rice husk ash. *Journal of Hazardous Materials, 163*(2-3), 1254–1264. doi:10.1016/j.jhazmat.2008.07.119 PMID:18783880

Naiya, T. K., Chowdhuary, P., Bhattacharya, A. K., & Das, S. K. (2009c). Saw dust and neem bark as low-cost natural biosorbent for adsorptive removal of Zn(II) and Cd(II) ions from aqueous solutions. *Chemical Engineering Journal, 148*(1), 68–79. doi:10.1016/j.cej.2008.08.002

Nakasaki, K., Yaguchi, H., Sasaki, M., & Kubota, H. (1993). Effects of pH control on composting of garbage. *Waste Management & Research, 11*(2), 117–125. doi:10.1177/0734242X9301100204

Narbón-Perpiñá, I., & Prior, D. (2020). Management of Waste Electrical and Electronic Equipment in European Union Countries: A Comparison. In *E-waste Recycling and Management* (pp. 127–141). Springer. doi:10.1007/978-3-030-14184-4_7

National Treasury. (2006). *A framework for considering market-based instruments to support environmental fiscal reform in SA*. National Treasury.

Nayak, A., & Bhushan, B. (2019). An overview of the recent trends on the waste valorization techniques for food wastes. *Journal of Environmental Management, 233*, 352–370. doi:10.1016/j.jenvman.2018.12.041 PMID:30590265

Naz, T., Khan, M. D., Ahmed, I., Rehman, S. U., Rha, E. S., & Malook, I. (2015). Biosorption of heavy metals by Pseudomonas species isolated from sugar industry. *Toxicology and Industrial Health, 10*, 525–534. PMID:25739395

Negi, S., Dhar, H., Hussain, A., & Kumar, S. (2018). Biomethanation potential for co-digestion of municipal solid waste and rice straw: A batch study. *Bioresource Technology, 254*, 139–144. doi:10.1016/j.biortech.2018.01.070 PMID:29413914

Nelms, S., Brownlow, B., Davidson, N., Deaville, R., Galloway, T., & ... . (2019). Microplastics in marine mammals stranded around the British coast: Ubiquitous but transitory? *Scientific Reports, 9*(1), 1075. doi:10.103841598-018-37428-3 PMID:30705316

Nemerow, N. L. (2009). *Environmental Engineering: Environmental Health and Safety for Municipal Infrastructure, Land Use and Planning, and Industry* (6th ed.). Wiley.

Neves, L., Oliveira, R., & Alves, M. M. (2006). Anaerobic co-digestion of coffee waste and sewage sludge. *Waste Management (New York, N.Y.)*, *26*(2), 176–181. doi:10.1016/j.wasman.2004.12.022 PMID:16310117

Neves, L., Ribeiro, R., Oliveira, R., & Alves, M. M. (2006). Enhancement of methane production from barley waste. *Biomass and Bioenergy*, *30*(6), 599–603. doi:10.1016/j.biombioe.2005.12.003

Neyens, E., & Baeyens, J. (2003). A review of classic Fenton's peroxidation as an advanced oxidation technique. *Journal of Hazardous Materials*, *98*(1–3), 33–50. doi:10.1016/S0304-3894(02)00282-0 PMID:12628776

Ng, M., Liu, S., Chow, C. W. K., Drikas, M., Amal, R., & Lim, M. (2013). Understanding effects of water characteristics on natural organic matter treatability by PACl and a novel PACl-chitosan coagulants. *Journal of Hazardous Materials*, *263*, 718–725. doi:10.1016/j.jhazmat.2013.10.036 PMID:24220196

Nhat, P. V. H., Ngo, H. H., Guo, W. S., Chang, S. W., Nguyen, D. D., Nguyen, P. D., Bui, X. D., Zhang, X. B., & Guo, J. B. (2018). Can algae-based technologies be an affordable green process for biofuel production and wastewater remediation? *Bioresource Technology*, *256*, 491–501. doi:10.1016/j.biortech.2018.02.031 PMID:29472123

Nico, G., Knee, P., Ye, Q., & Tiedje, O. (2020). Initial droplet conditions in numerical spray painting by electrostatic rotary bell sprayers. A framework for optimization of injection model coefficients. *Journal of Coatings Technology and Research*, *17*(5), 1091–1104. doi:10.100711998-020-00352-1

Niekerk, S., & Weghmann, V. (2019). *Municipal solid waste management services in Africa*. Working Paper, Public Service International, Johannesburg, RSA.

Nintendo, E., Nintendo, E., Miyamoto, S., Tezuka, T., & Kondo, K. (2019). *The Legend of Zelda.* Paper presented at the Spin.

Nissim, I., & Inbar, S. (2005). From dumping to sanitary landfills-solid waste management in Israel. *Waste Management (New York, N.Y.)*, *25*(3), 323–327. doi:10.1016/j.wasman.2004.06.004 PMID:15823747

Nkosi, N., Muzenda, E., Zvimba, J., & Pilusa, J. (2013). *The current waste generation and management trends in SA: A review. International Conference on Integrated Waste Management and Green Energy Engineering*, Johannesburg, SA.

Nourani, M., Baghdadi, M., Javan, M., & Bidhendi, G. N. (2016). Production of a biodegradable flocculant from cotton and evaluation of its performance in coagulation-flocculation of kaolin clay suspension : Optimization through response surface methodology (RSM). *Biochemical Pharmacology*. Advance online publication. doi:10.1016/j.jece.2016.03.028

Nowicki, M. A., Castro, N. J., Plesniak, M. W., & Zhang, L. G. (2016). 3D printing of novel osteochondral scaffolds with graded microstructure. *Nanotechnology*, *27*(41), 414001. doi:10.1088/0957-4484/27/41/414001 PMID:27606933

Nyika, J., Onyari, E., Mishra, S., & Dinka, M. (2020). Waste management in South Africa. In A. Pariatamby, F. Shahul, & M. Bhatti (Eds.), *Sustainable waste management challenges in developing countries*. IGI Global. doi:10.4018/978-1-7998-0198-6.ch014

Ob, A., & Muchie, M. (2010). *Remediation of heavy metals in drinking water and wastewater treatment systems: Processes and applications*. Academic Press.

Oelofse, S. (2018). Opportunities and challenges in waste management. *CSIR*. Available Online: http://sustainabilityweek.co.za/assets/files/Day%203%20%20CSIR%20-%20Waste.pdf

Ogata, T., & Nakano, Y. (2005). Mechanisms of gold recovery from aqueous solutions using a novel tannin gel adsorbent synthesized from natural condensed tannin. *Water Research*, *39*(18), 4281–4286. doi:10.1016/j.watres.2005.06.036 PMID:16221485

Öhman, F. (2006). *Precipitation and separation of lignin from kraft black liquor* (PhD Thesis). Chalmers University of Technology, Gothenburg, Sweden.

Ojoawo, S., Agbede, O., & Sangodoyin, A. (2011). On the Physical Composition of Solid Wastes in Selected Dumpsites of Ogbomosoland, South-Western Nigeria. *Journal of Water Resource and Protection*, *3*(09), 661–666. doi:10.4236/jwarp.2011.39076

Okot-Okumu, J. (2015). Solid waste management in Uganda; challenges and options. In T. Simelane & R. Mohee (Eds.), *Future directions of municipal solid waste management in Africa* (pp. 107–135). Africa Institute of South Africa. doi:10.2307/j.ctvh8r2sj.14

Okot-Okumu, J., & Nyenje, R. (2011). Municipal solid waste management under decentralisation in Uganda. *Habitat International*, *35*(4), 537–543. doi:10.1016/j.habitatint.2011.03.003

Oliveira, J. T. S., Hellmeister, J. C., Simões, J. W., & Tomazello Filho, M. (1999). Caracterização da madeira de sete espécies de eucaliptos para a construção civil: 1- avaliações dendrométricas das árvores. *Scientia Forestalis*, *56*, 113–124.

Oliveira, R. C. P., Mateus, M., & Santos, D. M. F. (2018). Chronoamperometric and chronopotentiometric investigation of kraft black liquor. *International Journal of Hydrogen Energy*, *43*(35), 16817–16823. doi:10.1016/j.ijhydene.2018.01.046

Olper, M., Maccagni, M., & Cossali, S. (2004). *U.S. Patent Application No. 10/798,143*. Washington, DC: US Patent Office.

Omez-Flores, R., Hernández-Martínez, H., Tamez-Guerra, P., Tamez-Guerra, R., & Quintanilla-Licea, R., Monreal-Cuevas, R., & Rodríguez-Padilla, C. (2010). Antitumor and immunomodulating potential of *Coriandrum sativum, Piper nigrum* and *Cinnamomum zeylanicum. Journal of Natural Products*, *3*, 54–63.

Omil, F., Garrido, J. M., Arrojo, B., & Méndez, R. (2003). Anaerobic filter reactor performance for the treatment of complex dairy wastewater at industrial scale. *Water Research*, *37*(17), 4099–4108. doi:10.1016/S0043-1354(03)00346-4 PMID:12946891

Onwosi, C., Igbokwe, V., Odimba, J., Eke, I., Nwankwoala, M., Iroh, I., & Ezeogu, L. (2017). Composting technology in waste stabilization: On the methods, challenges and future prospects. *Journal of Environmental Management*, *190*, 140–157. doi:10.1016/j.jenvman.2016.12.051 PMID:28040590

Oppelt, E. T. (1987). Incineration of hazardous waste. *JAPCA*, *37*(5), 558–586. doi:10.1080/08940630.1987.10466245 PMID:3328776

Orr, I. G., Hadar, Y., & Sivan, A. (2004). Colonization, biofilm formation and biodegradation of polyethylene by a strain of Rhodococcus ruber. *Applied Microbiology and Biotechnology*, *65*(1), 97–104. PMID:15221232

Oslan, A. M. R., Ahmad, M., & Mohd, K. (2013). *Investigation of oil palm frond properties for use as biomaterials and biofuels investigation of oil palm frond properties for use as biomaterials and biofuels*. Academic Press.

Ostrem, K. (2004). *Greening Waste: Anaerobic Digestion For Treating The Organic Fraction Of Municipal Solid Wastes*. Earth Engineering Center Columbia University.

Oswald, W., & Gotass, H. (1957). *Photosynthesis in sewage treatment*. Trans. Amer. Soc. Civil Engrs.

Otaru, A. J., Ameh, C. U., Okafor, J. O., Odigure, J. O., Abdulkareem, A. S., & Ibrahim, S. (2013). Study on the effectiveness of phytoremediation in the removal of heavy metals from soil using Corn. *International Journal of Computational Engineering Research*, *3*(4), 87–93.

Oteng-Ababio, M. (2010). Private sector involvement in solid waste management in Ghana: The case of the Greater Accra Metropolitan Area (GAMA). *Waste Management & Research*, *28*(4), 322–329. doi:10.1177/0734242X09350247 PMID:19854814

Oteng-Ababio, M. (2011). Governance crisis or attitudinal challenges: collection, storage and transportation of solid waste in Ghana. *Integrated Waste Management, 1, 3-22.*

Oteng-Ababio, M., Arguello, J., & Gabbay, O. (2013). Solid waste management in African countries: Sorting the facts from the fads in Accra, Ghana. *Habitat International*, *39*, 96–104. doi:10.1016/j.habitatint.2012.10.010

Otieno, N. E., Dai, X., Barba, D. De, Bahman, A., Smedbol, E., Rajeb, M., & Jaton, L. (2016). *Palm oil production in Malaysia : An analytical systems model for balancing economic prosperity, forest conservation and social welfare.* Academic Press.

Otti, V. I., Ifeanyichukwu, H. I., Nwaorum, F. C., & Ogbuagu, F. U. (2014). *Sustainable oil palm waste management in engineering development.* Academic Press.

Padalia, R. C., Karki, N., Sah, A. N., & Verma, R. S. (2011). Volatile constituents of leaf and seed essential oil of *Coriandrum sativum* L. *Journal of Essential Oil-Bearing Plants*, *14*(5), 610–616. doi:10.1080/0972060X.2011.10643979

Pan, Y., Shi, B., & Zhang, Y. (2014). *Research on Flocculation Property of Bioflocculant PG.* doi:10.5539/mas.v3n6p106

Pandey, S., Saha, P., Biswas, S., & Maiti, T. K. (2011). Characterization of two heavy metal resistant strains isolated from slag disposal site at Burnpur. *Indian J. Environ Biol.*, *32*, 773–779. PMID:22471215

Pangestuti, R., & Kim, S. K. (2011). Biological activities and health benefit effects of natural pigments derived from marine algae. *Journal of Functional Foods*, *3*(4), 255–266. doi:10.1016/j.jff.2011.07.001

Pansuk, J., Junpen, A., & Garivait, S. (2018). Assessment of air pollution from household solid waste open burning in Thailand. *Sustainability*, *10*(7), 2553. doi:10.3390u10072553

Parthasarathy, S., Mohammed, R. R., Fong, C. M., Gomes, R. L., & Manickam, S. (2016). A novel hybrid approach of activated carbon and ultrasound cavitation for the intensification of palm oil mill effluent (POME) polishing. *Journal of Cleaner Production*, *112*, 1218–1226. doi:10.1016/j.jclepro.2015.05.125

Pathak, P., Srivastava, R. R., & Ojasvi. (2017). Assessment of legislation and practices for the sustainable management of waste electrical and electronic equipment in India. *Renewable & Sustainable Energy Reviews*, *78*, 220–232. doi:10.1016/j.rser.2017.04.062

Patil, S. A., & Sharma, N. M. (2015). Electronic Waste-A Literature Review. *International Journal of Science and Research.*

Patterson, S. C., & Jancuk, W. A. (1977). Cementation treatment of copper in wastewater. In *Proceedings of 32nd Industrial Waste Conference* (pp. 85-865), Perdue University.

Pavankumar, S. B., Sachin, K. R., Shankar, R., Thyagaraja, B., & Madhusudhan, T. (2018). *Design and fabrication of organic waste shredding machine*. Academic Press.

Pearce, D., & Turner, R. K. (1994). *Economics and solid waste management in the developing world.* CSERGE working paper WM 1994-05. Centre for Social and Economic Research on the Global Environment.

Peng, K., Li, X., Luo, C., & Shen, Z. (2006). Vegetation composition and heavy metal uptake by wild plants at three contaminated sites in Xiangxi area, China. *Journal of Environmental Science and Health. Part A, Environmental Science and Engineering & Toxic and Hazardous Substance Control*, *40*, 65–76. PMID:16401571

Peralta-Videa, J. R., Gardea-Torresdey, J. L., Gomez, E., Tiemann, K. J., Parsons, J. G., Rosa, G. D., & Carrillo, G. (2002). Potential of alfalfa plant to phytoremediate individually contaminated montmorillonite-soils with cadmium (ii), chromium (vi), copper (ii), nickel (ii), and zinc (ii). *Bulletin of Environmental Contamination and Toxicology*, *69*(1), 74–81. doi:10.100700128-002-0012-y PMID:12053260

Pereira, E. G., Martins, M. A., Pecenka, R., & Carneiro, A. C. O. (2016). Pyrolysis gases burners: Sustainability for integrated production of charcoal, heat and electricity. *Renewable & Sustainable Energy Reviews*, *75*, 592–600. doi:10.1016/j.rser.2016.11.028

Permana, A., Towolioe, S., Aziz, N., & Ho, C. (2015). Sustainable solid waste management practices and perceived cleanliness in a low-income city. *Habitat International*, *49*, 197–205. doi:10.1016/j.habitatint.2015.05.028

Perman, R., Ma, Y., McGilvray, J., & Common, M. (2003). *Natural resource and environmental economics*. Pearson Education.

Pickin, J. (2008). Representations of environmental concerns in cost–benefit analyses of solid waste recycling. *Resources, Conservation and Recycling*, *53*(1), 79–85. doi:10.1016/j.resconrec.2008.09.003

Pienkos, P. T., & Darzins, A. L. (2009). The promise and challenges of microalgal-derived biofuels. *Biofuels, Bioproducts and Biorefining: Innovation for a Sustainable Economy, 3*(4), 431-440.

Pietro, L. C., Nicola, S., Lacobellis, A., Adriana, D. M., Francesco, C., & Felice, S. (2004). Antibacterial activity of *Coriandrum sativum* L. and *Foeniculum vulgare* Miller Var. *vulgare* (Miller) essential oils. *Journal of Agricultural and Food Chemistry*, *52*(26), 7862–7866. doi:10.1021/jf0493122 PMID:15612768

Piltz, B., & Melkonian, M. (2018). Immobilized microalgae for nutrient recovery from source-separated human urine. *Journal of Applied Phycology*, *30*(1), 421–429. doi:10.100710811-017-1266-4

Pires, A., Martinho, G., & Chang, N. B. (2011). Solid waste management in European countries: A review of systems analysis techniques. *Journal of Environmental Management*, *92*(4), 1033–1050. doi:10.1016/j.jenvman.2010.11.024 PMID:21194829

Pistek, D., Merinska, D., & Tupy, M. (2011). The Influence of Plasticizers on Mechanical and Optical Properties of Recycled Polyvinyl Butyral. *Proceedings of 13thWSEAS International Conference on Mathematical Methods and Techniques in Engineering and Environmental Science*, 305-307.

Poh, P. E., Ong, W. Y. J., Lau, E. V, & Chong, M. N. (2014). *Journal of Environmental Chemical Engineering Investigation on micro-bubble flotation and coagulation for the treatment of anaerobically treated palm oil mill effluent (POME)*. Elsevier. doi:10.1016/j.jece.2014.04.018

Polanyi, M. (1932). Theories of the adsorption of gases. A general survey and some additional remarks. *Transactions of the Faraday Society*, *28*(0), 316–333. doi:10.1039/TF9322800316

Polyakova, V., Degaev, E., & El Haddad, P. (2018). *Reduction of Ecological and Economic Risks in Utilization of Solid Domestic Wastes and Construction Waste*. Paper presented at the MATEC Web of Conferences. 10.1051/matecconf/201825106017

Pometto, A. L. III, Lee, B. T., & Johnson, K. E. (1992). Production of an extracellular polyethylene-degrading enzyme (s) by Streptomyces species. *Applied and Environmental Microbiology*, *58*(2), 731–733. doi:10.1128/AEM.58.2.731-733.1992 PMID:1610196

Popova, E. (2016). Accumulation of heavy metals in the "soil-plant" system. *AIP Conference Proceedings*, *1772*, 050006. doi:10.1063/1.4964576

Posten, C. (2009). Design principles of photo-bioreactors for cultivation of microalgae. *Engineering in Life Sciences*, *9*(3), 165–177. doi:10.1002/elsc.200900003

Post, J. (1999). The problems and potentials of privatizing solid waste management in Kumasi, Ghana. *Habitat International*, *23*(2), 201–216. doi:10.1016/S0197-3975(98)00045-9

Pradhan, J. K., & Kumar, S. (2014). Informal e-waste recycling: Environmental risk assessment of heavy metal contamination in Mandoli industrial area, Delhi, India. *Environmental Science and Pollution Research International*, *21*(13), 7913–7928. doi:10.100711356-014-2713-2 PMID:24652574

Pramendra, K., Nida, K., & Kumar, D. (2016). Polyvinyl Butyral (PVB), Versatile Template for Designing Nanocomposite/ Composite Materials: A Review. *Green Chemistry and Technology Letters*, *2*(4), 185–194. doi:10.18510/gctl.2016.244

Prasad, A., Chand, S., Kumar, S., Chattopadhyay, A., & Patra, D. D. (2012). Heavy metals affect yield, essential oil compound and Rhizosphere Microflora of Vetiver (*Vetiverial zizanioides* L. Nash) grass. *Communications in Soil Science and Plant Analysis*, *45*(11), 1511–1522. doi:10.1080/00103624.2014.904334

Psomopoulos, C. S., Bourka, A., & Themelis, N. J. (2009). Waste-to-energy: A review of the status and benefits in USA. *Waste Management (New York, N.Y.)*, *29*(5), 1718–1724. doi:10.1016/j.wasman.2008.11.020 PMID:19153036

Puckett, J., Byster, L., & Westervelt, S. (2002). *Exporting Harm: The High-Tech Trashing of Asia, The Basel Action Network (BAN) and Silicon Valley Toxics Coalition (SVTC), 2002*. https://ban. org/E-waste/technotrashfinalcomp.pdf

Pui, K., Shak, Y., & Wu, T. Y. (2014). Coagulation – flocculation treatment of high-strength agro-industrial wastewater using natural Cassia obtusifolia seed gum : Treatment efficiencies and flocs characterization. *Chemical Engineering Journal*, *256*, 293–305. doi:10.1016/j.cej.2014.06.093

Punetha, D., Pande, C., Tewari, G., & Tewari, K. (2014). Qualitative variation in Coriandrum sativum L. seed oil with manure applications. *Indian Perfumer*, *58*(2), 27–31.

Punetha, D., Tewari, G., Pande, C., Kharkwal, G. C., & Tewari, K. (2015). Investigation on heavy metal content in common grown vegetables from polluted sites of Moradabad district, India. *Journal of the Indian Chemical Society*, *92*, 97–103.

Punetha, D., Tewari, G., Pande, C., Rana, L., & Tripathi, S. (2019). *Effect of manure on the metal efficiency of Coriandrum sativum. In Zero Waste: Management Practices for Environmental Sustainability*. Taylor & Francis.

Quinet, P., Proost, J., & Van Lierde, A. (2005). Recovery of precious metals from electronic scrap by hydrometallurgical processing routes. *Mining, Metallurgy & Exploration*, *22*(1), 17–22. doi:10.1007/BF03403191

Quintas, P. Y., Dotto, G. L., Da Silva, S. M., & Escudero, L. B. (2019). Removal of thallium from environmental samples using a raw and chemically modified biosorbent derived from domestic wastes. *Environmental Science and Pollution Research International*, *26*(31), 32285–32297. doi:10.100711356-019-06442-x PMID:31598928

Raaman, N., Rajitha, N., Jayshree, A., & Jegadeesh, R. (2012). Biodegradation of plastic by Aspergillus spp. isolated from polythene polluted sites around Chennai. *Journal of Academia and Industrial Research*, *1*(6), 313–316.

Rachael, E., & Marshall, K. F. (2013). Systems approaches to integrated solid waste management in developing countries. *Waste Management (New York, N.Y.)*, *33*(4), 988–1003. doi:10.1016/j.wasman.2012.12.023 PMID:23360772

Ragaert, K., Delva, L., & Van Geem, K. (2017). Mechanical and chemical recycling of solid plastic waste. *Waste Management (New York, N.Y.)*, *69*, 24–58. doi:10.1016/j.wasman.2017.07.044 PMID:28823699

Rajinipriya, M., Nagalakshmaiah, M., Robert, M., & Elkoun, S. (2018). Importance of Agricultural and Industrial Waste in the Field of Nanocellulose and Recent Industrial Developments of Wood Based Nanocellulose: A Review. *ACS Sustainable Chemistry & Engineering*, *6*(3), 2807–2828. doi:10.1021/acssuschemeng.7b03437

Ramachandra, T., & Bachamanda, S. (2007). Environmental audit of Municipal Solid Waste Management. *International Journal of Environmental Technology and Management*, *7*(3/4), 369–391. doi:10.1504/IJETM.2007.015152

Ramos, L. P. (2003). The chemistry involved in the steam treatment of lignocellulosic materials. *Quimica Nova*, *26*(6), 863–871. doi:10.1590/S0100-40422003000600015

Rani, A., Ray, K., & Goel, R. (2010). *Strategies for crop improvement in contaminated soils using metaltolerant bioinoculants*. Springer.

Raposo, F., De La Rubia, M. A., Fernández-Cegrí, V., & Borja, R. (2012). Anaerobic digestion of solid organic substrates in batch mode: An overview relating to methane yields and experimental procedures. *Renewable & Sustainable Energy Reviews*, *16*(1), 861–877. doi:10.1016/j.rser.2011.09.008

Rasmeni, Z., & Madyira, D. (2019). A review of the current municipal solid waste management practices in Johannesburg city townships. *Procedia Manufacturing*, *35*, 1025–1031. doi:10.1016/j.promfg.2019.06.052

Rasouli, Z., Valverde-Pérez, B., D'Este, M., De Francisci, D., & Angelidaki, I. (2018). Nutrient recovery from industrial wastewater as single cell protein by a co-culture of green microalgae and methanotrophs. *Biochemical Engineering Journal*, *134*, 129–135. doi:10.1016/j.bej.2018.03.010

Rattanachaikunsopon, P., & Phumkhachorn, P. (2010). Potential of coriander (*Coriandrum sativum*) oil as a natural antimicrobial compound in controlling *Campylobacter jejuni* in raw meat. *Bioscience, Biotechnology, and Biochemistry*, *74*(1), 31–35. doi:10.1271/bbb.90409 PMID:20057155

Raven, J. A. (2010). Inorganic carbon acquisition by eukaryotic algae: Four current questions. *Photosynthesis Research*, *106*(1-2), 123–134. doi:10.100711120-010-9563-7 PMID:20524069

Ravi Kumar, M. N. (2000). A review of chitin and chitosan applications. *Reactive & Functional Polymers*, *46*(1), 1–27. doi:10.1016/S1381-5148(00)00038-9

Ray. (2012). Economic Performance of Indian Automobile Industry: An Econometric Appraisal. *Business Intelligence Journal,* 151-153.

Raychaudhuri, S., Raychaudhuri, M., Rautaray, S. K., Parida, S., & Kumar, A. (2017). *Urban wastewater–A potential irrigation source for summer paddy (Oryza sativa L.) in India*. Academic Press.

Raziyafathima, M., Praseetha, P. K., & Rimal, I. R. S. (2016). Microbial degradation of plastic waste: A review. *Chemical and Biological Sciences*, *4*, 231–242.

Razzak, S. A., Hossain, M. M., Lucky, R. A., Bassi, A. S., & de Lasa, H. (2013). Integrated CO_2 capture, wastewater treatment and biofuel production by microalgae culturing—A review. *Renewable & Sustainable Energy Reviews*, *27*, 622–653. doi:10.1016/j.rser.2013.05.063

Reddy, D. H. K., Harinatha, Y., Seshaiah, K., & Reddy, A. V. R. (2010). Biosorption of Pb(II) from aqueous solutions using chemically modified *Moringa oleifera* tree leaves. *Chemical Engineering Journal*, *162*(2), 626–634. doi:10.1016/j.cej.2010.06.010

Reichenberg, D. (1953). Properties of ion-exchange resins in relation to their structure. III. Kinetics of exchange. *Journal of the American Chemical Society*, *75*(3), 589–597. doi:10.1021/ja01099a022

Rengaraj, S., Moon, S. H., Sivabalan, R., Arabindoo, B., & Murugesan, V. (2002). Agricultural solid waste for the removal of organics: adsorption of phenol from water and wastewater by palm seed coat activated carbon. *Waste Management, 22*(5), 543-548.

Ren, L. J., Ji, X. J., Huang, H., Qu, L., Feng, Y., Tong, Q. Q., & Ouyang, P. K. (2010). Development of a stepwise aeration control strategy for efficient docosahexaenoic acid production by *Schizochytrium* sp. *Applied Microbiology and Biotechnology, 87*(5), 1649–1656. doi:10.100700253-010-2639-7 PMID:20445973

Reputation Matters. (2014). *Buffalo City Metropolitan Municipality calls for all hands on waste.* Available Online: http://reputationmatters.co.za/2014/02/17/buffalo-city-metropolitan-municipality-calls-for-all-hands-on-waste/

Reschovsky, J., & Stone, S. (1994). Market incentives to encourage household waste recycling: Paying for what you throw away. *Journal of Policy Analysis and Management, 13*(1), 120–139. doi:10.2307/3325093

Reyes, M. R., Reyes-Esparza, J., Angeles, O. T., & Rodríguez-Fragoso, L. (2010). Mutagenicity and safety evaluation of water extract of *Coriandrum sativum* leaves. *Journal of Food Science, 75*(1), T6–T12. doi:10.1111/j.1750-3841.2009.01403.x PMID:20492211

Richard, A., & Margaritis, A. (n.d.). *Poly (glutamic Acid) for Biomedical Applications.* Academic Press.

Rich, C., Gronow, J., & Voulvoulis, N. (2008). The potential for aeration of MSW landfills to accelerate completion. *Waste Management (New York, N.Y.), 28*(6), 1039–1048. doi:10.1016/j.wasman.2007.03.022 PMID:17531464

Riddin, T. (2018). *BCM solid waste operates on 10 of 26 refuse trucks.* Available Online: https://www.dispatchlive.co.za/news/2018-04-10-bcm-solid-waste-operates-on-10-of-26-refuse-trucks/

Ridzuan, M., Ali, M., Shirai, Y., Samsu, A., Amiruddin, A., Ali, M., & Idris, J. (2013). Treatment of effluents from palm oil mill process to achieve river water quality for reuse as recycled water in a zero emission system. *Journal of Cleaner Production, 10–13.* Advance online publication. doi:10.1016/j.jclepro.2013.12.004

Ritcey, G. M. (2006). Solvent extraction in hydrometallurgy: Present and future. *Tsinghua Science and Technology, 11*(2), 137–152. doi:10.1016/S1007-0214(06)70168-7

RitchieH.RoserM. (2018). *Plastic pollution.* https://ourworldindata.org/plastic-pollution

Robinson, B. H. (2009). E-waste: An assessment of global production and environmental impacts. *The Science of the Total Environment, 408*(2), 183–191. doi:10.1016/j.scitotenv.2009.09.044 PMID:19846207

Roda, A., De Faveri, D. M., Dordoni, R., & Lambri, M. (2014). Vinegar production from pineapple wastes- Preliminary Saccharification Trials. *Chemical Engineering Transactions, 37*, 607–612.

Rodolfi, L., Chini Zittelli, G., Bassi, N., Padovani, G., Biondi, N., Bonini, G., & Tredici, M. R. (2009). Microalgae for oil: Strain selection, induction of lipid synthesis and outdoor mass cultivation in a low-cost photobioreactor. *Biotechnology and Bioengineering, 102*(1), 100–112. doi:10.1002/bit.22033 PMID:18683258

Rodrigues, A. C., Boscov, M. E., & Günther, W. M. (2020). Domestic flow of e-waste in São Paulo, Brazil: Characterization to support public policies. *Waste Management (New York, N.Y.), 102*, 474–485. doi:10.1016/j.wasman.2019.10.052 PMID:31751920

Rojas, J., Bedoya, M., & Ciro, Y. Current Trends in the Production of Cellulose Nanoparticles and Nanocomposites for Biomedical Applications. In M. Poletto (Ed.), *Cellulose – Fundamental aspects and current trends* (pp. 193–228). IntechOpen. doi:10.5772/61334

Rosli, W., Daud, W., & Law, K. (2011). *Oil palm fibers as papermaking material: potentials and challenges*. Academic Press.

Rotich, H. K., Yongsheng, Z., & Jun, D. (2006). Municipal solid waste management challenges in developing countries e Kenyan case study. *Waste Management (New York, N.Y.)*, *26*(1), 92–100. doi:10.1016/j.wasman.2005.03.007 PMID:16006111

Roychoudhuri, R., & Debnath, B. (2020). Risk Assessment of E-Waste Recycling with a Focus on ISO 9000 Standard. In *Sustainable Waste Management: Policies and Case Studies* (pp. 431–440). Springer. doi:10.1007/978-981-13-7071-7_38

Roy, S. D., Thakur, S., Negi, A., Kumari, M., Sutar, N. Z., & Jana, G. K. (2010). *In vitro* antibiotic activity of volatile oils of *Carum carvi* & *Coriandrum sativim. International Journal of Chemical and Analytical Science*, *1*(7), 149–150.

Ruiz, J., Álvarez, P., Arbib, Z., Garrido, C., Barragán, J., & Perales, J. A. (2011). Effect of nitrogen and phosphorus concentration on their removal kinetic in treated urban wastewater by Chlorella vulgaris. *International Journal of Phytoremediation*, *13*(9), 884–896. doi:10.1080/15226514.2011.573823 PMID:21972511

Rupassara, S. I., Larson, R. A. A., Sims, G. K., & Marley, K. A. (2002). Degradation of Atrazine by hornwort in aquatic systems. *Bioremediation Journal*, *6*(3), 217–224. doi:10.1080/10889860290777576

Russell, C. S., & Vaughan, W. J. (2003). *The choice of pollution control policy instruments in developing countries: arguments, evidence and suggestions*. Edward Elgar.

Russell, J., Huang, J., Anand, P., Kucera, K., Sandoval, A., Dantzler, K., Hickman, D. S., Jee, J., Kimovec, F. M., Koppstein, D., Marks, D. H., Mittermiller, P. A., Núñez, S. J., Santiago, M., Townes, M. A., Vishnevetsky, M., Williams, N. E., Vargas, M. P. N., Boulanger, L.-A., ... Strobel, S. A. (2011). Biodegradation of polyester polyurethane by endophytic fungi. *Applied and Environmental Microbiology*, *77*(17), 6076–6084. doi:10.1128/AEM.00521-11 PMID:21764951

Saeeda, A., Iqbala, M., & Akhtarb, M. W. (2005). Removal and recovery of lead(II) from single and multimetal (Cd, Cu, Ni, Zn) solutions by crop milling waste (black gram husk). *Journal of Hazardous Materials*, *B117*(1), 65–73. doi:10.1016/j.jhazmat.2004.09.008 PMID:15621354

Safarzadeh, M. S., Bafghi, M. S., Moradkhani, D., & Ilkhchi, M. O. (2007). A review on hydrometallurgical extraction and recovery of cadmium from various resources. *Minerals Engineering*, *20*(3), 211–220. doi:10.1016/j.mineng.2006.07.001

Said, M., Hasan, H. A., Tusirin, M., & Nor, M. (2015). *Removal of COD, TSS and colour from palm oil mill effluent (POME) using montmorillonite*. doi:10.1080/19443994.2015.1036778

Said, M., Ahmad, A., Wahab, A., Tusirin, M., & Nor, M. (2014). Journal of Industrial and Engineering Chemistry Blocking mechanism of PES membrane during ultrafiltration of POME. *Journal of Industrial and Engineering Chemistry*, 1–7. doi:10.1016/j.jiec.2014.02.023

Sakthivel, M., Rajeshkannan, G., Naveenkumar, M., & Muralimanokar, D. M. (2017). Design and analysis of twin shaft shreader using Pro-E and Hyperworks software. *International Journal of Advanced Research in Basic Engineering Sciences and Technology (IJARBEST)*, *3*(24), 836–858.

Salama, E.-S., Saha, S., Kurade, M. B., Dev, S., Chang, S. W., & Jeon, B.-H. (2019). Recent trends in anaerobic codigestion: Fat, oil, and grease (FOG) for enhanced biomethanation. *Progress in Energy and Combustion Science*, *70*, 22–42. doi:10.1016/j.pecs.2018.08.002

Saldías, C., Speelman, S., Amerasinghe, P., & Van Huylenbroeck, G. (2015). Institutional and policy analysis of wastewater (re) use for agriculture: Case study Hyderabad, India. *Water Science and Technology*, *72*(2), 322–331. doi:10.2166/wst.2015.220 PMID:26177416

Salles, A., Wolff, D., & Silveira, G. (2012). Solid wastes drained in an urban river sub-basin. *Urban Water Journal*, *9*(1), 21–28. doi:10.1080/1573062X.2011.633612

Salvadori, M. R., Ando, R. A., Oller Do Nascimento, C. A., & Correa, B. (2014). Bioremediation from wastewater and extracellular synthesis of copper nanoparticles by the fungus Trichoderma koningiopsis. *Journal of Environmental Science and Health. Part A, Toxic/Hazardous Substances & Environmental Engineering*, *49*(11), 1286–1295. doi:10.1080/10934529.2014.910067 PMID:24967562

Samson, M. (2010). *Organising reclaimers in Tshwane, SA-Lessons from Topdown and bottom up experiences. Women in informal employment: Globalizing and Organizing*. WIEGO.

Sanlisoy, A., & Carpinlioglu, M. O. (2017). A review on plasma gasification for solid waste disposal. *International Journal of Hydrogen Energy*, *42*(2), 1361–1365. doi:10.1016/j.ijhydene.2016.06.008

Santini, J. M., Stolz, J. F., & Macy, J. M. (2002). Isolation of a new arsenate respiring bacterium –physiological and Phylogenetic studies. *Geomicrobiology Journal*, *19*, 41–52.

Santos, F., Queiroz, J. H. & Queiroz, J. H. (2013). *Bioenergia & Biorrefinaria: Cana-de-Açúcar & Espécies Florestais*. Viçosa: UFV.

Santos, A. F., Guimarães, J. B. Junior, Martins, E. H., Lima, J. V., & Protásio, T. P. (2015). Rendimento efetivo em laminação de madeira de *Pinus oocarpa* cultivada no estado de Minas Gerais. *Enciclopédia Biosfera*, *11*(21), 1662–1670.

Sarangi, Bano, & Pant. (2014). Future Trend in Indian Automobile Industry: A Statistical Approach. *Apeejay-Journal of Management Sciences and Technology*, *2*(1), 28–32.

Sarkar, N., Ghosh, S. K., Bannerjee, S., & Aikat, K. (2012). Bioethanol production from agricultural wastes: An overview. *Renewable Energy*, *37*(1), 19–27. doi:10.1016/j.renene.2011.06.045

Sawalha, M. F., Peralta-Videa, J. R., Romero-Gonzalez, J., & Gardea-Torresdey, J. L. (2006). Biosorption of Cd(II), Cr(III), and Cr(VI) by saltbush (*Atriplex canescens*) biomass: Thermodynamic and isotherm studies. *Journal of Colloid and Interface Science*, *300*(1), 100–104. doi:10.1016/j.jcis.2006.03.029 PMID:16600278

Schieber, A., Stintzing, F. C., & Carle, R. (2001). By-products of plant food processing as a source of functional compounds - recent developments. *Trends in Food Science & Technology*, *12*(11), 401–413. doi:10.1016/S0924-2244(02)00012-2

Schnoor, J. L. (2000). Degradation by Plants-Phytoremediation. *Biotechnology: An Indian Journal*, *11b*, 372–384.

Schübeler, P. (1996). Conceptual framework for municipal solid waste management in low-income countries. In K. Wehrle & J. Christen (Eds.). UNDP/UNCHS/World Bank/SDC Collaborative Programme on Municipal Solid Waste Management in Low-Income Countries. Academic Press.

Schulz-Berendt, V. (2000). Bioremediation with Heap Technique. *Biotechnology: An Indian Journal*, *11b*, 320–328.

Seeber, G., Buchmeiser, M. R., Bonn, G. K., & Bertsch, T. (1998). Determination of airborne, volatile amines from polyurethane foams by sorption onto a high-capacity cation-exchange resin based on poly (succinic acid). *Journal of Chromatography. A*, *809*(1-2), 121–129. doi:10.1016/S0021-9673(98)00146-0 PMID:9677714

Sehaqui, H., Alliais, M., Zhou, Q., & Berglund, L. A. (2011). Wood cellulose biocomposites with fibrous structures at micro and nanoscale. *Composites Science and Technology*, *71*(3), 382–387. doi:10.1016/j.compscitech.2010.12.007

Seixas, F., Bauch, S. C., & Oliveira Júnior, E. D. (2005). Energetic balance and economical study of two debarking alternatives for *Eucalypts* wood. *Scientia Forestalis*, *67*, 37–43.

Sekabira, K., Oryem–Origa, H., Mutumba, G., Kakudidi, E., & Basamba, T. A. (2011). *Heavy metal phytoremediation by Commelina benghalensis (L) and Cynodon dactylon (L) growing in urban stream sediments*. Academic Press.

Senkoro, H. (2003). *Solid Waste Management in Africa: A WHO/AFRO Perspective*. Paper 1, Presented in Dares Salaam at the CWG Workshop.

Sen, S. K., & Raut, S. (2015). Microbial degradation of low density polyethylene (LDPE): A review. *Journal of Environmental Chemical Engineering*, *3*(1), 462–473. doi:10.1016/j.jece.2015.01.003

Sentime, K. (2011). Profiling solid waste pickers: A case study of Braamfontein-Greater Johannesburg. *Africanus*, *41*, 96–111.

Şentürk, E., İnce, M., & Engin, G. O. (2012). The effect of transient loading on the performance of a mesophilic anaerobic contact reactor at constant feed strength. *Journal of Biotechnology*, *164*(2), 232–237. doi:10.1016/j.jbiotec.2012.08.016 PMID:22960727

Senzige, J., Makinde, D., Njau, K., & Nkansah-Gyeke, Y. (2014). Factors influencing solid waste generation and composition in urban areas of Tanzania: The case of Dar-es- Salaam. *American Journal of Environmental Protection*, *3*(4), 172–178. doi:10.11648/j.ajep.20140304.11

Sepulveda, A., Schluep, M., Renaud, F. G., Streicher, M., Kuehr, R., Christian, H., & Gerecke, A. C. (2010). A review of the environmental fate and effects of hazardous substances released from electrical and electronic equipment during recycling: Examples from China and India. *Environmental Impact Assessment Review*, *30*(1), 28–41. doi:10.1016/j.eiar.2009.04.001

Seymour, R. B. (1989). Polymer science before and after 1899: Notable developments during the lifetime of Maurits Dekker. *Journal of Macromolecular Science. Chemistry*, *26*(8), 1023–1032. doi:10.1080/00222338908052032

Seyram, S. K., Yaovi, A., Simplice, K. D., & Comlan, D. S. (2009). Study of pineapple peelings processing into vinegar by biotechnologies. *Pakistan Journal of Biological Sciences*, *12*(11), 859–865. doi:10.3923/pjbs.2009.859.865 PMID:19803120

Shah, A., Hasan, F., Hameed, A., & Ahmed, S. (2008). Biological degradation of plastics: A comprehensive review. *Biotechnology Advances*, *26*(3), 246–265. doi:10.1016/j.biotechadv.2007.12.005 PMID:18337047

Shah, M., Hashmi, H. N., Ali, A., & Ghumman, A. R. (2014). Performance assessment of aquatic macrophytes for treatment of municipal wastewater. *Journal of Environmental Health Science & Engineering*, *12*(1), 106. doi:10.1186/2052-336X-12-106 PMID:25089203

Shahriari, H., Warith, M., Hamoda, M., & Kennedy, K. J. (2012). Anaerobic digestion of organic fraction of municipal solid waste combining two pre-treatment modalities, high temperature microwave and hydrogen peroxide. *Waste Management (New York, N.Y.)*, *32*(1), 41–52. doi:10.1016/j.wasman.2011.08.012 PMID:21945550

Shannon, M. J. R., & Unterman, R. (2000). Evaluation Bioremediation: Distinguishing Fact from Fiction. *Annual Review of Microbiology*, *47*, 715–738. PMID:8257113

Sharafi, S. M., Rasooli, I., & Beheshti-Maal, K. (2010). Isolation, characterization and optimixation of indigenous acetic acid bacteria and evaluation of their preservation methods. *Iranian Journal of Microbiology*, *2*(1), 41–48. PMID:22347549

Sharholy, M., Ahmad, K., Mahmood, G., & Trivedi, R. (2008). Municipal solid waste management in Indian cities-A review. *Waste Management (New York, N.Y.)*, *28*(2), 459–467. doi:10.1016/j.wasman.2007.02.008 PMID:17433664

Sharma, J., Gurung, T., Upadhyay, A., Nandy, K., Agnihotri, P., & Mitra, A. K. (2015). Isolation and characterization of plastic degrading bacteria from soil collected from the dumping grounds of an industrial area. *International Journal of Advance & Innovative Research*, *3*, 2278–7844.

Sharma, M., Sharma, P., Sharma, A., & Chandra, S. (2015). Microbial Degradation of Plastic—A Brief Review. *CIBTech Journal of Microbiology*, *4*(1), 85–89.

Sharma, R. K., Agrawal, M., & Marshall, F. M. (2009). Heavy metals in vegetables collected from production and market sites of a tropical urban area of India. *Food and Chemical Toxicology*, *47*(3), 583–591. doi:10.1016/j.fct.2008.12.016 PMID:19138719

Sharma, S., & Malaviya, P. (2016). Bioremediation of tannery wastewater by chromium resistant novel fungal consortium. *Ecological Engineering*, *91*, 419–425. doi:10.1016/j.ecoleng.2016.03.005

Sharma, V., Testa, C., Lastella, G., Cornacchia, G., & Comparato, M. (2000). Inclined-plug-flow type reactor for anaerobic digestion of semi-solid waste. *Applied Energy*, *65*(1-4), 173–185. doi:10.1016/S0306-2619(99)00084-7

Shavandi, M. A., Haddadian, Z., Ismail, M. H. S., Abdullah, N., & Abidin, Z. Z. (2012). *Removal of Residual Oils from Palm Oil Mill Effluent by Adsorption on Natural Zeolite*. doi:10.100711270-012-1169-6

Sheehan, J., Dunahay, T., Benemann, J., & Roessler, P. (1998). A look back at the US Department of Energy's aquatic species program: Biodiesel from algae. *National Renewable Energy Laboratory*, *328*, 1–294.

Sheng, P. P., & Etsell, T. H. (2007). Recovery of gold from computer circuit board scrap using aqua regia. *Waste Management & Research*, *25*(4), 380–383. doi:10.1177/0734242X07076946 PMID:17874665

Sheth, M., Kwartler, S., Schmaltz, E., Hoskinson, S., Martz, E., Dunphy-Daly, M., Schultz, T. F., Read, A. J., Eward, W. C., & Somarelli, J. A. (2019). Bioengineering a future free of marine plastic waste. *Frontiers in Marine Science*, *6*, 1–10. doi:10.3389/fmars.2019.00624

Shibata, J., & Matsumoto, S. (1999). *Development of environmentally friendly leaching and recovery process of gold and silver from wasted electronic parts*. Available from: http://www. environmentalcenter. com/articles/article320/ article320. htm

Shimao, M. (2001). Biodegradation of plastics. *Current Opinion in Biotechnology*, *12*(3), 242–247. doi:10.1016/S0958-1669(00)00206-8 PMID:11404101

Shi, X., Luo, C., Lu, H., & Yu, K. (2019). Primary recycling of anhydride-cured engineering epoxy using alcohol solvent. *Polymer Engineering and Science*, *59*(s2), E111–E119. doi:10.1002/pen.24997

Siddiqui, J., & Pandey, G. (2013). A review of plastic waste management strategies. *International Research Journal of Environmental Sciences*, *2*(12), 84–88.

Silva, G. F., Mendonça, A. R., Hoffmann, R. G., Zaneti, L. Z., Chichorro, J. F. & Ferreira, R. L. C. (2015). *Rendimento em laminação de madeira de parica na região de Paragominas*. Academic Press.

Simatele, D., & Etambakonga, C. L. (2015). Scavenging for solid waste in Kinshasa: A livelihood strategy for the urban poor in the Democratic Republic of Congo. *Habitat International*, *49*, 266–274. doi:10.1016/j.habitatint.2015.05.029

Simelane, T., & Mohee, R. (2012). Future directions of municipal solid waste management in Africa. *Policy Brief. Briefing*, *81*, 1–5.

Simelane, T., & Mohee, R. (2015). *Future directions of municipal solid waste management in Africa*. Africa Institute of SA.

Singh, Singh, Arora, Bharti, Rathi, Tyagi, & Kaicker. (2019). Utilization of PVB windshield waste for making protective primer and peelable paint. Poster presentation in DST – Plastic Waste Management, Swachhta Hi Seva (SHS), Delhi, India.

Singh, A., Agrawal, M., & Marshall, F. M. (2010). The role of organic vs. inorganic fertilizers in reducing phytoavailability of heavy metals in a wastewater-irrigated area. *Ecological Engineering*, *36*(12), 1733–1740. doi:10.1016/j.ecoleng.2010.07.021

Singha, B., & Das, S. K. (2011). Biosorption of Cr(VI) ions from aqueous solutions: Kinetics, equilibrium, thermodynamics and desorption studies. *Colloids and Surfaces. B, Biointerfaces*, *84*(1), 221–232. doi:10.1016/j.colsurfb.2011.01.004 PMID:21282045

Singha, B., & Das, S. K. (2013). Adsorptive removal of Cu(II) from aqueous solution and industrial effluent using natural/agricultural wastes. *Colloids and Surfaces. B, Biointerfaces*, *107*, 97–106. doi:10.1016/j.colsurfb.2013.01.060 PMID:23466548

Singhabhandhu, A., & Tezuka, T. (2010). The waste-to-energy framework for integrated multi-waste utilization: Waste cooking oil, waste lubricating oil, and waste plastics. *Energy*, *35*(6), 2544–2551. doi:10.1016/j.energy.2010.03.001

Singh, H., Chauhan, G., Jain, A. K., & Sharma, S. K. (2017). Adsorptive potential of agricultural wastes for removal of dyes from aqueous solutions. *Journal of Environmental Chemical Engineering*, *5*(1), 122–135. doi:10.1016/j.jece.2016.11.030

Singh, J., & Gu, S. (2010). Commercialization potential of microalgae for biofuels production. *Renewable & Sustainable Energy Reviews*, *14*(9), 2596–2610. doi:10.1016/j.rser.2010.06.014

Singh, M., & Srivastava, R. (2011). Sequencing batch reactor technology for biological wastewater treatment: A review. *Asia-Pacific Journal of Chemical Engineering*, *6*(1), 3–13. doi:10.1002/apj.490

Singh, R. N., & Sharma, S. (2012). Development of suitable photobioreactor for algae production–A review. *Renewable & Sustainable Energy Reviews*, *16*(4), 2347–2353. doi:10.1016/j.rser.2012.01.026

Singh, S., & Prerna, P. (2009). Review of recent advances in anaerobic packed-bed biogas reactors. *Renewable & Sustainable Energy Reviews*, *13*(6-7), 1569–1575. doi:10.1016/j.rser.2008.08.006

Sinha-Khetriwal, D. (2002). *The management of electronic waste: A comparative study on India and Switzerland* (MS thesis). St. Gallen, Switzerland: University of St. Gallen.

Sinha, V., Patel, M. R., & Patel, J. V. (2010). PET waste management by chemical recycling: A review. *Journal of Polymers and the Environment*, *18*(1), 8–25. doi:10.100710924-008-0106-7

Sivakumar, G. D., Barnabas, S. G., & Anatharam, S. (2013). Indian Automobile Material Recycling. Management. *International Journal of Innovative Research in Science, Engineering and Technology*, *3*(3), 2347–2671.

Sivian, A. (2011). New perspectives in plastic biodegradation. *Current Opinion in Biotechnology*, *22*(3), 422–426. doi:10.1016/j.copbio.2011.01.013 PMID:21356588

Slade, R., & Bauen, A. (2013). Micro-algae cultivation for biofuels: Cost, energy balance, environmental impacts and future prospects. *Biomass and Bioenergy*, *53*, 29–38. doi:10.1016/j.biombioe.2012.12.019

Smith, W. M. (Ed.). (1964). *Manufacture of plastic* (Vol. 1). Reinhold Pub. Corp.

Smook, G. (2016). *Handbook for pulp and paper technologists* (4th ed.). TAPPI Press.

Sokollek, S. J., Hertel, C., & Hammes, W. P. (1998). Cultivation and preservation of vinegar bacteria. *Journal of Biotechnology*, *60*(3), 195–206. doi:10.1016/S0168-1656(98)00014-5

Solomons, T. W. G., Fryhle, C. B., & Snyder, S. A. (2016). *Organic Chemistry* (12th ed.). Wilwy.

Sonashree, R., Bhavana, J., Rashmi, R. H., Shamsiya, R., & Praveena, B. (2020). Isolation and Characterization of *Acetobacter aceti* sugar cane- A review. *Indo American Journal of Pharmaceutical Sciences, 7*(3), 478-481.

Sônego, Costab, & Ambrósio. (2016). Dynamic and Static Crosslinking of Polyvinyl Butyral with a Silane Agent. *Plastic Research Online*, 1-3.

Sônego, M., Costab, L. C., & Ambrósio, J. D. (2016). Polyvinyl Butyral Chemically Modified With a Silane Agent in the Molten State. *Polymer Engineering and Science, 56*(9), 971–979. doi:10.1002/pen.24326

Sônego, M., Cristina, L. C., & Ambrósio, J. D. (2015). Flexible Thermoplastic Composite of Polyvinyl Butyral (PVB) and Waste of Rigid Polyurethane Foam. *Polímeros, 25*(2), 175–180. doi:10.1590/0104-1428.1944

Song, S., Duan, Y., Zhang, T., Zhang, B., Zhao, Z., Bai, X., Xie, L., He, Y., Ouyang, J., Huang, X., & Sun, H. (2019). Serum concentrations of bisphenol A and its alternatives in elderly population living around e-waste recycling facilities in China: Associations with fasting blood glucose. *Ecotoxicology and Environmental Safety, 169*, 822–828. doi:10.1016/j.ecoenv.2018.11.101 PMID:30597781

Soon, A. N., & Hameed, B. H. (2011). Heterogeneous catalytic treatment of synthetic dyes in aqueous media using Fenton and photo-assisted Fenton process. *DES, 269*(1–3), 1–16. doi:10.1016/j.desal.2010.11.002

South African Cities Network (SACN). (2014). State of waste management in cities-Phase 2: Modelling the effects of landfilling as a disposal method. SACN.

South African Institute of Race Relations (SAIRR). (2013). *SA 'two-thirds urbanised.* Available at: www.southafrica.info/news/urbanisation-240113.htm

Sparks, S. (2012). *Nanotechnology: business applications and commercialization.* CRC Press.

Spolaore, P., Joannis-Cassan, C., Duran, E., & Isambert, A. (2006). Commercial applications of microalgae. *Journal of Bioscience and Bioengineering, 101*(2), 87–96. doi:10.1263/jbb.101.87 PMID:16569602

Sridhar, M. K. C. (2000). *Nature of Solid Wastes and Their Management.* Paper Presented at *a one Day Seminar Organized by NINNAFEH*, Ibadan, Nigeria.

Srivastava, R., Krishna, V., & Sonkar, I. (2014). Characterization and management of municipal solid waste: A case study of Varanasi city, India. *International Journal of Current Research and Academic Review, 2*(8), 10–16.

Srivastava, V. C., Mall, I. D., & Mishra, I. M. (2006). Characterization of mesoporous rice husk ash (RHA) and adsorption kinetics of metal ions from aqueous solution onto RHA. *Journal of Hazardous Materials, B134*(1-3), 257–267. doi:10.1016/j.jhazmat.2005.11.052 PMID:16386363

Srivastava, V. C., Mall, I. D., & Mishra, I. M. (2009). Competitive adsorption of cadmium(II) and nickel(II) ions from aqueous solution onto rice husk ash. *Chemical Engineering and Processing, 48*(1), 370–379. doi:10.1016/j.cep.2008.05.001

Statistics, S. A. (2003). *Investigations into appropriate definitions of urban and rural in SA.* Report No. 03-02-21.

Statistics, S. A. (2011). *Buffalo City Municipality, Census 2011.* Available Online: http://beta2.statssa.gov.za/?page_id=1021andid=buffalo-city-municipality

Statistics, S. A. (2009). *Non-financial municipal census of municipalities for the year ending June 2008.* Statistics SA.

Steiner, C., Das, K. C., Melea, N., & Lakly, D. (2010). Reducing nitrogen loss during poultry litter composting using biochar. *Journal of Environmental Quality, 39*(4), 1236–1242. doi:10.2134/jeq2009.0337 PMID:20830911

Stevens, E. (2003). What makes green plastics green? *BioCycle*, *44*, 24–27.

Stottmeister, U., Wießner, A., Kuschk, P., Kappelmeyer, U., Kästner, M., Bederski, O., Müller, R. A., & Moormann, H. (2003). Effects of plants and microorganisms in constructed wetlands for wastewater treatment. *Biotechnology Advances*, *22*(1-2), 93–117. doi:10.1016/j.biotechadv.2003.08.010 PMID:14623046

Sujauddin, M., Huda, S., & Hoque, A. R. (2008). Household solid waste characteristics and management in Chittagong, Bangladesh. *Waste Management (New York, N.Y.)*, *28*(9), 1688–1695. doi:10.1016/j.wasman.2007.06.013 PMID:17845843

Sukiran, M. A., Abnisa, F., Wan Daud, W. M. A., Abu Bakar, N., & Loh, S. K. (2017). A review of torrefaction of oil palm solid wastes for biofuel production. *Energy Conversion and Management*, *149*, 101–120. doi:10.1016/j.enconman.2017.07.011

Sulaiman, F. R., & Hamzah, H. A. (2018). Heavy metals accumulation in suburban roadside plants of a tropical area (Jengka, Malaysia). *Ecological Processes*, *7*(28), 1–11. doi:10.118613717-018-0139-3

Sum, E. Y. (1991). The recovery of metals from electronic scrap. *JOM*, *43*(4), 53–61. doi:10.1007/BF03220549

Sung, M. H., Park, C., Kim, C. J., Poo, H., Soda, K., & Ashiuchi, M. (2005). Natural and edible biopolymer poly-γ-glutamic acid: Synthesis, production, and applications. *Chemical Record (New York, N.Y.)*, *5*(6), 352–366. doi:10.1002/tcr.20061 PMID:16278834

Svarstad, H., Petersen, L. K., Rothman, D., Siepel, H., & Watzold, F. (2008). Discursive biases of the environmental research framework DPSIR. *Land Use Policy*, *25*(1), 116–125. doi:10.1016/j.landusepol.2007.03.005

Swain, B., Park, J. R., Shin, D. Y., Park, K.-S., Hong, M. H., & Lee, C. G. (2015). Recycling of Waste Automotive Laminated Glass and Valorization of Polyvinyl Butyral through Mechanochemical Separation. *Environmental Research*, *142*, 615–623. doi:10.1016/j.envres.2015.08.017 PMID:26318256

Swift, G. (1997). *Non-medical biodegradable polymers: environmentally degradable polymers. Handbook of biodegradable polymers*. Hardwood Academic.

Switzenbaum, M. A. (1983). Comparison of the anaerobic filter and the anaerobic expanded/fluidized bed processes. *Water Science and Technology*, *15*(8-9), 345–358. doi:10.2166/wst.1983.0177

Taljaard, A. (2011). *Critical perspectives on the definition of wastes in SA- experiences within the steelmaking industry in environmental management*. North West University, SA.

Tan, Y. H., Goh, P. S., Lai, G. S., Lau, W. J., & Ismail, A. F. (2014). *Jurnal Teknologi Full paper Treatment of Aerobic Treated Palm Oil Mill Effluent (AT-POME) by using TiO 2 Photocatalytic Process*. Academic Press.

Tan, G., Chen, C., Ge, L., Li, L., Tan, S., & Wang, J. (2015). Bioconversion of styrene to poly (hydroxyalkanoate) (PHA) by the new bacterial strain *Pseudomonas putida* NBUS12. *Microbes and Environments*, *30*(1), 76–85. doi:10.1264/jsme2.ME14138 PMID:25740622

Tang, D., Han, W., Li, P., Miao, X., & Zhong, J. (2011). CO_2 biofixation and fatty acid composition of *Scenedesmus obliquus* and *Chlorella pyrenoidosa* in response to different CO_2 levels. *Bioresource Technology*, *102*(3), 3071–3076. doi:10.1016/j.biortech.2010.10.047 PMID:21041075

Tang, S., Zou, P., Xiong, H., & Tang, H. (2008). Effect of nano-SiO2 on the performance of starch/polyvinyl alcohol blend films. *Carbohydrate Polymers*, *72*(3), 521–526. doi:10.1016/j.carbpol.2007.09.019

Tan, S. C. (2005). *Vinegar Fermentation*. Department of Food Science, Louisiana State University, Agricultural and Mechanical College.

Tansel, B. (2017). From electronic consumer products to e-wastes: Global outlook, waste quantities, recycling challenges. *Environment International*, *98*, 35–45. doi:10.1016/j.envint.2016.10.002 PMID:27726897

Tanskanen, P. (2013). Management and recycling of electronic waste. *Acta Materialia*, *61*(3), 1001–1011. doi:10.1016/j.actamat.2012.11.005

Tan, X., Chu, H., Zhang, Y., Yang, L., Zhao, F., & Zhou, X. (2014). *Chlorella pyrenoidosa* cultivation using anaerobic digested starch processing wastewater in an airlift circulation photobioreactor. *Bioresource Technology*, *170*, 538–548. doi:10.1016/j.biortech.2014.07.086 PMID:25164347

Tatsuno, H., Nishina, K., Shibayama, Y., Osaka, T., & Hirakawa, M. (2019). *Focus adjustment system for an image display device*. Google Patents.

Tavlarides, L. L., Bae, J. H., & Lee, C. K. (1987). Solvent extraction, membranes, and ion exchange in hydrometallurgical dilute metals separation. *Separation Science and Technology*, *22*(2-3), 581–617. doi:10.1080/01496398708068970

Taxier, A. C., Andrès, Y., Brasquet, C. F., & Cloirec, P. I. (2002). Fixed-bed study for anthanide (La, Eu, Yb) in removal from aqueous solution by immobilized Pseudomonas aeruginosa: Experiental data and modelization. *Chemosphere*, *47*(3), 333–342. doi:10.1016/S0045-6535(01)00244-2 PMID:11996155

Taylor, P., Malakahmad, A., & Chuan, S. Y. (2013). *Desalination and Water Treatment Application of response surface methodology to optimize coagulation – flocculation treatment of anaerobically digested palm oil mill effluent using alum*. doi:10.1080/19443994.2013.791778

Teller, M. (2006). Recycling of electronic waste material. In *Sustainable metals management* (pp. 563–576). Springer. doi:10.1007/1-4020-4539-5_23

Temkin, M. J., & Pyzhev, V. (1940). Recent modifications of Langmuir isotherms. *Acta Physiochim URSS*, *12*, 217–222.

Temoor, A., Shahid, M., Azeem, F., Rasul, I., Shah, A., Noman, M., & … . (2017). Biodegradation of plastics: Current scenario and future prospects for environmental safety. *Environmental Science and Pollution Research International*, *25*(8), 7287–7298. PMID:29332271

ten Brummeler, E., & Koster, I. W. (1990). Enhancement of Dry Anaerobic Batch Digestion of the Organic Fraction of Municipal Solid Waste by an Aerobic Pre-treatment Step. *Biological Wastes*, *31*(3), 199–210. doi:10.1016/0269-7483(90)90159-P

Teng, H., & Hsieh, C. T. (1999). Activation energy for oxygen chemisorptions on carbon at low temperature. *Industrial & Engineering Chemistry Research*, *38*(1), 292–297. doi:10.1021/ie980107j

Teoh, J. (2018). *Oil palm fibre*. Academic Press.

Teotia, Meenu, & Soni. (2014). Polymer Interlayers for Glass Lamination-A Review. *International Journal of Science and Research*, *3*(8), 1264-1270.

Thandasherry, S. (2018, Jan.). High Fives for Solar Powered Ferry Ops. *Ship & Boat International*, 18-24.

The Star. (2017). *SMEs growth outpaces country's GDP growth.* Star Media Group Berhad.

Themelis, N. J., & Ulloa, P. A. (2007). Methane generation in landfills. *Renewable Energy*, *32*(7), 1243–1257. doi:10.1016/j.renene.2006.04.020

Theo, L. (1998, May). Integrated recycling of non-ferrous metals at Boliden Ltd. In *Proceedings of the 1998 IEEE International Symposium on Electronics and the Environment. ISEE-1998* (Cat. No. 98CH36145) (pp. 42-47). IEEE.

Thoemen, H., Irle, M., & Sernek, M. (2010). *Wood-based panels: an introduction for specialists*. Brunel University Press.

Thomas, H. G. (1948). Chromatography: A problem in kinetics. *Annals of the New York Academy of Sciences*, *49*(2), 161–182. doi:10.1111/j.1749-6632.1948.tb35248.x PMID:18911539

Tian, X., Fang, Z., Smith, R. L. Junior, Wu, Z., & Liu, M. (2016). Properties, Chemical Characteristics and Application of Lignin and Its Derivatives. In Z. Fang & R. L. Smith Junior., (Eds.), *Production of Biofuels and Chemicals from Lignin* (pp. 3–34). Springer. doi:10.1007/978-981-10-1965-4_1

Tiehm, A., Nickel, K., & Neis, U. (1997). The use of ultrasound to accelerate the anaerobic digestion of sewage sludge. *Water Science and Technology*, *36*(11), 121–128. doi:10.2166/wst.1997.0402

Tiehm, A., Nickel, K., Zellhorn, M., & Neis, U. (2001). Ultrasonic waste activated sludge disintegration for improving anaerobic stabilization. *Water Research*, *35*(8), 2003–2009. doi:10.1016/S0043-1354(00)00468-1 PMID:11337847

Tıkız, İ., & Pehlivan, H. (2018). Experimental Investigation for Alternative Use of Agricultural and Domestic Wastes. *Kırklareli Üniversitesi Mühendislik ve Fen Bilimleri Dergisi*, *4*(1), 142–152.

Ting, H., Haifeng, L., Shanshan, M., Zhang, Y., Zhidan, L., & Na, D. (2017). Progress in microalgae cultivation photobioreactors and applications in wastewater treatment: A review. *International Journal of Agricultural and Biological Engineering*, *10*(1), 1–29.

Tiquia, S. M., & Tam, N. F. Y. (2000). Microbiological and chemical parameters for compost maturity evaluation of spent pig litter disposed from the pig-on litter system. In *Proceedings of the International Composting Symposium (ICS 1999)*. CBA Press.

Tiwari, D. P., Singh, D. K., & Saksena, D. N. (1995). Hg(II) adsorption from aqueous solutions using rice-husk ash. *Journal of Environmental Engineering*, *121*(6), 479–481. doi:10.1061/(ASCE)0733-9372(1995)121:6(479)

Tokiwa, Y., & Calabia, B. (2004). Review degradation of microbial polyesters. *Biotechnology Letters*, *26*(15), 1181–1189. doi:10.1023/B:BILE.0000036599.15302.e5 PMID:15289671

Tokiwa, Y., & Calabia, B. P. (2007). Biodegradability and biodegradation of polyesters. *Journal of Polymers and the Environment*, *15*(4), 259–267. doi:10.100710924-007-0066-3

Tokiwa, Y., Calabia, B., Ugwu, C., & Aiba, S. (2009). Biodegradability of plastics. *International Journal of Molecular Sciences*, *10*(9), 3722–3742. doi:10.3390/ijms10093722 PMID:19865515

Toledo-Cervantes, A., Morales, M., Novelo, E., & Revah, S. (2013). Carbon dioxide fixation and lipid storage by *Scenedesmus obtusiusculus*. *Bioresource Technology*, *130*, 652–658. doi:10.1016/j.biortech.2012.12.081 PMID:23334023

Tomita, K., Kuroki, Y., Hayashi, N., & Komukai, Y. (2000). Isolation of a thermophile degrading poly (butylene succinate-co-butylene adipate). *Journal of Bioscience and Bioengineering*, *90*(3), 350–352. doi:10.1016/S1389-1723(00)80096-1 PMID:16232871

Tong, K. S., Kassim, M. J., & Azraa, A. (2011). Adsorption of copper ion from its aqueous solution by a novel biosorbent *Uncaria gambir*: Equilibrium, kinetics, and thermodynamic studies. *Chemical Engineering Journal*, *170*(1), 145–153. doi:10.1016/j.cej.2011.03.044

Tredici, M. R. (2004). Mass production of microalgae: photobioreactors. Handbook of Microalgal culture: Biotechnology and Applied Phycology, 1, 178-214.

Tredici, M. R., Margheri, M. C., Zittelli, G. C., Biagiolini, S., Capolino, E., & Natali, M. (1992). Nitrogen and phosphorus reclamation from municipal wastewater through an artificial food-chain system. *Bioresource Technology, 42*(3), 247–253. doi:10.1016/0960-8524(92)90028-V

Tribedi, P., & Sil, A. (2014). Cell surface hydrophobicity: A key component in the degradation of polyethylene succinate by *Pseudomonas sp.* AKS2. *Journal of Applied Microbiology, 116*(2), 295–303. doi:10.1111/jam.12375 PMID:24165295

Trisakti, B., Mhardela, P., Trisakti, B., Lubis, J., Hanum, H., & Tantawi, A. R. (2018). *Production of oil palm empty fruit bunch compost for ornamental plant cultivation Production of oil palm empty fruit bunch compost for ornamental plant cultivation.* doi:10.1088/1757-899X/309/1/012094

Trivedi, P., & Fardim, P. (2019). Recent advances in cellulose chemistry and potential applications. In Z. Fang, J. R. Smith, & X. F. Tian (Eds.), *Production of Materials from Sustainable Biomass Resources. Biofuels and Biorefineries* (pp. 99–115). Springer. doi:10.1007/978-981-13-3768-0_4

Troschinetz, A. M., & Mihelcic, J. R. (2009). Sustainable recycling of municipal solid waste in developing countries. *Waste Management (New York, N.Y.), 29*(2), 915–923. doi:10.1016/j.wasman.2008.04.016 PMID:18657963

Trovaslet, M., Enaud, E., Guiavarc'h, Y., Corbisier, A. M., & Vanhulle, S. (2007). Potential of a Pycnoporus sanguineus laccase in bioremediation of wastewater and kinetic activation in the presence of an anthraquinonic acid dye. *Enzyme and Microbial Technology, 41*(3), 368–376. doi:10.1016/j.enzmictec.2007.03.007

Trudel, D., Scheringer, M., von Goetz, N., & Hungerbühler, K. (2011). Total consumer exposure to polybrominated diphenyl ethers in North America and Europe. *Environmental Science & Technology, 45*(6), 2391–2397. doi:10.1021/es1035046 PMID:21348481

Tukahirwa, J. T., Mol, A., & Oosterveer, P. (2011). Access of urban poor to NGO/CBO-supplied sanitation and solid waste services in Uganda: The role of social proximity. *Habitat International, 35*(4), 582–591. doi:10.1016/j.habitatint.2011.03.006

Tuomela, M., Vikman, M., Hatakka, A., & Itävaara, M. (2000). Biodegradation of lignin in a compost environment: A review. *Bioresource Technology, 72*(2), 169–183. doi:10.1016/S0960-8524(99)00104-2

Tupý, M., Měřínská, D., & Kašpárková, V. (2015). *PVB Sheet Recycling and Degradation.* Retrieved from www.intechopen.com

Tupy, M., Merinska, D., Tesarikova, A., & Peternek, V. (2014). Mechanical Properties of Recycled Plasticized PVB / PVC Blend. World Academy of Science, Engineering and Technology. *International Journal of Chemical and Molecular Engineering, 8*(9), 981-986. Retrieved from www.waset.org

Tupý, M., & Petránek, V. (2014). Use of Recycled PVB as a Protection against Carbonation. *International Journal of Chemical, Molecular, Nuclear, Materials and Metaullargical Egg., 8*(6), 514–541.

Uçkun Kiran, E., Stamatelatou, K., Antonopoulou, G., & Lyberatos, G. (2016). Production of biogas via anaerobic digestion. Handbook of Biofuels Production, 259–301.

Uçkun Kiran, E., Trzcinski, A. P., Ng, W. J., & Liu, Y. (2014). Bioconversion of food waste to energy: A review. *Fuel, 134*, 389–399. doi:10.1016/j.fuel.2014.05.074

UNEP-United Nations Environment Programme. (2005). *Integrated Waste Management Scoreboard: a tool to measure performance in municipal solid waste management.* UNEP.

UN-Habitat. (2010). Solid waste management in the world's cities e Water and sanitation in the world's cities 2010. London: United Nations Human Settlements Programme (UN-Habitat), Earthscan.

United Nations Environment Program (UNEP). (2005). *Solid waste management*. Available Online: http://www.unep.or.jp/ietc/Publications/spc/SolidWaste Management/index.asp

United Nations Secretariat. (2010). *Population division of the department of economic and social affairs: World population prospects*. Author.

United States Environmental Protection Agency. (2013). *Wastes-non-hazardous waste-municipal solid waste*. Author.

Urase, T., Okumura, H., Panyosaranya, S., & Inamura, A. (2008). Emission of volatile organic compounds from solid waste disposal sites and importance of heat management. *Waste Management & Research*, *26*(6), 534–538. doi:10.1177/0734242X07084321 PMID:19039069

US20080268206A1. (2006). *A process for making heat resistant adhesive*. United States Patent.

US2358355. (1944). *Treatment of polyvinyl butyral resins*. United States Patent.

US2430053. (1942). *Polyvinyl butyral-resorcinol-formaldehyde adhesive*. United States Patent.

US2798015 A. (1954). *Coating composition for Metals*. United States Patent.

US3035005. (1956). *Method for preparing a corrosion inhibiting resinous coating composition*. United States Patent.

US3282865 A. (1963). *Composition of hexamethoxymethyl-melamine with viscous plasticizers*. United States Patent.

US3320203A. (1963). *A sprayable Polyvinyl butyral wood primer for polyurethane or polyester finish*. United States Patent.

US3325432 A. (1965). *Sprayable weldable primer containing polyvinyl butyral, chromium trioxide and aluminum powder*. United States Patent.

US3392130 A. (1963). *Coating compositions comprising polyvinyl butyral-hydrolyzed alkyl silicate reaction product and zinc dust*. United States Patent.

US4251591. (1979). *A process for making laminates*. United States Patent.

US5130370. (1991). *Molding compositions of polyvinylbutyral blend*. United States Patent.

US5972176A. (1997). *A process for making adhesive using corona treatment*. United States Patent.

US7521107. (2003). *Recycled polyvinyl butyral compositions and uses*. United States Patent.

USEPA. (1997). *Cleaning Up the Nation's Waste Sites: Markets and Technology Trends, United States Environmental Protection agency EPA/542/R-96/005*. Office of Solid Waste and Emergency Response.

USEPA. (2012). *Industrial food processing waste analyses*. USEPA.

Utomo, R. P., Chang, Y. R., Lee, D. J., & Chang, J. S. (2013). Lutein recovery from *Chlorella* sp. ESP-6 with coagulants. *Bioresource Technology*, *139*, 176–180. doi:10.1016/j.biortech.2013.04.025 PMID:23660381

Vallejos, M. E., Zambon, M. D., Area, M. C., & Curvelo, A. A. S. (2015). Low liquid-solid ratio fractionation of sugarcane bagasse by hot water autohydrolysis and organosolv delignification. *Industrial Crops and Products*, *65*, 349–353. doi:10.1016/j.indcrop.2014.11.018

Van Den Hende, S., Vervaeren, H., & Boon, N. (2012). Flue gas compounds and microalgae: (bio-) chemical interactions leading to biotechnological opportunities. *Biotechnology Advances*, *30*(6), 1405–1424. doi:10.1016/j.biotechadv.2012.02.015 PMID:22425735

Van Den Hende, S., Vervaeren, H., Desmet, S., & Boon, N. (2011). Bioflocculation of microalgae and bacteria combined with flue gas to improve sewage treatment. *New Biotechnology*, *29*(1), 23–31. doi:10.1016/j.nbt.2011.04.009 PMID:21565287

Van Deuren, J., Lloyd, T., Chhetry, S., Raycharn, L., & Peck, J. (2002). Remediation Technologies Screening Matrix and Reference Guide. *Federal Remediation Technologies Roundtable, 5*, 1-4.

Vara, S., Karnena, M. K., Dwarapureddi, B. K., & Chintalapudi, B. (2019). Will Single Use Products Lead to Sustainability? *International Journal of Social Ecology and Sustainable Development*, *10*(2), 37–52. doi:10.4018/IJSESD.2019040104

Vargas, I. M., & Wiebeck, H. (2007). *Formulation of Primers to Wood Floorings Covering Using PVB Films from the Recycling of Windshields.* Paper Presented at Ninth Conferences Brasilo De Polymers. Retrieved from www.iepn.org

Varghese, J., Saritha, V., Bhavya Kavitha, D., Yasaswitha, O., & Karnena, M. K. (2016). Journal of Global Biosciences. *Journal of Global Biosciences*, *5*(3), 3776–3784.

Vasumathi, K. K., Premalatha, M., & Subramanian, P. (2012). Parameters influencing the design of photobioreactor for the growth of microalgae. *Renewable & Sustainable Energy Reviews*, *16*(7), 5443–5450. doi:10.1016/j.rser.2012.06.013

Vegas, E., Mateo, W., Tesfaye, A. B., Cerezo, R. M., Callejón, M., Poblet, J. M., Guillamón, A., & Mas, M. J. (2010). Population dynamics of acetic acid bacteria during traditional wine vinegar production. *International Journal of Food Microbiology*, *138*(1-2), 130–136. doi:10.1016/j.ijfoodmicro.2010.01.006 PMID:20117853

Veit, H. M., Bernardes, A. M., Ferreira, J. Z., Tenório, J. A. S., & de Fraga Malfatti, C. (2006). Recovery of copper from printed circuit boards scraps by mechanical processing and electrometallurgy. *Journal of Hazardous Materials*, *137*(3), 1704–1709. doi:10.1016/j.jhazmat.2006.05.010 PMID:16757116

Vekaria, R. H., Patel, M. N., Bhalodiya, P. N., Patel, V., Desai, T. R., & Tirgar, P. R. (2012). Evaluation of neuroprotective effect of *Coriandrum sativum* Linn. against ischemic - reperfusion insult in brain. *International Journal of Phytopharmacology*, *3*(2), 186–193.

Veldhuizen, H., & Sippel, B. (1994). Mining discarded electronics. *Ind. Environ.(Switzerland)*, *17*(3), 7–11.

Velmurugan, B., & Arathy, E. C. (2010). Anaerobic co-digestion of fruit and vegetables wastes and primary sewage sludge. *Environ. Sc. And Engg*, *52*, 19–22. PMID:21114101

Venkatesh, S., & Kukreti, I. (2018). *An Indian consumes 11 kg plastic every year and an average American 109 kg. Down to Earth.* https://www.downtoeart h.org.in/news/waste /an-indian-consumes-11-kg-plastic-every-year-and-an-average-american-109-kg-60745

Vermeulen, T. (1953). Theory for irreversible and constant pattern solid diffusion. *Industrial & Engineering Chemistry*, *45*(8), 1664–1670. doi:10.1021/ie50524a025

Vieira, R. S., Lima, J. T., Silva, J. R. M., Hein, P. R. G., Bailleres, H., & Baraúna, E. E. P. (2010). Small wooden objects using eucalypt sawmill wood waste. *BioResources*, *5*, 1463–1472.

Vignesh, R., Deepika, R. C., Manigandan, P., & Janani, R. (2016). Screening of plastic degrading microbes from various dumped soil samples. *Int Res J Eng Tech*, *3*(4), 2493–2498.

Vikas, O.V., & Umesh, M. (2014). Bioconversion of papaya peel waste into vinegar using Acetobacter aceti. *Int. J. Sci. Res, 3*(11), 409-411.

Vulić, M., & Tomović, A. (2015). Recycling of Auto Glass from Aspect of Technological Impacts on Some Parts of Recycling. *Proceedings of 9th International Quality Conference held at University of Karagujevac.*

Waldron, K. (2014). *Advances in Biorefineries: Biomass and Waste Supply Chain Exploitation.* Cambridge: Woodhead.

Wallberg, O., Linde, M., & Jonsson, A. (2006). Extraction of lignin and hemiceluloses from black liquor. *Desalination, 199*(1-3), 413–414. doi:10.1016/j.desal.2006.03.094

Wang, D. H. (2016). *Patent landscape report : palm oil production and waste treatment technologies.* Academic Press.

Wang, D., Cai, Z., Jiang, G., Leung, A., Wong, M. H., & Wong, W. K. (2005). Determination of polybrominated diphenyl ethers in soil and sediment from an electronic waste recycling facility. *Chemosphere, 60*(6), 810–816. doi:10.1016/j.chemosphere.2005.04.025 PMID:15946727

Wang, G., Qin, L., Li, G., & Chen, L. (2009). Landfill site selection using spatial in- formation technologies and AHP: A case study in Beijing, China. *Journal of Environmental Management, 90*(8), 2414–2421. doi:10.1016/j.jenvman.2008.12.008 PMID:19375842

Wang, H. M. D., Chen, C. C., Huynh, P., & Chang, J. S. (2015). Exploring the potential of using algae in cosmetics. *Bioresource Technology, 184*, 355–362. doi:10.1016/j.biortech.2014.12.001 PMID:25537136

Wang, J. Y., Stabnikova, O., Tay, S. T., Ivanov, V., & Tay, J. H. (2004). Biotechnology of intensive aerobic conversion of sewage sludge and food waste into fertilizer. *Water Science and Technology, 49*(10), 147–154. doi:10.2166/wst.2004.0631 PMID:15259949

Wang, M., Mao, M., Zhang, M., Wen, G., Yang, Q., Su, B., & Ren, Q. (2019). Highly efficient treatment of textile dyeing sludge by CO2 thermal plasma gasification. *Waste Management (New York, N.Y.), 90*, 29–36. doi:10.1016/j.wasman.2019.04.025 PMID:31088671

Wang, S. K., Stiles, A. R., Guo, C., & Liu, C. Z. (2014). Microalgae cultivation in photobioreactors: An overview of light characteristics. *Engineering in Life Sciences, 14*(6), 550–559. doi:10.1002/elsc.201300170

Wang, W., Zeng, C., Sardans, J., Zeng, D., Wang, C., Bartrons, M., & Penuelas, J. (2018). Industrial and agricultural wastes decreased greenhouse-gas emissions and increased rice grain yield in a subtropical paddy field. *Experimental Agriculture, 54*(4), 623–640. doi:10.1017/S001447971700031X

Wang, X., Liu, Y., Zeng, G., Chai, L., Xiao, X., Song, X., & Min, Z. (2008). Pedological characteristics of Mn mine tailings and metal accumulation by native plants. *Chemosphere, 72*(9), 1260–1266. doi:10.1016/j.chemosphere.2008.05.001 PMID:18555510

Wang, Y., Geng, S., Zhao, P., Du, H., He, Y., & Crittenden, J. (2016). Cost–benefit analysis of GHG emission reduction in waste to energy projects of China under clean development mechanism. *Resources, Conservation and Recycling, 109*, 90–95. doi:10.1016/j.resconrec.2016.02.010

Wang, Y., Ho, S. H., Cheng, C. L., Guo, W. Q., Nagarajan, D., Ren, N. Q., Lee, D. J., & Chang, J. S. (2016). Perspectives on the feasibility of using microalgae for industrial wastewater treatment. *Bioresource Technology, 222*, 485–497. doi:10.1016/j.biortech.2016.09.106 PMID:27765375

Ward, A. J., Hobbs, P. J., Holliman, P. J., & Jones, D. L. (2008). Optimization of the anaerobic digestion of agricultural resources. *Bioresource Technology, 99*(17), 7928–7940. doi:10.1016/j.biortech.2008.02.044 PMID:18406612

Wastowski, A. D. (2018). *Química da madeira.* Interciência.

Waterman, P. G. (1993). The chemistry of volatile oils. In R. K. M. Hay & P. G. Waterman (Eds.), *Volatile oil crops* (pp. 47–62). Logman Scientific & Technical.

Way, D. A., & Oren, R. (2010). Differential responses to changes in growth temperature between trees from different functional groups and biomes: A review and synthesis of data. *Tree Physiology*, *30*(6), 669–688. doi:10.1093/treephys/tpq015 PMID:20368338

Webb, H., Arnott, J., Crawford, R., & Ivanova, E. (2012). Plastic degradation and its environmental implications with special reference to poly(ethylene terephthalate). *Polymers*, *5*(1), 1–18. doi:10.3390/polym5010001

Weber, W. J., & Morris, J. C. (1963). Kinetics of adsorption on carbon from solution. *Journal of the Sanitary Engineering Division*, *89*, 31–60.

Webi, T. W., & Chakravort, R. K. (1974). Pore and solid diffusion models for fixed bed absorbers. *AIChE Journal. American Institute of Chemical Engineers*, *20*(2), 228–238. doi:10.1002/aic.690200204

Wegner, T., Skog, K. E., Ince, P. J., & Michler, C. J. (2010). Uses and desirable properties of wood in the 21st century. *Journal of Forestry*, *108*(4), 165–173.

Wertz, J. L., Deleu, M., Coppée, S., & Richel, A. H. (2019). *Hemicelluloses and lignin in biorefinerires*. CRC Press.

WHO. (1993). *Guidelines for drinking water quality*. Retrieved from https://www.who.int/water_sanitation_health/publications/gdwq2v1/en/

Wiemes, L., Pawlowsky, U., & Mymrin, V. (2017). Incorporation of industrial wastes as raw materials in brick's formulation. *Journal of Cleaner Production*, *142*, 69–77. doi:10.1016/j.jclepro.2016.06.174

Wierckx, N., Narancic, T., Eberlein, C., Wei, R., Drzyzga, O., Magnin, A., & … . (2018). *Plastic biodegradation: challenges and opportunities*. Springer International Publishing AG.

Wijffels, R. H., & Barbosa, M. J. (2010). An outlook on microalgal biofuels. *Science*, *329*(5993), 796–799. doi:10.1126cience.1189003 PMID:20705853

Wills, B. A., & Finch, J. (2015). *Wills' mineral processing technology: an introduction to the practical aspects of ore treatment and mineral recovery*. Butterworth-Heinemann.

Wilson, D. C., Velis, C. A., & Rodic, L. (2013). Integrated sustainable waste management in developing countries. *Proceedings of the Institution of Civil Engineers - Waste and Resource Management*, *166*(2), 52–68. 10.1680/warm.12.00005

Wilson, D. C. (2007). Development drivers for waste management. *Waste Management & Research*, *25*(3), 198–207. doi:10.1177/0734242X07079149 PMID:17612318

WO1993002141 A1. (1992). *Recycling of Polyvinyl butyral*. World Intellectual Property Organization Patent.

Wolborska, A. (1989). Adsorption on activated carbon of p-nitrophenol from aqueous solution. *Water Research*, *23*(1), 85–91. doi:10.1016/0043-1354(89)90066-3

Wolborska, A., & Pustelnik, P. (1996). A simplified method for determination of breakthrough time of an adsorbent layer. *Water Research*, *30*(11), 2643–2650. doi:10.1016/S0043-1354(96)00166-2

Wolfville, P., & Scotia, N. (2002). *Study on solid waste management options for Africa. Project Report Final Draft Version*. The African Development Bank.

Wong, P. Y., & Kitts, D. D. (2006). Studies on the dual antioxidant and antibacterial properties of parsley (*Petroselinum crispum*) and cilantro (*Coriandrum sativum*) extracts. *Food Chemistry*, *97*(3), 505–515. doi:10.1016/j.foodchem.2005.05.031

World Bank. (2019). *From Waste to Resource: Shifting paradigms for smarter wastewater interventions in Latin America and the Caribbean*. World Bank Group.

Worm, B., Lotze, H. K., Jubinville, I., Wilcox, C., & Jambeck, J. (2017). Plastic as a Persistent Marine Pollutant. *Annual Review of Environment and Resources, 42*(1), 1–26. doi:10.1146/annurev-environ-102016-060700

Wosten, J. H. M. (1997). Pedotransfer functions to evaluation soil quality. In E. G. Gregorich & M. R. Carter (Eds.), *Soil quality for crop production and ecosystem health* (pp. 221–245). Elsevier science Ltd. doi:10.1016/S0166-2481(97)80037-2

Wu, T Y, Mohammad, A. W., Jahim, J., & Anuar, N. (2007). *Palm oil mill effluent (POME) treatment and bioresources recovery using ultrafiltration membrane : Effect of pressure on membrane fouling*. doi:10.1016/j.bej.2007.01.029

Wu, T. Y., Mohammad, A. W., Jahim, J. M., & Anuar, N. (2010). Pollution control technologies for the treatment of palm oil mill effluent (POME) through end-of-pipe processes. *Journal of Environmental Management, 91*(7), 1467–1490. doi:10.1016/j.jenvman.2010.02.008 PMID:20231054

Xu, F., Li, Y., Ge, X., Yang, L., & Li, Y. (2017). Anaerobic digestion of food waste – challenges and opportunities. *Bioresour Technol.*

Yadav, B. K., Siebel, M. A., & van Bruggen, J. J. (2011). Rhizofiltration of a heavy metal (lead) containing wastewater using the wetland plant Carex pendula. CLEAN–Soil, Air. *Water (Basel), 39*(5), 467–474.

Yadav, G., Karemore, A., Dash, S. K., & Sen, R. (2015). Performance evaluation of a green process for microalgal CO2 sequestration in closed photobioreactor using flue gas generated in-situ. *Bioresource Technology, 191*, 399–406. doi:10.1016/j.biortech.2015.04.040 PMID:25921786

Yamada-Onodera, K., Mukumoto, H., Katsuyaya, Y., Saiganji, A., & Tani, Y. (2001). Degradation of polyethylene by a fungus, Penicillium simplicissimum YK. *Polymer Degradation & Stability, 72*(2), 323–327. doi:10.1016/S0141-3910(01)00027-1

Yan, M., Khoshnevisana, B., Tsapekos, P., Fotidis, I., & Angelidaki, I. (2019). *Continuous biomethanation of organic fraction of municipal solid waste under extreme ammonia levels.* Paper presented at the 16th World Congress on Anaerobic Digestion.

Yanqiao, P., Zhang, D., Singh, P., & Ragauskas, A. (2008). The new forestry biofuels sector. *Biofuels, Bioproducts & Biorefining, 2*(1), 58–73. doi:10.1002/bbb.48

Yeh, K. L., Chen, C. Y., & Chang, J. S. (2012). pH-stat photoheterotrophic cultivation of indigenous Chlorella vulgaris ESP-31 for biomass and lipid production using acetic acid as the carbon source. *Biochemical Engineering Journal, 64*, 1–7. doi:10.1016/j.bej.2012.02.006

Yener & Jirsak. (2011). Improving Performance of Polyvinyl Butyral Electrospinning. *Nanocon 2011*, 21–23.

Yen, H. W., Ho, S. H., Chen, C. Y., & Chang, J. S. (2015). CO_2, NO_x and SO_x removal from flue gas via microalgae cultivation: A critical review. *Biotechnology Journal, 10*(6), 829–839. doi:10.1002/biot.201400707 PMID:25931246

Yen, H. W., Hu, I. C., Chen, C. Y., Ho, S. H., Lee, D. J., & Chang, J. S. (2013). Microalgae-based biorefinery–from biofuels to natural products. *Bioresource Technology, 135*, 166–174. doi:10.1016/j.biortech.2012.10.099 PMID:23206809

Yilmaz, E., Wzorek, M., & Akçay, S. (2018). Co-pelletization of sewage sludge and agricultural wastes. *Journal of Environmental Management, 216*, 169–175. doi:10.1016/j.jenvman.2017.09.012 PMID:28943060

Ying, G. G. (2018). Ecological risk assessment of pesticides used in agriculture. In *Integrated Analytical Approaches for Pesticide Management* (pp. 67–79). Academic Press. doi:10.1016/B978-0-12-816155-5.00005-1

YKL Engineering Sdn. Bhd. (2002). *MY-142874-A*. Malaysia: MyIPO.

Yoon, J., Cao, X., Zhou, Q., & Ma, Q. L. (2006). Accumulation of Pb, Cu and Zn in native plants growing on a contaminated Florida site. *The Science of the Total Environment*, *368*(2-3), 456–464. doi:10.1016/j.scitotenv.2006.01.016 PMID:16600337

Yoon, Y. H., & Nelson, J. H. (1984). Application of gas adsorption kinetics II. A theoretical model for respirator cartridge service life. *American Industrial Hygiene Association Journal*, *45*(8), 509–516. doi:10.1080/15298668491400197 PMID:6475759

Yoshida, S., Hiraga, K., Takehana, T., Taniguchi, I., Yamaji, H., Maeda, Y., Toyohara, K., Miyamoto, K., Kimura, Y., & Oda, K. (2016). A bacterium that degrades and assimilates poly (ethylene terephthalate). *Science*, *351*(6278), 1196–1199. doi:10.1126cience.aad6359 PMID:26965627

Yousif, E., & Haddad, R. (2013). Photodegradation and photostabilization of polymers, especially polystyrene. *SpringerPlus*, *2*(1), 398. doi:10.1186/2193-1801-2-398 PMID:25674392

Yuan, X., Wen, B., Ma, X., Zhu, W., Wang, X., Chen, S., & Cui, Z. (2014). Enhancing the anaerobic digestion of lignocellulose of municipal solid waste using a microbial pre-treatment method. *Bioresource Technology*, *154*, 1–9. doi:10.1016/j.biortech.2013.11.090 PMID:24365784

Zahoor, T., Siddique, F., & Farooq, U. (2006). Isolation and characterization of Vinegar Culture (*Acetobacter aceti)* from indigenous sources. *British Food Journal*, *108*(6), 429–439. doi:10.1108/00070700610668405

Zahrim, A. Y., Nasimah, A., & Hilal, N. (2014). Journal of Water Process Engineering Pollutants analysis during conventional palm oil mill effluent (POME) ponding system and decolourisation of anaerobically treated POME via calcium lactate-polyacrylamide. *Journal of Water Process Engineering*, *4*, 159–165. doi:10.1016/j.jwpe.2014.09.005

Zainal, N. H., Jalani, N. F., Mamat, R., & Astimar, A. A. (2017). A review on the development of palm oil mill effluent (POME) final discharge polishing treatments. *Journal of Oil Palm Research*, *29*(4), 528–540. doi:10.21894/jopr.2017.00012

Zainul, I. F. (2013). *SMEs in palm oil urged to venture into refined downstream activities*. Star Media Group Berhad.

Zaki, A. B., El-Sheikh, M. Y., Evans, J., & El-Safty, S. A. (2000). Kinetics and mechanism of the sorption of some aromatic amines onto amberlite IRA-904 anion-exchange resin. *Journal of Colloid and Interface Science*, *221*(1), 58–63. doi:10.1006/jcis.1999.6553 PMID:10623452

Zakil, F. A., Ali, N., Hassan, O., & Pretreatment, A. (2008). *The usage of empty fruit bunch (EFB) and palm pressed fibre (PPF) as substrates for the cultivation of Pleurotus ostreatus*. Academic Press.

Zakrewski, J., Chamer, R., & Koscielniak, A. (1992). *Method for recovering gold and nickel from electronic scrap and waste*. Polish Patent, PL158889B (C22B7/00).

Zayed, A., Pilon-Smits, E., deSouza, M., Lin, Z.-Q., & Terry, N. (1999). *Remediation of Selenium-Polluted Soils and Waters by Phytovolatilization. In Phytoremediation of contaminated soil and water*. CRC Press.

Zeng, X., Duan, H., Wang, F., & Li, J. (2017). Examining environmental management of e-waste: China's experience and lessons. *Renewable & Sustainable Energy Reviews*, *72*, 1076–1082. doi:10.1016/j.rser.2016.10.015

Zeng, X., Yang, C., Chiang, J. F., & Li, J. (2017). Innovating e-waste management: From macroscopic to microscopic scales. *The Science of the Total Environment*, *575*, 1–5. doi:10.1016/j.scitotenv.2016.09.078 PMID:27723459

Zhang, B., He, Y., Zhu, H., Huang, X., Bai, X., Kannan, K., & Zhang, T. (2020). Concentrations of bisphenol A and its alternatives in paired maternal–fetal urine, serum and amniotic fluid from an e-waste dismantling area in China. *Environment International*, *136*, 105407. doi:10.1016/j.envint.2019.105407 PMID:31955035

Zhang, B., Li, S., Michel, F., Li, G., Luo, Y., & Zhang, D. (2016). Effects of mix ratio, moisture content and aeration rate on sulfurodor emissions during pig manure composting. *Waste Management (New York, N.Y.)*, *56*, 498–505. doi:10.1016/j.wasman.2016.06.026 PMID:27363616

Zhang, C., Xiao, G., Peng, L., Su, H., & Tan, T. (2013). The anaerobic co-digestion of food waste and cattle manure. *Bioresource Technology*, *129*, 170–176. doi:10.1016/j.biortech.2012.10.138 PMID:23246757

Zhang, J., Lü, F., Shao, L., & He, P. (2014). The use of biochar-amended composting to improve the humification and degradation of sewage sludge. *Bioresource Technology*, *168*, 252–258. doi:10.1016/j.biortech.2014.02.080 PMID:24656550

Zhang, L., Lee, Y. W., & Jahng, D. (2011). Anaerobic co-digestion of food waste and piggery wastewater: Focusing on the role of trace elements. *BioresTech*, *102*(8), 5048–5059. doi:10.1016/j.biortech.2011.01.082 PMID:21349706

Zhang, P., Chen, Y., & Zhou, Q. (2009). Waste activated sludge hydrolysis and short-chain fatty acids accumulation under mesophilic and thermophilic conditions: Effect of pH. *Water Research*, *43*(15), 3735–3742. doi:10.1016/j.watres.2009.05.036 PMID:19555988

Zhang, S., & Forssberg, E. (1997). Electronic scrap characterization for materials recycling. *Journal of Waste Management and Resource Recovery*, *3*, 157–168.

Zhang, W., Yang, J., Wu, X., Hu, Y., Yu, W., Wang, J., Dong, J., Li, M., Liang, S., Hu, J., & Kumar, R. V. (2016). A critical review on secondary lead recycling technology and its prospect. *Renewable & Sustainable Energy Reviews*, *61*, 108–122. doi:10.1016/j.rser.2016.03.046

Zhang, Z., Zhu, Y., Yang, T., Li, L., Zhu, H., & Wang, H. (2017). Conversion of local industrial wastes into greener cement through geopolymer technology: A case study of high-magnesium nickel slag. *Journal of Cleaner Production*, *141*, 463–471. doi:10.1016/j.jclepro.2016.09.147

Zhao, X., Jiang, G., Li, A., & Wang, L. (2016). Economic analysis of waste-to-energy industry in China. *Waste Management (New York, N.Y.)*, *48*, 604–618. doi:10.1016/j.wasman.2015.10.014 PMID:26514312

Zhao, X., Xia, Y., Zhan, L., Xie, B., Gao, B., & Wang, J. (2018). Hydrothermal treatment of e-waste plastics for tertiary recycling: Product slate and decomposition mechanisms. *ACS Sustainable Chemistry & Engineering*, *7*(1), 1464–1473. doi:10.1021/acssuschemeng.8b05147

Zheljazkov, V. D., Craker, L. E., & Xing, B. (2006). Effect of Cu, Pb and Cu on growth and essential oil contents in dill, peppermint and basil. *Environmental and Experimental Botany*, *58*(1-3), 9–16. doi:10.1016/j.envexpbot.2005.06.008

Zheljazkov, V. D., Craker, L. E., Xing, B., Nielson, N. E., & Wilcox, A. (2008). Aromatic plant production on metal contaminated soils. *The Science of the Total Environment*, *395*(2-3), 51–62. doi:10.1016/j.scitotenv.2008.01.041 PMID:18353428

Zheljazkov, V. D., & Warman, P. R. (2003). Application of high Cu compost to Swiss chard and basil. *The Science of the Total Environment*, *302*(1-3), 13–26. doi:10.1016/S0048-9697(02)00390-X PMID:12526895

Zheljazkov, V., & Nielsen, N. E. (1996). Studies on the effect of heavy metals (Cd, Pb, Cu, Mn, Zn and Fe) upon the growth, productivity and quality of lavender (*Lavandula angustifolia* Mill) production. *The Journal of Essential Oil Research*, *8*(3), 259–274. doi:10.1080/10412905.1996.9700612

Zhiqi, H., Reiske, H. R., & Wilson, D. B. (1999). Chatacterization of cadmium uptake in Lacto plantarum and isolation of cadmium and manganese uptake system. *Applied and Environmental Microbiology*, *65*, 4741–4745. PMID:10543780

Zhou, P., Zheng, Z., & Tie, J. (2005). *Technological process for extracting gold, silver and palladium from electronic industry waste*. Chinese Patent, CN1603432A (C22B 11/00).

Zhou, C., Gong, Z., Hu, J., Cao, A., & Liang, H. (2015). A cost-benefit analysis of landfill mining and material recycling in China. *Waste Management (New York, N.Y.)*, *35*, 191–198. doi:10.1016/j.wasman.2014.09.029 PMID:25453315

Zhu, B., Gikas, P., Zhang, R., Lord, J., Jenkins, B., & Li, X. (2009). Characteristics and biogas production potential of municipal solid wastes pretreated with a rotary drum reactor. *Bioresource Technology*, *100*(3), 1122–1129. doi:10.1016/j.biortech.2008.08.024 PMID:18849162

Zhu, C., Tian, H., Cheng, K., Liu, K., Wang, K., Hua, S., Gao, J., & Zhou, J. (2016). Potentials of whole process control of heavy metals emissions from coal-fired power plants in China. *Journal of Cleaner Production*, *114*, 343–351. doi:10.1016/j.jclepro.2015.05.008

Zink, T., Geyer, R., & Startz, R. (2018). Toward estimating displaced primary production from recycling: A case study of US aluminum. *Journal of Industrial Ecology*, *22*(2), 314–326. doi:10.1111/jiec.12557

Zorpas, A. A., Arapoglou, D., & Panagiotis, K. (2003). Waste paper and clinoptilolite as a bulking material with dewatered anaerobically stabilized primary sewage sludge (DASPSS) for compost production. *Waste Management (New York, N.Y.)*, *23*(1), 27–35. doi:10.1016/S0956-053X(02)00042-9 PMID:12623099

Zoubiri, S., & Baaliouamer, A. (2010). Essential oil composition of *Coriandrum sativum* seed cultivated in Algeria as food grains protectant. *Food Chemistry*, *122*(4), 1226–1228. doi:10.1016/j.foodchem.2010.03.119

Zubris, K., & Richards, B. (2005). Synthetic fibres as an indicator of land application of sludge. *Environmental Pollution*, *138*(2), 201–211. doi:10.1016/j.envpol.2005.04.013 PMID:15967553

Zucconi, F., Pera, M.A., & Forte, M., & Bertoldi, M. d. (1981). Evaluating toxicity of immature compost. *BioCycle*, *22*, 54–57.

Zurbrugg, C. (2002). Urban solid waste management in low income countries of Asia: How to cope with the garbage crisis: scientific committee on problems of the environment (SCOPE). Urban solid waste management review session, Durban, SA.

Zurbrügg, C., Caniato, M., & Vaccari, M. (2014). How assessment methods can support solid waste management in developing countries—A critical review. *Sustainability*, *6*(2), 545–570. doi:10.3390u6020545

About the Contributors

Ashok K. Rathoure holds doctorate degree in Bioremediation/Biotechnology from Central University (HNBGU Srinagar UK India) and M.Tech. Biotechnology. Currently, he is associated with M/s Akone Services Lucknow, India and also working as Managing Director, Biohm Consultare Pvt. Ltd. Surat India. Previously he was associated with Eco Group of Companies Surat Gujarat; Vardan Environet Gurgaon; En-vision Group Surat (En-vision Environmental Services and En-vision Enviro Engineers Pvt. Ltd.) for EIA studies; Himachal Institute of Life Sciences Paonta and Beehive College of Ad. Studies Dehradun for teaching to Biotechnology, Microbiology, Biochemistry and other biosciences subjects. His area of research is environmental biotechnology and publication includes 90 full length research papers in international and national journals of repute, more than 45 books from reputed publishers in India & abroad and more than 30 Book chapters. He is also associated as Editor-in-Chief for Octa Journal of Environmental Research, Managing Editor for Octa Journal of Biosciences and Executive Editor for Scientific India Magazine. He also holds PG Diploma in Human Resource Management.

* * *

Shujahat Ali is Assistant Professor in MUST Business School, Mirpur University of Science and Technology, Azad Kashmir.

Priya Banerjee is presently serving as Assistant Professor, Environmental Studies, Directorate of Distance Education, Rabindra Bharati University, Kolkata. She has completed her PhD research from the Department of Environmental Science, University of Calcutta, Kolkata. She has several research articles to her credit. She has also participated in various national and international seminars till date and has received awards for best platform and poster presentations including the best poster presentation award at the 100th Indian Science Congress, held at Kolkata. Her current research interests include integrated waste management strategies and environmental policies.

Pawan Kumar Bharti, MSc, PhD, NET, FASEA, FANSF, has written/edited 125 Books and more than 150 articles and filed 4 patents. Dr. Bharti is the fellow, founder and life member of several academic societies and received Bharat Excellence Award (2013); Best Research Award (2015), Limca Book Record (2016), Young Scientist Award (2017), Research Excellence Award (2018) and few other awards from different agencies/societies. He is in the editorial/advisory board of 36 international journals. Dr. Bharti was the member of 30th ISEA of MoES, Gov. of India during 2010-11. His biography has been published in 'Who's Who in the World' in 2012. He has visited South Africa, Antarctica, UAE,

Bhutan and Nepal. Presently, working as Scientist in Shriram Institute, Delhi, India and several PhD/MSc scholars are working under his supervision. He is QCI/NABET approved FAE in AP, NV, SC, RH for EIA projects.

Brij Bhushan is a researcher involved in recovery of resources by integration of membrane technology from wastewater, activated carbon development and its use in wastewater treatment. Actively involved in research since the last 8 years. PhD from IIT Roorkee and postdoc from UPC, Barcelona, Spain. Currently working as Associate Professor at Graphic Era University, Dehradun.

Arbind Choudhary is presently working as the Principal Scientist (Plant Breeding), and have been leading the quality seed production program of the institute under the aegis of NFSM funded national project "Creation of seed hubs for increasing indigenous production of pulses in India".

Abhishek Dubey is working as Scientist (Plant Pathology) at ICAR-Research Complex for Eastern Region, Patna.

Rachana Dubey is a scientist, ICAR-Research Complex for Eastern Region, Patna working on the aspect of greenhouse gas emission grom agriculture, conservation agriculture and rice-fallow area.

Sharad Kumar Dwivedi is scientist in ICAR having expertise in Plant physiology.

Zia Ur Rahman Farooqi is Doctoral Student in University of Agriculture, Faisalabad.

Abdul Kareem is doing postgraduation in Soil Science from University of Agriculture, Faisalabad.

Manoj Kumar Karnena is a Researcher and Teaching Assistant in the Department of Environmental Science. He has many years of research experience working on various lab-scale projects. He has publications in reputed international and national journals. His research interests include Water and Wastewater Treatment, Natural Coagulants, Nanocatalysis, Storage Technologies, and membrane technology.

Amit Kumar is working as an Assistant Professor, Department of Biotechnology, Keral Verma Subharti College of Science, Swami Vivekanand Subharti University, Meerut.

Man Djun Lee was born in Malaysia on the 17th of October 1989. He graduated with his first-class degree from Universiti Malaysia Sarawak in Mechanical and Manufacturing Engineering in 2013. He later pursued his Ph.D. in Advanced Manufacturing Engineering at the same institution and received his Ph.D. in 2017. His greatest strengths are his research, and communication skills. He started teaching and working in technical work since 2009 in various backgrounds. He is currently working as a lecturer at the University College of Technology Sarawak. Man Djun Lee has written more than 20 papers since he started writing in 2014. He has written several book chapters on the topic within his research interest. His research interest is in desalination, water treatment, wave energy, and waste management. Man Djun Lee had won several engineering design awards in waste management and wave energy projects.

Pui San Lee was born in Malaysia on the 27th of May 1999. She has written several book chapters and papers related to wastewater treatment.

Elaine Lengowski is a wood industrial engineer, D.sc, Wood Science and Technology Professor at Federal University of Mato Grosso.

Aniruddha Mukhopadhayay is presently the head/in charge of Department of Environmental Science, University of Calcutta and has been teaching for more than the last two decades. After completion of his graduation and master's from University of Calcutta he was awarded the D. N. Roy Chowdhury Gold Medal for his results in post graduation. He had also completed his M. Phil. in Environmental Science from University of Calcutta. He had carried out his post doctoral research in USA. He has published around a hundred national and international research publications and edited five books to his credit. He has been elected as FELLOW of West Bengal Academy of Science and Technology (FAScT). Currently he is the Chairman of the UG board of studies in Environmental Science, University of Calcutta and member of several other Board of Studies of other Universities, in state and outside. He has supervised doctoral students and been in research collaborations with research institutes and other Universities both at national and international level. He is a member of different committees instituted by the Government Agencies. His research interests include areas like environmental biotechnology, toxicology and ecology.

Joan Mwihaki Nyika is an Environmental Science PhD student at the University of South Africa and a Tutorial Fellow at the Technical University of Kenya, Geoscience and the Environment Department. She obtained a Bsc. degree in Biochemistry and Molecular Biology from the Jomo Kenyatta University of Agriculture and Technology in 2011. She also obtained an Msc degree in Land and Water Management from the University of Nairobi and, Kenya in 2017. Her PhD work revolves around solid waste management and the effects of landfill leachate on soil and groundwater. She has also done research on the use of microbes for pollutant bioremediation and phytoremediation. She has interests in sustainable environmental management, water management, ecological and ecosystem management. Joan has published more than 15 articles and 6 book chapters in peer reviewed journals and books, respectively. She has and is mentoring both undergraduate and diploma students and has more than 5 years experience teaching and training in environmental science.

Ednah Onyari is currently an Associate Proffessor at the University of South Africa in the civil engineering department. She has a PhD in Civil & Environmental engineering from the University of the Witwatersrand and a MSc degree in Water Engineering from UNESCO-IHE in the Netherlands. Dr Onyari has contributed to knowledge and has several publications in the field of water resources and environmental engineering.

Faizan Rafi is doing postgraduation in Soil Science from University of Agriculture, Faisalabad.

Vyshnavi Rao is a highly motivated researcher and academician by choice with a teaching experience of 4 years and an all India ranking of 16 in CSIR-NET Examination. Have an industrial Clinical research experience of 1 year with around 15 Research article and Review articles published in National and international peer reviewed journals. Hope to seek newer opportunities to evolve and learn interests in the process of knowledge sharing.

Biswajit Singha is Assistant Professor of Chemistry & Head, Department of Basic Science and Humanities, Camellia Institute of Technology, received B. Sc. (Hons.), M.Sc. and Ph.D. degree from University of Calcutta, working on the different research area like water pollution, toxic effect of heavy metals on plants, ANN, IOT etc., published 17 research papers in International journals which received more than 685 citations. Presented more than 35 research papers in International/National Conferences. Received young scientist award from Indian Chemical Society on 2011. Supervising two research scholars for their Ph.D. work. Life Member of Indian Science Congress.

Ashutosh Upadhyaya is presently working as Principal Scientist and Head, Division of Land and Water Management at ICAR Research Complex for Eastern Region, Patna. He has more than 30 years experience on addressing various issues related to land and water management in India. His area of interest is canal water management, ground water management, rainwater water management, on-farm water management, waste water utilization and management of Natural Resources for getting optimum agricultural output.

Index

O

P

R

S

T

U

V

W

Z

www.ingramcontent.com/pod-product-compliance
Lightning Source LLC
LaVergne TN
LVHW060354300926
849162LV00006B/1263
9781799849216